HYDRAULIC HANDBOOK

8TH EDITION

by R.H.WARRING

TRADE & TECHNICAL PRESS LTD.,
MORDEN, SURREY, SM4 5EW,
ENGLAND.

ISBN 85461–094–4

Printed in Great Britain by TTP (Printers) Limited, East Molesey, Surrey.

PREFACE

Many methods in the engineering discipline of oil-hydraulics which have become accepted practice during the intervening years since the Second World War are now subject to radical change. The use of interface technology, miniaturization, computerization, super durable materials, automatic monitoring and control are just some of the factors responsible for the current impositions on oil-hydraulic power transmission. To keep abreast of change and anticipate the future, whilst appreciating the past, is a problem for all engineers, not least hydraulicians. For this reason alone, the 8th Edition HYDRAULIC HANDBOOK is a vital reference work for the design engineer, hydraulic technician, chief engineer, plant engineer, or anyone concerned with the selection, installation, operation or maintenance of oil hydraulic equipment. It embraces a wealth of essential and useful information and data interspersed with numerous diagrammatic illustrations, charts, references, and ready-reference tables. It is probably the most comprehensive and authoritative work ever published on this specialized subject — a veritable tome of technology — the 'bible' of oil-hydraulics.

THE PUBLISHERS

ACKNOWLEDGMENTS

Aeroquip (UK) Limited
Automotive Products
K. G. Beresford
Benmar (UK) Limited
Burmah-Castrol and Verdol International
R. A. Burnison
R. Cakebread
Dowty Hydraulic Units Limited
Dowty Seals Limited
R. G. Eggleton
Enerpac Limited
Fenner Fluid Power
H. Huhnke Limited
HML Engineering Limited
Intec Limited
J. R. Loder
A. L. Mills
Sperry Vickers
Unimation (Europe) Limited
Uniroyal
R. H. Warring
J. R. Wearing

CONTENTS

SECTION 3 – SYSTEMS

SECTION 4 – MACHINES, ETC.

SECTION 5 – INSTRUMENTATION AND TESTING

SECTION 6 – APPLICATIONS

SECTION 7 – SURVEYS

SECTION 8

SECTION 1

Properties of Fluids

BASIC FLUID parameters of main significance in hydraulics are *density* (or alternatively *specific gravity*), *viscosity, specific heat* and *compressibility*.

The *density* of a fluid is defined as the mass per unit volume, *viz*

$$\text{density } (\rho) = \frac{\text{(mass) lb}}{\text{volume}} = \frac{\text{lbf}}{\text{volume x g}} = \frac{\text{kgf}}{\text{volume x g}}$$

in consistent units

Although rendered obsolete by modern unit definitions, *specific weight* remains a convenient engineering unit, defined as weight per unit volume, or

$$\text{specific weight (w)} = \frac{\text{lbf}}{\text{volume}} = \frac{\text{kgf}}{\text{volume}}$$

in consistent units

Specific gravity is a dimensionless quantity and is the ratio of the density (or specific weight) of a fluid to the density (or specific weight) of water. In the case of equations for engineering calculations it is often desirable to render density or specific weight, where it appears as a factor, in terms of specific gravity, thus avoiding any possible confusion between the true numerical values of density and specific weight which are to be employed in the formula.

The significance of specific gravity as a hydraulic fluid parameter is that it gives an indication of the weight of the fluid in the system, or more directly a comparison of fluid weights for a given system where different fluids may be considered. Also the higher the specific gravity of the fluid the more difficult to lift the fluid in the suction part of the system. The design of the suction side may therefore need particular care in order to avoid the possibility of cavitation and erratic pump operation.

Viscosity

The viscosity of a fluid is a measure of its internal resistance. *Dynamic* viscosity (μ) is defined in terms of the force in dynes between two parallel laminae or layers of fluid each 1 cm^2 in area with a slip velocity of 1 cm/sec between them, the corresponding unit of viscosity being the poise. Because the dynamic viscosity of real fluids determined in poises is invariably a fractional quantity, the more usual unit employed for expressing dynamic viscosity is the centipoise, or one hundredth of a poise. The significance of dynamic viscosity is that it is effectively a *friction coefficient.*

For engineering calculations it is usually more convenient to employ *kinematic viscosity (ν)* rather than dynamic viscosity, this being determined as the absolute dynamic viscosity divided by the mass density of the fluid. The standard unit is the stoke – (St) but for the same reason as above the practical unit is invariably taken as a centistoke or one hundredth of a stoke (cSt).

Kinematic viscosity is used for the calculation of flow characteristics, and thus dynamic pressure.

Practical Viscosity Values

Logically all fluid viscosity values should now be quoted in centistokes. Actual measurement of fluid viscosity, however, continues to be made in viscometers yielding values in arbitrary scales – Redwood No 1 seconds, Saybolt Universal Seconds (SUS), or Engler degrees.

There are no exact conversions between these scales, nor for conversion of arbitrary viscosities in seconds or degrees to kinematic viscosities in centistokes, ft^2/sec, or in^2/sec units. Close approximate conversions can be made by reference to conversion scales or conversion tables. It should be noted that such conversions apply only at the same temperature as the original measurement.

In the case of a non-Newtonian fluid the instantaneous viscosity of the fluid is dependent on the shear stress in that fluid at that particular moment. If necessary, a specific viscosity figure can be obtained with a viscometer which ensures a uniform shear rate throughout measurement. Such a figure will, however, have limited practical value, unless the shear stability characteristics of the fluid are also known.

The variation of viscosity with temperature is one of the most significant parameters with hydraulic fluids, affecting both the performance and selection of a fluid. This can be fully expressed by plotting a characteristic curve for the fluid on an ASTM chart – Fig 1. The scales of an ASTM chart are so designed that the characteristic curve for most fluids is linear. Given a number of spot readings for viscosity and temperature, a close approximation to the viscosity-temperature characteristics of that fluid at intermediate temperatures can be obtained by joining these points with a straight line, and for temperature outside the range covered by the spot values, by extending this line in either direction.

This chart (Fig 1) also shows equivalent kinematic viscosity values in four different scales; also the viscosity ranges covered by SAE oil number ratings – these are still widely used for general classification of oil viscosities.

ISO/BS Viscosity Classification

Viscosity classification for industrial liquid lubricants defined by BS 4231 establishes 18 viscosity grades in the range 2–1500 centistokes covering approximately (in the case of mineral oils) the range from kerosene to cylinder oils and thus also embracing the normal range of mineral-oil-based hydraulic fluids. Classification is based on the principle that the mid-point kinematic viscosity of each grade should be approximately 50% higher than that of the preceding one. Using this numbering system, oil viscosities are quoted as ISO viscosity grade (number), or ISO VG (number) – see Table I. Table II gives corresponding kinematic viscosities at various temperatures (related also to Viscosity Index). VDA classifications for hydraulic fluids are given in Table III.

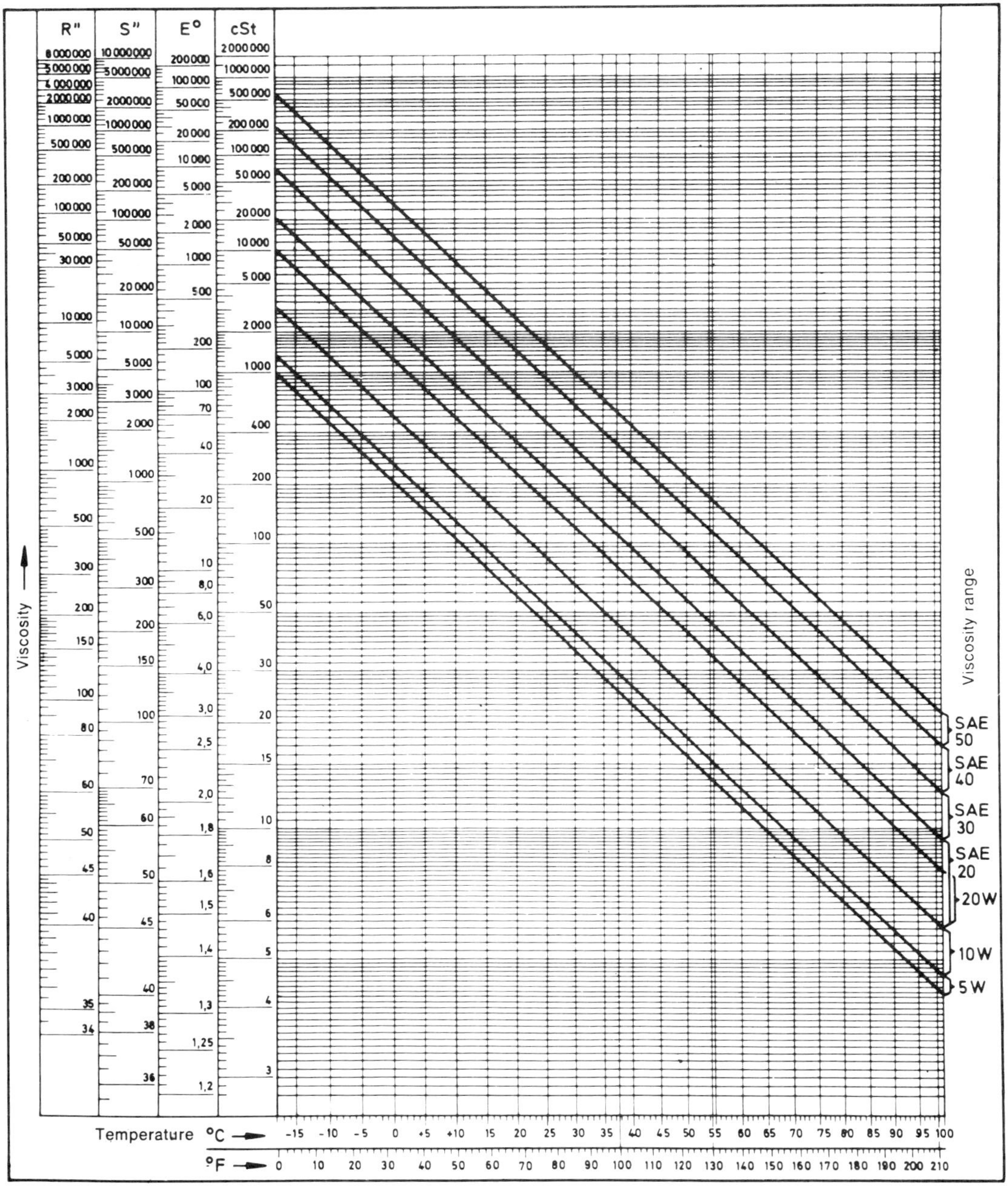

Fig 1 Viscosity—temperature chart.
(Bosch).

TABLE I – ISO VISCOSITY CLASSIFICATION

ISO viscosity grade	Mid-point kinematic viscosity cSt at 40.0 °C	Kinematic viscosity limits cSt at 40.0 °C Minimum	Maximum	ISO viscosity grade	Mid-point kinematic viscosity cSt at 40.0 °C	Kinematic viscosity limits cSt at 40.0 °C Minimum	Maximum
ISO VG 2	2.2	1.98	2.42	ISO VG 68	68	61.2	74.8
ISO VG 3	3.2	2.88	3.52	ISO VG 100	100	90.0	110
ISO VG 5	4.6	4.14	5.06	ISO VG 150	150	135	165
ISO VG 7	6.8	6.12	7.48	ISO VG 220	220	198	242
ISO VG 10	10	9.00	11.0	ISO VG 320	320	288	352
ISO VG 15	15	13.5	16.5	ISO VG 460	460	414	506
ISO VG 22	22	19.8	24.2	ISO VG 680	680	612	748
ISO VG 32	32	28.8	35.2	ISO VG 1000	1000	900	1100
ISO VG 46	46	41.4	50.6	ISO VG 1500	1500	1350	1650

TABLE II – ISO VISCOSITY CLASSIFICATION WITH CORRESPONDING KINEMATIC VISCOSITIES AT VARIOUS TEMPERATURES FOR DIFFERING VISCOSITY INDICES (BS 4231 : 1975)

ISO viscosity grade	Kinematic viscosity range	Approximate kinematic viscosity at other temperatures for differing values of viscosity index					
		Viscosity index = 50			Viscosity index = 95		
ISO VG	cSt at 40 °C	cSt at 20 °C	cSt at 37.8 °C	cSt at 50 °C	cSt at 20 °C	cSt at 37.8 °C	cSt at 50 °C
2	1.98–2.42	(2.87–3.69)	(2.05–2.52)	(1.69–2.03)	(2.92–3.71)	(2.06–2.52)	(1.69–2.03)
3	2.88–3.52	(4.59–5.92)	(3.02–3.70)	(2.38–2.84)	(4.58–5.83)	(3.01–3.69)	(2.39–2.86)
5	4.14–5.06	(7.25–9.35)	(4.37–5.37)	(3.29–3.95)	(7.09–9.03)	(4.36–5.35)	(3.32–3.99)
7	6.12–7.48	(11.9–15.3)	(6.52–8.01)	(4.68–5.61)	(11.4–14.4)	(6.50–7.98)	(4.76–5.72)
10	9.00–11.0	19.1–24.5	9.68–11.9	6.65–7.99	18.1–23.1	9.64–11.8	6.78–8.14
15	13.5–16.5	31.6–40.6	14.7–18.0	9.62–11.5	29.8–38.3	14.6–17.9	9.80–11.8
22	19.8–24.2	51.0–65.8	21.7–26.6	13.6–16.3	48.0–61.7	21.6–26.5	13.9–16.6
32	28.8–35.2	82.6–108	31.9–39.2	19.0–22.6	76.9–98.7	31.7–38.9	19.4–23.3
46	41.4–50.6	133–172	46.3–56.9	26.1–31.3	120–153	45.9–56.3	27.0–32.5
68	61.2–74.8	219–283	69.2–85.0	37.1–44.4	193–244	68.4–83.9	38.7–46.6
100	90.0–110	356–454	103–126	52.4–63.0	303–383	101–124	55.3–66.6
150	135–165	583–743	155–191	75.9–91.2	486–614	153–188	80.6–97.1
220	198–242	927–1180	230–282	108–129	761–964	226–277	115–138
320	288–352	1460–1870	337–414	151–182	1180–1500	331–406	163–196
460	414–506	2290–2930	488–599	210–252	1810–2300	478–587	228–274
680	612–748	3700–4740	728–894	300–360	2880–3650	712–874	326–393
1000	900–1100	5960–7640	1080–1330	425–509	4550–5780	1050–1290	466–560
1500	1350–1650	9850–12600	1640–2010	613–734	8390–9400	1590–1960	676–812

TABLE III – VDMA FLUID CLASSIFICATION

Hydraulic Fluids		Approximate viscosity at 50°C (122°F) cSt
With additives for increasing the resistance to ageing and for improving the protection against corrosion	HL 16	12 to 20
	HL 25	21 to 29
	HL 36	32 to 40
	HL 49	44 to 54
	HL 68	62 to 74
With extra additives for improving the behaviour in the boundary lubrication range	HLP 16	12 to 20
	HLP 25	21 to 29
	HLP 36	32 to 40
	HLP 49	44 to 54
	HLP 68	62 to 74

TABLE IV – SPECIFIC HEAT OF TYPICAL FLUIDS cal/g degC

Fluid	Temperature – °C				
	15	20	25	60	80
Mineral oil (typical)	–	1.0	–	–	–
Water-glycol	–	–	0.718	0.774	0.785
Phosphate ester	0.43	–	–	–	–
Silicone	0.39	–	–	–	–
Water	1.0011	1.000	0.9992	1.0000	1.0033

Viscosity Index

Viscosity Index (VI) is a single number representation of the viscosity-temperature characteristics of a fluid. The higher the viscosity index the smaller the change in viscosity unit temperature, and *vice versa*, although this is only a rough guide as to actual change. VI values may extend beyond 100, when the correct designation is VI_E. (Extended Viscosity Index).

Specific Heat

The specific heat of a fluid is defined as the ratio of the heat required to raise the temperature of a given volume of fluid by one degree to that required to raise the temperature of the same volume of water by the same amount. It can thus be specified either in Btu or calorie-gm-cm-units. The specific heat is not constant but varies with temperature, although for practical calculation a constant value is often assumed based on a nominal temperature range.

In the absence of specific figures the following semi-empirical formula can be used to calculate the specific heat of typical hydraulic oils. (See also Table IV).

$$\text{specific heat} = \frac{0.388 \times 0.00045t}{\sqrt{sg}}$$

where sg = specific gravity of the fluid at 60°F

t = temperature

Specific heat is then given in units of Btu/lb per deg F with temperature in °F; or cal/g per deg C with temperature in °C

The actual specific heat of a hydraulic oil may be appreciably modified by the presence of additives, and also of contaminants in the fluid.

Compressibility of Fluids

Specifically, compressibility (usually designated by the symbol β) is the reciprocal of the fluid bulk modulus (K). However, the bulk modulus is not constant, tending to increase with temperature and decrease non-linearly with pressure. The instantaneous value of the bulk modulus at any pressure is called the *tangent bulk modulus* (K_t), and the mean value of the bulk modulus from atmospheric pressure to any pressure P is called the *secant bulk modulus* (K_s)

$$K_t = -V \frac{dP}{dV}$$

where V is the volume at pressure P

$$K_s = -\frac{V_oP}{V_o - V}$$

where V_o is the volume at atmospheric pressure

The values of the tangent and secant moduli will tend to coincide at lower pressures (*ie* as P approaches atmospheric pressure), and for pressures up to about 70 bar (1 000 lb/in^2) the difference can usually be ignored. For general engineering calculations a 'typical' bulk modulus value may be quoted and used for pressures up to 700 bar (10 000 lb/in^2). Logically this should be the secant modulus. For working at specific high pressures, however, the tangent modulus should be used, if known. (See also Table V).

TABLE V – COMPRESSIBILITY OF TYPICAL FLUIDS
(Compressibility expressed as percentage reduction in volume)

Fluid	Pressure	
	70 bar (1 000 lb/in^2)	700 bar (10 000 lb/in^2)
Water	0.34	3.3
Water-in-oil emulsions	0.35	3.5
Water-glycol	0.26	2.6
Mineral oils (typical)	0.35	3.4
Phosphate ester	0.25	2.5
Chlorinated hydro carbon	0.24	2.4

Typical working figures are a reduction in initial volume of 0.5% per 1 000 lb/in^2 (0.00735% per atmosphere) for hydraulic oils and 0.4% per 1 000 lb/in^2 (0.0059% per atmosphere) for water. Such figures are reasonably valid for pressures up to 700 bar (10 000 lb/in^2) and over a temperature range of 10–100°C (50–200°F).

The compressibility of a normal hydraulic oil at 20°C (68°F) and 70 bar (1 000 lb/in^2) can also be estimated quite accurately from its kinematic viscosity at 22°C (72°F), using the empirical formula

$$\text{'compressibility'} = 3.5 - 0.2 \log \nu$$

where ν = viscosity in centistokes

and compressibility is given in inch^3/lb x 10^{-6} units

Alternatively, 'compressibility' may be expressed in terms of relative density, *viz*

$$\text{'compressibility'} = (1 - \frac{\Delta_{1000}}{\Delta_0}) \times 100\% \text{ per } 1\,000 \text{ lb/in}^2$$

where Δ_0 = fluid density at atmospheric pressure

Δ_{1000} = fluid density at 70 bar (1 000 lb/in^2)

As a direct result of compressibility, the density of any real fluid will increase with pressure. In very high pressure systems this may be more significant than the volumetric change.

Vapour Pressure

The vapour pressure of a fluid is the pressure exerted by the saturated vapour in contact with the surface of the fluid at a specified temperature. The higher the vapour pressure the more volatile the fluid, and/or the nearer it is to boiling point. Fluids with a high vapour pressure, either due to their volatile nature or because of a high operating temperature, are therefore prone to 'flash' into vapour under suction conditions, thus setting a specific limit to the net positive suction head a pump can accommodate without cavitating.

Additionally, as the boiling point of a fluid approaches, the more volatile fractions will come off first, progressively changing the nature of the fluid. This is seldom significant in oil fluids at normal working temperatures, but with water fluids progressive loss of water may be experienced at quite moderate working temperatures.

Aniline Point

The aniline point of a mineral oil is the lowest temperature at which the oil is completely miscible with an equal volume of freshly distilled aniline. It is a general indication of the aromatic content of the oil (paraffinic oils having a high aniline point and aromatic oils a low aniline point), and because of this is sometimes used as a form of compatibility index.

Surface Tension

Surface tension may be significant as affecting:

(i) Foaming characteristics at the free liquid surface or interface between two non-miscible fluids.

(ii) The ability of the fluid to 'wet' a metal surface.

(iii) Inherent leakage past seals and at joints, *etc.*

Characteristics (i) and (ii) can be adequately controlled by additives if necessary. Item (iii) is not normally significant in the case of oil fluids which have adequate surface tension to make sealing relatively easy. Fluids which have a surface tension of less than 30 dynes/cm^2, however, are troublesome to seal, needing particular attention to joints, and one must often accept that some leakage will be inevitable with practical designs of seals.

Cloud Point

The cloud point of a mineral oil is that temperature at which waxes or other solids normally present in solution tend to crystallize out, or come out of solution. This can lead to clogging or partial choking of the system.

Pour Point

This is the temperature at which the thickening action of separation of the waxy constituents is so marked that the fluid ceases to flow. On the ASTM curve this would be marked by an abrupt rise in viscosity. The pour point (temperature) is determined with regard to specific flow conditions.

Thermal Expansion

The coefficient of volumetric expansion of an oil remains practically constant over the usual range of working temperatures encountered in hydraulic systems. Specific values are largely related to the specific gravity of the oil. The coefficient of volumetric expansion, however, decreases rapidly with increasing pressure, *eg* of the order of 0.000025 per degC per 70 bar (1 000 lb/in^2). The relative volume of a pressurized fluid is thus less than that predicted on the basis of volume correction coefficients alone.

Thermal Conductivity (See also Table VI)

The thermal conductivity is a measure of the ability of a fluid to dissipate heat. In a practical system heat dissipation may be hindered by the formation of boundary layer films and thus fluids which do not 'wet' the internal surfaces tend to have lower thermal conductivities. The lower the thermal conductivity of the fluid the higher its working temperature will tend to be, under similar operating conditions. Thus mineral oils, having a generally low thermal conductivity, will tend to run at higher working temperatures than water-based or water-glycol fluids used in a similar system.

TABLE VI – THERMAL CONDUCTIVITY OF FLUIDS
cal/cm sec degC x 10^{-3}

Fluid	Nominal	Temperature			
		10°C	50°C	70°C	80°C
Mineral oil	3.24	–	–	–	–
Water-glycol	9.61	–	–	–	–
Phosphate ester	5.30	–	–	–	–
Glycerine	–	–	–	6.8	–
Water	–	14.7	15.4	–	16.0

The thermal conductivity of a hydraulic oil can be estimated from the following semi-empirical formula:

$$\text{thermal conductivity} = \frac{0.821 - 0.000244t}{sg} \text{ Btu in/ft}^2 \text{ h}$$

$$= \frac{(1.2 - 0.00035t) \times 10^{-3}}{sg} \text{ cal gm/cm}^2 \text{ h}$$

where t = temperature in °C or °F, respectively
sg = specific gravity of fluid

By comparison the thermal conductivity of water is some 4.5 times greater, and that of water-glycol solutions some 3 times greater.

Flash Point

The flash point broadly defines the relative fire hazard of a fluid. The flash point may also serve as an indication of the type of an oil or blend, since the more volatile the oil the lower the flash point and *vice versa.* The flash point may be determined by various standard 'open' or 'closed' cup tests. These tests measure the tendency of the fluid to ignite when brought into contact with a hot surface. They are particularly relevant where spills may leak on to hot exhaust pipes, *etc,* but give similar results to auto-ignition temperature values (AIT).

Spontaneous Ignition Temperature

A steel block, with a thermometer set into it to allow its temperature to be read, is heated strongly from below. Droplets of the fluid under test are allowed to fall into a cavity in the block.

Conventional hydraulic oils vaporize at relatively low temperatures and ignite with a sharp report. The water containing types of water-in-oil emulsions and water-glycol mixtures form a steam blanket which must be dispersed so that extremely high temperatures are needed for spontaneous ignition.

Phosphate ester fluids, on the other hand, with their innate resistance to fire, cannot support combustion but will ignite spontaneously at very high temperatures. The behaviour of the various fluids in this test differs; interpretation of the results is difficult and not amenable to tabulation.

Auto-Ignition Temperature (AIT) (ASTM D 2155–66)

This is the temperature at which a fluid will ignite spontaneously. It is a measure of the tendency of the fluid to withstand overheating and contact with hot surfaces. This is considered to be one of the relevant tests for hydraulic fluids.

Basic Hydraulic Theory

THE MAJORITY of simple systems are *hydrostatic* in character, *ie* transmitting power by *pressure energy.* However, most practical systems involve fluid flow through pipes and related components, which involves forces generated by motion and thus *hydrodynamic* phenomena. Thus, basically, actuators work on a hydrostatic principle, whilst flow through the pipelines, *etc,* to the actuators conform to hydrodynamic laws.

Hydrostatics

In a purely hydrostatic system a static pressure is produced by virtue of a mass of fluid supported by a base area, this pressure (P) being dependent only on the height of the column of liquid (H) and its density, (ρ)

$$P = H \cdot \rho \cdot g$$

where g is the acceleration due to gravity,

In engineering terms:

$$P = H \times w$$

where w is the specific weight of the fluid

In the case of fluid in a container (*eg* a tank or reservoir), hydrostatic pressure exerts a force F on the base area:

$$F = A \cdot \rho \text{ (in consistent units)}$$

where A is the base area

The pressure at any point within the fluid is determined by its depth below the surface of the liquid, or head (h) — Fig 1. A head, so determined, is called the *potential head* and may be less than the total head available. Thus total head represented by h + z in Fig 1, is called the *piezometric head.* It follows that piezometric head equals the potential head plus any additional head available or:

$$\text{piezometric head} = P/w + z$$

The piezometric head is a constant for all fluids at rest. This is a special case of the Bernoulli equation where the velocity terms are zero. In fact, hydrodynamic equations are directly applicable to fluid statics by rendering all the terms involving fluid motion as zero. Conversely, the

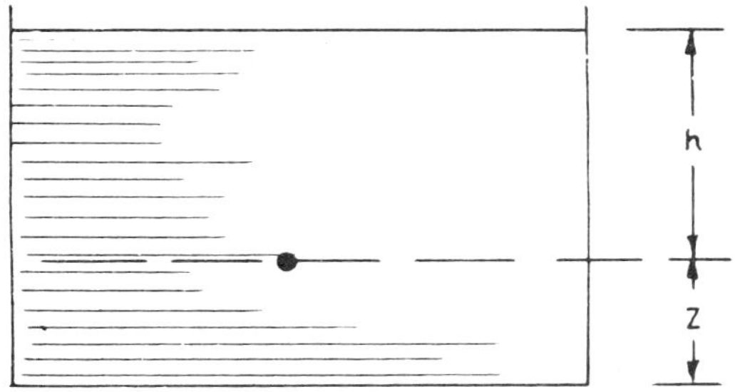

Fig 1

application of hydrostatic principles to the transmission of force through a fluid must take velocity into account if fluid movement is involved, although in practice this may be small enough to be negligible. This applies generally when flow rates are small, fluid performs a 'passive' role in the transmission of pressure and the small amount of fluid movement involved has no significant effect on the performance of the system.

The pressure at any point in a static fluid is the same in every direction. The pressure exerted on any *surface* immersed in the fluid is thus equal to the product of the fluid pressure and the surface area. Similarly, pressure exerted on a fluid in a closed system is transmitted equally to all parts of the system and acts perpendicularly to all surfaces in contact with the fluid. For practical hydrostatic application the weight of fluid can be ignored since the potential head involved is very small compared with the applied pressure.

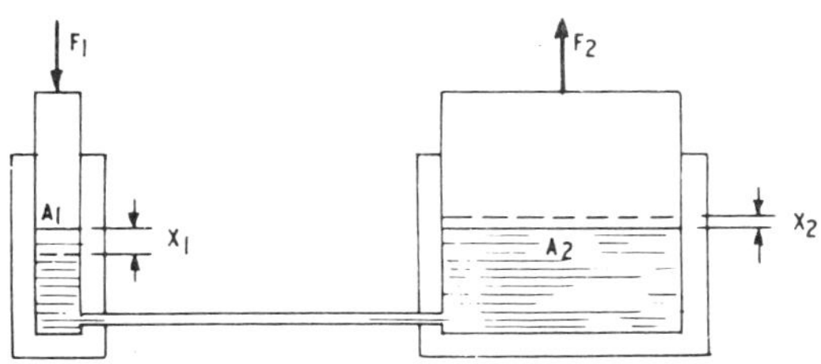

Fig 2

Since forces exerted on surfaces by hydrostatic pressure are proportional to areas, force multiplication is possible — *eg* by utilizing pistons of different areas (see Fig 2). Here F_1 applied to piston A_1 develops a pressure of F_1/A_1 which is transmitted throughout the fluid. Thus the resulting pressure on piston area A_2 is (F_1/A_1), or:

$$\text{output force } (F_2) = \text{input force } (F_1) \times \frac{A_2}{A_1}$$

This force multiplication is achieved in the ratio A_2/A_1.

This is a valid hydrostatic system since the displacement of fluid is small enough for velocity loss to be neglected. In a practical system, the modifying factors are the weights of the respective pistons (W_1 and W_2) and the friction of the piston seals (f_1 and f_2) — see Fig 3. The true output can then be calculated as:

$$\text{output force } (F_2) = (F_1 + W_1 - f_1) \quad \frac{{D_2}^2}{{D_1}^2} \quad - (W_2 - f_2)$$

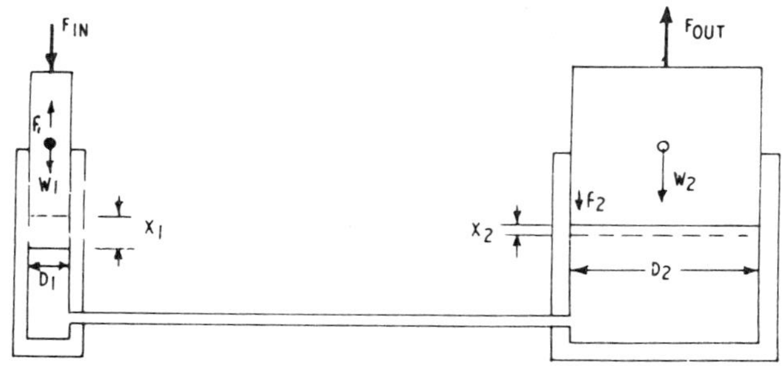

Fig 3

Respective piston travels are not modified in any way and are:

$$X_2 = X_1 \frac{D_1{}^2}{D_2{}^2}$$

Similarly the work W_1 done on piston 1 is equal to the work W_2 done by the loaded piston.

$$W_1 = F_1 \cdot X_1$$
$$W_2 = F_2 \cdot X_2$$
$$\text{or } F_1 \cdot X_1 = F_2 \cdot X_2$$

These equations define the performance in a basic hydrostatic system of force multiplication over pressure ranges where the compressibility of the fluid is negligible. Not all such systems are necessarily force multipliers, however. They are equally suitable for force transfer with high efficiency. Thus the typical automobile brake system may, in fact, employ slave cylinders of smaller diameter than the master cylinder. All the force multiplication requires is applied at the mechanical output, where a maximum pedal pressure of the order of 45 kgf (100 lbf) with a master cylinder of 25 mm (1 in) diameter can yield a system pressure of the order of 42–56 bar (600–800 lb/in^2), with virtually 100% force transfer for the slave cylinders.

Pumps in Hydrostatic Systems

Where a hydraulic pump acts as the transmitter in a force transfer system, the ram is 'sized' on the basis of the output force required and the fluid pressure available from the pump, *ie*

$$\text{ram diameter (d)} = \sqrt{\frac{4F}{\pi P}}$$

where F = output force required
P = pressure available from pump

The *delivery* required from the pump is governed by the speed of working required; or *vice versa,* thus:

$$\text{delivery (Q)} = \frac{D^2}{4} \times \frac{S}{t}$$

where S = stroke
t = time to complete stroke

Pump output force required then follows as:

$$\text{hp out} = \frac{F}{k} \cdot \frac{L}{t}$$

where k is a constant (depending on units employed)

In engineering units:

$$\text{hp out} = \frac{F \times L}{6600 \times t}$$

where F is in lbf
and L is in inches

$$\text{hp out} = \frac{F \times L}{3700 \times t}$$

where F is in kgf
and L is in centimetres

The aforementioned relationships assume that fluid velocity components are negligible (*ie* the system is truly hydrostatic). In practical systems it may be necessary to take into account velocity components and back-pressure effects; also seal friction and, in the case of large vertical rams, ram weight.

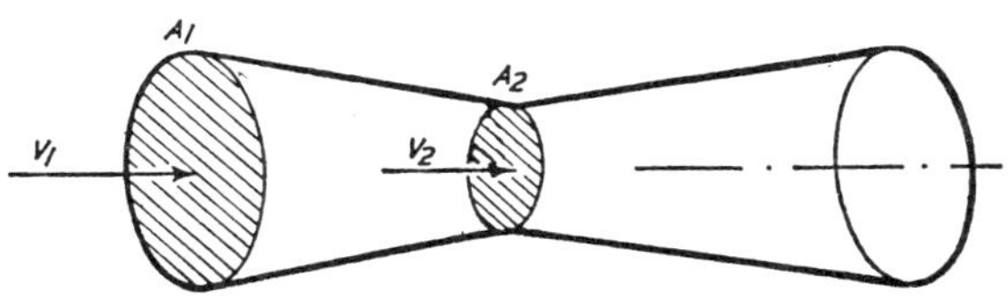

Fig 4

Hydrodynamics

Fluid under pressure will always tend to fill a pipe and for all practical purposes in engineering hydraulics, pipes will always be filled by pressurized flow. The *velocity* of flow through pipes of changing or variable cross section will then automatically adjust to maintain continuity of flow. Thus at a change of section (Fig 4):

$$Q = V_1 \cdot A_1 = V_2 \cdot A_2$$

where Q = flow rate (constant)
A_1 and A_2 are pipe cross sections
V_1 = flow velocity in section A_1
V_2 = flow velocity in section A_2

Flow *velocity* will also result in a *dynamic pressure* being generated, proportional to the energy of motion of the fluid. Dynamic pressure or velocity-pressure (P_V) has the theoretical value:

$$P_V = \rho \frac{V^2}{2}$$

where ρ = fluid density.

Equally the fluid will also be subject to some static pressure generating the flow (unless it is flowing down an inclined pipe under gravity); and possibly potential pressure (due to a change in weight along the length of flow).

Since the total energy of a fluid is constant (neglecting frictional losses) it follows that the sum of the static pressure, dynamic pressure and potential pressure is constant for any flow line cross section.

In most practical systems potential pressure can be neglected, assuming simply that static pressure falls with increasing dynamic pressure (*ie* increasing flow velocity) and *vice versa.*

Energy Equations

Energy equations for steady flow are derived from the first law of thermodynamics which states that the amount of heat added to a fluid as it passes through a system, is equal to the change in energy content of the fluid, plus any work done by the fluid.

For a closed system:

$$q = \Delta\ \text{(kinetic energy)} + \Delta\ \text{(displacement energy)} + \Delta\ \text{(potential energy)}.$$

For an open system:

$$q = \Delta\ \text{(kinetic energy)} + \Delta\ \text{(displacement energy)} + \Delta\ \text{(potential energy)} + \Delta\ \text{(internal energy)} + \text{(work done by fluid)}.$$

The kinetic energy of any mass m is mV^2. In the case of a fluid in other than one-dimensional flow, the true kinetic energy is always greater than that calculated on the mean velocity. It can be found by integrating the product of the mean flow rate through an infinitesimal area and the kinetic energy per unit mass over the entire flow area. The ratio of the true kinetic energy per unit time to that based on the average velocity, is called the kinetic energy correction factor (α).

Thus:

$$\text{true kinetic energy} = \frac{\alpha\, mV^2}{2}$$

For laminar flow in a pipe $\alpha = 2$; and
For turbulent flow in a pipe $\alpha = 1.1$ (approx).

The potential energy is simply defined as the work required to bring a fluid to any elevation other than the datum, and is thus numerically equal to gh per unit mass of fluid.

The internal energy is a function of the pressure and temperature of the fluid, but is more generally applicable to gases than liquids. Thermal terms are generally ignored when dealing with liquids, when an additional term is introduced to take into account the conversion of some mechanical energy into thermal energy due to viscous friction. A general expression of the energy equation then becomes:

$$\frac{\alpha_1 V_1{}^2}{2} + \frac{P_1}{\rho} + gh_1 - W = \frac{\alpha_2 V_2{}^2}{2} + \frac{-P_2}{\rho} + gh_2 + h_f$$

or

$$\frac{\alpha_1 V_1{}^2}{2g} + \frac{P_1}{w} + h_1 \frac{-W}{g} = \frac{\alpha_2 V_2{}^2}{2g} + \frac{P_2}{w} + h_2 + \frac{h_f}{g}$$

where h_f is the amount of mechanical energy converted into thermal energy, or head loss.

W = work done by, or on fluid, expressed in terms of head

w = specific weight of fluid.

In a practical system the head loss is always positive and in the direction of flow. In a hydrostatic system work is added to the system (*eg* by the pump). In the case of a hydrokinetic system work is both added (by the pump) and removed (by the hydrokinetic machine).

Friction

In any practical hydrodynamic system friction is produced in the fluid so that some energy is lost in the form of heat. Frictional losses derive from the fact that a real fluid has *viscosity*. In the case of laminar flow (Reynolds' Number below 2 000), viscous forces predominate and a simple friction factor applies which is independent of the condition of the surfaces (*ie* pipe bore) within which the fluid is contained.

$$\text{friction factor } (f_{lam}) = \frac{64\nu}{VD}$$

$$= \frac{64}{R_e}$$

where ν = kinematic viscosity of the fluid

V = flow velocity

D = pipe bore

R_e = Reynolds Number

Pressure drop then follows as:

$$\Delta P = f_{lam} \frac{\rho V^2}{2D} \cdot L$$

where ρ = density of fluid

L = length of pipe

For energy calculations it is more conventional to reach the pressure drop equation directly in terms of flow rate (Q), fluid kinematic viscosity (ν) and specific gravity of the fluid (sg), *viz*

$$\Delta P/L = K \cdot \frac{Q\nu}{D^4} \times sg$$

where K is a constant depending on the units employed — see Table I.

TABLE I – CONSTANT K FOR LAMINAR FLOW

ΔP	L	Q	D	ν	K =
lb/in^2	feet	in^3/sec	inches	centistokes	1/14 030
lb/in^2	feet	gallons/min	inches	centistokes	1/3 635
kg/cm^2	metres	litres/min	millimetres	centistokes	6.52

In the case of turbulent flow the friction factor is less clearly defined since radial components of velocity exist and there is an inter-change of momentum between adjacent layers of fluid. With modern small bore hydraulics employing drawn tubing, the bore condition is consistent with 'smooth turbulence' (*ie* surface roughness does not protrude beyond the laminar sub-layer of the boundary flow), when the friction factor approaches closely to $0.3164/R_e^{0.25}$.

The basic expression for pressure drop:

$$\Delta P = f_t \frac{\rho V^2}{2D} \cdot L$$

$$= \frac{0.3164}{R_e^{0.25}} \cdot \frac{\rho V^2}{2D} \cdot L$$

can be expressed in terms of flow rate and energy units as:

$$\frac{\Delta P}{L} = \frac{Q^x \nu^y}{K_t D^z} \times sg$$

where exponential values are: $x = 1.75$

$y = 0.25$

$z = 4.75$

The constant K_t depends on the units employed (see Table II).

TABLE II – CONSTANT K_t FOR TURBULENT FLOW

ΔP	L	Q	D	ν	K_t
lb/in^2	feet	in^3/sec	inches	centistokes	18 750
lb/in^2	feet	gallons/min	inches	centistokes	1 270
kg/cm^2	metres	litres/min	millimetres	centistokes	358 000

In practice, solutions for pressure drop are most conveniently obtained from charts or nomograms. Formula solutions generally give sufficiently accurate results up to Reynolds Numbers of 10 000, and reasonably consistent results up to Reynolds Numbers of 100 000. For higher Reynolds Numbers more accurate solutions are obtained by using modified friction factors based on Prandtl's law of pipe friction for smooth pipes or Nikuradse's empirical data for rough pipes or Colebrook/White equations.

The modified friction factor for 'smooth turbulence' is:

$$\frac{1}{\sqrt{f_t}} = 2 \log (R_e \sqrt{f_t}) - 0.8$$

The corresponding Colebrook/White equation for all turbulent flow conditions is:

$$\frac{1}{\sqrt{f_t}} = 1.74 - 2 \log \left(\frac{2k}{D} + \frac{18.7}{R_e \sqrt{f_t}}\right)$$

where k/D is the relative roughness of the pipe bore (*ie* ratio of roughness dimension to pipe diameter)

For smooth pipes k approaches zero and so the quantity 2 k/D can be eliminated from the formula.

Flow Through Changing Cross Section

Losses in flow through changing cross sections are largely due to separation, although frictional losses will also be present. Such losses are most conveniently expressed in terms of the velocity head ($V^2/2g$) at the downstream end. For engineering calculations the loss is considered as occurring at the loss-producing section, when expressing the loss as a head (h_L):

$$h_L = \frac{V^2}{2g}$$

where V is the downstream velocity

or

$$\Delta P = \frac{V^2 \rho}{2g}$$

where ρ is the density of the fluid

To account for different geometry (and the friction loss component in practical flow), an empirical coefficient (K_L) is introduced:

$$h_L = K_L \frac{V^2}{2g}$$

$$\Delta P = K_L \frac{V^2 \rho}{2g} = K_L \frac{V^2 w}{2}$$

where w is the specific weight of the fluid

This is valid for all general cases except that for expansion of cross section the effective velocity value is the difference between upstream and downstream flow velocities.

Typical values of K_L for entries and exits are given in Fig 5. In the case of a sudden contraction (Fig 6) the loss coefficient is a function of the diameter ratio D_2/D_1, as shown. This can also be derived mathematically as:

$$K_L = \frac{1 - \left(\frac{D_2}{D_1}\right)^2}{2}$$

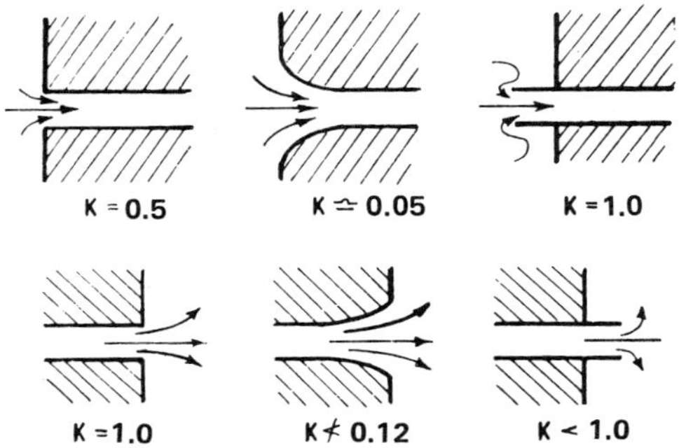

$\delta\Delta$

ΔP

$$\Delta P = K \cdot \frac{\rho V^2}{2}$$

Fig 5

where K has the typical values given in the diagram.

Fig 6

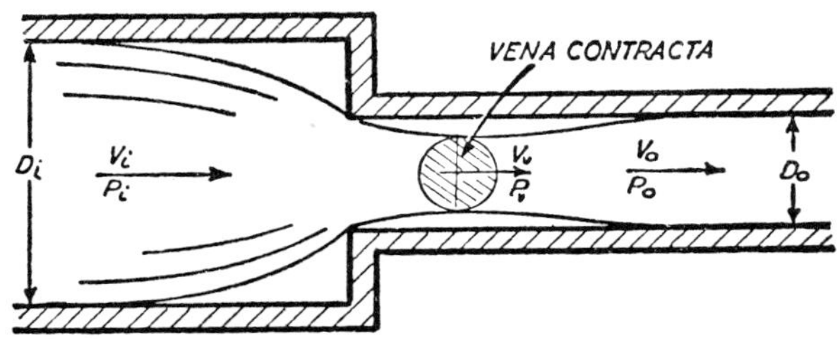

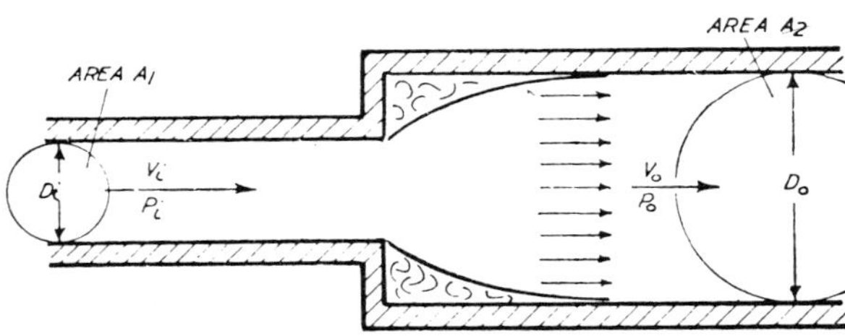

Fig 7

Fig 8

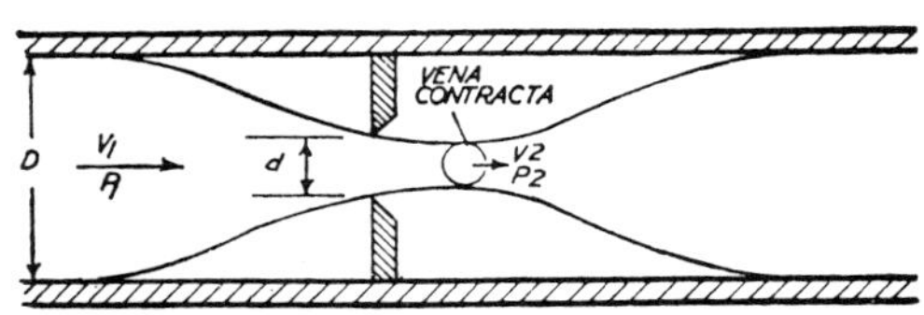

In the case of a sudden enlargement (Fig 7) the loss coefficient is determined by the area ratios:

$$K_L = \frac{\frac{A_2}{A_1}}{\frac{A_2}{A_1} - 1}$$

Sharp-Edged Orifices

Flow through a sharp-edged orifice is an individual case — Fig 8. The streamlines converge on approaching the orifice and continue to converge after passing through the orifice, reaching a minimum cross sectional area, known as the *vena contracta,* downstream of the orifice before diverging again. For small circular orifices the downstream position of the vena contracta is of the order of half the diameter of the orifice. At the vena contracta all the streamlines are perpendicular to the plane of the orifice.

The theoretical solution for downstream velocity is:

$$V_2 = \frac{1}{\sqrt{1 - \frac{A_2^2}{A_1^2}}} \sqrt{\left[\frac{2g}{\rho} (P_1 - P_2)\right]}$$

The flow rate (Q) thus follows as:

$$Q = \frac{A_2}{\sqrt{1 - \frac{A_2^2}{A_1^2}}} \sqrt{\left[\frac{2g}{\rho} (P_1 - P_2)\right]}$$

Introducing an empirical constant (K_o) to account for friction losses:

$$Q = \frac{K_o \cdot A_2}{\sqrt{1 - \frac{A_2^2}{A_1^2}}} \sqrt{\left[\frac{2g}{\rho} (P_1 - P_2)\right]}$$

The quantity $\dfrac{K_o \cdot A_2}{1 - \frac{A_2^2}{A_1^2}}$ is most commonly expressed as a single *orifice coefficient*.

(C_o), when:

$$Q = C_o A_2 \sqrt{\frac{2g}{\rho} (P_1 - P_2)}$$

whence pressure drop (ΔP) = ($P_1 - P_2$)

$$= \frac{\rho Q^2}{C_o^2 A_2 \cdot 2g}$$

Again this is more conveniently expressed in terms of orifice diameter (d_o):

$$\Delta P = \frac{8 \rho Q^2}{\eta^2 C_o^2 \cdot d_o^4 \cdot g}$$

or in engineering terms:

$$\Delta P = \frac{K_o Q^2}{C_o^4 d_o^4} \times \text{sg of fluid}$$

where K_o is a constant depending on the units employed. — (see Table III).

TABLE III — VALUES OF K_o FOR ORIFICE FORMULA

Q	d	K_o
in^3/sec	inches	7.78×10^{-5}
gallons/min	inches	3.64×10^{-4}
litres/min	millimetres	32.33

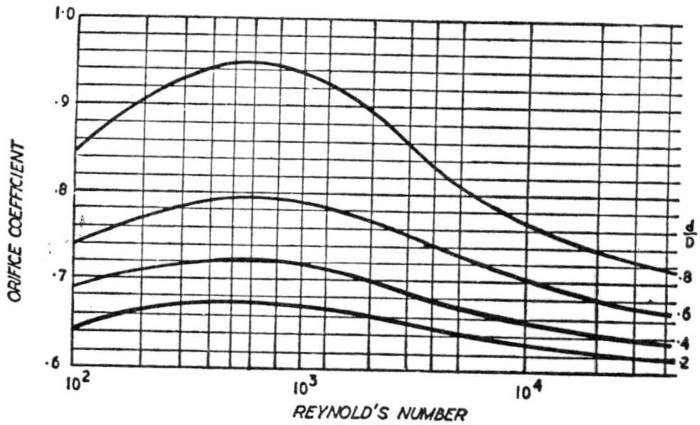

Fig 9

Practical values of orifice coefficients range from about 0.6 to 0.8, an average value of 0.65 commonly being adopted for sharp-edged orifices. The orifice coefficient varies both with geometry and Reynolds Number (Fig 9). The smaller the orifice diameter relative to pipe diameter the smaller the variation with Reynolds Number and the closer the value to 0.65 for sharp-edged orifices.

For rapid working the engineering formulas can be simplified still further by rendering the orifice coefficient as a conductance coefficient embracing the fluid density, *eg*

$$\text{conductance coefficient (Y)} = \frac{\pi}{4} \cdot C_o \sqrt{\frac{2g}{\rho}}$$

Two formulas are then available for quick calculation:

$$Q = Y d^2 \sqrt{\Delta P}$$

$$\Delta P = \frac{Q^2}{Y^2 d^4}$$

In this case 'Y' is calculated for a given value of orifice coefficient (*eg* 0.65), applicable to a given fluid at a given temperature (*ie* employing the appropriate fluid density figure at that temperature). Alternatively, solutions may be expressed in the form of charts, either for different orifice diameters or for a specific fluid at a specific temperature.

Annular Orifices

Viscous flow through the annular space formed by two concentric cylinders is essentially similar to that between two flat, parallel plates, provided in the first instance that annular thickness, b, is small compared with the cylinder radii. The same solution will therefore apply in both cases — Fig 10.

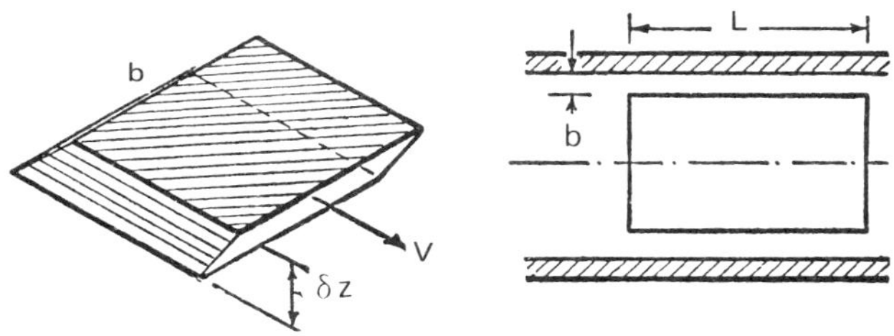

Fig 10

The shear force, resultant on viscosity of the fluid and velocity on any plane or area A, due to velocity V is:

$$S = A\nu\rho\,\delta v/\delta z$$

The shear force on a plane dx distance away is then:

$$S + \delta S = S + \frac{dS}{dz}\ \delta z$$

Hence:

$$\delta S = A\nu\rho\ \frac{d^2 V}{dz^2}\ \cdot dx \cdot dz$$

The difference in normal pressure at each end is:

$$P = b\,g\,\delta p\,\delta z$$

which equals the difference in shear forces δS

Hence:

$$V = -\ \frac{g}{2\nu\rho} \cdot \frac{dp}{dx}\ (cZ - Z^2)$$

where the clearance c is in the plane of Z

For a uniform clearance (over a distance L), and rendering the expression in engineering terms:

$$P = \frac{12\nu\rho\,L\,Q}{Dc^3\,g}$$

Relating this formula for pressure drop (ΔP) to an annular orifice where the clearance space is small, the total thickness of the flow path approximates to πD, where D is the diameter of the inner cylinder. The standard equation for the flow through an orifice formed by concentric, circular cylinders then becomes:

$$Q = \frac{\pi D b^3 \Delta P g}{12 \nu \rho L}$$

where b is the thickness of the annulus and
L is the length of the flow path

or

$$\Delta P = \frac{12 \nu \rho L Q}{\pi D b^3 g}$$

It will be appreciated from this formula that since flow rate, or escape of fluid through a small annular orifice is proportional to the cube of the clearance space, this radial dimension is particularly critical as regards sealing — far more so than the length of the flow path provided by the annulus.

See also chapter on *Pipework Calculations.*

Pipework Calculations

FLOW VELOCITY (V) is the primary factor governing friction in flow through pipes, and thus pressure drop. Formulas for determining flow velocities are:

$$V\ (\text{m/s}) = \frac{Q}{4.71d^2} \times 100$$

where Q is flow rate in lit/min
d is pipe bore in mm

$$V\ (\text{ft/sec}) = \frac{Q}{2.04d^2}$$

where Q is flow rate in Imperial gal/min
d is pipe bore in inches

The flow rate (Q) is established by the delivery of the pump and the take-off or demand of the system. Simple formula calculation is used to establish the pipe size necessary to provide a specific flow rate at an acceptable flow velocity. The following flow rates are usually adopted as generally giving moderate to low frictional losses:

	minimum	**average**	**maximum**
High-pressure (delivery) lines	2 m/s 7 ft/s	3 m/s 10 ft/s	4.5 m/s 15 ft/s
Intake lines	0.6 m/s 2 ft/s	0.9 m/s 3 ft/s	1.2 m/s 4 ft/s
Return lines	0.6 m/s 2 ft/s	0.9 m/s 3 ft/s	1.2 m/s 4 ft/s

Calculated flow rate follows from a re-arrangement of the flow velocity formula, *viz*

$$Q = K \cdot V \cdot d^2$$

where K is a constant depending on the units employed for Q, V and d (see Table I).

TABLE I – CONSTANT (K) FOR FLOW RATE FORMULA

Unit for Q	K for V in ft/sec, d in inches	K for V in m/sec, d in mm
in^3/sec	9.42	0.048
in^3/min	565	2.875
ft^3/min	0.33	0.00166
Imperial gallons/min	2.04	0.104
US gallons/min	2.45	0.125
litres/sec	0.154	0.0008
litres/min	9.24	0.047

Pipe Sizing

Optimum pipe sizing then follows using the following formulas to determine equivalent bore sizes:

$$d\ (\text{mm}) = 4.6\sqrt{\frac{Q}{V_d}}$$

where Q is in lit/min

and V_d is the design flow velocity in m/sec

$$d\ (\text{inches}) = 0.7\sqrt{\frac{Q}{V_d}}$$

where Q is in Imperial gallons/min

V_d is the design flow velocity in ft/sec

Critical Systems

For critical systems designed for high performance within narrow limits, pipe size may be determined directly from the design pressure drop figure. The latter would include total pressure drop through lines and fittings. The following basic formula then applies:

$$d^x = \frac{Q^a \nu^b L}{k \Delta P} \times sg$$

If flow is laminar:

$$d\ (\text{inches}) = \sqrt[4]{\frac{Q \nu L}{14030 \Delta P} \times \text{sg of fluid}}$$

where d = bore in inches

Q = flow rate in in^3/sec

ν = fluid viscosity in centistokes at working temperature

L = equivalent length of pipes in feet

ΔP = design pressure drop, lb/in^2

$$d\ (mm) = \sqrt[4]{\frac{Q\,\nu\,L}{7.85\Delta P} \times \text{sg of fluid}}$$

for Q in cm^3/sec
L in metres
ΔP in bar

If the flow is turbulent:

$$d\ (inches) = 4.75\sqrt{\frac{Q^{1.75}\,\nu^{0.25}\,L}{18750\Delta P} \times \text{sg of fluid}}$$

$$d\ (mm) = 4.75\sqrt{\frac{Q^{1.75}\,\nu^{0.25}\,L}{10.5\Delta P} \times \text{sg of fluid}}$$

This method can be particularly tedious in the case of turbulent flow and also additionally requires a fairly exact knowledge of fitting losses, *etc.* Line sizing by pressure drop, however, has the advantage that system performance can be more closely predicted than with line sizing by arbitrary flow velocity limits. It may also be found that using fully acceptable pressure drop values as a design figure the resultant line size is considerably smaller than that given by flow velocity selection, usually showing a saving in bulk and weight in the pipework and fittings.

Reynolds Number

Fluid friction (and thus pressure drop) is dependent on the *type* of flow, determined by the Reynolds Number (R_e). This is a dimensionless quantity given by:

$$R_e = \frac{dV}{\nu} \quad \text{in consistent units}$$

or

$$R_e = \frac{kdV}{\nu}$$

where ν is the kinematic viscosity of the fluid
k is a factor dependent on units employed — see Table II.

TABLE II — CONSTANT (k) FOR REYNOLDS NUMBER FORMULA (VISCOSITY ν IN CENTISTOKES)

d	V	Q	k
in	ft/sec	—	7750
mm	m/sec	—	1000
in	ft/sec	in^3/sec	820
in	ft/sec	Imperial gal/min	3800
in	ft/sec	US gal/min	4560
mm	m/sec	litres/min	21200

Flow is *laminar* with Reynolds Number up to about 1 500. With increasing Reynolds Number flow becomes transitional, turning to fully *turbulent* flow at R_e = 3 200. Frictional effects are impossible to predict accurately in the transitional range and so, for safety, only flow with a Reynolds Number less than 1 500 can positively be accepted as laminar. In both cases pressure drop (through frictional losses) can be expressed by the equations:

$$\text{pressure drop } (\Delta P) = f \cdot \frac{\rho V^2}{gd} \times L$$

where f is a friction factor dependent on Reynolds Number
L = length of pipe

This is more usually written in the form:

$$\frac{\Delta P}{L} = f \cdot \frac{\rho V^2}{gd}$$

The pressure drop is then normally calculated per m or 100 m of pipe length; or 100 ft of pipe length.

Laminar Flow

With laminar flow pressure drop is independent of the pipe bore condition and the corresponding friction factor can be determined simply as:

$$f_{lam} = \frac{64}{\sqrt{R_e}}$$

Turbulent Flow

In the case of turbulent flow the friction factor is dependent both on the Reynolds Number and the smoothness of the pipe bore. For Reynolds Numbers below 10 000:

$$f_t = \frac{0.3164}{\sqrt{R_e}}$$

For Reynolds Numbers above 10 000 this modified value can be used:

$$f_t = 0.0032 + \frac{0.221}{R_e\ 0.237}$$

A more accurate factor is $f_t = 2 \log (R_e\sqrt{f_t}) - 0.8$

Where the surface of the pipe or tube is not smooth the friction factor is dependent on both Reynolds Number and the degree of surface roughness. The latter is usually expressed in terms of the relative roughness D/k (or k/D), where k is a linear measure of mean surface roughness.

Friction factors are then normally read from a chart. Three degrees of turbulent flow exist — (i) *smooth turbulence* consistent with smooth bore pipes and tubes and where the friction factor varies with Reynolds Number only; (ii) *rough turbulence* where fully rough turbulent flow is produced and the friction factor is substantially independent of Reynolds Number and varies with relative roughness; and (iii) a *transitional range* where the friction factor varies with both Reynolds Number and relative roughness. Graphical interpretation of the appropriate friction factor is usual for engineering purposes, although the following formulas apply:

smooth turbulence: $$\frac{1}{\sqrt{f_t}} = 2 \log_{10} R_e \frac{\sqrt{f_t}}{2.51}$$

rough turbulence: $$\frac{1}{\sqrt{f_t}} = 2 \log_{10} 3.7 \frac{D}{k}$$

transitional range: (turbulent flow) $$\frac{1}{\sqrt{f_t}} = 2 \log_{10} \frac{k}{3.7D} + \frac{2.5}{R_e\sqrt{f_t}}$$

Friction factors for turbulent flow are normally determined from graphs — *eg* Fig 1.

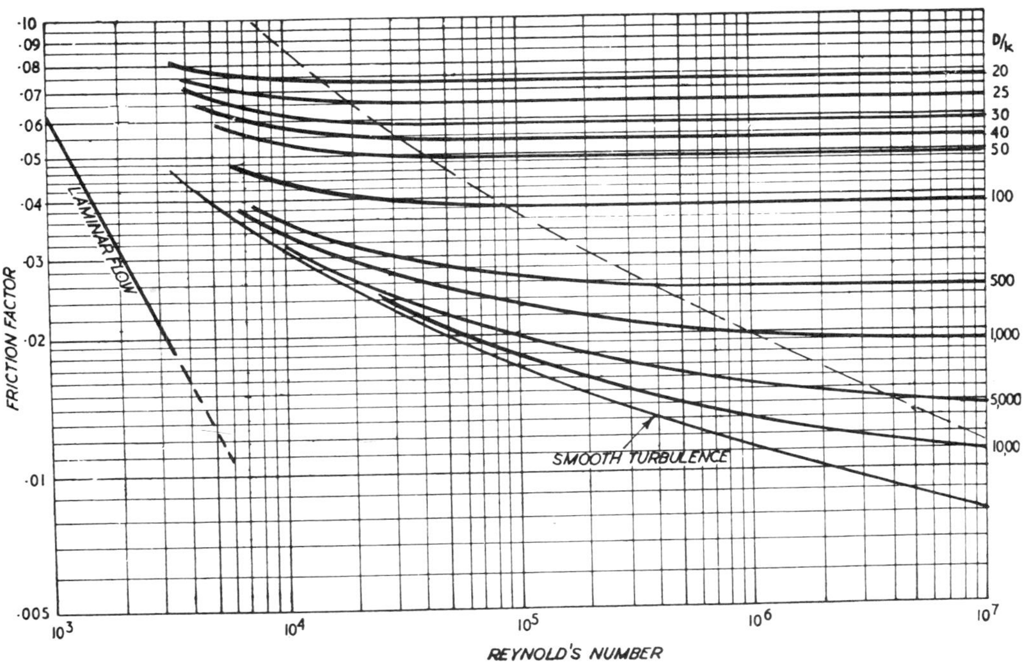

Fig 1 Friction factors for turbulent flow — curves designated D/k are for rough bore pipes where k is surface roughness.

Pressure Drop at Bends

The previous solutions apply only to flow through straight pipes of uniform cross section. Losses at bends do not lend themselves to complete mathematical analysis and solutions are invariably based on empirical data. As a generalization, there is no effective loss in bends of R/d of 8:1 or greater (where R = radius of bend) with turbulent flow. In the case of laminar flow, some loss will be experienced at values of R/d of 20:1 or less, increasing with decreasing R/d ratio.

At an R/d ratio of 3:1, approximate practical values are:

equivalent straight length of pipe = 2 x bend for turbulent flow
equivalent straight length of pipe = 5 x bend for laminar flow

Losses at Fittings

Losses at fittings are determined empirically and can be expressed in terms of 'resistance coefficient' or equivalent pipe length. In the former case an 'average' or 'typical' friction factor is assumed which does not take into account the variation with flow Reynolds Number and can lead to considerable error. In fact, there are three separate ways of rendering 'resistance coefficients' which can lead to serious errors in interpretation.

The true resistance coefficient (K_X) is a friction factor combined with the linear dimensions of the fitting which can be used in the basic pressure drop formula:

$$\Delta P = K_X \frac{\rho}{2g} V^2$$

$$= A K_X V^2 \times sg$$

where A is an arithmetical constant depending on the units employed.

It follows that the resistance coefficient (K_X) is the equivalent of fL/d.

Alternative criteria are the *conductance coefficient* (Y_C) or the *valve coefficient* (K_V), defined specifically as:

$$Y_C = \sqrt{1.35 \times 10^{-2}\ fL/d \times sg}$$

$$K_V = \sqrt{1.35 \times 10^{-2}\ fL/d^5 \times sg}$$

In practice it is generally satisfactory, and considerably easier, to work with equivalent pipe length — *ie* the length of straight pipe of the same diameter as the fitting which would have the same resistance or pressure drop as the fitting. Again, such data can only be determined empirically, but can be expressed as a typical L/d value for various types of fitting although such values are subject to scale effect as well as minor variations in geometry with different sizes.

Total Pressure Drop

The total pressure drop in a complete pipework system can then be estimated by rendering all the bends, fittings, *etc* in terms of equivalent pipe lengths, adding these to the actual lengths of pipe involved, *viz*

$$L_T = L + l_1 + l_2 + l_3 + \ldots\ldots$$

where L_T = equivalent straight length of pipe
L = length of straight pipe
l_1, l_2, l_3 = equivalent pipe lengths for individual fittings, *etc.*

This will then give a single value (L_T) to enter in the appropriate formula for calculating pressure drop. Note, however, that a further pressure drop may be present at the working point(s) of the system, *eg* pressure drop through the inlet port to a cylinder; and pressure drop through control valve(s), unless the latter has already been included as 'equivalent straight length(s)'.

Pressure Rating of Pipes

Pressure ratings for pipes and tubes are normally taken from manufacturers' figures, but can also be calculated from first principles. In the case of homogeneous metal tubes the following simple formula can be used:

$$P_W = 2S_{max} \cdot \frac{t}{D}$$

where P_W = maximum permissible working pressure
S_{max} = maximum permissible material stress
t = tube wall thickness
D = tube overall diameter

Maximum permissible material stress is normally taken as one third (or 0.3) of the ultimate tensile stress of the material – see Table III.

Provided the maximum material stress figure is taken within the limit of proportionality of the material, this simple formula is valid. It does not hold true for higher stress values, and thus will not accurately predict bursting pressures, *eg* using S_{ult} in place of S_{max}. It is also not valid where the ratio D/t is 16:1 or less, as stress is then no longer uniformly distributed through the wall thickness but ranges from a maximum at the inner surface to a minimum at the outer surface.

The simple formula is thus restricted to thin-walled tubes (D/t greater than 16:1). It will over-estimate the pressure rating for thick-walled tubes (D/t 16:1 or less), and in such cases an alternative formula must be used. An alternative formula which can be applied in the case of thick-walled tubes is:

$$S_{max} = P \frac{R_i^2}{R_o^2 - R_i^2} \left(\frac{R_o^2}{R_i^2} + 1\right)$$

where R_i = inner radius of tube
R_o = outer radius of tube
P = internal pressure

TABLE III – MAXIMUM PERMISSIBLE STRESS FOR TUBE CALCULATIONS
(Minimum UTS divided by 3)

Material	Condition	S_{max}	
		bar	lb/in^2
Low-carbon steel	As drawn	1 280	18 300
	Drawn and polished*	1 950	28 000
20-ton steel	As drawn	1 000	15 000
Stainless (304) steel	Annealed	2 350	33 300
	Half-hard	2 800	40 000
	Hard	3 500	50 000
Light alloy 61S-T6	As drawn	1 000	15 000
Copper	Annealed	480	6 800†
	Half-hard	630	9 000†
	Hard	800	11 300†
Tungum	Annealed	1 550	22 000
	Precipitation-hardened	1 550	22 000
Titanium 115/125		1 300	18 500
Titanium 150/160		2 100	30 000

*Cylinder tubes † Up to 65°C (150°F) only

Alternative formulas, written as a solution for wall thickness required are:

$$t = \frac{D}{2}\left(\sqrt{\frac{S+P}{S-P}} - 1\right)$$

$$t = \frac{D}{2}\left(\sqrt{\frac{3S+P}{3S-4P}} - 1\right)$$

for $S = S_{max}$ the corresponding value of P is P_w

for $S = S_{ult}$ the corresponding value of P is the bursting pressure

Modified formulas are used in the case of non-homogeneous tubes; and also for non-metallic tubes.

(i) In the case of welded tubes, a correction factor may be introduced; or alternatively, a higher divisor may be used to establish P_{max} from the ultimate tensile strength of the material. This is not necessarily the invariable rule, as welded tubes can have the same working strength as drawn tubes. Corrections may be applied to tubes with welded connections on a similar basis, however.

(ii) In the case of cast tubes, a nominal (and substantially lower) value may be adopted for P_{max}. Cast tubes are associated with older hydraulic systems and large pipe sizes, where pressure rating is established on empirical lines, permitting fairly large tolerances in wall thickness.

(iii) In the case of copper pipes and tubes intended for brazed or soldered connections, the standard thin-walled formula is de-rated:

$$P_W = \frac{2S_{max}t}{D - 0.8t}$$

(iv) In the case of metallic tubes intended for threaded connections, an allowance is made for the reduction in tube strength due to threading. The following formula can then be used for thin-walled tubes:

$$P_W = \frac{2S_{max}(t - C)}{D - 0.8t\ (t - C)}$$

where C is taken as equal to the depth of thread cut, with a minimum value of 1.25 mm (0.05 in).

(v) In the case of plastic tubes, an allowance is made for the higher elastic moduli of such materials, when a suitable formula is:

$$P_W = \frac{2S_{max}t}{D - t}$$

Actuators

THE BROAD definition hydraulic *actuators* covers linear actuators (single- and double-acting cylinders); semi-rotary actuators; pulse motors; and (full) rotary actuators (hydraulic motors) used specifically for actuator duties as opposed to motor drives. Components normally involved are hydraulic cylinders and semi-rotary actuators, although pulse motors offer particular advantages for precision positioning movements.

Linear Actuators

The ideal operating conditions of a linear actuator (hydraulic cylinder) are represented by a constant load moved with constant velocity over the whole of the stroke, with the resulting load-travel diagram shown in Fig 1. This is seldom realized in practice since the usual motion is acceleration from rest which may or may not be followed by a period of uniform velocity and possibly deceleration approaching the end of the stroke (*eg* if the cylinder is fitted with a cushion).

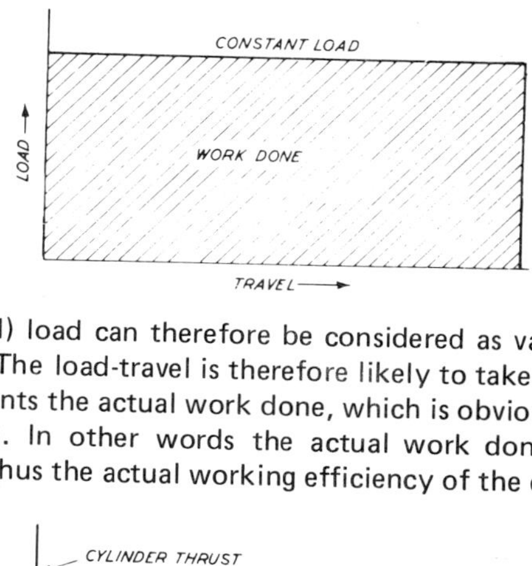

Fig 1

The actual (practical) load can therefore be considered as variable, with the work done by the actuator also variable. The load-travel is therefore likely to take the form shown in Fig 2. The area under this curve represents the actual work done, which is obviously less than that under ideal conditions (constant load). In other words the actual work done by the cylinder is less than its potential capacity and thus the actual working efficiency of the cylinder is reduced.

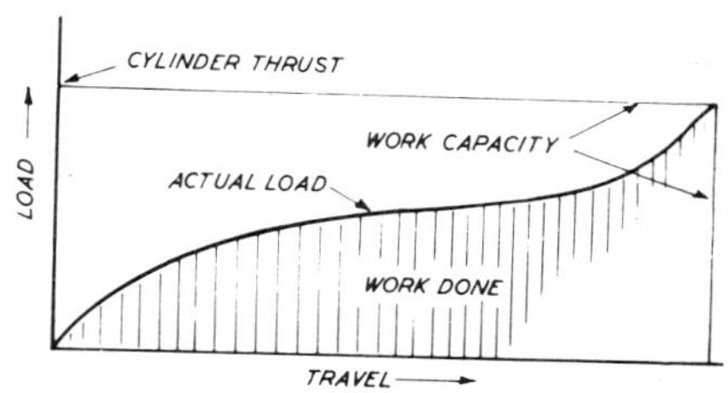

Fig 2

Specifically, the force necessary to accelerate (or decelerate) a load is equal to the load divided by the acceleration due to gravity, multiplied by the resulting acceleration (a), *viz*

$$F = \frac{W \cdot a}{g}$$

This can be expressed in terms of an *acceleration factor,* (g_f) defined as $g_f = a/g$. Acceleration can be defined in terms of resultant velocity (V) produced over the stroke (S), when:

$$a = \frac{V^2}{2S}$$

Friction will also normally be present; which must be added to the total force required to accelerate a load, or subtracted from the total force required to decelerate a load. In general the frictional forces inherent in hydraulic cylinders is negligible and only the frictional forces associated with movement of the load itself are significant. The final force equation for horizontal movements thus becomes:

$$F = W \cdot g_f \pm W \cdot f$$
$$= W \cdot (g_f \pm f)$$

Where the load is moved vertically, the load itself represents a further force to be overcome (for upward movement) or assisting the actuator force (for downward movement). Thus for vertical movement:

$$F = W\,(g_f \pm f) \pm W$$

Since the friction force may or may not be of the same sign as W, the four possible conditions are detailed below for clarity –

Load movement	*Force value*
accelerating upwards	$F = W(g_f + f) + W$
decelerating upwards	$F = W(g_f - f) - W$
accelerating downwards	$F = W(g_f + f) - W$
decelerating downwards	$F = W(g_f - f) + W$

One further case needs to be considered. That is where hydraulic force generated by the actuator continues to be applied during deceleration of the load. This is the case with a cushioned cylinder approaching the end of its stroke or where separate cushioning devices are used to decelerate the moving load. For horizontal movements, the total force involved during deceleration then becomes:

$$F = W(g_f - f) + F_h$$

where F_h is the hydraulic force present during deceleration (*eg* normal cylinder force)

For vertical movements this is modified by the load, *ie* by –W for upward movement deceleration; or +W for downward movement deceleration.

This total force represents the force present in the cushioned side of the cylinder.

Load-Travel Diagrams

Where the load-travel diagram shows that the work done is substantially less than the theoretical work capacity of the cylinder, some modification of the geometry may be beneficial so that the load may be spread more evenly over the travel, increasing the area under the curve and enabling the maximum force requirement to be reduced. This could enable the same duty to be performed by a smaller cylinder. This is particularly so in the case of a load-travel curve with rising characteristics, when the actual efficiency of cylinder operation achieved may be less than 25%. If a pre-load can be applied so that the resultant of pre-load and normal load is more nearly horizontal, as in Fig 3, not only is the maximum force required substantially reduced, but the area under the load-travel curve now shows very much greater efficiency. Despite the fact that the cylinder now has to work against an artificial load as well as the normal load, a much smaller cylinder can be used for the duty required, because it is being worked so much more efficiently.

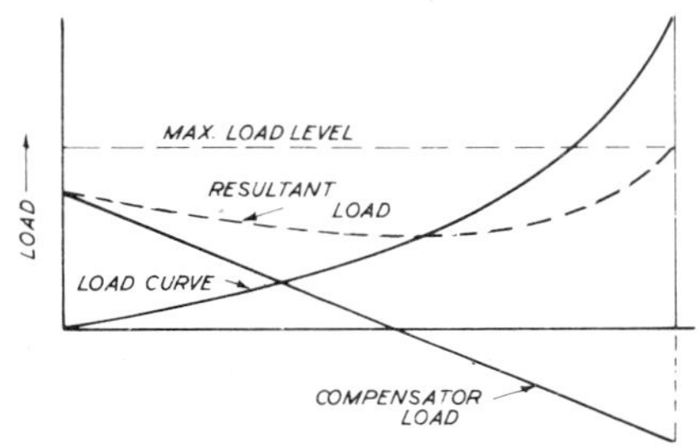

Fig 3

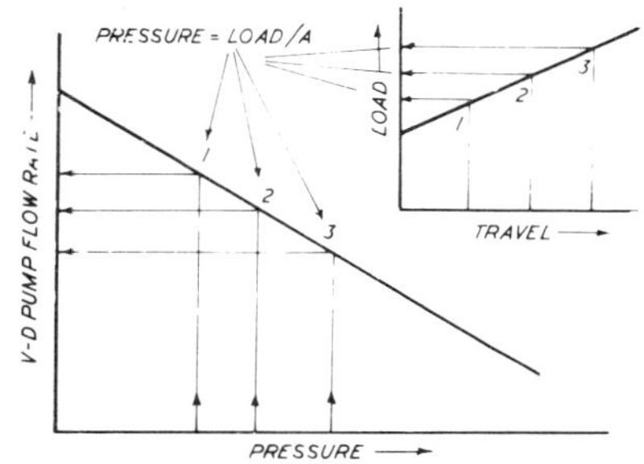

Fig 4

Pre-load devices normally take the form of springs or auxiliary cylinders. In the latter case, the auxiliary cylinder may be connected to the main circuit, or energized separately from an accumulator, or even a gas bottle. They are generally referred to as compensators. Compensators can, quite obviously, only be used in specific cases, but their possibilities should not be overlooked.

The load-travel curve also provides a means of estimating the operating time of a cylinder with a good degree of accuracy, particularly where a variable delivery pump is being employed. Specific points can be taken on the load-travel curve, from which the pressure required at each point can be determined by first dividing the load by the cylinder bore area. These values can be transferred to the pump characteristic curve to read delivery available at each pressure – Fig 4.

In practice this will over-estimate performance, since the pressure calculation is derived directly from the load only. For a true value, the total pressure drop in the lines (taking into account back-pressure effects, if significant) must also be added to this value to obtain the true pump delivery – Fig 5. For most purposes an estimate of pressure drop may be sufficient, unless the speed of operation is to be determined very accurately.

The actual velocity of the piston at each point can then be calculated by dividing the delivery available by the cylinder area, which can be plotted as a further curve of the *reciprocal* piston

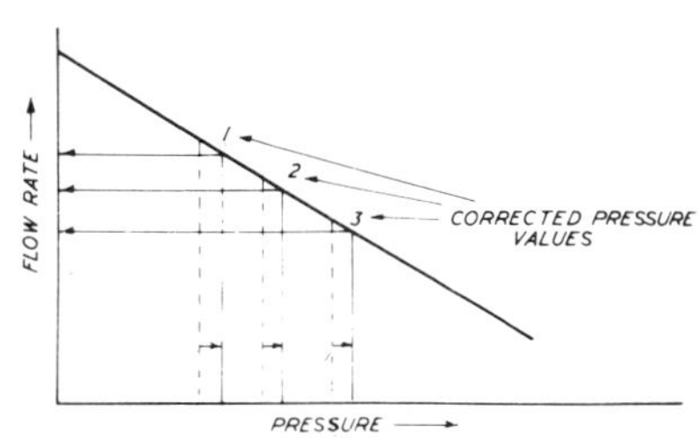

Fig 5

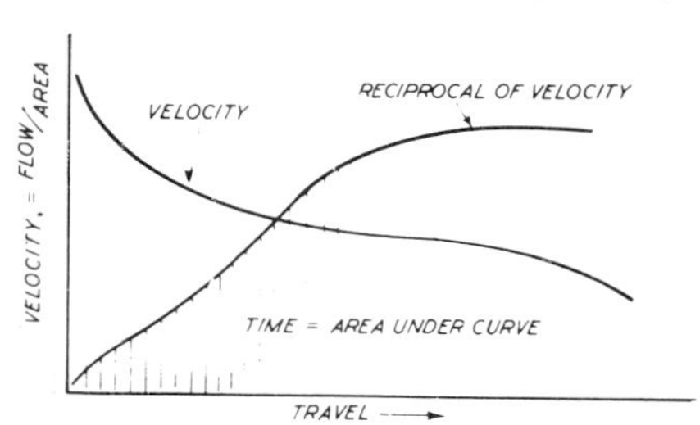

Fig 6

velocity against travel — Fig 6. Enough points should be taken on the initial curve to enable a realistic velocity-travel curve to be plotted, paying particular attention to any irregular regions of the original load-travel curve. The area under this curve will then give the full time of operation. If, for any reason, piston velocity needs to be known at particular points of the stroke, this can be plotted separately as velocity against travel.

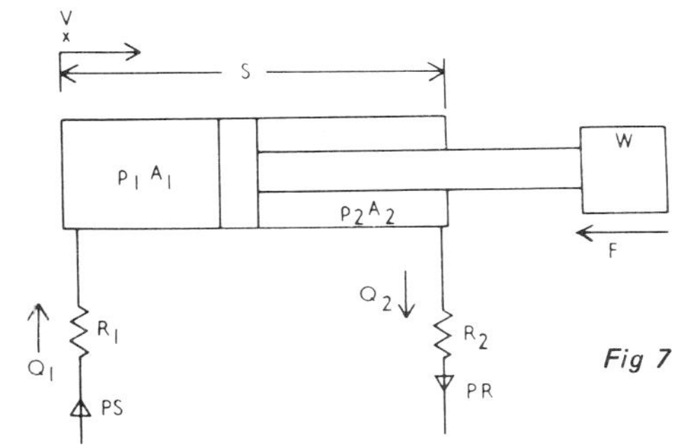

Fig 7 Theoretical circuit model.

Theoretical analysis of the same subject can be based on the parameters defined in Fig 7. The basic energy relationship is:

$$P_S - P_1 = R_1 \cdot Q_1^2$$

$$P_2 - P_R = R_2 \cdot Q_2^2$$

which, because P_R is usually small enough to be negligible can be taken as:

$$P_2 = R_2 \cdot Q_2^2$$

Also, in the absence of compressibility, $V = Q_1/A_1 = Q_2/A_2$

The dynamic equation for the cylinder and mass extending can be shown to be:

$$P_S - R_e\, Q_1^2 - P_f = \frac{P_W}{gA_1} \cdot \frac{dQ_1}{dt}$$

where R_e = equivalent resistance $= R_1 + \dfrac{R_2}{\delta\eta}\left(\dfrac{A_2}{A_1}\right)^3$

P_f = static pressure coefficient

$$= \frac{F}{\eta A_1}$$

where η is the efficiency of the cylinder

p_W = dynamic pressure coefficient

$$= \frac{W}{\eta A_1}$$

where W is the weight moved by the cylinder

If fluid motion is taken in account the weight (W) is replaced by equivalent weight (W_e) where:

$$W_e = W + \left(\frac{l_1}{a_1} A_1^2 + \frac{l_2}{a_2} A_2^2\right)$$

where l_1 is the length of inlet pipe
a_1 is the cross sectional area of inlet pipe
l_2 is the length of outlet pipe
a_2 is the cross sectional area of outlet pipe

If X defines the position of the rod less than the total stroke, the following 'working' equations apply:

$$\frac{X}{Y} = \log_e\left(\frac{1}{1-\delta^2}\right)$$

$$\text{where } \delta = \frac{\text{velocity at point X}}{\text{mean velocity}} = \frac{V_x}{V_m}$$

$$Y = \frac{V_m^{\,2}}{g}\quad\frac{P_w}{P_s - P_f} = \frac{W}{\eta\,\alpha\,A_1^{\,3}\,g}$$

$$\frac{T}{\gamma} = \log_e\left(\frac{1+\delta}{1-\delta}\right)$$

$$\text{where } \gamma = \frac{V_m}{g}\left(\frac{P_w}{P_s - P_f}\right) = \frac{Y}{V_m}$$

Numerical values for $\frac{T}{\gamma}$ and $\frac{X}{Y}$ are given in Fig 8 related to δ. Fig 9 shows the relationship between $\frac{T}{\gamma}$ and $\frac{X}{Y}$. This particular diagram shows that the gradient of the curve approaches 1, so that approximations may be made as below.

(i) $\frac{X}{Y} > \delta\ T \approx \frac{X}{V_m}$

(ii) $\delta > \frac{X}{Y} > 0.3\ T \approx \frac{1.1X}{V_m} + \frac{V_m}{2g}\quad\frac{P_w}{P_s - P_f}$

(iii) $0.3 > \frac{X}{Y} > 0$ T obtained from Figs 9 and 10.

Use of Figs 8 and 9

In the application of Figs 8 and 9, it is necessary to consider two cases:

(i) The steady state velocity is less than that available from the pump flow or pressure-compensated flow control valve, *ie*

$Q_m < Q_p$ or $V_m < V_p$

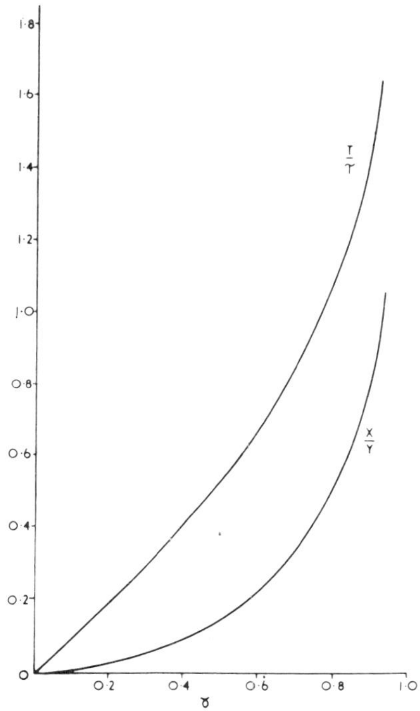

Fig 8

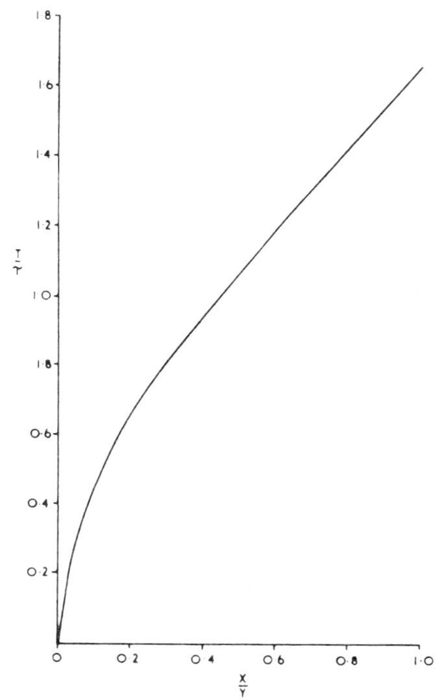

Fig 9

(The theoretical steady-state flow is that caused by the maximum available pressure difference applied across the circuit resistances).

In this case the equations and hence the graphs can be used to find the time to complete the stroke either to the limit stops or to the cushion.

The various quantities are calculated together with the appropriate value of X.

and $\frac{X}{Y}$ is calculated.

From Fig 9 the value of $\frac{T}{\gamma}$ and hence T is determined.

Fig 10(a) shows the pressures and velocity along the stroke.

(ii) The flow available is less than the steady-state value, *ie*

$Q_p < Q_m$ and $V_p < V_m$

The maximum value of δ possible is:

$$\delta = \frac{Q_p}{Q_m}$$

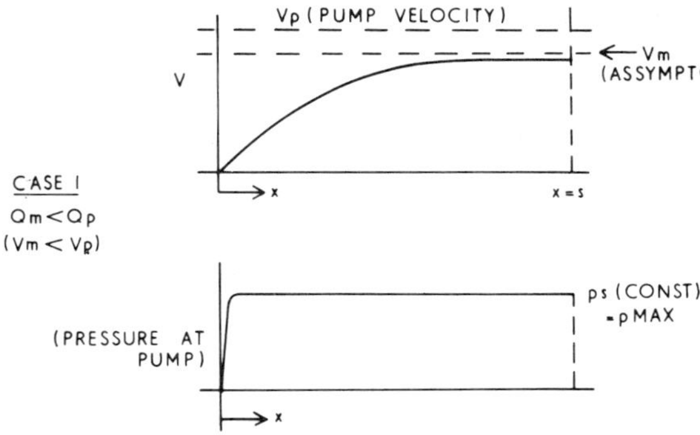

Fig 10(a)

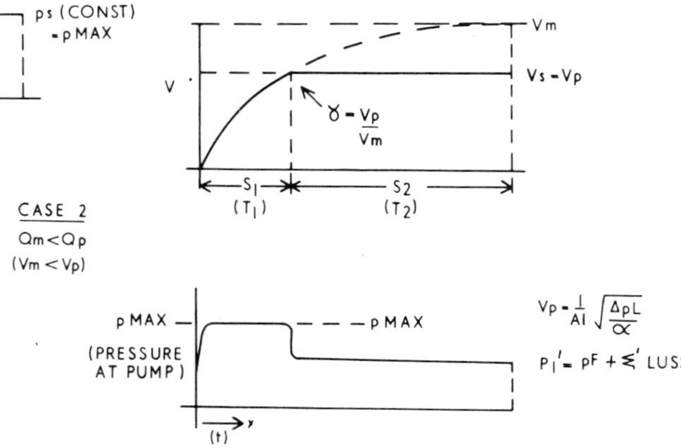

Fig 10(b)

$V_p = \frac{1}{A_1}\sqrt{\frac{\Delta p L}{\alpha}}$

$P_1' = p_F + \xi'$ LUSS

This value is now used, with Fig 10, and the various calculated factors to determine the distance (S_1) at which this occurs. Then the time to reach this point can be derived from Fig 8, *ie* T_1.

From this point the ram will travel at a uniform velocity of V_p and hence the time to complete the remaining part of the stroke can be calculated.

$$T_2 = \frac{S_2}{V_p}$$

The total stroke $S = S_1 + S_2$

Total time $T = T_1 + T_2$

The actual pressure at the pump will fall once the point of maximum velocity is reached, because

$$P_p - P_f = Q_p^2 \, \alpha \text{ instead of } Q_m^2$$

Therefore, for any fixed value of P_f, P_p will be less than the maximum pump pressure of P_s. Fig 10(b) shows how the velocity and pressure change vary throughout the stroke.

For the case where P_f is zero it is possible to determine an area A, for minimum operating time, assuming a given ratio of A_1/A_2 and that $Q_p > Q_m$, *viz*

$$A_1 \quad \frac{1.145W}{g\,\alpha\,S}^{1/3}$$

giving

$$T_{min} = \frac{1.182W}{\eta\, g\, A_1 \sqrt{\alpha\, P_s}}$$

Compressibility 'Delay'

Theoretically, at least, a constant delivery pump will take a finite time to build up pressure although in fact this is very small and can be disregarded for most practical applications. If very rapid transit times are involved, then it may be necessary to determine the time needed to compress the fluid, which is:

$$t_C = \frac{P_S V}{Q_p N}$$

where P_S = is the supply pressure at the pump outlet
V = is the pump volume
Q_p = pump delivery
$N = V \dfrac{dp}{dv}$

Compressibility Effects

The performance of a cylinder can also be modified by the compressibility of the fluid, this being generally known as *compliance.* In practice this effect is negligible at low to moderate pressure, but can become significant in critical applications at pressures of the order of 140 bar (2 000 lb/in^2) or greater.

Cylinder compliance is related to the fluid volumes on each side of the piston and the bulk modulus of the fluid.

$$\text{compliance } (\lambda) = \frac{1}{\beta\left(\dfrac{1}{V_1} + \dfrac{1}{V_2}\right)}$$

where β = fluid bulk modulus
V_1, V_2 are the fluid volumes on each side of the piston

Compliance will be a maximum when $V_1 = V_2$, and the position of the piston giving maximum compliance is thus:

$$L_1 A_1 = L_2 A_2$$

where L_1 and L_2 are the effective cylinder port lengths and
A_1 and A_2 are the effective piston areas

In the case of a through-rod cylinder:

$$\text{maximum compliance} = \frac{L}{4A}$$

where L is the cylinder stroke

For critical applications, cylinder compliance can be determined over the full stroke range, as in Fig 11. The compliance ratio in this case is the ratio of the actual compliance at any particular stroke position to the maximum compliance.

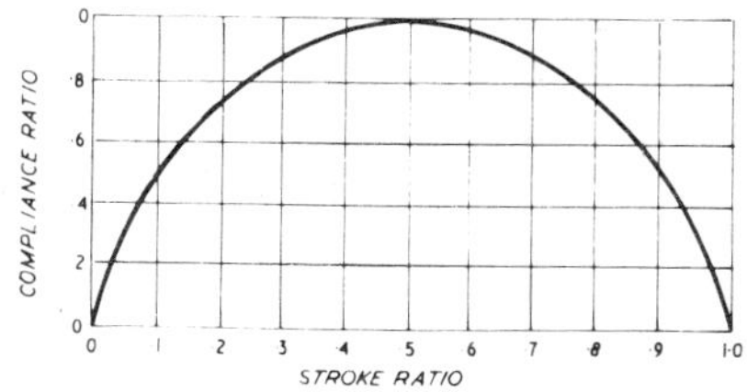

Fig 11

Strictly speaking, for accurate analysis, the compliance of the fluid column lengths in the lines to the cylinder should also be taken into account.

The significance of compliance is that a cylinder is essentially a non-linear actuator except for very small displacements at around the point of maximum compliance, and departs mostly from linear characteristics at each end of the stroke. In practice, however, the actual compliance present usually renders the non-linear characteristics negligible. Thus, for general applications, a cylinder can be considered to be a true linear actuator.

Semi-Rotary Actuators

Semi-rotary actuators are normally based on cylinders, or a special form of limited-movement vane 'motor'. The force available is in the form of *torque* produced on the output shaft. In the case of cylinder types this is directly proportional to the cylinder force modified by the form of internal gearing or linkage and the resultant frictional losses in the mechanical system. In the case of vane-type semi-rotary actuators, the output torque is proportional to the effective vane area modified by internal leakage (*ie* the effective pressure acting on the vane(s) is reduced by leakage). The only frictional losses are those of the seals (usually negligible) and the output shaft bearing(s).

Performance characteristics of semi-rotary actuators are thus largely determined on empirical grounds.

Motors

The theoretical torque available from a hydraulic motor is given by:

$$\text{Torque (T)} = K_1 \frac{P \cdot Q}{N}$$

P = system pressure
Q = fluid delivery rate
N = rev/min

$$\text{also } \frac{Q}{N} = K_2 V_m$$

where V_m is the volume displacement of the motor per revolution.

It follows that:

$$\text{Torque (T)} = K_3 \cdot P \cdot V_m$$

where K_1, K_2 and K_3, are constant factors depending on the units employed for P, Q and V_m

In practice a hydraulic motor is subject to both fluid losses and mechanical losses, so actual performance is related both to its *volumetric efficiency* and *mechanical efficiency.* Volumetric efficiency is readily determined as a percentage by:

$$\eta_v = K_4 \frac{V_m \cdot N}{Q_a} \times 100$$

where Q_a = actual volume flow rate applied to the motor and
K_4 is the 'unit' factor

Overall efficiency (η_o) can be determined by measuring the actual power output and comparing with the theoretical power. Mechanical efficiency then follows as:

$$\eta_m = \frac{\eta_o}{\eta_v} \times 100$$

where all efficiencies are expressed as percentages.

Pulse Motors

Pulse motors are essentially electro-hydraulic devices, correctly described as electro-hydraulic pulse motors (EHPMs). Basically they comprise a low-power digital electric stepping motor combined with a precision high-torque hydraulic motor. Square-wave digital pulses are applied as input signals to the stepping motor which then controls the direction, speed and position of the hydraulic motor without the need for feedback. Thus an EHPM operates on an open-loop basis. The basis of control is that shaft speed is proportional to the signal pulse *rate;* position to the *number* of input pulses; and direction of rotation depends on the pulse *sequence.*

See also chapters on *Cylinders, Rotary Actuators* and *Motors.*

System Design and Performance

A FLUID can be pressurized directly by the application of a force over the whole of one free surface or indirectly by being pumped. In the former case the unit pressure applied (force per unit area) is transmitted through the fluid without loss in a purely hydrostatic system, and is independent of any loading. In the case of a pump, mechanical energy is converted into fluid energy, using the power input to the pump to change the pressure level of the fluid passing through the pump, and thence into the system. The pump used is almost invariably of positive displacement type and hence the discharge pressure is largely independent of load or 'head'; in favourable cases pump efficiency can be in excess of 95%.

Pump volumetric efficiency can readily be determined by taking two readings at the intended operating speed, one at zero delivery pressure and one at the required pressure. This simple test is valid only if the pump is not cavitating under zero delivery pressure. If so, the pump will have to be pressurized to the extent necessary to ensure full displacement.

The mechanical efficiency of a pump can be calculated on a theoretical basis, namely by comparing the calculated input power required to raise the theoretical delivery of the fluid to a given pressure with the actual power input required. Again this is only true if the pump is not cavitating.

The pressurized output from a pump is not necessarily constant. Depending on the type of pump, valve timing (and the number of cylinders in the case of a piston pump), pressure may be in a series of pulses. Such cyclic variations usually occur at a frequency equal to the product of the pump speed and the number of pump elements involved, which is normally too high to cause jerky movement. In certain applications, however, this pump 'ripple' may be objectionable, or be emphasized by system elasticity and require damping. Pressure fluctuations may also be produced by variations in delivery rates, and these can be more objectionable than pump ripple.

In the majority of cases of force-transmitting systems, the pressure transmitter is a circulating pump working in a closed circuit. The delivery is then determined by the rated pump performance, *ie* the quantity of fluid it can deliver at the specified pressure.

Pressure and Head

System performance is normally calculated on the basis of system pressure; however, equivalent *head* of fluid is sometimes used. The relationship between the two is:

Pressure (P) = Head (H) x specific weight of fluid (w).

Working formulas are:

$$P\ (\text{lb/in}^2) = 0.434\ H\ (\text{feet}) \times \text{sg of fluid}$$
$$= 0.37\ H\ (\text{feet})\ \text{for typical hydraulic oils}$$

$$H \text{ (feet)} = \frac{2.3 \times P \text{ (lb/in}^2\text{)}}{\text{sg of fluid}}$$

$$= 2.67 \times P \text{ (lb/in}^2\text{) for typical hydraulic oils}$$

$$P \text{ (bar)} = H \text{ (metres)} \times \text{sg of fluid}$$

$$= 0.86\ H \text{ (metres) for typical hydraulic oils}$$

$$H \text{ (metres)} = \frac{10 \times P \text{ (bar)}}{\text{sg of fluid}}$$

$$= 11.63 \times P \text{ (bar) for typical hydraulic oils}$$

Velocity of Pressure Transmission

When a fluid is pressurized, the velocity of transmission of that pressure through the fluid is given by:

$$V_p = \frac{Kg}{\rho}$$

where K = bulk modulus of the fluid
ρ = density of fluid

For all practical purposes the velocity of pressure transmission can be considered instantaneous *ie* is of the order of 1 470 metres/sec (4 840 ft/sec) through a typical hydraulic oil.

Flow Velocity

Pipes are always assumed to flow full, when the following basic relationship applies:

$$\text{flow rate } (Q) = \frac{\pi d^2 V}{4} = 0.785 d^2 V \quad \text{in consistent units}$$

$$\text{or } (V) = \frac{4Q}{\pi d^2} = \frac{1.273Q}{d^2} \quad \text{in consistent units}$$

where d = bore size
V = velocity.

Working formulas are:

$$Q \text{ (lit/min)} = 0.047 d^2 V$$

where d is in mm
V is in m/sec

$$Q \text{ (Imp gal/min)} = 2.04 d^2 V$$

$$Q \text{ (US gal/min)} = 2.45 d^2 V$$

where d is in inches
V is in ft/sec

Flow velocity (V) is calculated on a mean basis, the effect of velocity gradient due to boundary friction and fluid velocity being negligible in most practical cases. Localized higher or lower (mean) velocities, however, may well occur at bends, abrupt changes in pipe section, or in flow through ports and valves, *etc.*

Any change of pipe section will initiate some disturbance of flow, subsequently settling down with a new (mean) velocity V_2, when:

$$V_2 = \frac{1.273Q}{d_2^{\ 2}}$$

where d_2 = new bore size.

Pipe Sizing

Pipe sizes are normally chosen on the basis of arbitrary limits for flow velocity, or more rarely on the basis of design pressure drop. Velocity is significant only as a parameter affecting flow losses (although excessive flow velocities can have other adverse effects). For engineering purposes it is usually more convenient to work in terms of flow rate (Q), or the demands of the system.

In principle any size of pipe is capable of providing any delivery within practical limits of flow velocity. In practice pipe size will govern the time required to meet a given volumetric demand, as flow velocities are necessarily limited to avoid excessive frictional losses.

For a majority of systems a suitable pipe size can be determined by adopting arbitrary values for flow velocity which ensure moderate and acceptable pressure drop. These are:

3 m/sec (10 ft/sec) for delivery lines,

1 m/sec (3 ft/sec) for intake and return lines.

For further details of line sizing see chapter on *Pipework Calculations.*

Hydraulic Horsepower

Hydraulic horsepower is represented by the rate at which useful energy is delivered by, or to, the fluid.

$$\text{Hydraulic hp} = \frac{Q\,w\,H}{K_h}$$

where Q = flow rate

w = specific weight of fluid

K_h is an energy/time constant consistent with the units employed (see Table I).

Hydraulic horsepower can also be expressed directly in terms of pressure (P) rather than head, when it is independent of the specific weight of the fluid.

$$\text{Hydraulic hp} = \frac{Q\,P}{K_p}$$

where K_p is an energy/time constant consistent with the units employed (see Table I).

TABLE I – CONSTANT K_h FOR HYDRAULIC HORSEPOWER

Q	P	w	H	K_h	K_p
in^3/sec		lb/ft^3	feet	940 400	
in^3/sec	lb/in^2				5 500
gallons/min		lb/ft^3	feet	205 600	
gallons/min	lb/in^2				1 428
litres/min		kg/litre	metres	363 000	
litres/min	bar				92 000

Cylinder Performance

The thrust generated by a hydraulic cylinder is directly related to its (effective) piston area and actual pressure of the fluid entering the cylinders (*ie* system pressure less pressure drop through pipework and valves to the cylinder, and pressure drop through the inlet port). Full theoretical values are normally assumed, less a nominal allowance for frictional losses – see chapter on *Hydraulic Cylinders*.

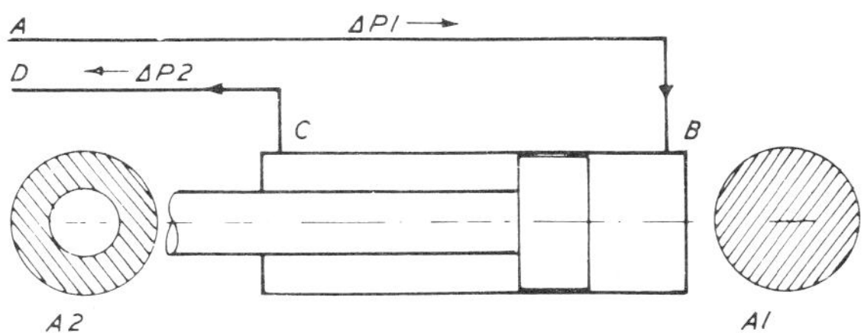

Fig 1

Actual cylinder performance will, however, be modified by *back-pressure*. The effect of back-pressure can be followed with reference to Fig 1. Assuming (for the sake of simplicity) that the lines to the cylinder are of equal length and size, pressure drop in either line will be proportional to the flow rate in each line (for different line lengths and/or line sizes, exact pressure drops can be calculated separately if back-pressure is likely to be significant).

With the cylinder extending, the flow rate in the inlet line AB will be proportional to the piston area A_1 and the flow rate in the outlet line CD will be proportional to the annulus area A_2. The pressure drops in the lines AB and CD will thus be in the ratio $A_1 : A_2$.

The pressure drop in line AB will represent an initial pressure drop ΔP_1, from the system pressure applied at the far end of the line. The pressure drop in line CD will represent a *back-pressure* on the piston, opposing movement, which can only be overcome by further pressure on the other side of the piston. This additional pressure represents a further loss of effective pressure. However, since the piston area is greater than the annulus area (to which the back-pressure is applied), this additional loss will be less than the actual back-pressure (P_2) in the ratio $A_2 : A_1$. In other words:

pressure drop in the inlet line = ΔP_1

additional loss of effective pressure due to overcoming back-pressure in return line

$$= \frac{A_2}{A_1} \times \Delta P_2$$

Therefore true effective pressure (P_e) equals system pressure (P) less each of these losses:

$$P_e = P - P_1 - \frac{A_2}{A_1} \cdot \Delta P_2$$

Also:

$$P_2 = \frac{A_2}{A_1} \cdot P_1$$

because of the flow rate relationship previously established.

This can be rendered more conveniently in the form:

$$\text{Total pressure drop} = P\left(1 + \frac{1}{r}\right)$$

where P is the calculated pressure drop in the inlet line to the cylinder
r is the ratio of effective piston areas or $A_1 : A_2$

With the reverse direction of operation, line DC becomes the inlet and line BA, the outlet. Similar analysis will show that in this case:

$$\text{Total pressure drop} = P\left(r + \frac{1}{r}\right)$$

Such a method of calculation enables an accurate estimate of pressure drop to be made, which may be necessary for a critical application. It will show how performance can be modified if a large differential exists between the two back-pressures, as could occur with a large area ratio (*ie* larger than usual rod diameter). Such loss of performance will generally be apparent by one or another of the cylinder movements slowing up to adjust to the considerable difference in total pressure drops. In normal applications, however, back-pressure effects are usually ignored, and their possible effect on performance is usually negligible provided the ratio of areas is 4:3 or less (4:3 being the usual proportion for cylinders with standard rod sizes).

Cylinders — Speed of Operation

Speed of operation is related to cylinder displacement and actual delivery rate achieved by the pump, *viz*

$$\text{time (seconds)} = \frac{A_e L}{qP} = \frac{A_e L}{Q} \quad \text{in consistent units}$$

where A_e is the effective piston area (*ie* $0.2854 \times D^2$ or $0.2854 \times (D^2 - d^2)$)
Q is the pump delivery

It follows that this relationship also provides a means of calculating the pump delivery required to achieve a given speed of operation. Working formulas which can be used to calculate the speed of operation of a cylinder (or pump delivery necessary to achieve a required operating speed) are given in Table II. It should be noted, however, that these are theoretical figures and may be modified in practice by throttling effect of the inlet and outlet ports (see section on Speed Control later) and by back-pressure.

In these formulas D, d and L are specified in inches and D_m, d_m and L_m in centimetres.

See also chapter on *Hydraulic Cylinders.*

TABLE II – OPERATING TIME OF CYLINDERS

	Single-rod	Through-rod
Single-acting cylinders	$\text{Time (secs)} = \dfrac{47.12\,D^2\,L}{Q}$	$= \dfrac{47.12\,(D^2 - d^2)\,L}{Q}$
	where Q is in in^3/min	
	$\text{Time (secs)} = \dfrac{0.17\,D^2\,L}{Q}$	$= \dfrac{0.17\,(D^2 - d^2)\,L}{Q}$
	where Q is in gallons/min	
	$\text{Time (secs)} = \dfrac{0.774\,D^2\,L}{Q}$	$= \dfrac{0.774\,(D^2 - d^2)\,L}{Q}$
	where Q is in litres/min	
	$\text{Time (secs)} = \dfrac{D_m^2\,L_m}{1\,000\,Q}$	$= \dfrac{(D_m^2 - d_m^2)\,L_m}{1\,000\,Q}$
	where Q is in litres/min	
Double-acting cylinders Extending:	$\text{Time (secs)} = \dfrac{0.7854\,D^2\,L}{Q}$	$= \dfrac{0.7854\,(D^2 - d^2)\,L}{Q}$
	where Q is in in^3/sec	
	$\text{Time (secs)} = \dfrac{47.12\,D^2\,L}{Q}$	$= \dfrac{47.12\,(D^2 - d^2)\,L}{Q}$
	where Q is in in^3/min	
	$\text{Time (secs)} = \dfrac{0.17\,D^2\,L}{Q}$	$= \dfrac{0.17\,(D^2 - d^2)\,L}{Q}$
	where Q is in gallons/min	
	$\text{Time (secs)}\ (0.) = \dfrac{0.774\,D^2\,L}{Q}$	$= \dfrac{0.774\,(D^2 - d^2)\,L}{Q}$
	where Q is in litres/min	
	$\text{Time (secs)} = \dfrac{D_m^2\,L_m}{1\,000\,Q}$	$= \dfrac{(D_m^2 - d_m^2)\,L_m}{1\,000\,Q}$
	where Q is in litres/min	
Retracting:	$\text{Time (secs)} = \dfrac{0.7854\,(D^2 - d^2)\,L}{Q}$	$= \dfrac{0.7854\,(D^2 - d^2)\,L}{Q}$
	where Q is in in^3/sec	
	$\text{Time (secs)} = \dfrac{47.12\,(D^2 - d^2)\,L}{Q}$	$= \dfrac{47.12\,(D^2 - d^2)\,L}{Q}$
	where Q is in in^3/min	
	$\text{Time (secs)} = \dfrac{0.17\,(D^2 - d^2)\,L}{Q}$	$= \dfrac{0.17\,(D^2 - d^2)\,L}{Q}$
	where Q is in gallons/min	
	$\text{Time (secs)} = \dfrac{0.774\,(D^2 - d^2)\,L}{Q}$	$= \dfrac{0.774\,(D^2 - d^2)\,L}{Q}$
	where Q is in litres/min	
	$\text{Time (secs)} = \dfrac{(D_m^2 - d_m^2)\,L_m}{1\,000\,Q}$	$= \dfrac{(D_m^2 - d_m^2)\,L_m}{1\,000\,Q}$
	where Q is in litres/min	

Velocity Limits

In practice it is generally recommended that flow velocities in inlet lines should be limited to a maximum of 4.5 metres/sec (15 ft/sec) to minimize turbulence and pressure drop and hydraulic shock. Table III is then a general guide to the piston speeds likely to be achieved with typical standard and oversize cylinder ports. Where higher speeds are required, the inlet line size can be increased to increase delivery whilst maintaining a maximum flow velocity of 4.5 metres/sec (15 ft/sec), with delivery to the cylinder through two (or more) ports.

Velocity limits may also be imposed by the construction of the cylinder. This is to say, excessive velocities may be potentially damaging, or may affect cushioning suitability. Thus, in general, if the intended operating velocity of a hydraulic cylinder exceeds 35 metres/min (120 ft/min), the supplier should be consulted regarding the suitability of the cylinder (and cushion, where applicable) for working at such piston speeds.

The operating speed from a pump of given (fixed) delivery can, of course, also be altered by varying the size of the cylinder used, if this is possible within the output force requirements of the system. Thus the use of a small cylinder will give increased speed of operation at the expense of reduced output.

For particularly accurate determination of the speed of operation of a hydraulic cylinder the load-travel curve should be plotted — see chapter on *Actuators.*

Speed Control

The speed of a cylinder can be controlled by means of a restrictor fitted either in the inlet line (metering-in) or the outlet line (metering-out) — Fig 2. Of the two, metering-out is generally preferred. A restrictor can, of course, be placed in both lines when it is necessary to control the speed in both directions of operation.

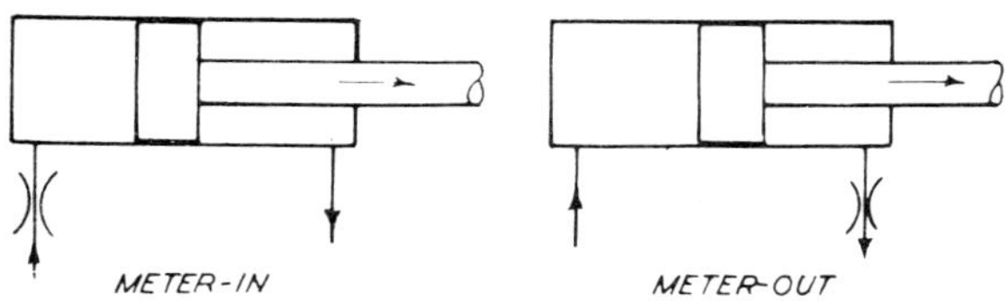

Fig 2

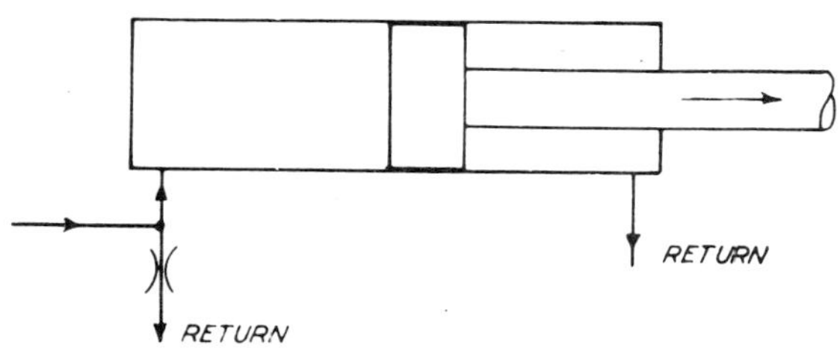

Fig 3

A further method is bleed-off control — Fig 3. General recommendations for the application of these alternative methods are given in Table III.

It is generally necessary to associate the restrictor with a non-return valve in a bypass loop, when speed is controlled in one direction only, to allow full flow through that line in the reverse direction of motion. However, it is possible to place the restrictors on the pump side of the control valve, which eliminates the need for non-return valves. In this case, a 5/4 valve is necessary if speed control is required in both directions in order to separate the return flows — Fig 4.

TABLE III – SPEED CONTROL APPLIED TO HYDRAULIC MACHINE TOOLS

Type	Recommended Applications
Meter-in	Feeds on grinder tables, welding machines, milling machines, *etc,* rotary hydraulic motor drives.
Meter-out	Boring, drilling, reaming, turning, threading, tapping, cut-off and cold swaging machines.
Bleed-off	Reciprocating grinding tables, broaching machines, honing machines, *etc*, rotary hydraulic motor drives.

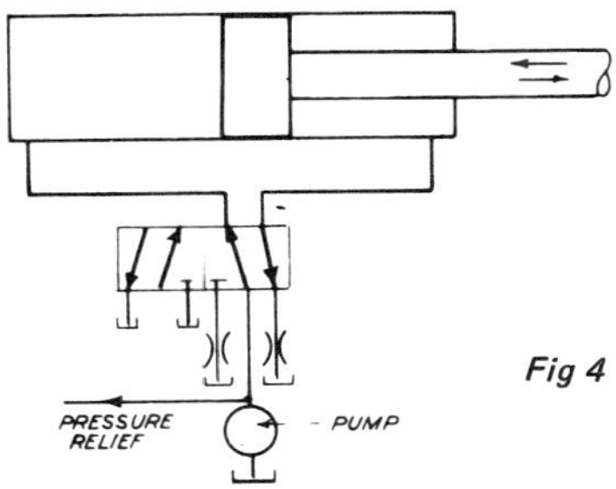

Fig 4

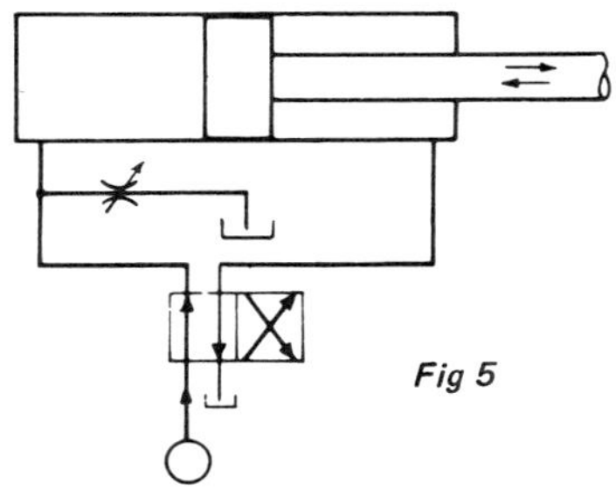

Fig 5

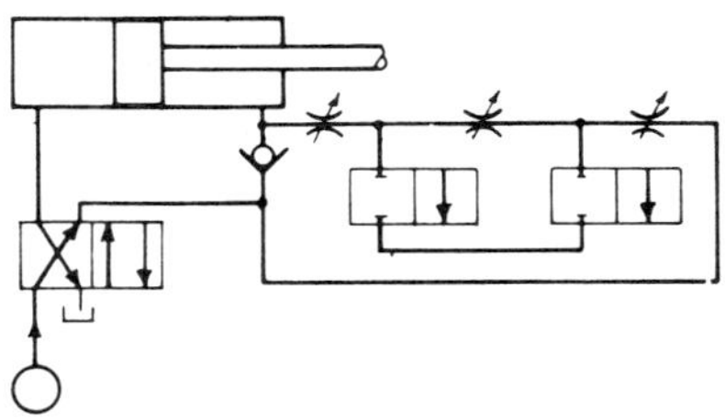

Fig 6

An alternative method is shown in Fig 5, using a restrictor in parallel with the cylinder instead of in series. This has the advantage that full system pressure is available across the cylinder. In this case the restrictor cannot be bypassed by a non-return valve and so the flow through the restrictor is effective in both directions. If it is required to eliminate the effect of the bleed (through the restrictor) in one direction, this can only be done by incorporating an additional 2-way valve in series with the restrictor to shut off the restrictor line when that direction of motion is initiated.

Where multiple speed control is required over a single stroke, two or more restrictors may be used in series (Fig 6) or in parallel. The former is by far the more critical arrangement and needs greater care in setting up and adjustment.

It should be noted that if simple restrictors are used, the speed obtained will only be constant with the piston under constant load. Any variations in load or flow conditions will produce variable speed of operation. If positive speed control is required, independent of load, then a pressure-compensated flow restrictor must be used. Requirements differ depending on whether metering-out or metering-in speed control is used. In the former case, the pressure-compensated flow restrictor comprises, basically, a restrictor in series with a reducing valve. For metering-in operation, downstream pressure needs feeding back to the reducing valve to act as a base reference for the reducing section. It is important also that the flow restrictor be designed to cope with a suitable flow ratio, as operating outside the design range will result in loss of compensation. A flow rate ratio of 3:1 is fairly commonplace, but very much higher figures can be achieved with special designs.

Output Control

Since the output force for a given size of cylinder depends only on the pressure (or, more specifically, the pressure drop between the inlet and outlet ports), it can be controlled by controlling the pressure, or the pressure drop. The former is simpler, since only the inlet pressure needs to be controlled. For more exact control the back-pressure should also be regulated, although this is usually small enough for the effect of such control to be negligible.

Inlet pressure may be regulated quite simply by fitting a pressure-control valve in the inlet line – Fig 7 – although this will be wasteful of power. There is also the possibility of an over-centre load 'running away', although this risk can be eliminated by fitting a pressure-relief valve in the outlet line. This will act as a counter-balance valve, maintaining any back-pressure, which will limit the force available and any chance of running away. Relief valves can also be used in the outlet line to relieve excessive back-pressure, which could apply undue braking effect. Where controlled output is required, therefore, due consideration should be given to other aspects of the working stroke, as well as simple pressure control.

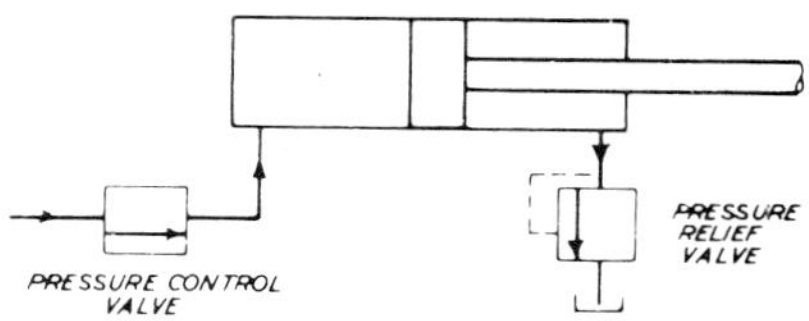

Fig 7

Pump Heating

In a practical hydraulic circuit the pump may be called upon to operate continuously against an intermittent demand. If the pump was left pumping its full output through a relief valve when there was no system demand, the bulk of the output power would be converted into heat. The resulting temperature rise would be independent of the flow rate and would depend primarily on the pressure energy.

$$\text{Temperature rise } (^\circ\text{C}) = \frac{\text{P (lb/in}^2\text{) x 0.00165}}{\text{SG x Sh}}$$

$$= \frac{\text{P (bar) x 0.0001155}}{\text{SG x Sh}}$$

where SG = specific gravity of fluid

Sh = specific heat of fluid

Where the pump has to be run continuously, this condition could be avoided in practice by off-loading the pump when there is no system demand. This can be done by 'pressure off-loading' or 'flow off-loading'. In the former case the pump circulates fluid at nominal zero pressure until there is a demand from the system. In the latter case pump displacement is reduced to a nominal figure to balance internal leakage only until there is a demand from the system. In both cases the pump input horsepower is substantially reduced and only moderate heating is experienced. The actual temperature rise is given by:

$$\text{Temperature rise } (^\circ\text{C}) = \frac{\text{K x pump input hp}}{\text{SG x Sh x } Q_O}$$

where Q_O = offload flow through the pump
K = 2.35 for offload flow in gallons/min
= 0.515 for offload flow in litres/min

Operating Temperatures

Average working temperatures for hydraulic fluids in industrial systems are of the order of 38–50°C (100–120°F). The nominal viscosity range is from about 20 to 100 centistokes, the description 'nominal' being given to indicate that these viscosities refer to fluids at normal ambient temperatures. Ideally — and logically — the fluid viscosity is selected so as to be at an optimum value at the working temperature of the system. The fact that viscosity changes with temperature imposes a number of practical difficulties.

The working temperature of any system is controllable. That is to say, the heating effect of friction (or, more specifically, the heat input generated when the system is working) can be offset by cooling by natural ventilation, forced cooling or intercooling, so as to avoid an excessive working temperature, or to limit the fluid temperature rise to a specific maximum above the ambient temperature. Controlled cooling can also be introduced — the simplest method being to switch an intercooler in or out of the circuit — to maintain a specific fluid temperature with changes of ambient temperature. Such precise control of temperature can be employed to give constant fluid characteristics, although it is seldom needed on the majority of systems.

Such conditions can only apply where a heat input is available, *ie* the system is working and part of the power input is converted into heat energy. Systems which work only intermittently will be subject to cooling down when idle, and if the idle period is long enough, will assume the ambient temperature. If the ambient temperature falls drastically, this will substantially increase the viscosity of the fluid, which may present difficulties in starting up. If this possibility is avoided by the use of a lower viscosity fluid, then the loss of viscosity on warming up may render the fluid characteristics unsuitable for the working of the system. Equally, in the case of a system which has to operate at a high temperature (*eg* in a higher ambient temperature), selection of a suitable viscosity for that temperature may well yield an excessive fluid viscosity on standing.

The overall problems involved can best be referred to a viscosity temperature graph from which limits can be found. These limits are necessarily broad for general issues, but can be quite specific for individual requirements.

The lowest limit for the viscosity of oil fluids is normally set by lubrication requirements and is of the order of 10 centistokes. A corresponding upper limit is more difficult to define; it could probably be set at around 800 centistokes, although a much lower figure would normally be preferred to minimize power output on starting up a cold system — see Fig 8.

Between these limits viscosity-temperature characteristics of individual fluids would then indicate their suitability, *ie* the range of temperatures which would fall within the extreme limits of acceptable viscosity. Requirements are, however, likely to be more specific, particularly where the hydraulic pump is concerned. The ideal or design velocity in this case is set by the viscosity of the fluid at working temperature, giving a spot point on the chart. The 'recommended' oil viscosity

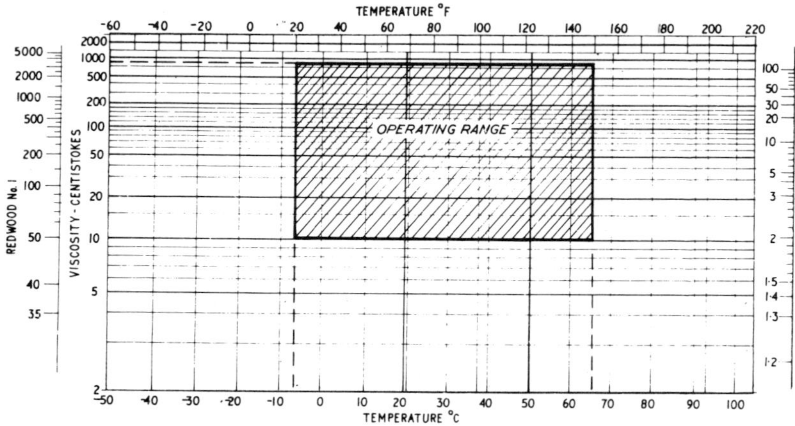

Fig 8

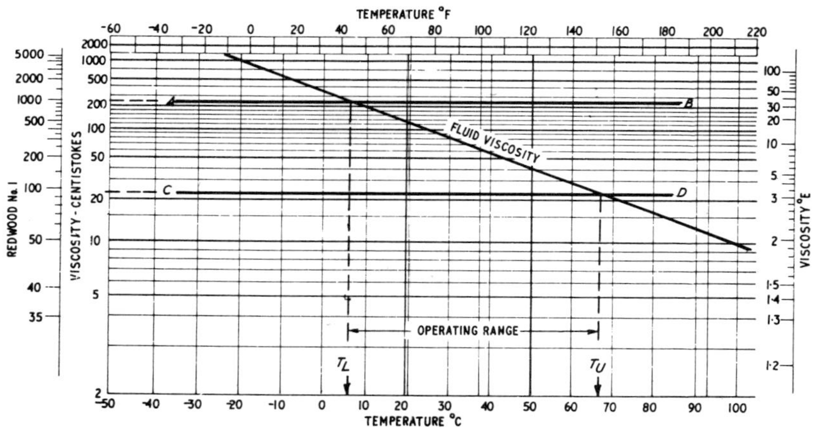

Fig 9

may well be quoted as the viscosity at normal temperatures, in which case a fluid with the same viscosity index as the design fluid would have to be employed to achieve identical characteristics at the working temperature. It is thus necessary to follow closely fluid recommendations for individual pumps which will vary both with the type of pump and individual designs.

Where viscosity requirements can be defined more explicitly in terms of upper and lower limits, as in AB, CD in Fig 9, the usable temperature range of a particular fluid is also clearly defined, by the intersection of AB with the left hand side of the fluid curve, and CD with the right hand side of the fluid curve. Any fluid temperature between T_U and T_L will be suitable. Such limits will be specific to individual fluids, *ie* individual fluid curves.

The actual working temperature range will be from the lowest ambient temperature (cold start) up to $T_U = T_a + \Delta t$, where T_a is the ambient temperature and Δt, the rise in temperature due to energy conversion into heat. The latter can be limited by cooling, but this does not necessarily limit the value of T_U, for with a constant value of Δt, T_U can shift up or down with changes in the ambient temperature T_a. However, if Δt is based on the maximum ambient temperature likely to be experienced, so that $T_{a(max)} + t = T_U$ this will ensure that the system will never operate at a temperature above T_U.

No similar control over T_L is possible, however, since this is a natural (ambient) temperature, or $T_{a(min)}$. If $T_{a(min)}$ is lower than T_L as indicated on the graph, then either another fluid must be chosen which provides the required maximum viscosity at this temperature, or artificial heating must be employed to make up the difference between $T_{a(min)}$ and T_L. An alternative choice of fluid may not be practical, and in any case will yield a different value of T_U and possibly more stringent cooling requirements. A low viscosity oil, in fact, would normally only be selected for systems which have to work continuously at very low temperatures, or intermittently at low temperatures. Fluid heating, on the other hand, is fairly straightforward and suitable for general applications, and can be accomplished quite simply with an electric immersion heater in the reservoir.

The capacity of such a heater can be readily calculated from the weight of fluid in the reservoir, the specific heat of the fluid and the rate of temperature rise. In the case of typical hydraulic oils, 1 kilowatt is required to raise 400 litres (88 gallons) of oil through 5 degC (9 degF) in one hour. The capacity for other heating requirements can be derived by simple proportion. Certain practical limitations are placed on the current density of electric heaters used in oil tanks. This should not exceed 230 watts/dm^2 (15 kW/in^2) for static oil and 300 watts/dm^2 (19.5 kW/in^2) for moving oil.

Miniature Hydraulics

MINIATURIZATION OF hydraulic components is attractive from the point of space and weight saving. Miniaturized valves, too, can form the working *logic* elements in a control system, operating at very low power levels. Miniaturization of the main circuit, however, inevitably leads to restriction on power levels, mainly on account of practical limits to the flow velocities which can be accommodated without excessive pressure drop losses. For that reason reduction in size as such offers little advantage except in applications requiring power levels up to about 1 kilowatt (or a little over 1 horsepower). Here miniature hydraulic systems are directly competitive with, and can offer specific advantages over, mechanical, pneumatic and electrical power systems. In particular, they embrace applications involving the instrumentation and control of mechanisms, and the transfer of control signals of limited output potential.

TABLE I – THEORETICAL MAXIMUM THRUST OF MINIATURE HYDRAULIC CYLINDERS

Cylinder Bore mm	SYSTEM PRESSURE – bar						
	35	50	75	100	150	200	210
5	7	10	15	20	30	40	41
7.5	15.5	22	33	44	66	88	92
10	27.5	39	59	78.5	118	157	165
15	62	88	132	176	263	352	370
20	110	157	235	314	470	628	660

Thrust in kgf

Cylinder Bore in	SYSTEM PRESSURE – lb/in^2								
	100	200	300	400	500	1000	1500	2000	3000
0.3125	7	1							
(8 mm)	7	14	21	28	34	65	105	130	210
3/8	7.5	15	22	30	37.5	70	112.5	140	220
1/2	17.5	35	52.5	70	87.5	170	262.5	340	525
3/4	40	80	120	160	200	380	600	750	1200
1	70	140	210	280	350	675	1050	1300	2100

Thrust in lbf

Cylinder Sizes and Forces Available

Specifically, to be described as miniature, a hydraulic cylinder needs to be smaller than the smallest standard hydraulic cylinder, *ie* 25 mm or 1 in bore. Logically other components, especially valves, should likewise be reduced to a matching size. Cylinder forces available should then be of the order shown in Table I.

Power Levels

The power level obtained from any hydraulic system is directly proportional to the product of fluid pressure and flow rate. Theoretically, at least, the flow level for any size of system can be increased by increasing the pressure. Practical limits are set by the size of clearances necessary to maintain satisfactory leakage levels. Material stress levels are less important and become less significant still with miniaturization. Optimum maximum pressure set by technological demands is of the order of 210 bar (3 000 lb/in^2).

The sizes associated with miniaturized hydraulic systems are usually within the range of 1.8 mm (0.07 in) to 4 mm (0.16 in) bore — or say a nominal size of 3 mm (0.125 in). The upper limit of flow rate which can be accommodated through such a line size is of the order of 3 litres/min (0.66 gallons/min or 3 in^3/sec).

These parameters, *ie* P = 210 bar max and Q = 3 litres/min max set the output power levels obtainable, *viz*

$$P\ max = \frac{3 \times 210}{612}$$

$$= 1\ \text{kilowatt}$$

$$= 1.34\ \text{horsepower}$$

Corresponding cylinder sizes commonly used in such systems range from 8 mm (0.3125 in) to 20 mm (0.75 in) bore, although larger sizes could also be accommodated. There is little advantage in making much smaller cylinder sizes as lower output requirements can be supplied more efficiently by working with reduced pressure.

Maximum thrust force from the 20 mm (0.75 in) size cylinder is thus 660 kg (1 450 lbs) at a maximum pressure of 210 bar (3 000 lb/in^2).

Cylinders may be single- or double-acting. In the case of double-acting cylinders miniaturization imposes certain limits on the output force available for the outward stroke because of the size of the rod required to accommodate the inward stroke force. In other words, the rod diameter may need to be larger in proportion to the cylinder bore than in conventional hydraulic cylinders. Typically, for example, the rod diameter used may be as much as 50% of the bore.

Miniature Valves

Control valve dimensions are largely governed by the method of actuation — *eg* the size of the magnet in the case of electro-magnetic controls.

A particular problem which has to be solved in the design of miniature valves is how to reduce the hydrodynamic force. This force originates in the change of fluid stream momentum when flowing into and out of the valve spool ports, and acts on the spool in such a way as to tend to lock it in the closed-flow position. It is especially unfavourable in miniaturized flow distribution valves with small spool travel and high pressure drop on the distributing edges. The need to overcome this force may substantially increase the dimensions of the control electromagnet and make the miniaturization of the valves useless.

The basic requirement for efficient valve operation is that the change of flow momentum does not exert a closing force on the valve spool as long as the fluid stream does not act on a movable part of the valve. Thus the valve port geometry has to be designed to provide a substantial reduction in the 'initial hydrodynamic' force present. In the case of hydrodynamically-operated valves, both inlet and return flows must be considered.

Miniature Tube Sizes

Typical tube sizes employed for a 200 bar system are 3 mm overall diameter, although larger sizes (*eg* up to 6 mm) may be employed for high pressure lines. Typical bore sizes for a similar pressure rating would be 1.7 mm for 3 mm o.d. and 4 mm for 6 mm o.d. tubes. The main reason for selecting the larger tube size would be to reduce flow velocity (and thus pressure drop) for a system with a fairly high demand from the pressure line, *eg* for rapid cylinder movement. In general, maximum piston speed likely to be achieved with a 200 bar system is of the order of 200 mm/sec (8 inches/sec).

Whilst circuit design follows that of conventional hydraulics it is often possible with miniature hydraulics to simplify the control system. Thus Fig 1 shows a circuit where a cylinder can be controlled by a single-acting directional control valve.

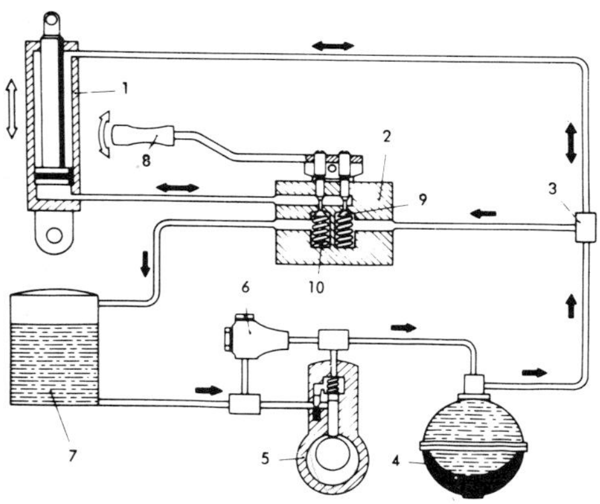

1. Hydraulic cylinder, double-acting.
2. Hydraulic directional-control valve.
3. Flow distributor.
4. Accumulator.
5. Hydraulic pump.
6. Pressure relief valve.
7. Hydraulic reservoir.
8. Lever.
9. Valve.
10. Valve spring.

Fig 1 Typical miniature hydraulic circuit diagram. (Bosch).

The system is shown in the neutral position with the valve closed. The hydraulic fluid is pumped from the hydraulic pump through the accumulator to the directional control valve inlet and to the rod end of the hydraulic cylinder. The rod end of the cylinder is therefore constantly under operating pressure. To initiate an operation, the valve on the supply side is opened with the lever by means of a thrust pin. The hydraulic fluid then flows into the cylinder head end. Upon releasing the lever the valve is closed by its spring.

When the other valve is opened with the same lever, piston return begins. Upon releasing the lever the valve is closed by its spring. The stroke speed of the piston can be controlled by means of a throttle screw in the housing. The entire hydraulic system is protected from overloading during operation by a pressure relief valve in the pump.

Should the permissible maximum pressure be exceeded (*eg* through external forces acting on the cylinder, or the like), the directional control valve connected with the return flow acts as a safety valve for the cylinder in question.

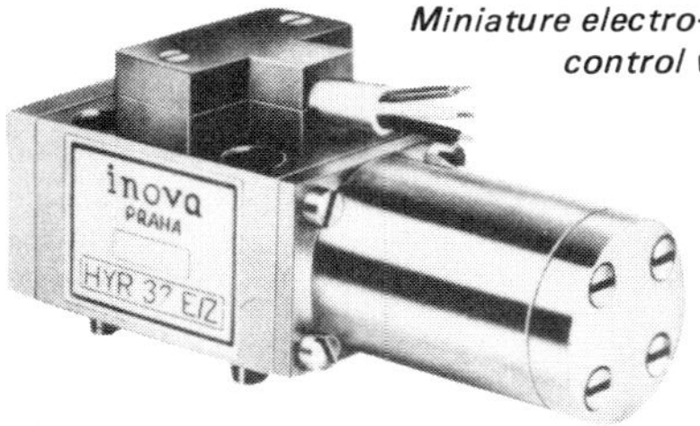

Miniature electro-magnetic flow control valve.

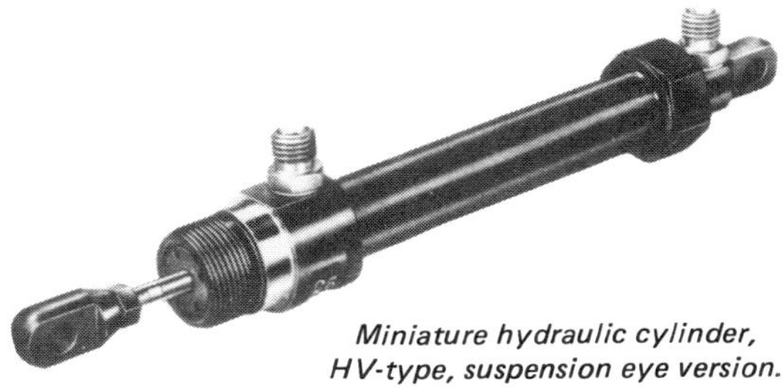

Miniature hydraulic cylinder, HV-type, suspension eye version.

Miniature cam-operated flow control valve.

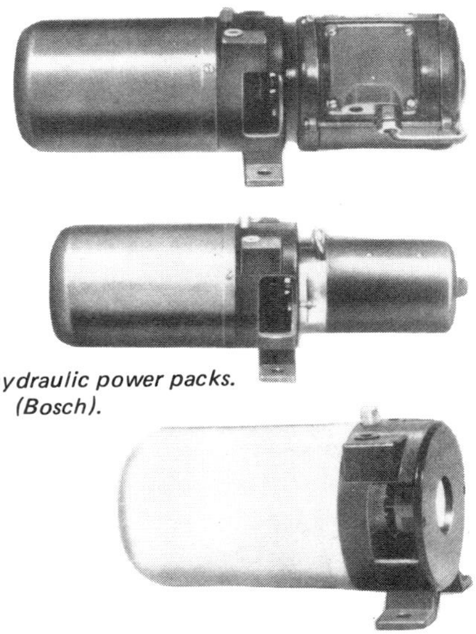

Miniature hydraulic power packs. (Bosch).

Miniature hydro-operated flow control valve.

Fields of Application

The main fields of application for miniaturized hydraulics are:

(1) Utilization of the advantages of hydrostatic drives in instrumentation techniques, and of their compatibility with electric control systems.

(2) Replacement of mechanical parts of low power output in both machines and various instruments.

(3) Replacement of electrical or pneumatic components where it is advantageous from both the technical and the economic point of view.

Practical fields of application are numerous, and include:

(i) operation of windows, sun roofs, seat adjustments, *etc*, on automobiles.

(ii) operation of windows, doors, blinds in houses and commercial buildings.

(iii) operation of dentists' chairs, operating tables, hospital beds, *etc*.

(iv) operation of feed clamping devices and similar mechanisms associated with machine tools.

(v) automation of controls.

(vi) source of servo power in radio control systems.

Towler the ultimate performer...
Towler capabilities cover every requirement of the design, manufacture and installation of hydraulic and electronic control systems. No system is too large or too small. Applications include extrusion and forging presses, plate bending and powder compacting presses, injection moulding and continuous casting machines, metal and woodworking machinery, marine and civil engineering equipment.
Extensive Product Range
Installed within Towler's own systems these products are engineered to exacting requirements and give high operational efficiency and long service life. The current range includes axial and radial piston pumps, fixed and variable displacement port plate pumps, inline plunger pumps, cam rotor pumps and a comprehensive range of hydraulic valves, operating through the pressure range 10 to 6000 bar.
95/5 Soluble Water Oil Micro Emulsion
Towler supports an ongoing development programme into system and component design, with great emphasis on future environmental and safety demands.
Extensive testing of established designs have shown that Towler can now offer a range of pumps from 3 to 187 litres per min, with pressures up to 350 bar, using a 95/5 Water oil emulsion, giving a better safety factor in hazardous conditions.
Service
Big or small, Towler is internationally renowned for high manufacturing standards, reliability and dedicated customer service.
Call us for details on our range of hydraulic components and system design capability.
THORN EMI Hydraulics
THORN EMI
Towler Hydraulics (U.K.) Limited Oaklands Road Leeds LS13 1LG England telephone (0532) 577721 telex 55413
TOWLER
A THORN EMI Comp

SECTION 2A

Hydraulic Pumps

A MULTIPLICITY of types are used as hydraulic pumps, with the usual choice being a positive displacement pump. Centrifugal pumps are used only in low-pressure hydraulic systems which require very high delivery rates; although they do find some application in small sizes because of their simplicity of construction. Apart from non-positive delivery characteristics, smaller centrifugal pumps have the disadvantage of low volumetric efficiency. Thus even where relatively large volumes of oil fluid may be involved a screw pump may be a more attractive solution.

Pressure performance is most readily predictable with reciprocating piston pumps. These embrace in-line, vee and other multi-cylinder configurations; radial piston pumps with cylinders disposed radially around a cam crankshaft; rotary piston pumps where the cylinders are rotated around a fixed shaft; and axial piston pumps with cylinder reciprocating motion being derived *via* a swash-plate. Other types of positive displacement pump which have found wide application as hydraulic pumps include vane pumps and, to a lesser extent, screw pumps.

A summary of hydraulic pump types and operating parameters is given in Table I.

Piston Pumps

Piston pumps offer high volumetric efficiencies together with virtually no limit on capacity, and thus cover a wide range of delivery requirements. Because of the greater complexity of construction, however, they are seldom competitive in smaller sizes with gear or vane pumps unless high system pressures are required. In this respect they are superior to all other types of pump, although the pressure rating of a piston pump is governed by the types of valve which can be employed with the design. In general, the highest pressures can only be achieved with seated valves. Configurations relying on porting or sliding or rotary valves are limited in the maximum pressures they can develop.

In-line Piston Pumps

Multi-cylinder pumps of in-line configuration are generally robust, large units designed for applications requiring high pressures and large deliveries, and in such sizes are invariably of horizontal configuration. Apart from the fact that they are capable of developing the highest pressures of any type of hydraulic pump, volumetric efficiencies in excess of 97% can be realized, an important consideration when large volumes of fluid are being handled. Speeds are generally restricted to 100–600 rev/min, although so-called high-speed pumps of this type are rated for continuous running up to 1 500 rev/min. Smaller in-line pumps are either of horizontal or vertical configuration, and may be designed for higher operating speeds.

TABLE I – HYDRAULIC PUMP TYPES

Type	Pressure Rating	Capacities Available	Speeds up to rev/min	Remarks
External gear	G/P types up to 35 bar (500 lb/in^2) Precision types up to 210 bar (3000 lb/in^2)	Up to 400 lit/min (90 gal/min) at 1000 rev/min	3000–6000 (smaller units up to 10000) Special designs up to 50 000	Efficiencies may range from 40–90%, depending on precision of manufacture. Fixed displacement only. Special designs up to 315 bar (4500 lb/in^2).
Internal gear (crescent)	High			Large displacement for given pump size. Volumetric efficiency depends largely on precision of manufacture. Generally fixed displacement.
Internal gear (lobe-rotor)	Moderate to high	Up to 200 lit/min (45 gal/min)	2000–3600	Ripple-free delivery. Generally fixed displacement.
Rotary abutment	Moderate			Fixed or variable displacement.
Sliding vane	Up to 70–100 bar (1000–1500 lb/in^2) or 175 bar (2500 lb/in^2) with two-stage units		2500 maximum (also limited by size)	Hydraulically balanced two-cell unit usual. Numerous differences in design detail – efficiencies *circa* 90%. Fixed and variable displacement types. Minimum speed 200–450 rev/min.
Cam-vane (Deri-Sine)	Up to 140 bar (2000 lb/in^2)	Up to 275 lit/min 1–60 gal/min		Ripple-free delivery.
Multi-piston in-line	Up to 1050 bar (15000 lb/in^2)	Large		Usually large, horizontal, fixed displacement units for heavy duties and continuous running at high pressures and high delivery rates.
Multi-piston rotary (Hele-Shaw type)	Up to 210 bar (3000 lb/in^2) or higher if required	Moderate	1000–3000	Fixed or variable displacement. Many variations on this basic design.

Bosch Gear Pumps Bosch Directional Valves Bosch Accumulators Bosch Variable Piston Pumps Bosch AC & DC Power Packs Bosch Proportional Valves Bosch

The foremost name behind all these products.

The Bosch name is synonymous with quality and reliability and this reputation is further reinforced with our wide range of hydraulic products. We are one of the world's leading hydraulic companies with over 30 years experience in industrial and mobile hydraulics.

The Products

Quiet running gear pumps; high flow-rated directional valves; bladder and diaphragm type accumulators; long service life variable piston pumps; directional, pressure, and flow control proportional valves–just some of the comprehensive range of Bosch hydraulics equipment.

The Offer

Write today for detailed literature on our range of hydraulic products. Specify and purchase Bosch, for consistently high quality at competitive prices.

Off the shelf availability in the U.K.

BOSCH Hydraulics

Robert Bosch Ltd, P.O. Box 166, Rhodes Way, Watford, Herts WD2 4LB.
Tel: Watford (0923) 44233. Telex No: 8812298.

Rotary piston	Up to 140 bar (2000 lb/in^2) with Berry pump			Revived interest in this Carey rotary ball piston design. Some use for aircraft hydraulics and now being further developed for hydro-static drives and low speed, high torque motors. Berry rotary piston is another type.
Multi-piston radial	Up to 700 bar (10000 lb/in^2)	Large	Up to 4000	Variable or fixed capacity. Mainly used for earlier aircraft and marine hydraulics.
Axial piston rotary (Williams-Janey type)	Up to 210 bar (3000 lb/in^2)	Moderate to large		Variable delivery type (usually) with wide-spread application for aircraft and specialized hydraulic systems.
Axial piston swash-plate	Up to 700 bar (10000 lb/in^2)	Moderate		Fixed-displacement type – mainly used for high-pressure systems. Typical maximum rating for continuous running 350 bar (5000 lb/in^2).
Axial piston tilted-body (Thoma type)		Moderate		Most widely favoured variable-delivery piston pump.
Annular piston	Up to 100 bar (1500 lb/in^2)	Small		Small capacity pump with high volumetric efficiency and high power:weight ratio.
Screw	Up to 140 bar (2000 lb/in^2); 210 bar (3000 lb/in^2) in some designs.	Up to 18000 lit/min (1–4000 gal/min)	1800	Quiet running, ripple-free output. Generally superior to gear pumps or vane pumps for handling large volumes of oil.
Centrifugal	Low	Small to very large	Up to 10000	Limited applications; mainly for high volume, low pressure systems.

Individual designs may differ in detail, notably in the method of reciprocating the cylinders. These may be directly driven by a crankshaft and connecting rods, or displaced by cams or push rods and returned by springs. Pistons are plain, working with fine clearances in the bores. Valves are invariably of the seated type (*eg* poppet valves) and normally operated automatically by pressure difference.

In-line pumps lend themselves to adaptation for handling fluids with low lubricity. In this case a separate pressure lubrication system may be incorporated to supply the requirements of the pump unit, particularly the bearings.

Radial Piston Pumps

The radial piston pump is a more compact type capable of a similar performance to an in-line pump. It also lends itself to variable-delivery performance. Pumps of this class embrace both radial (fixed cylinder, rotating cam plate) and rotary (fixed cam plate, rotating cylinder block).

As with the in-line pump, the fine clearances and long leakage path provided by the pistons make high pressures readily obtainable, although this is limited in designs where port-type rather than seating valves have to be employed, because of the movements involved.

Where the cylinder block as a whole rotates about a stationary cam, porting has to be employed, limiting the maximum pressure which can be developed to about 210 bar (3 000 lb/in^2). Speed is also limited by the mass of the rotating cylinder block.

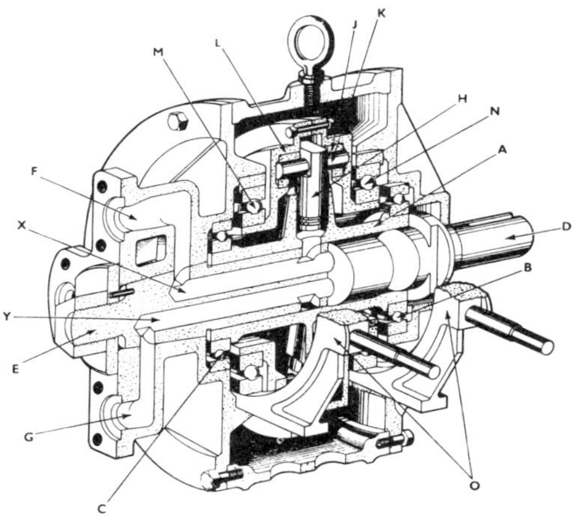

A —cylinder body
B,C —ball bearings
D —driving shaft
E —pintle
F, G —suction or delivery branches
H —piston
J —slipper
K —gudgeon pin
L —floating ring
M, N —ball bearings
O —guides
X, Y —suction or delivery ports

Fig 1 Hele-Shaw pump.

A typical configuration is shown in Fig 1, where the rotating cylinder block is mounted on a stationary eccentrically-positioned pintle carrying diametrically-opposed inlet and outlet ports, the whole cylinder block being surrounded by the casing of the pump or a tracking ring. Variable-delivery performance is readily achieved by making the eccentricity of the tracking ring variable and thus the design is particularly versatile in this respect — see Fig 2.

The rotary type radial piston has found its main uses as an aircraft hydraulic pump and for marine applications, but is less favoured than axial piston types for general industrial hydraulics. It is not competitive with the in-line piston pump for higher pressure services.

The alternative configuration, or true radial, employs a fixed cylinder block with the cylinders radially disposed and a rotating cam or eccentric driving the pistons. Piston return is accomplished either by suction or spring loading. In the case of higher speed pumps, positive-return drives may

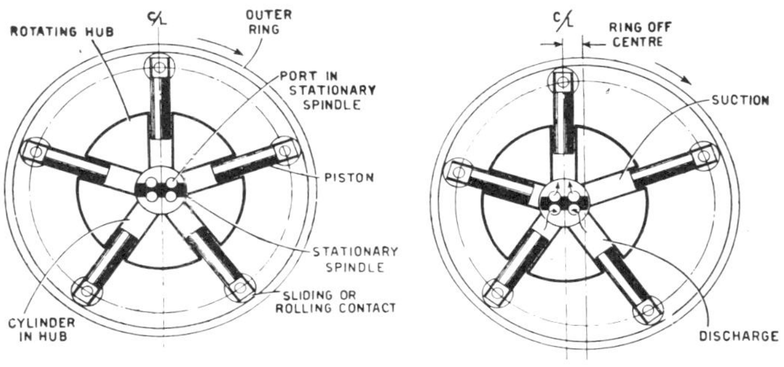

Fig 2

be provided to ensure that the pistons return satisfactorily during the suction stroke. Higher speeds, and very much higher pressures, can be realized than with rotary piston pumps, and the type can also have a variable delivery by providing axial movement of the cam. It also lends itself admirably to tapping off separate deliveries from individual cylinders or combinations of cylinders, to obtain multiple outlets at different pressures, if required — see Fig 3.

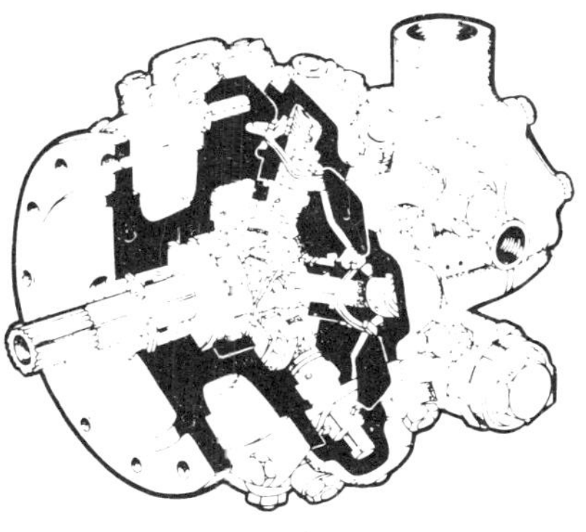

Fig 3 Radial piston pump with stationary cylinder block.

Fig 4

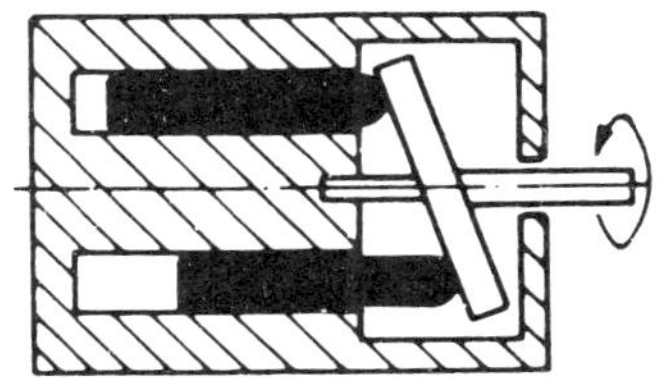

(a) Wobble-plate

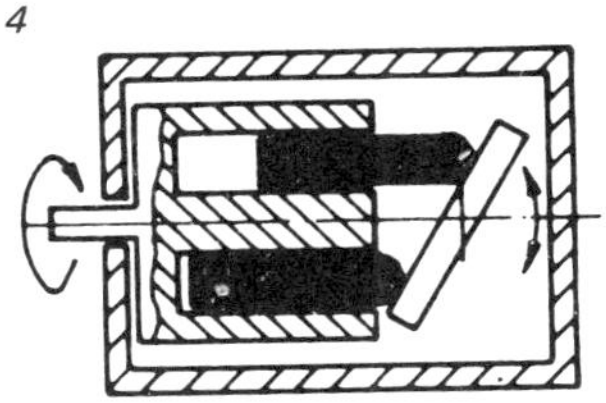

(b) Inclined swashplate.

Axial Piston Pumps

In the axial piston pump the cylinders are disposed axially (parallel to the crankshaft) and reciprocated by an inclined swash plate. If the cylinder block is fixed and the swash plate oscillates the type is known as a wobble-plate pump, (Fig 4(a)). If the swash plate angle is fixed, then the pump is a fixed-capacity type (Fig 4(b)). It is, however, a relatively simple matter to arrange for the swash-plate angle to be adjustable, thus giving the pump variable delivery characteristics.

The rotating swash-plate pump is more usually produced as a fixed-displacement type and also differs from the rotating-block type in that the direction of flow is independent of the direction of rotation. There are numerous variations in design detail, one of the main problems being to minimize wear produced by the sliding contact of the pistons. Instead of a plain swash plate the pistons often butt against a separate reaction plate or wobble plate carried on bearings in the swash plate — Fig 5. The wobble plate may then be constrained to remain stationary, or in some designs may be driven at a relatively slow speed, *via* gearing, to produce even wear on its surface.

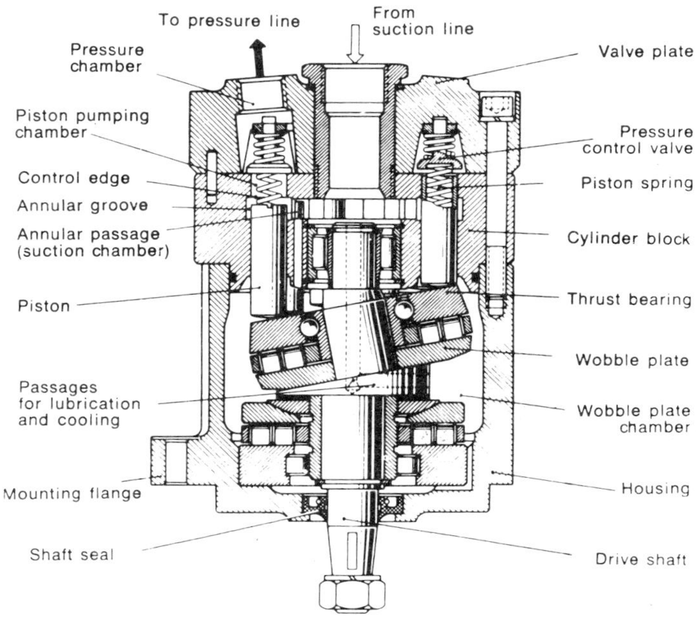

Fig 5 Axial piston pump —wobble plate type. (Bosch).

The rotating-block configuration has the advantage that it is easy to arrange for the fixed swash-plate to be tiltable, to provide infinitely variable flow characteristics, and it is thus a favoured type for specialized applications such as hydrostatic drives. On the other hand, porting has to be employed, which limits the maximum pressure capacity of the pump, and the design can become complicated when positive drive methods are used to ensure that the pistons always remain in contact with the swash plate. Often further complications are added by devices, such as floating port plates, incorporated to reduce internal leakage at higher pressures — Fig 6.

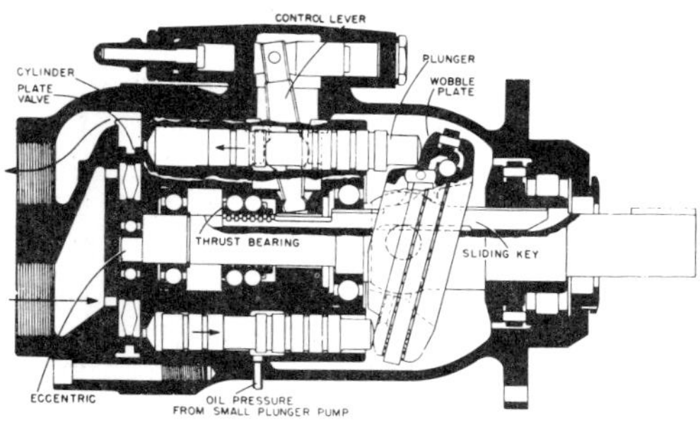

Fig 6 Sundstrand axial-piston pump.

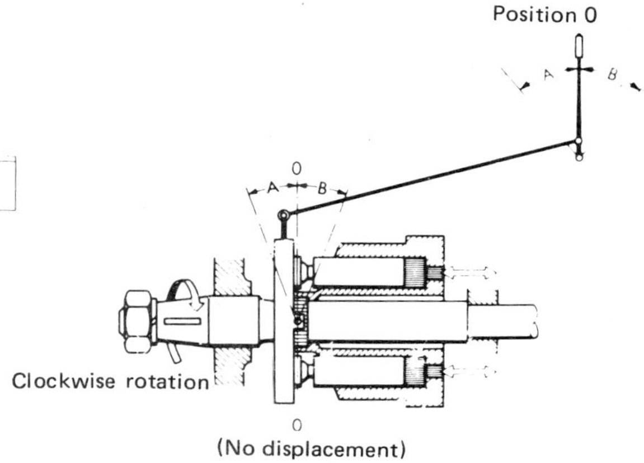

Fig 7 Variable capacity pump control. (Bosch).

Variable-Capacity Pumps

The principle of variable-capacity control by adjusting the angle of the swash plate on an axial-piston pump is shown in Fig 7. The angle of inclination of the swash plate is varied by the movement of the control rod E, aided by the servo-control. In this way the stroke of the pistons, and hence the displacement, is varied.

In the initial position the face of the inclinable swash-plate is vertical to the drive axis (neutral position, *ie* swash plate angle = 0). The pistons execute no stroke, the displacement and hence the delivery rate are therefore = 0. By varying the swash plate angle the displacement can be infinitely varied. With the maximum swash plate angle of ±20° the pistons execute their longest stroke. This results in maximum displacement. The delivery rate is directly proportional to the speed of rotation of the drive and the control rod travel. An important consideration is that when passing through the neutral position (from angle +20° to −20° or *vice versa*) the direction of flow reverses in spite of the direction of rotation remaining constant, *ie* the hydraulic fluid is delivered in the opposite direction.

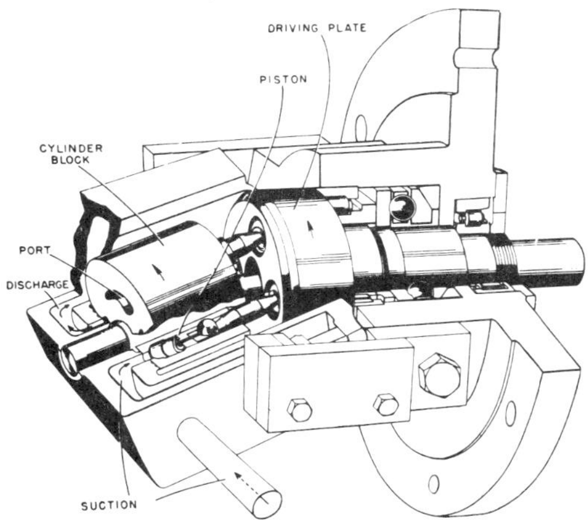

Fig 8 Vickers axial-piston pump.

Bent-Axis Pumps

The bent-axis or tilted-body pump employs a driving disc mounted square with the shaft, and the driven member and pump body rotate together, *eg* see Fig 8. Angular displacement of the driven member and body will then drive the pistons with a reciprocating motion, the stroke being proportional to the angular displacement. Piston drive is normally through ball-ended connecting rods providing a simple form of universal joint coupling which also serves to rotate the cylinder block drive from the disc end. The block itself must, of course, be suitably supported so that it is free to rotate.

As with all rotating-block pumps, seating type valves cannot be employed, so pressure rating is somewhat restricted. The configuration does, however, readily provide for infinitely variable flow, simply by altering the angle of the driven member relative to the block. It is also suitable for running at relatively high speeds, and is superior to the rotating block axial piston pump in this respect. Port connections do, however, call for a more complicated design of internal porting.

Annular Piston Pump

The annular piston or so-called orbital pump, is a lesser known design, although it appears to be returning to favour for specific applications. Its chief virtue is its relative simplicity coupled with

TABLE II – PISTON PUMPS

Type	Sub-Type(s)	Pressures	Speeds rev/min	Application and/or Remarks
Single piston	(i) Rotary or sliding valves			Virtually no application as performance provided by such pumps is better met by vane or gear pumps.
	(ii) Seated valves	Up to 700 bar (10 000 lb/in^2)		Possibly some limited application for high pressure low demand services requiring minimum pump leakage.
Multiple piston in-line	(i) Fixed-displacement (a) Horizontal	Up to 1 000 bar (15 000 lb/in^2)	100–1 500	Large, robust units – capable of generating the highest hydraulic pressures and large volumes. Also used for large-demand power packs. Suitable for use with all types of hydraulic fluids.
	(b) Vertical	Up to 1 000 bar (15 000 lb/in^2)	500–3 000	Smaller units. Both (a) and (b) are suitable for continuous heavy-duty use.
	(ii) Variable-displacement			Suitable for heavy-duty use, but not widely produced in this form. Most in-line pumps are of fixed-displacement type.
Radial piston	(i) Rotary rotating block	Limited to 210 bar (3 000 lb/in^2) maximum	Up to 3 000 but usually lower	Easy to render in variable-delivery form as well as fixed-displacement. Variable-delivery pumps readily adapted to give pressure-compensated characteristics.
	(ii) Radial rotating cam	Up to 700 bar (10 000 lb/in^2)	Up to 4 000	Versatile type suitable for both variable delivery and fixed displacement. Valves can be flow-operated.
Axial piston	(i) Rotating block	Up to 210 bar (3 000 lb/in^2); 140 bar (2 000 lb/in^2) for continuous operation	Moderate	Readily adaptable to variable-delivery by varying swash-plate angle and usually made in this form. Maximum pressure largely limited by porting employed.
	(ii) Rotating swash-plate	Up to 700 bar (10 000 lb/in^2); 350 bar (5 000 lb/in^2) for continuous operation	Moderate	Basically fixed-displacement units but adaptable to variable-delivery. Poppet valves used, hence higher pressure rating.
	(iii) Tilted-body	Up to 280 bar (4 000 lb/in^2)	High	Driving member and cylinder block rotate together, pumping action being obtained by tilting the body. Readily rendered in variable-delivery. Maximum pressure rating depends on porting.
	(iv) Semi-axial			Angled cylinders rotating about a fixed swash-plate. Virtually a variation of type (i).
Annular piston orbital		Up to 100 bar (1 500 lb/in^2)		Simplest form of piston pump with high power:weight ratio. Generally small capacity units – either fixed-displacement or variable-capacity.

a high power-weight ratio. The pump is also hydraulically balanced, so that bearing loads are low. Pressure rating is somewhat limited, however, unless multiple units are employed. Most units of this type are of 3-piston configuration, and relatively small in size.

The pistons are in the form of rectangular section sectors, driven round an annular channel by an eccentrically mounted rotor, pumping action being produced by the pistons alternately approaching and retracting from each other. The delivery is also controlled by the degree of eccentricity, making the design readily adaptable for variable delivery. The pumping action is smooth, and maintained down to very low pressures. Volumetric efficiency is also high.

Characteristics of Piston Pumps are summarized in Table II.

Vane Pumps

Vane pumps are particularly suited to medium-pressure, medium-speed duties and hence have the advantage over gear pumps that the rotor can be hydraulically balanced, thus minimizing bearing loads. Their main application is for low- and medium-pressure systems requiring a compact low-cost pump (*eg* machine-tool hydraulic systems), their versatility being an attractive feature.

Modern vane pumps are capable of developing pressures up to 175 bar (2 500 lb/in^2), with maximum deliveries up to 410–455 litres/min (90–100 gal/min) and speeds up to 2 500 rev/min or higher. In the case of variable delivery pumps, maximum pressure rating is usually 70 bar (1 000 lb/in^2), Minimum speed for vane pumps is of the order of 200–450 rev/min.

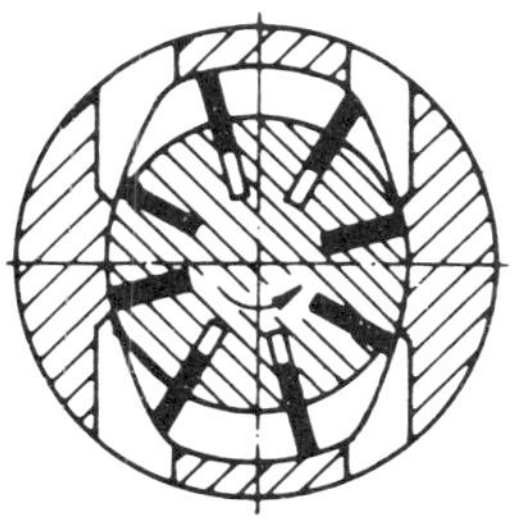

Fig 9(a) Fixed capacity vane pump.

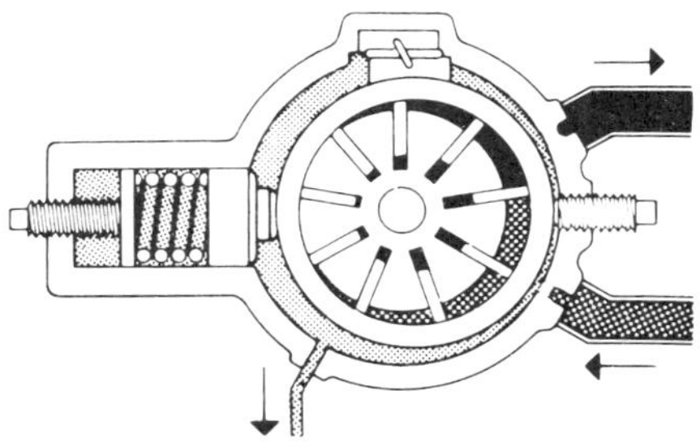

Fig 9(b) Variable capacity vane pump.

The simple single-cell vane pump (Fig 9(a)) is unbalanced hydraulically and tends to suffer from low volumetric efficiency because of the difficulty in controlling internal leakage. The geometry does, however, readily lend itself to providing variable delivery characteristics, simply by adjustment of the relative positions of the rotor spindle and outer ring modifying the capacity of the 'pockets' swept during revolution – Fig 9(b). The casing can also readily be fitted with a replacement liner to take wear.

A degree of balance can be achieved by adopting two-cell or three-cell configuration with a single rotor although this complicates the design and can compromise the pumping performance. The usual form of balanced design incorporates two inlet ports and two outlet ports on opposite sides of the motor, with an elliptic shaped cam ring. Both sides of the rotor then have equal pressure excited in opposite directions, cancelling out side-thrust on the rotor shaft and its bearings which now have only to carry the external load.

Other forms of balanced vane pumps are the intra-vane type, and dual-vane designs. As well as being hydraulically balanced both have higher volumetric efficiency than simple vane pumps and can generate higher pressures without slippage.

Variable-delivery vane pumps are normally unbalanced designs. The displacement of the pump is changed by movement of the cam ring relative to the rotor, effectively modifying the 'pocket' volumes. Ring movement in such cases is usually spring controlled with a pre-determined spring pre-compression opposing the horizontal force developed on the ring. Vertical movement of the ring is constrained by a thrust block. The spring rate then determines the change in pressure necessary to move the ring from full flow rate to zero. Delivery remains directly proportional to the ring setting or effective 'stroke', between the range of full eccentricity and concentricity, unless the drop-off 'knee' is reached, when the ring will move automatically to reduce the stroke to compensate for increasing pressure. The pump is thus pressure-compensated and does not need a relief valve, as an over-pressurization will simply reduce the delivery to zero by moving the ring to the concentric position. The main requirement is to establish a suitable delivery/eccentricity gradient so that the pump cannot become unstable and attempt to oscillate over its normal working range.

The Cam-Vane Pump

The Deri-Sine cam-vane pump is shown in Fig 10. Whilst this is basically a vane pump, only two sliding vanes are employed, mounted in slots in the casing. The rotor is cam shaped, rotation producing a sinusoidal-type wave form displacement. Two such cells are mounted in tandem, keyed to a common shaft, but with the rotors 90° apart circumferentially. The combined output from such a double unit provides a constant-delivery flow without the pulsations normally present in a conventional vane pump output.

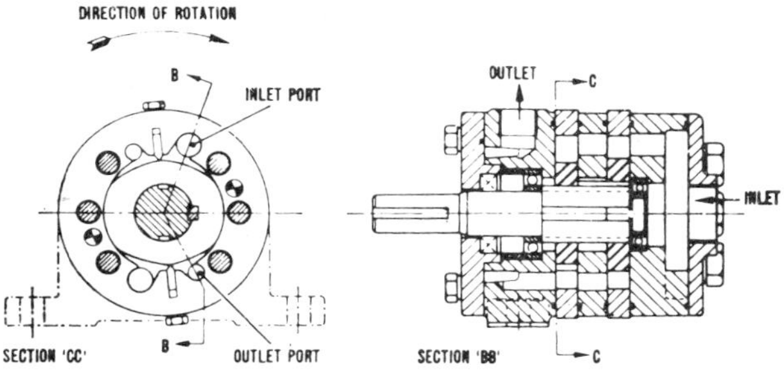

Fig 10

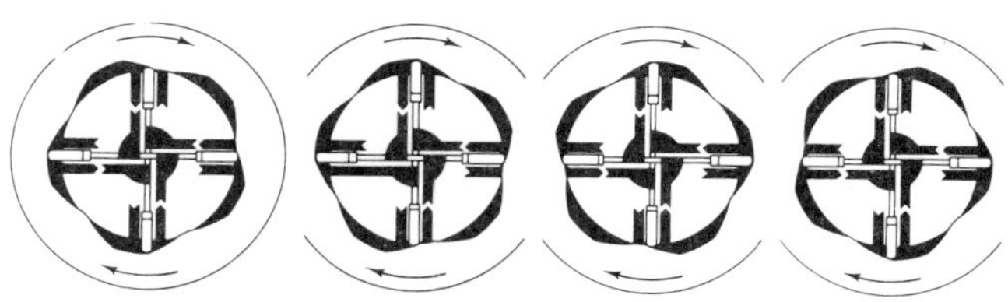

Fig 11 Pushrod vane pump.

Pushrod Vane Pump

The pushrod vane pump — shown in Fig 11 — employs pairs of diametrically opposed vanes linked by pushrods which ride in a contained outer casing which forms the rotor. Vane movement also controls the opening and closing of the inlet and outlet ports.

Vane pumps of this type are normally limited to low speeds but can develop relatively high pressures for high torques when used as a motor.

Gear Pumps

The external gear pump is capable of developing higher fluid pressures than a vane pump and can also be run at higher speeds. Its original limitation was high internal leakage which has been overcome by the introduction of pressure balancing methods. Modern gear pumps are rated for up to 210 bar (3 000–6 000 rev/min). Again in special designs speeds of up to 50 000 rev/min may be achieved. Average speed for most gear pumps is 400–500 rev/min.

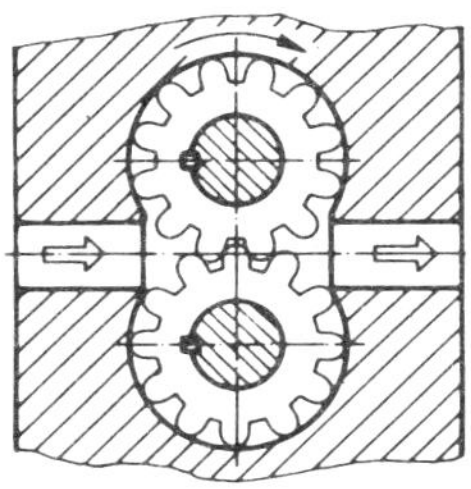

Simple gear pump

Fig 12 Diagram showing gear arrangement and internal leakage paths in an external gear pump. A = Inlet. B = Outlet.

The basic elements of an unbalanced external gear pump are shown in Fig 12. Two intermeshing gears of the same diameter and form are mounted on separate spindles and housed in a close fitting casing. Inlet and outlet ports are formed directly in the sides of the casing, in line with the point of meshing. One gear shaft is driven whilst the other idles. Both shafts are carried in low-friction bearings (usually rolling bearings).

During rotation, as each pair of teeth inter-mesh on the inlet side, the volume on that side is reduced by the volume of two tooth spaces, providing a suction effect. Oil flowing into the suction space is then trapped on each side by a tooth crest approaching the bore of the housing and carried round to the delivery side by the 'pockets' between adjacent pairs of teeth. On the delivery side, the oil is displaced from the delivery port under pressure.

For maximum volumetric efficiency there should be no leakage between the teeth, and no leakage across the end faces of the gears. Even if such internal leakages are reduced to a practical minimum, there may be a further penalty to pay in the high bearing loads and unbalancing loads introduced by fluid trapped in the teeth pockets, which can result in increased friction and lowered mechanical efficiency. Thus a compromise may have to be reached in order to achieve maximum overall efficiency, although usually attention to increasing volumetric efficiency results in higher gains than any corresponding reduction in mechanical efficiency.

As a consequence a premium is placed on detail design, precision workmanship and rigidity, in order to produce a high-performance gear pump with a high overall efficiency. Overall efficiencies in excess of 90% are attainable, but only by increasing the cost of production. The relatively inexpensive general-purpose gear pump will have lower efficiencies, although it may be perfectly adequate for certain duties. Development of the high-efficiency gear pump took place primarily for aircraft hydraulic systems, where the design problems were further aggravated by the high system pressures required, *eg* 140 and 210 bar (2 000 and 3 000 lb/in^2). Lower-cost gear pumps of virtually comparable rating are now available for industrial hydraulics, whereas previously industrial hydraulic gear pumps were of simpler design and lower efficiencies and used mainly for delivery pressures below 35 bar (500 lb/in^2).

Pumps of this type may be balanced by tooth venting although on low pressure relief may be provided by allowing the gears to run with a small backlash. On high-pressure pumps, grooves may be cut in the side plates to provide communication between the inter-tooth spaces (see also Fig 13).

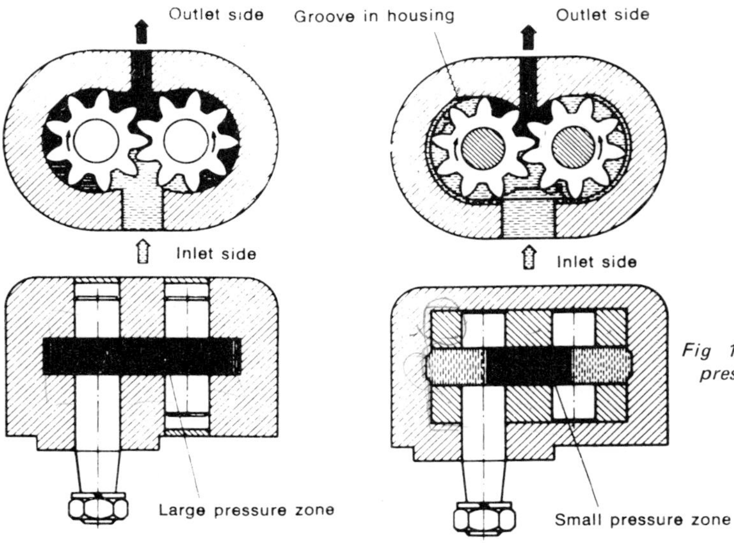

Fig 13 Simple gear pump (left) and pressure-compensated (double-gland) gear pump (right). (Bosch).

Different tooth forms will also provide different degrees of sealing. Various gear forms have been tried, but the stub gear remains the favourite and about the best for sealing, even though the displacement per revolution may be less than that which can be achieved with other tooth forms. Single- and double-helical forms are also used, but mainly where a quieter running pump is required. Apart from the doubtful possibility of achieving any real performance gains, the use of more complex tooth forms is usually undesirable from the point of view of increased complexity and cost of manufacture. The tendency, therefore, is to use simple stub tooth forms and use the smallest possible number of teeth consistent with smooth continuous drive. Reducing the number of teeth has the advantage of increasing the displacement per revolution, regardless of gear form.

The more or less standard method of producing a range of gear pumps of different capacities is to adopt one or more gear diameter sizes and then render each diameter size in several different face widths – *ie* extend the capacity by increasing the face width of the gears in lengthened casings. This enables the same detail design features to be preserved throughout the series at minimum cost, although in the case of high-pressure pumps, altering the length of the gears will modify the casing stress and bearing loads. Increasing the gear length will decrease the casing stress but increase bearing loads, and *vice versa.* Geometrically, the optimum form for a high-pressure pump is given by a gear width equal to the pitch circle diameter of the gear.

Methods of detail design, aimed at reducing internal leakage across the end faces of the gears, are based on pressure-loading techniques. One such method is to fit side plates between the gear faces and end covers, with high-pressure oil fed to specific areas on the outside of these plates. The plates are thus pressurized to bring them into contact with the gear ends and prevent end leakage, with a force proportional to the speed and internal pressure of the pump. Numerous variations on this theme have been developed by individual manufacturers and although this involves additional components, the overall cost is usually less than the cost of the more precise manufacture necessary to achieve similar end-leakage control with fixed side clearances.

The delivery of a gear pump is directly proportional to its speed, and the delivery pressure to the external load. With increasing load the pressure will continue to rise, up to limits set by either

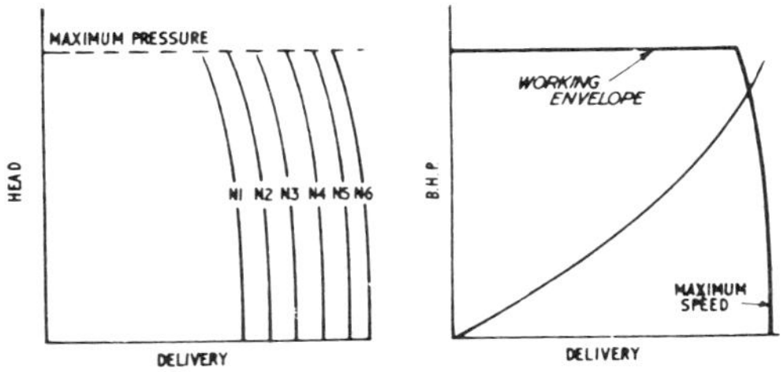

Fig 14

a relief valve or the strength of the casing itself. Some slip is inevitable, and hence, for a given speed, delivery will decrease with increasing resistance – Fig 14. Slip is virtually independent of speed, except at very low speeds where it will normally tend to increase and volumetric efficiency decrease quite rapidly. As a general rule, high efficiencies can only be achieved with gear pumps by operating them at relatively high speeds. Where a pump is rated for operation over a range of speeds, operating at the highest speed will give the highest volumetric efficiency. A maximum speed of 3 000 rev/min for continuous running is general for most modern high-performance gear pumps, although this will also depend on size. Some smaller units may be rated for running at speeds of up to 8 000 rev/min, but 3 500 to 4 000 rev/min is a more usual maximum.

Rated delivery may be expressed in terms of displacement per revolution, or actual delivery given for some specified speed (often maximum speed for continuous running). In the former case, delivery at any particular speed can be obtained by multiplying by the speed in rev/min. In the latter case, delivery at any other speed follows from simple proportion. Note, however, that delivery is normally determined empirically, using a low- or medium-viscosity oil discharging freely (*ie* zero discharge pressure). Actual delivery will be modified by a different fluid viscosity, efficiency tending to fall with increasing viscosity, and will be reduced slightly when working against an output load.

Gear pumps should not normally be run at speeds above the manufacturer's rated maximum (although little harm is likely to result from such running for short periods under no-load conditions); nor should they be run continuously at high pressures and very low speeds. Otherwise they can be regarded as a particularly versatile type of pump which can be run at any speed. Variable-delivery operation would not normally be obtained by varying the speed, however. It would be better to use two or more gear pumps, with off-loading of individual pumps as necessary. Multiple units can then be mounted in tandem, with a common drive.

Internal-Gear Pumps

In an internal gear pump one (internal) gear is located within an outer gear ring, the tooth form being chosen so that each internal gear tooth is always in contact with the inner surface of the

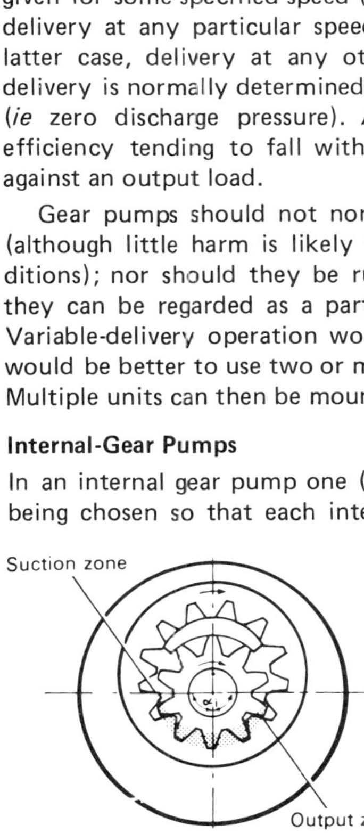

Fig 15

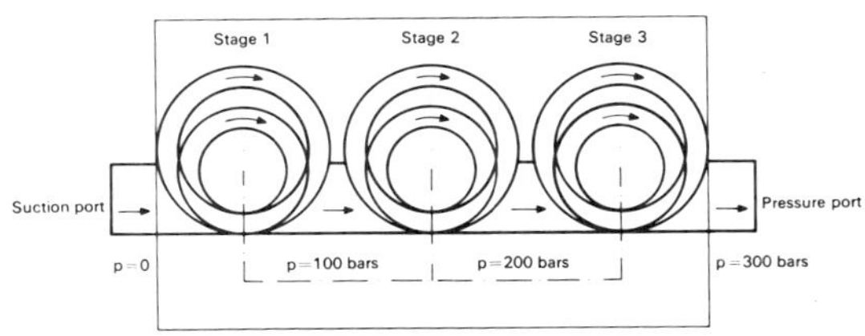

Fig 16

outer ring – Fig 15. Rotation produces a series of contracting and expanding 'pockets' transferring oil from the inlet side to the outlet side. This geometry has some specific advantages over the external gear pump, notably the lower localized fluid pressures generated and the lower shearing forces on the fluid. It can also show much lower operating noise levels than a comparable external gear pump, and readily lends itself to multi-staging – Fig 16.

In the simplest form of the internal-gear pump, the eccentrically mounted inner gear drives the outer gear, and some means of blocking the high-pressure side from the low-pressure side is needed in order to prevent back-flow. The simplest way of achieving this is by the introduction of a crescent shaped 'wedge', filling the clearance space on the non-meshing side of the internal gear. Precision manufacture is called for to give the close clearances necessary to minimize internal leakage, but an internal gear pump of this type is notable for the large displacement which it can offer relative to its overall size.

The 'Gearotor' pump is another form of internal-gear pump which is now produced in designs capable of developing 70–140 bar (1 000–2 000 lb/in^2) pressure per stage. In the main, however, all internal-gear pumps are low-pressure types.

Screw Pumps

The screw pump has positive-displacement characteristics and is well suited for handling large volumes of oil fluids; thus it can have a particular application as a hydraulic pump in systems requiring high deliveries.

Both single and double helical screws may be used. Single screws are unbalanced hydraulically and generate considerable end thrust, unless suitably compensated. One method of doing this is by applying pressure to hydraulic pistons integral with the rotor – see Fig 17. Double screws (tandem configuration) are balanced axially and do not require compensation.

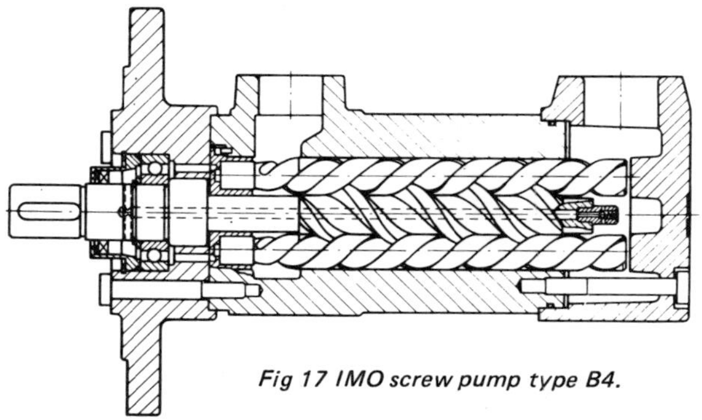

Fig 17 IMO screw pump type B4.

Screw pumps provide a smooth output, free from pump ripple. They are free from vibration and inherently quiet running, as well as being particularly suitable for driving at high speeds. They may, therefore, be considered as hydraulic pumps where such parameters are significant. Capacities available range from about 5 litres/min to 7 200 litres/min (1 gal/min to 2 000 gal/min), with pressures up to 140 bar (2 000 lb/in^2). They can also be produced in variable-delivery configuration, usually by the incorporation of a movable blanking sleeve over the screws.

Hydraulic Hand Pumps

Manually-operated hydraulic pumps – generally known as hand pumps – are almost invariably of piston type. Limited use has also been made of gear pumps hand-driven *via* a winding handle *via* step-up gearing, but much higher deliveries (and pressures) can be achieved by piston pumps for the same effort.

Hand-operated piston pumps may be either single-acting or double-acting. Two-stage, or two-speed, pumps provide for high delivery and minimum pumping time at light loads, the pump output being reduced at a pre-determined level in order to reduce the load required to build up to maximum pressure. The change-over is normally based on a maximum effort required to the order of 27 kg (60 lb).

Alternative methods of providing two-stage operation include mechanical systems to alter the leverage (and thus the stroke of the pump) and the use of two separate cylinders, one providing high output at low to moderate pressures and the other low output at high pressure. Change-over is accomplished by unloading the low-pressure cylinder either manually or automatically.

A typical two-stage design would incorporate two double-acting cylinders, one having twice the stroke of the other. When the pressure rises to a given point, the load automatically cuts out the longer stroke cylinder, allowing pumping to proceed with reduced output at the higher pressure.

Pump Performance

THE DELIVERY of a positive displacement pump is directly proportional to its displacement and cycling speed (*ie* rev/min in the case of rotary pumps and strokes/min in the case of piston pumps). Displacement for any pump is normally expressed as volume per rev, or volume per stroke. Maximum delivery is thus governed by the maximum speed at which the pump can be operated, although the majority of pumps are designed for direct coupling to electric motors and thus operating rev/min is fixed by the motor design.

Delivery available governs the speed of operation of the actuator(s) in the system, true power output then being the product of actuator speed and the load against which it is operating. It is generally more convenient to determine system power in terms of hydraulic power which is directly proportional to the product of pressure and delivery. This directly represents the work capacity or *maximum power rating* of the pump, mechanical output power equivalent being obtained by factoring by the actuator efficiency. The maximum power rating of a pump is determined both by its design (*eg* pump type) and mechanical considerations (*eg* elements of construction). Typical values in this respect are given in Table I.

TABLE I – TYPICAL MAXIMUM POWER RATINGS

Pump Type	Maximum Power Rating hp
Multi-piston, rotary valve	400–500
Multi-piston, seated valves	150
Gear	50–150 (but usually nearer the lower figure)
Vane	50

Pressure Rating

Maximum pressure developed by a pump depends both on the pump type and its design and construction. With certain types (*eg* vane and external gear pumps), practical maximum pressures are limited. With others, such as piston pumps, there is no absolute limit to maximum pressure, only that related to construction strength and other practical parameters. Table II summarizes normal maximum pressure ratings for various types of hydraulic pump.

All positive displacement pumps need some form of protection by pressure relief against excessive and damaging pressures being developed by accidental over-load on the output side. This is normally provided by a relief valve, the operation of which must lie within or on the power envelope. Certain designs, however, permit pressure-compensation to be introduced – this being a general characteristic of variable delivery pumps.

TABLE II – TYPICAL PRESSURE RATINGS FOR HYDRAULIC PUMPS

Type	Maximum Pressure Rating bar	Maximum Pressure Rating lb/in^2
Multi-piston, in-line	700 to 1 000	10 000 to 15 000
Multi-piston, radial	700	10 000
Axial-piston, swashplate	700	10 000
Axial-piston, tilted-body	280	4 000
Precision gear	210	3 000
Cam-vane	210	3 000
Two-stage vane	140 to 175	2 000 to 2 500
Screw	140	2 000
Rotary abutment	100	1 500
Annular piston	100	1 500
Vane	100	1 000 to 1 500
Simple external gear	35	500

Spenborough axial piston pump with rated flows up to 57 lit/min and rated pressures up to 200 bar (3 000 lb/in^2).

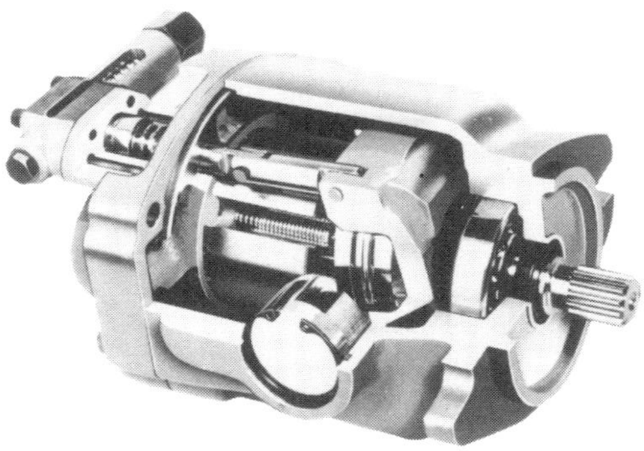

Webster PA series pressure compensated, axial piston pump.

An example from the range of Lucas internal gear pumps with exceptionally low noise levels and continuous pressure ratings up to 300 bar (4 300 lb/in^2).

TABLE III – TYPICAL EFFICIENCIES OF HYDRAULIC PUMPS

Type	Maximum volumetric efficiency	Maximum overall efficiency
Multi-piston, in-line	up to 99%	up to 95%
Multi-piston, radial	over 95%	over 90%
Multi-piston, axial	over 95%	over 90%
Precision gear	can approach 98%	can approach 95%
Screw		75–85%
Vane	85–90%	75–80%
External gear (non-precision)		20–60%

Efficiency

Overall efficiency of a pump is expressed as the ratio of hydraulic power output to power input (normally multiplied by 100 to give efficiency as a percentage value). Overall efficiency also includes mechanical losses (*ie* friction), and so actual hydraulic efficiency or *volumetric efficiency* may be quoted separately. Table III summarizes some typical values for different pump types. It should be noted, however, that the actual pump efficiency achieved in a system can be affected by many operating factors.

As a general rule, piston pumps have higher overall efficiencies than rotary pumps, although pressure balancing or precision manufacture of gear pumps with fixed end clearances has resulted in comparable efficiencies being achieved with gear pumps. On the other hand, the overall efficiency of a simple, inexpensive gear pump can be quite low.

It should be noted that such figures are a general guide only. Actual efficiencies achieved depend very much on the detail design and manufacture of an individual pump, and also to a large extent on the size of the pump (smaller pumps tend to be less efficient than larger ones of the same type).

The question of the duty cycle involved must also be taken into account. Thus where there is considerable variation in demand it is usually more efficient as regards both operating and running costs to use a variable-delivery pump, although the initial cost of the pump will be higher. In many cases a variable-delivery pump may be virtually essential, as this form of delivery regulation is more efficient and easier to arrange than a variable-speed drive; however, in certain pump designs (*eg* multi-stage units of in-line piston pumps) it may be possible to tap off different deliveries and different pressures from a single constant-speed pump.

Efficiency may be very important (*eg* where large volumes of fluid are being pumped); or relatively insignificant (*eg* in a light duty system where ample input power is available from an inexpensive driver, or an 'over-size' pump is to be used for a particular reason).

Power Regulation

Power regulation may be necessary in systems with fluctuating loads so that the power output of the driver running at constant speed can be fully utilized. This can be done with a horsepower regulator which automatically reduces delivery as the load increases, and *vice versa,* to maintain a constant value of hydraulic horsepower in the system.

A theoretical power curve for a positive displacement pump is shown in Fig 1. The pressure level is determined by the external load, the maximum value being set by the mechanical strength of the pump itself. Maximum flow is established by the maximum permissible running speed. Maximum power is thus developed at maximum pressure and delivery, the corresponding point on the curve being called the corner horsepower.

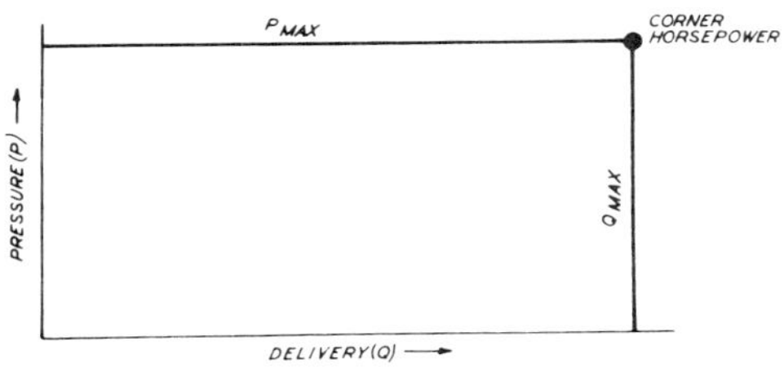

Fig 1

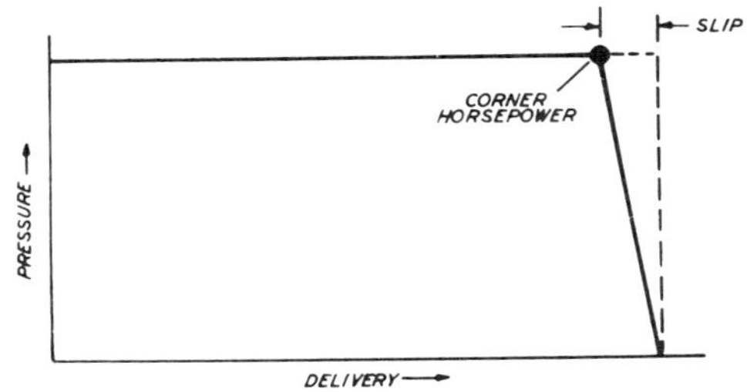

Fig 2

In practice the envelope is not quite rectangular, since increasing load will produce increasing slip, modifying the envelope to the form shown in Fig 2. The value of slip depends to a large extent on the precision with which the pump is manufactured and may be as low as 3% to 4% at maximum pressure. The effect of slip is, of course, to reduce the overall efficiency achieved.

Pumps capable of operating continuously at maximum pressure and maximum delivery can be fully defined for power rating by the form of curve shown in Fig 2. They can obviously be operated anywhere within the envelope, or right up to the corner horsepower provided the loading is steady.

Many duty cycles will, however, involve transient loads, in which case these must be taken into account, otherwise an immediate load may place a demand on the pump beyond either its available horsepower or its maximum pressure rating. The usual method of rating in such cases is to plot the envelope as a constant horsepower line, as in Fig 3. This now allows for transient loads to be accepted within the corner horsepower, and also defines the safe limit for transient loads. The over-load on the driver can also be determined, relative to any momentary displacements above the constant horsepower line.

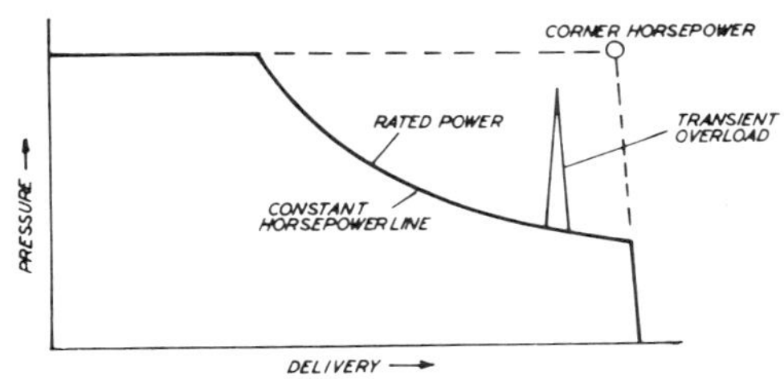

Fig 3

Power:Weight Ratio

The weight of a hydraulic pump is generally comparable with that of any other power generator of the same horsepower rating. In certain applications – *eg* aircraft hydraulic systems – installed weight and bulk can be an important factor. In such cases possible pump types may be assessed on a power:weight ratio basis. The aircraft field in particular has been responsible for developing a number of pumps with exceptionally high power:weight ratios. These are invariably of precision construction with pressure ratings of 210 bar (3 000 lb/in^2), and expensive productions. Industrial counterparts are becoming more and more available, somewhat de-rated in performance, but considerably less costly.

See also chapter on *Pump Drivers.*

Hydraulic Pump Selection

OPERATIONAL PARAMETERS to be considered in the selection of a hydraulic pump are:

(i) System pressure requirements.

(ii) Delivery volume requirements.

(iii) Circuit requirements.

(iv) Drive speed.

(v) Cost-effectiveness.

Items (i), (ii) and (iii) effectively define the duty requirements.

System Pressure

In the case of positive displacement types the pump operating under steady conditions cannot develop a pressure greater than the resistance offered by the system to which it is connected — *ie* the pressure developed is extremely dependent on the load. In a closed system this load can (theoretically) approach infinity (or the failure pressure of individual components) so that maximum working pressure is normally set by a relief valve.

In general, high pressures can only be achieved by pumps with seated valves, although the inherent limitations of rotary valves or ports can be overcome to some extent by refinements in design detail. For the very high pressures needed for heavy presses and similar applications, only multi-piston pumps with seated valves are likely to be suitable, when pressures of the order of 700 bar (10 000 lb/in^2) can be achieved continuously.

For lower system pressures the choice of pump type becomes much wider. Thus, for general industrial hydraulics, where system pressures may range from 30—35 bar (400—500 lb/in^2) up to about 100 bar (1 500 lb/in^2), virtually any type of positive displacement pump can be used, selection of type then being based on other specific requirements, or power rating. However, higher system pressures are becoming commonplace, so that 140 bar (2 000 lb/in^2) can also be considered an 'industrial' rating; this is beyond the limit of simple vane pumps unless they are coupled as two-stage units. The general tendency to uprate system pressures for higher working efficiency has extended the performance requirements beyond the capabilities of certain basic (and simple) types of pump, calling either for the use of alternative pumps or further development of such basic types to meet new pressure requirements. Either solution has generally resulted in a more efficient pump, although a more costly one.

Delivery

Basically for any one type of pump, the delivery volume required governs the size of pump necessary. However there are practical limits to volumes achievable with different pump types. Where very large deliveries are required a screw pump (or even a centrifugal pump) may have to be considered. The alternative is to consider the relative cost-effectiveness of providing the demand from a number of smaller pumps.

Delivery is also related to the speed at which the pump is driven. For directly-coupled pumps this is normally the standard speed(s) of electric motors. Higher speeds may be obtained *via* belt drives or gearing, which may also prove advantageous in increasing the efficiency of the pump. However, it is usually more efficient and less costly to drive a pump by direct coupling rather than through gearing as this must involve some power loss. Also there are practical limits as to the maximum speed at which individual pumps can be driven.

An important parameter in this respect is the *maximum power rating* of the pump, as representing its effective work capacity — see chapter on *Pump Performance.*

Cost-Effectiveness

Cost-effectiveness embraces a study of initial costs, operating costs, system reliability and maintenance. This can finally decide the most suitable type of pump where there are alternatives which can equally well meet the duty requirements. Here, cost factors may be in opposition. Thus piston pumps tend to offer the highest overall efficiencies with high reliability, but higher initial cost. Gear pumps are generally less expensive than vane pumps and more tolerant of adverse operating conditions (*eg* contaminated fluid), but have lower efficiencies than vane pumps capable of the same duty.

TABLE I — VDMA FLUID RECOMMENDATIONS

Climatic conditions	Requisite hydraulic fluid types (viscosity) for hydraulic systems with:			
	Plate-type pumps	Double gland pumps Double gland motors	Bushing-type pumps Vane-type pumps	Piston pumps and motors
At low ambient temperatures below 0°C (32°F)	HL 16	HL 16	HL 16	HL 16
	HL 25	HL 25	HL 25	HL 25
	HL 36	HL 36	HL 36	
	HL 49	HL 49	HL 49	
At medium ambient temperatures between 0°C and + 30°C (32°F–87°F)	HL 36	HL 16	HL 36	HL 16
	HL 49	HL 25	HL 49	HL 25
	HL 68	HL 36	HL 68	HL 36
		HL 49		HL 49
		HL 68		
At high ambient temperatures above 30°C (87°F)	HL 68	HL 68	HL 36	HL 16
			HL 68	HL 25
				HL 36
				HL 49

Fluid

Mineral-oil fluids present no problems with any type of hydraulic pump, provided the viscosity is suitable for the design of pump or the system, and the fluid is free from contaminants. Viscosity requirements are usually specific, and determined by the pump manufacturer's recommendation. The majority of hydraulic pumps are, in fact, designed around a fluid of a particular viscosity (see also Table I). Any departure from this viscosity will modify the performance of the pump, resulting in lowered efficiency, and in extreme cases even calling for re-rating of the maximum speed.

All pumps are likely to be harmed by the presence of abrasive solid contaminants in the fluid, some types being more susceptible to damage than others — *eg* all precision-made pumps with close clearance and gear pumps.

The use of fluids other than oils may present further problems, mainly as regards adequate lubrication of rubbing surfaces within the pump. In particular, this can affect the load rating on the bearings, especially on those types of pump operating under hydraulic unbalance where bearing loads are high.

TABLE II – APPROXIMATE LOAD FACTORS FOR BEARINGS

Fluid	Comparative Lubricity	Bearing Load Factor	Equivalent Bearing Load
Mineral oil	100 (standard)	1.0	1.0
Water-in-oil emulsions	33–50	2.0–3.0	0.5–0.33
Water-glycol	25	4.0	0.25
Phosphate ester	70–80	1.3	0.7–0.8
Phosphate ester blends	80–85	1.2	0.8–0.85
Chlorinated hydrocarbons	80–85	1.2	0.85

An assessment of load factors for bearings for different types of fluids is given in Table II, taking mineral oil as a standard. Where high bearing load factors are involved this generally precludes the use of pumps with rolling bearings lubricated only by the pumped fluid. Note however that in some pump types, such as in-line piston pumps, adequate lubrication for the pump bearings can often be provided by a pressurized oil system, and they are thus independent of the lubricity of the fluid being pumped.

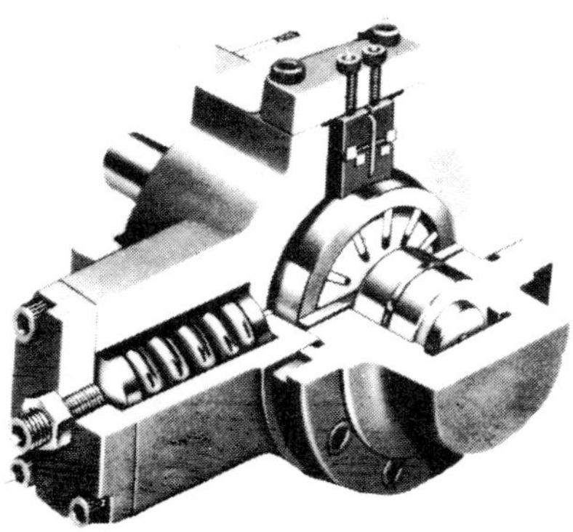

Racine variable capacity vane pump.

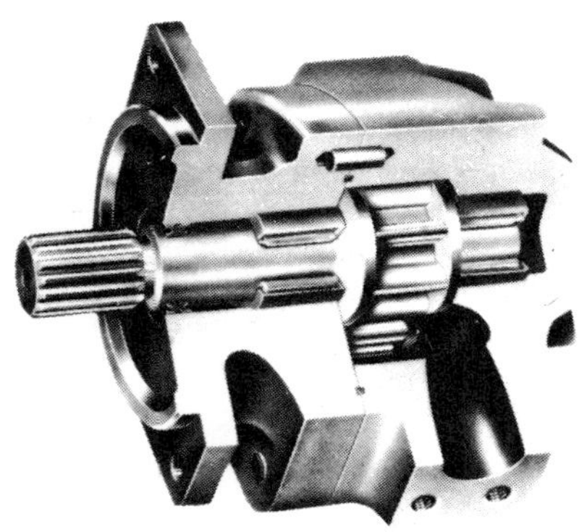

Hamworthy gear pump.

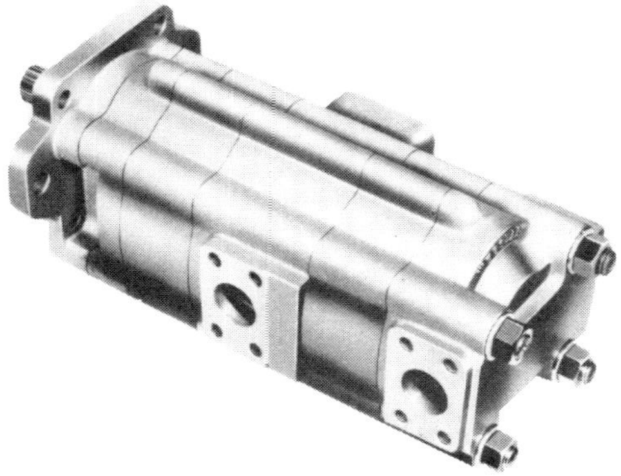

Webster G series dual gear pump.

Webster K series single-gear pump.

Bulk and Weight

No general comparisons can be drawn between power:weight ratios of different types of pump, since so much depends on the individual design and the constructional materials used. Comparative power:weight ratios can, therefore, only be extracted from manufacturers' specifications.

A further point here is that where weight is an important factor, it is usually the total system weight rather than the weight of the pump which is the major parameter. Optimum solutions are usually obtained by utilizing system pressures of the order of 210 to 280 bar (3 000 to 4 000 lb/in^2) using smaller sizes of components throughout and also minimizing fluid volume.

Noise and Vibrations

Noise and vibrations are generally limited, and whilst excessive noise may be disturbing, or a nuisance, excessive vibration can aggravate wear. Noise generated in pumps is largely the result of sudden pressure changes between the suction and outlet side. Thus pumps which produce high localized pressures are likely to be noisier than those providing a more gradual pressure change. No specific data can be given since the actual noise generated also depends on the mass and rigidity of the pump body, as well as the materials of construction, where and how it is mounted, and whether or not it is isolated from its mounting and delivery line.

In general, however, all piston pumps tend to be relatively noisy in operation, whilst screw pumps are normally the quietest. Gear pumps are noisier than vane pumps, although they can be quietened by using helical gear forms. Virtually all types of pump, in fact, can be made quieter by modification of design detail, but inevitably this is achieved only at the expense of a loss of efficiency. Again, as an alternative, a positive way of reducing pump noise is simply to use an over-size pump and run it at a much lower speed to give the same delivery.

See also chapters on *Hydraulic Pumps* and *Hydraulic Fluids*.

Pump Drivers

IN A majority of industrial applications hydraulic pumps are driven by electric motors as these are the most compact, efficient and quietest form of driver. Direct coupling is favoured with the pump being driven at a constant (motor) speed.

Except in the case of variable-capacity pumps, pump driven speed governs the actual delivery. Operating speed limits are set by the design of the pump, whilst desirable running speeds are set by the normal running speed of the driver — it is more efficient and less costly to drive the pump by direct coupling, rather than through gearing.

Normal maximum operating speeds for various types of pumps are summarized in Table I. Specific proprietary pumps will be rated for the maximum speed at which they can be run, either continuously or with separate speed ratings for continuous and intermittent operation. These recommendations should not be exceeded.

TABLE I — TYPICAL MAXIMUM SPEEDS FOR HYDRAULIC PUMPS

Type	Maximum speed, rev/min	Type	Maximum speed rev/min
Precision gear	5 000 (or higher)	Internal gear	3 000
Normal gear	3 500	Cam-vane	3 000
Radial piston	3 500	Rotary abutment	2 500
Axial piston	3 500	Vane	1 800
Screw	3 000 (or higher)	Multi-piston, in-line	1 500

Where direct coupling is used and the driver speed is lower than the maximum rated speed for the pump, then some sacrifice of potential capacity must be accepted and the pump size selected accordingly. Pump deliveries are usually expressed in terms of displacement per revolution, or displacement per 1 000 rev/min, so determination of delivery at any particular driven speed follows by simple calculation.

With variable-speed drives the delivery at any speed can be calculated on a similar basis.

Electric Motors

Most electric motors manufactured in Great Britain are produced to British Standard requirements for electric design and dimensions, the chief of which are:

BS 2048 : 1961 Dimensions of fractional horsepower motors

BS 2960 : 1961 Dimensions of 3-phase motors

BS 3979 : 1961 Dimensions of electric motors (metric series)

TABLE II – INSULATION AND OVER-LOAD CAPACITIES (BS 170 AND BS 2613)

Insulation Class	Type	Permissible Temp. Rise Measured °C	Permissible Temp. Rise Average Calculated from Increased Winding Resistance °C
Y	Cotton, silk, paper; not impregnated or immersed in oil	35	45
A	Cotton, silk, paper; impregnated or oil-immersed	55	60
E	Materials rated for operation up to 120°C	65	70
B	Asbestos, mica, glass fibre, *etc*; in built-up form with binder	75	80
F	Class B materials with binder, rated for up to 155°C	95	105
H	Mica, glass fibre, *etc*, silicone elastomer; impregnated; rated up to 180°C	115	125

TABLE III – INSULATION CLASSES AND MAXIMUM TEMPERATURE RISE (VDE 0530)

Component	Insulation Class – °C		
	B	F	H
Insulated windings	80	100	125
Commutators and slip-rings	80	80	80
Sleeve and rolling bearings	50	50	50
Rolling bearings with special grease	60	60	60

TABLE IV – TYPES OF ENCLOSURES FOR ELECTRIC MOTORS

Type of Enclosure	Duty
Protected or screen protected	General purposes, engineering workshops, *etc*.
Drip proof	Outdoor plant, 'wet' surroundings indoors.
Pipe ventilated	Mills, cement works, chemical works, paint plant, *etc*.
Forced draught	Where improved cooling is required.
Totally enclosed (TE)	Boiler house plant, foundries, steel plant.
Totally enclosed, fan-cooled (TEFC)	General duties requiring total enclosure.
Totally enclosed, flame-proof (TEFP)	Coal mines, gas works, oil plant, gaseous chemical plant, and for operation in hazardous atmospheres.

Electric motors are also specified by service ratings covered by BS and DW Standards, VPE specifications, and others. Various other authorities also lay down their own regulations and standards. Two general methods of rating are:

(i) Type of insulation — see Tables II and III.

(ii) Type of enclosure — see Table IV.

The majority of industrial electric motors are of the induction type, used for 3-phase supplies. Depending on the type and application, motors may be star-connected or delta-connected — Fig 1.

Fig 1

With slip ring rotor — *Star-connected*, *Delta-connected*

With squirrel-cage rotor — *Star-connected*, *Delta-connected*

The torque generated by 3-phase motors varies widely between zero speed and synchronous speed. For adequate acceleration starting torque will exceed the break-away torque of the hydraulic pump by a sufficiently large amount and during the rev-up to working speed the motor torque will always be greater than the load torque. Minimum requirements for pull-up types of fixed-speed motors are 0.5 times the starting torque and not lower than 0.9 times the rated torque for motors of less than 50 kW.

Limit of over-load capacity is determined by the break-down torque. Induction motors should be capable of delivering a torque up to 1.6 times the rated torque at rated voltage and frequency for 15 seconds (VDE 0530).

Permissible motor-on times are governed by the permitted loading, otherwise the motors may over-heat. Permitted loading is also related to the duty cycle. A 100% duty cycle corresponds to continuous running with constant-rated load, or a cyclic duration factor of 100%.

For periodic rather than continuous operation, the cyclic duration factor is given by:

$$\frac{\text{on-load time}}{\text{duty cycle time}} \times 100$$

or

$$\frac{t_B}{t_S} \times 100 \text{ (see Fig 2)}$$

Example : Cycle time is 10 min, with interval 6 min, (*ie* the motor is running for 4 mins, followed by a 6 minute idle period, then restarting); the cyclic duration time $= \frac{4}{10} \times 100$

$= 40\%$

With a low duty factor (*ie* t_S appreciably greater than t_B) the motor has sufficient time to cool down between the intervals of operation and can be rated accordingly. With a higher duty factor the rest intervals may be insufficient for the motor to cool off. Motor characteristic curves show the permitted cyclic duration factor for different duties.

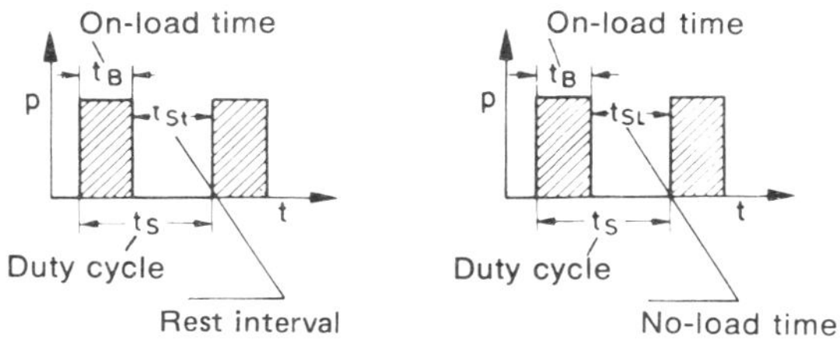

Fig 2

Internal Combustion Engines

Where a hydraulic system is associated with an engine-driven system or machine it is normally logical to drive the hydraulic pump from the main driver *via* a V-belt or gearing to match engine speed to optimum pump speed. Off-loading of the pump is essential with a continuously running engine. On some systems a clutch may be fitted to engage/disengage the hydraulic pump as required.

Where the hydraulic power demand is low it can be more advantageous to drive the hydraulic pump by an integrally mounted d.c. motor worked off a battery which is normally charged by an alternator associated with the main driver. Alternatively on larger installations – *eg* ship's hydraulics – the hydraulic pump(s) may be driven by electric motor(s) drawing current from the electrical services supply. The advantage of this is that individual pumps can be located close to actuator points.

See also chapter on *Pump Performance.*

Hydraulic Cylinders

HYDRAULIC CYLINDERS are the standard form for linear actuators, also described as rams or jacks. Specifically the description *ram* is applied to heavy-duty cylinders in some industries, or where the piston and rod are cylinders of the same diameter (also called a *plunger-type* or displacement cylinder). True plunger-type cylinders are usually single-acting and have relatively limited application. The term *jack* is fairly widely used to describe a single-acting cylinder, or for cylinders employed for lifting or 'jacking' actions. It is also used to describe double-acting cylinders in certain industries (*eg* in the aircraft industry).

The other variation in basic geometry is whether the rod terminates at the piston – *(single-rod)*; or passes right through the cylinder *(through-rod or double-ended rod (DER))*. The former provides maximum piston area for one direction of motion (and then maximum output force for a given system pressure). The through-rod cylinder provides equal piston wear and equal forces, and 'two-bearing' support for the rod.

These basic configurations are illustrated in Fig 1.

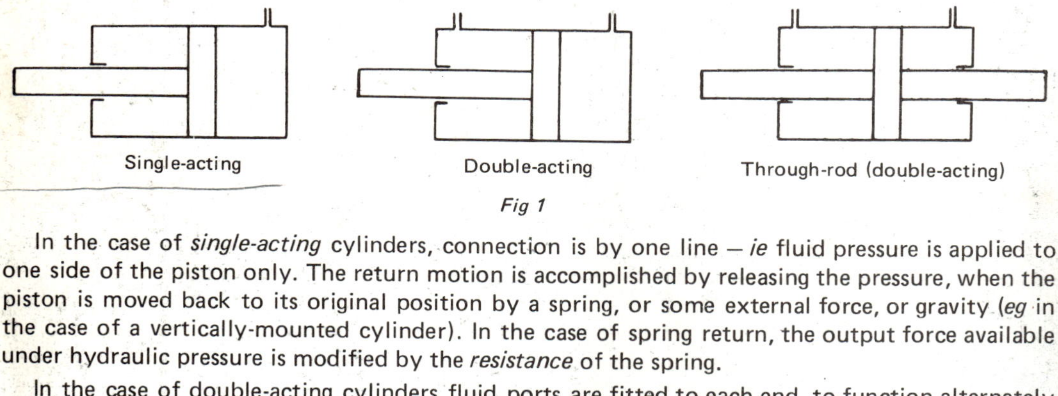

Fig 1

In the case of *single-acting* cylinders, connection is by one line – *ie* fluid pressure is applied to one side of the piston only. The return motion is accomplished by releasing the pressure, when the piston is moved back to its original position by a spring, or some external force, or gravity (*eg* in the case of a vertically-mounted cylinder). In the case of spring return, the output force available under hydraulic pressure is modified by the *resistance* of the spring.

In the case of double-acting cylinders fluid ports are fitted to each end, to function alternately as inlet and outlet ports, switched by a selector. The maximum output available is slightly less than that obtainable from a single-acting cylinder, since, when the fluid pressure is applied to the full piston area (outward stroke or *extending*), some back-pressure will be generated on the outlet side; also a rod seal will be required to prevent leakage when the piston is pressurized in the reverse direction, with consequently increased frictional resistance to motion.

In the reverse direction of motion (inward stroke or *retracting*), the force available will be lower since the effective piston area is reduced to that of the difference between the piston and rod

TABLE I – OTHER TYPES OF HYDRAULIC CYLINDERS

Description	Remarks
Telescopic 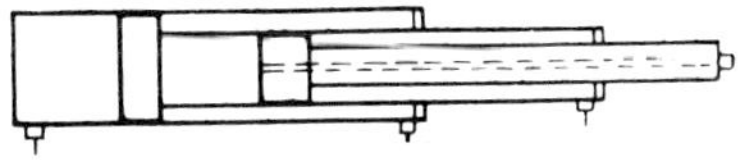	For very long strokes without excessive retracted length. Construction is normally based on using a hollow rod for the first section, which acts as a cylinder for the second section, and so on. Actual number of sections is usually limited to three or four. Return force is low with double-acting telescopic cylinders.
Differential 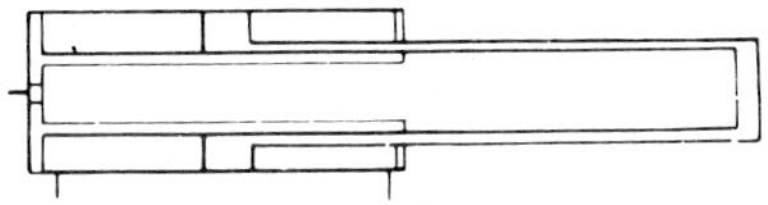	Telescopic cylinders designed to work as multi-volume devices. *Three-volume* differential cylinder gives inward and outward stroking of the piston and main hollow rod, with independent extension *via* pressurization of the third volume. This can be from same or an independent fluid source.
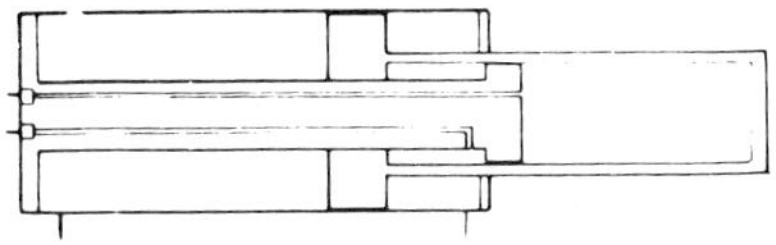	*Four-volume* differential cylinder provides double-acting working of the third (independent) movement. Basically comprises two double-acting cylinders mounted concentrically as an integral unit.
Duplex 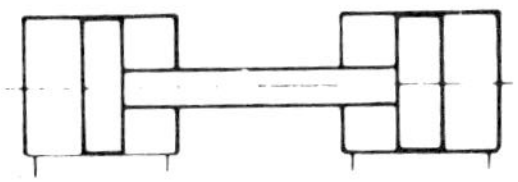	Two single- or double-acting cylinders opposed, with pistons mounted on a common rod. Capable of three-position movement.
Tandem 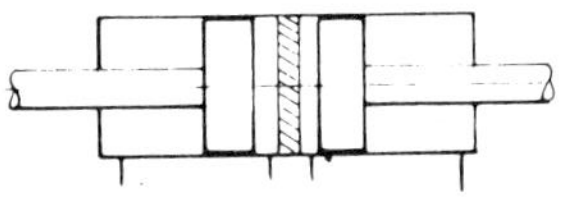	Two single- or double-acting cylinders mounted back-to-back as an integral unit. Provides three-position movement; or multi-position movements, as required.
Three-position tandem 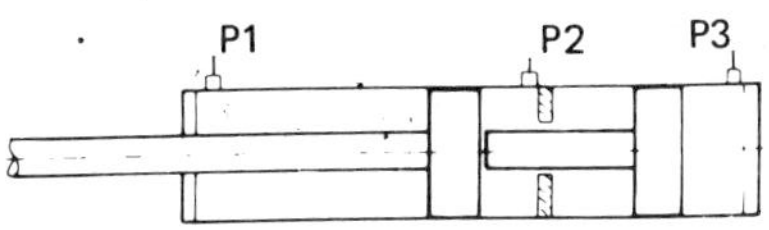	Tandem cylinder with equal or different cylinder lengths, and independent piston rods. With P1 or P2 pressurized, the main piston can complete a full stroke. With P1 and P3 pressurized, a half-stroke movement, or intermediate position can be held. Further movement is then possible by switching pressure from P1 to P2 (for further extension); or retracted half a stroke by switching P3 to exhaust. The system is not necessarily rigid if P3 only is pressurized.

areas. Back-pressure effects will, of course, also be present. Such performance losses may be small, but can appreciably modify the theoretical performance in practice, and it is usually on theoretical performance that cylinders are sized, with a nominal allowance for frictional losses.

Various 'multiple' configurations of cylinders are used, the chief types being summarized in Table I. Other types include the true *displacement cylinder* with rod diameter equal to the cylinder bore; and the *cable cylinder.* Displacement cylinders (Fig 2) are invariably single-acting. They are normally of short stroke and used for applications requiring high rigidity where the plunger can be returned by an external load or gravity. For this the cylinder must be mounted vertically. The plunger-type cylinder (Fig 3) is also employed in a similar fashion.

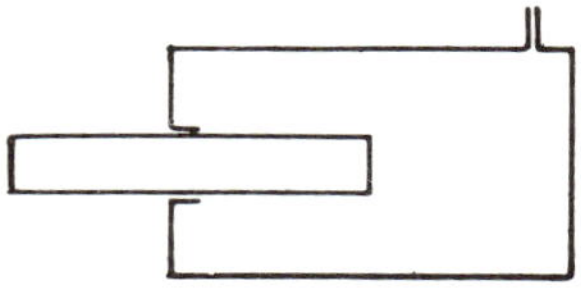

Fig 2 Displacement cylinder.

Fig 3 Plunger cylinder.

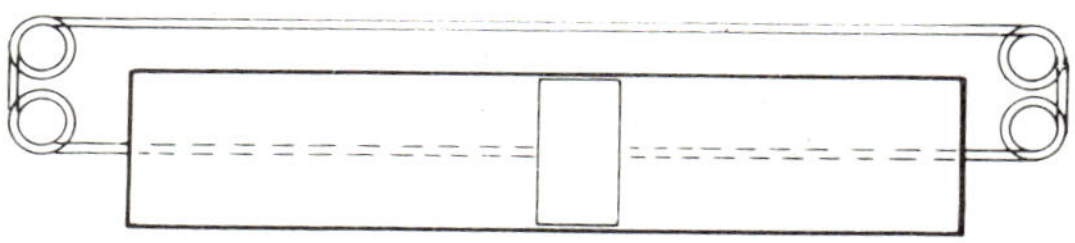

Fig 4 Cable cylinder.

A *cable cylinder* (Fig 4) is capable of providing a long stroke within the nominal overall length of the cylinder itself. This is accomplished by connecting the piston to an endless cable passing around external pulleys. Force output is taken by a suitable connection to the external cable.

Cable cylinders present sealing problems with high fluid pressure and are more usually rated as pneumatic than hydraulic cylinders. The same principle is, however, used in some designs of hydraulic rotary actuators.

Cylinder Performance

The theoretical thrust developed by a hydraulic cylinder is given by:

$$F = A_e P_e$$

where A_e = effective piston area, P_e = effective fluid pressure } in consistent units

The effective piston area is the full piston area in the case of a single-rod cylinder extending, the annulus area in the case of a single-rod cylinder retracting or either stroke of a through-rod cylinder.

$$F = 0.7854\ D^2\ P_e \text{ (single-rod, extending)}$$
$$F = 0.7854\ (D^2 - d^2)\ P_e \text{ (other cases)}$$

where D = cylinder bore
d = rod diameter
P_e = actual fluid pressure entering the cylinder

Pressure drop in the lines to the cylinder may be computed directly to arrive at a value of P_e. Strictly speaking the pressure drop through the cylinder ports should also be taken into account, but this is usually ignored because of the low velocities involved.

The following formulas can be used for quick calculation of cylinder sizes, allowing a correction for a typical 'average' pressure drop and frictional losses.

$$F = \frac{7 \times D^2 \times P}{10} \quad \text{or} \quad F = \frac{7 \times (D^2 - d^2) \times P}{10}$$

where P is the system pressure, or pump pressure at rated output

The output force is then given in kgf (kilograms) when D and d are in centimetres and P is in bar and in lbf (pounds force) when D and d are in inches, and P is in lb/in^2.

For more exact working the theoretical formulas should be used, with pressure drop correctly determined and back-pressure effects also considered in the case of double-acting cylinders. Friction is also involved, but an arbitrary loss of 2–5% will generally be adequate.

TABLE II – CYLINDER DEMAND FORMULAS

Single-acting Cylinders	**Single-rod**	**Through-rod**
Extending	$0.7854\ D^2 L\ in^3$ $0.00284\ D^2 L$ gallons $0.0129\ D^2 L$ litres $0.001\ D_m^2 L_m$ litres	$0.7854\ (D^2-d^2) L\ in^3$ $0.00284\ (D^2-d^2) L$ gallons $0.0129\ (D^2-d^2) L$ litres $0.001\ (D_m^2 - d_m^2) L$ litres
Retracting	Nil	Nil
Combined extension and retraction	$0.7854\ D^2 L\ in^3$ $0.00284\ D^2 L$ gallons $0.0192\ D^2 L$ litres $0.001\ D_m^2 L_m$ litres	$0.7854\ (D^2-d^2) L\ in^3$ $0.00284\ (D^2-d^2) L$ gallons $0.0129\ (D^2-d^2) L$ litres $0.001\ (D_m^2 - d_m^2) L$ litres
Double-acting Cylinders	**Single-rod**	**Through-rod**
Extending	$0.7854\ D^2 L\ in^3$ $0.00284\ D^2 L$ gallons $0.0129\ D^2 L$ litres $0.001\ D_m^2 L_m$ litres	$0.7854\ (D^2-d^2) L\ in^3$ $0.00284\ (D^2-d^2) L$ gallons $0.0129\ (D^2-d^2) L$ litres $0.001\ (D_m^2-d_m^2) L$ litres
Retracting	$0.7854\ (D^2-d^2) L\ in^3$ $0.00284\ (D^2-d^2) L$ gallons $0.0129\ (D^2-d^2) L$ litres $0.001\ (D_m^2-d_m^2) L_m$ litres	$0.7854\ (D^2-d^2) L\ in^3$ $0.00284\ (D^2-d^2) L$ gallons $0.0129\ (D^2-d^2) L$ litres $0.001\ (D_m^2-d_m^2) L$ litres
Combined extension and retraction	$0.7854\ (2\ D^2-d^2) L\ in^3$ $0.00284\ (2\ D^2-d^2) L$ gallons $0.0129\ (2\ D^2-d^2) L$ litres $0.001\ (2\ D_m^2 - d_m^2) L_m$ litres	$1.57\ (D^2-d^2) L\ in^3$ $0.00568\ (D^2-d^2) L$ gallons $0.0258\ (D^2-d^2) L$ litres $0.002\ (D_m^2 - d_m^2) L$ litres

Cylinder Demand

The volume of fluid required to operate the cylinder is equal to its displacement, thus fluid demand is related to both bore and stroke:

$$Q = A_e L$$

when A_e is the effective piston area
and L is the stroke

In working units:

$$Q\ (\text{Imp gal/min}) = \frac{A_e \times L}{227}$$

$$Q\ (\text{US gal/min}) = \frac{A_e \times L}{231}$$

where A_e is in in^2 and
L is in inches

$$Q\ (\text{lit/min}) = \frac{A_e \times L}{1\,000}$$

where A_e is in cm^2 and
L is in cm

Cylinder demand (or fluid volume per stroke) can be calculated directly from the working formulas given in Table II.

Power Output of Hydraulic Cylinders

The power output of a hydraulic cylinder can be derived directly from the output force and speed of operation, *viz*

$$\text{hp} = \frac{F \times S}{6600 \times \text{time (sec)}} \qquad = \frac{F \times S}{7614 \times \text{time (sec)}}$$

$$\text{Power (Watts)} = \frac{F \times S}{8.85 \times \text{time (sec)}} \qquad = \frac{F \times S}{10.2 \times \text{time (sec)}}$$

where F = force, lbf
S = stroke in in

where F = force in kgf
S = stroke in cm

or directly from the output force, cylinder diameter and pump delivery:

$$\text{hp} = \frac{F \times Q}{396 \times D^2} \qquad = \frac{F \times Q}{1652 \times D^2}$$

$$\text{Power (Watts)} = \frac{1\text{–}88\ F \times Q}{D^2} \qquad \frac{F \times Q}{2.2 \times D^2}$$

where F is in lbf
Q is the delivery, gallons/m
D is the bore, inches

where F is in lbf
Q is the delivery, lit/m
D is the bore in cm

In practice the speed of the operation of a cylinder is not constant, so such formulas are approximate. However, direct power costs can be estimated on this basis, allowing for a pump input power of, say, 10% greater than the calculated cylinder power.

For other aspects of cylinder performance, see chapter on *System Design and Performance.*

Cushioning

The speed of a hydraulic cylinder movement is governed directly by the pump delivery. Many cylinder applications demand fast movements, when the shock load, as the piston reaches the end of its movement, can be very high. To eliminate, or at least substantially reduce deceleration loads, the latter part of the piston movement can be 'cushioned'.

A typical cushion is based on providing dashpot damping on the final movement by throttling the fluid escaping from the outlet port. The cushion end provides an extension of the cylinder with substantially reduced bore. At the same time the leading face of the piston is fitted with a spigot of the same diameter which enters the cushion chamber as the piston approaches the end of the stroke, effectively sealing off fluid flow to the outlet port. The only escape for the fluid is then through a metered orifice, the throttling effect of which is usually adjustable by means of a needle valve – Fig 5. This adjustment may be preset (for fixed orifice throttling); or left adjustable for the user to set the degree of throttling required to suit a particular application (although this has certain dangers).

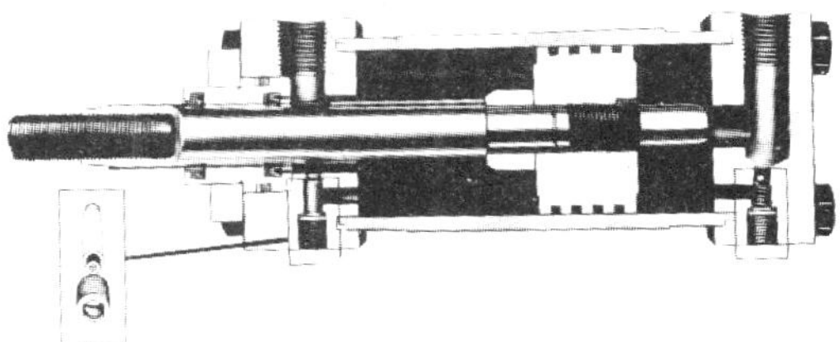

Fig 5 Typical design of cylinder cushion at tail ends of cylinder.

To initiate return movement, provision must be made to bypass the orifice, otherwise the fluid flow to the face of the piston would be so restricted that the necessary 'break-out' force might never be realized (or at least initial movement would be very sluggish). This usually takes the form of a spring-loaded non-return ball valve, which bypasses the return throttling orifice for the initial return flow, until the spigot of the piston has left the cushion chamber. Not until this motion has been accomplished will the full output force be available from the cylinder.

The basic type of cushion may have disadvantages for certain applications. Since the bleed orifice is fixed (either in design or by adjustment), a sudden change in piston velocity, which could affect machine operation, will occur at the beginning of cushioning. In such cases a relief valve may be employed instead for cushioning as the cushioning action will then be more moderate during the initial part of the cushioned stroke – Fig 6. However, cushioning will become progressively 'harder' as energy is dissipated and the cylinder may well stop before the end of its stroke. This can be overcome by using a restrictor in parallel with the relief valve.

Cushioning may be applied to one, or both, ends of a double-acting cylinder. Cushioning is mostly used on fast-acting cylinders where speeds may range up to 0.3 metres/sec (1 ft/sec) or more.

As a rough rule, the fitting of a cushion adds about one half of the cylinder diameter to the total length. However, there can be considerable differences in cushion lengths in individual manufacturers' designs, and different cushion lengths may also be available for standard sizes of

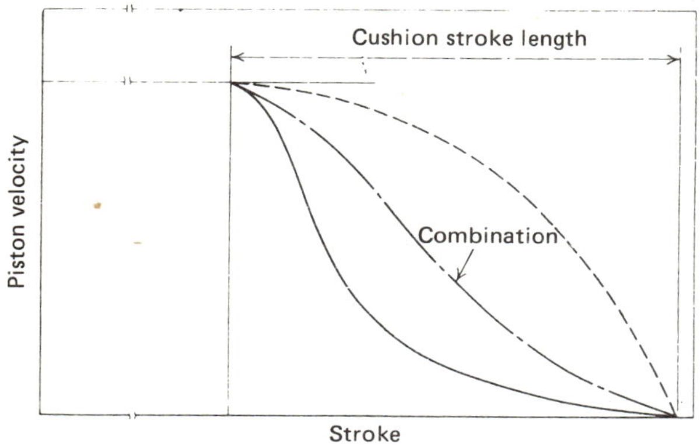

Fig 6 Full line shows effect of needle-valve cushioning. Dashed line shows gentler action introduced by C relief valve. Middle curve shows combination of restrictor (needle valve) and relief valve.

cylinders to suit specific requirements. Production sizes of hydraulic cylinders are commonly available with or without cushioning.

The most likely fault with cushions is leakage developing between the piston spigot and cushion chamber, which can modify the cushioning action. Dirt or mal-seating of the ball valve may also introduce leakage through this bypass port, substantially reducing the cushioning available. To be effective, the pressure developed in the cushion chamber must be higher than that of the fluid pressure on the piston. This pressure can be increased by increasing the throttling effect, but if this is taken to extremes, excessively high fluid pressures may be developed within the cushion chamber. This can be even more damaging than not having a cushion at all. Essentially therefore, cushion throttling adjustment is something which should only be attempted by someone with expert knowledge of the subject and the particular cylinder construction.

Rotating Cylinders

Rotating cylinders are usually mounted on machine spindles or similar rotating devices. Such cylinders are invariably of compact length and short stroke, piston and cylinder being locked together by means of an integral pin on which the piston slides, with the piston rod connected to the machine spindle. The blind end of the cylinder terminates in a distributor shaft, normally made of hardened steel, supported by a stationary distributor housing, which also acts as a bearing for the distributor shaft. The actual bearing surface may be plain or rolled bearings may be used.

The distributor shaft is drilled and ported to provide open passage to each end of the cylinder. The shaft ports correspond in position to distribution rings machined in the distributor bearing surface, these rings being connected to external ports. The design must also incorporate suitable seals, although some leakage is desirable both to maintain adequate lubrication of the rubbing surfaces and to conduct away frictional heat. A slight venting action may also be necessary in order to avoid excessive back-pressure on the pressure-energized seals. Provision is often made in the design for collecting leakage oil in a sump at the bottom of the distributor, from whence it can be returned to the reservoir – Fig 7.

Rotating cylinders are usually designed for lower working pressure than most other types of modern hydraulic cylinder. Typical pressure ratings are 17.5 to 35 bar (250 to 500 lb/in^2), although some designs may be capable of working at much higher pressures without excessive leakage developing. Rotational speeds may range up to 1 000 rev/min or more. Problems associated

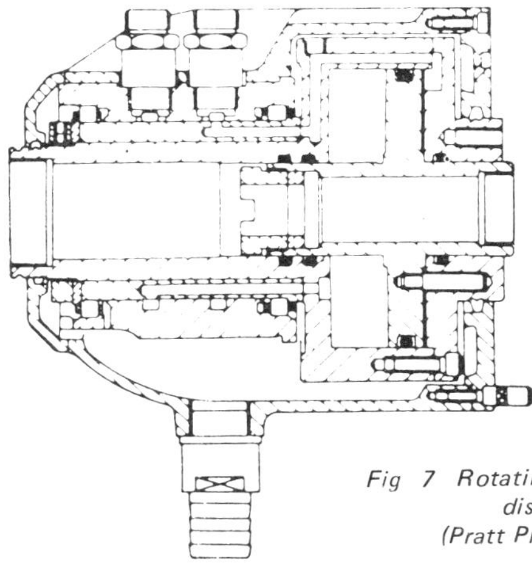

Fig 7 Rotating cylinder with special distributor gland. (Pratt Precision Hydraulics).

with the design of a suitable distributor gland or sealing system increase considerably with increasing rotational speed and cylinder diameter. In the latter respect hydraulic cylinders have an advantage over air cylinders for rotational applications because the higher fluid pressures available can result in a considerable reduction in cylinder diameter.

Seal design is always a primary problem, for the most effective types of rotating shaft seals are not usually those which are capable of withstanding high pressures. Packed glands may be preferred. Also the cooling effect of the lubricant film may be inadequate, in which case jacketing and water cooling may have to be employed.

Materials

Casting was the original method used for the production of high-pressure cylinders and is still used for larger high-pressure cylinders; however, forging may be used for even higher pressures. Cast (or forged) cylinders have the advantage that one end cover can be formed integral with the cylinder tube, thus saving one joint and a source of leakage. In the case of a cast cylinder the blind end is normally dome shaped, with generous radii and fillets for stress relief at abrupt changes in contour. Forged cylinder blind ends are usually flatter, but with similar attention to the elimination of stress raisers.

The disadvantage of this form of construction with an integral end cover is that it is more difficult to bore finish. It is easy to leave machining marks in the blind end of the tube, and honing is always more difficult applied to a blind tube. For this reason, and in view of the importance of surface finish for satisfactory seal performance, cast cylinders may be cast as open tubes so that they can be finished by through machining and polishing.

Bore surfaces may be protected by nickel or chromium plating (subsequently polished) or by other protective treatments. This is seldom necessary in the case of cylinders intended for use with oil fluids, although plating may be used to give a harder and more scratch resistant bore surface. In that case hard chrome plating is the usual choice. With water fluids some form of protective treatment may be considered essential to combat corrosion, even though the fluid may contain corrosion inhibitors.

For working in corrosive surroundings the outside of the cylinder tube may be protected, either by surface coating or treatment, or by jacketing in a corrosion resistant material. The use of plastics for tube construction is a possibility in such cases, provided the cylinder is reasonably small and the working pressure not too high. Plastic cylinders have been used successfully for marine applications, as an alternative to stainless steel or a resistant aluminium alloy, where the cylinder is exposed to salt spray.

Double-tube construction is also used to protect the main (inner) cylinder from impact damage, or where 'armouring' offers obvious advantages.

Conventional finishes normally represent a compromise between production costs and duty requirements, particularly as regards the type of seals employed. Basic requirements in this respect are summarized in Table III.

TABLE III – CYLINDER SURFACE FINISH

Tube Condition	Surface Finish		Application
	Microns Micromillimetres	Microinches	
As drawn (general production)	400–1200 (typical)	10–30 (typical up to 80)	Suitable for use with leather seals or heavy-duty fabricated seals.
As drawn (controlled production)	800–1200	20–30	Suitable for use with rubber-impregnated fabric seals.
Honed	800	20	Suitable for use with rubber-impregnated fabric seals.
	400–600	10–15	Suitable for use with elastomeric seals.
Honed and polished	better than 800	better than 20	Suitable for use with elastomeric seals at higher speeds.
Deep polished	400	10	Precision cylinders for high-pressure applications with any type of seal.
	200	5	Superior sealing at higher pressures.

A two-stage telescoping ram under test in the W.F.E. Hydraulics Limited factory in Sheffield.

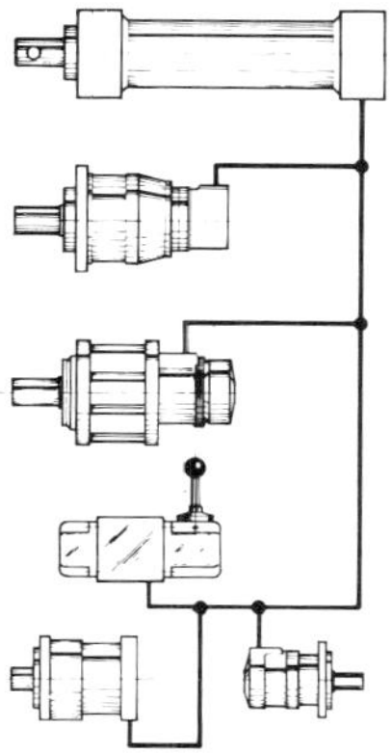

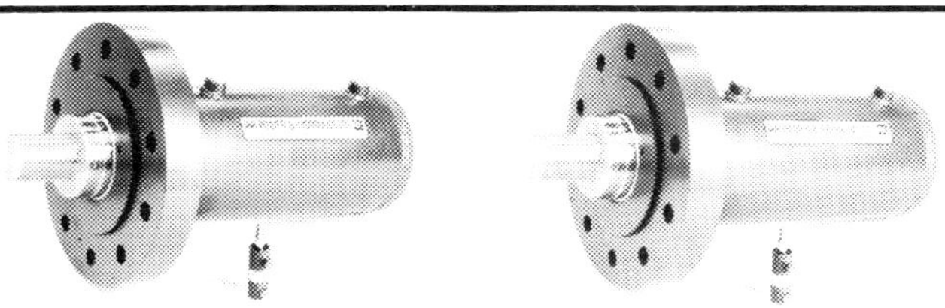

Construction

Various forms of cylinder construction are employed, *eg*

(i) Screwed-on ends

(ii) Tie-rod

(iii) Flanged-and-bolted covers

(iv) Welded construction

All have some advantages and limitations. All-welded cylinders are generally cheaper to produce, but basically non-serviceable. They are quite widely used on mobile equipment, however. Cylinders with screwed ends have a slight advantage in diametric size over other types (except welded), but usually require a thicker cylinder to accommodate cutting the screw threads without weakening or 'bellmouthing' the ends of the cylinder.

Tie-rod construction is now widely used and length-for-length compares favourably with other types. It is the standard form of construction accepted by the Joint Industry Council (JIC) of the Society of Automotive Engineers and specified in National Fluid Power Association Standard ASA – B93.3 1965. A particular advantage of tie-rod construction is that it allows a very wide range of alternative mounting styles and optional features from a standard range of components. It is now generally accepted as a standard for the machine tool industry, (see Fig 8).

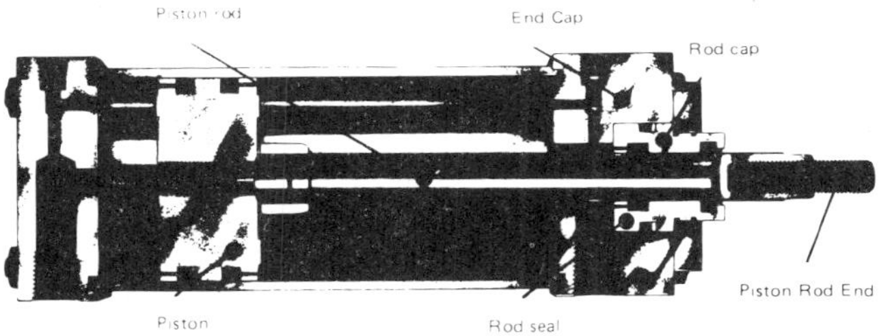

Fig 8 A standard JIC tie-rodded cylinder. (Double A Division, Brown & Sharpe Ltd).

The use of flanges welded to the cylinder ends with bolted-on covers is one of the strongest forms of construction (but not necessarily superior to tie-rod construction). Its disadvantage is that it is more costly to produce and extreme care has to be taken during welding to avoid distortion of the cylinder barrel. Cast cylinders, however, can be produced with cast-iron flanges. This type of construction is still used for some heavy-duty cylinders, particularly in larger sizes.

The cylinder wall thickness usually depends on the pressure rating of the cylinder but may also be influenced by the form of construction and/or the application. For heavy-duty use or use under arduous conditions a heavier wall cylinder may be the most economic proposition in the long run.

Strength of Cylinders

Where the diameter:thickness ratio of the cylinder tube is greater than 16:1, the stress produced in the wall material due to internal pressure can be determined from the simple formula for uniformly distributed hoop stress:

$$S = \frac{PD}{2t}$$

where S = hoop stress

P = internal pressure

D = internal diameter of tube

t = wall thickness

This can be rendered in the form of a working formula, written as a solution for wall thickness required:

$$t = \frac{P_w D}{2S_m} \times F + c$$

where P_w = design working pressure
S_m = maximum permissible material stress
F = design factor of safety
c = an additional thickness allowance for corrosion. This additional correction is normally ignored, the necessary margin being accommodated in the safety factor employed.

For thick-walled homogeneous tubes the stress is no longer uniformly distributed through the tube walls, when the maximum hoop stress is given more accurately by:

$$S = \frac{D^2 - 2t + 2t^2}{2t\ (D - t)} \times P$$

A simpler formula, written as a solution for wall thickness is:

$$t = \frac{D}{2}\left[\sqrt{\frac{S_m + P_w}{S_m - P_w}}\right] - 1 \quad \times F + c$$

Again the added value c may be ignored.

The stress produced in cast cylinders may also be determined from the aforementioned formulas, typical values for maximum permissible material stress being:

cast iron	–	280 bar (4 000 lb/in^2)
high duty cast iron	–	420–550 bar (6 000–8 000 lb/in^2)
cast steel	–	840 bar (12 000 lb/in^2)
cast aluminium alloy	–	550 bar (8 000 lb/in^2)
cast brass	–	420 bar (6 000 lb/in^2)
cast bronze	–	420 bar (6 000 lb/in^2)

The safety factor allowed in such cases is normally very generous.

The strength of various tube materials is summarized in Table IV. The strength of low carbon mild steel is substantially increased by cold working, with a potential ultimate tensile strength of the order of 6 300 bar (40 tons/in^2) after cold drawing and deep polishing. Cold working also tends to promote better machinability of the material, but at the same time increases the tendency towards distortion when machined. Stress-relieved tubes, with minimum distortion on machining, are more difficult to machine. On the other hand, stress relief does avoid ovality which might otherwise result from machining, and is particularly advisable in the case of thin-walled tubes (diameter:thickness ratio in excess of 16:1).

TABLE IV – DRAWN CYLINDER TUBE MATERIALS

Material	Condition or Type	Ultimate Tensile Strength (min)		0.1% Proof Stress		0.2% Proof Stress		Design Maximum Permissible Stress	
		bar	lb/in^2	bar	lb/in^2	bar	lb/in^2	bar	lb/in^2
Low carbon steel	As cold-drawn	3 550	55 000	3 200	45 000			1 250	18 000
	After deep polishing	6 000	85 000	5 000	70 000			1 750–2 100	25 000–30 000
Stainless steel (304)	Annealed	7 000	100 000			2 400	34 000	2 300	33 000
	Half-hard	8 800	125 000			6 000	85 000	3 000	42 000
	Hard	10 000	150 000			8 300	118 000	3 500	50 000
Tungum alloy	Annealed	4 650	66 000	2 600	37 000			1 550	22 000
	Precipitation-hardened	4 700	67 000	3 200	45 000			1 750	25 000
Aluminium alloy	61S–T6	3 200	45 000					1 000	15 000
Titanium	DTD 5013A	4 700	67 000†	2 350	33 500*			1 550	22 000
	DTD 5063	6 300–7 700	90 000–110 000	4 700	67 000*			2 100	30 000

† Maximum * Minimum

All cold-drawn tubing is subject to ovality. However very close tolerances can be held by controlled production, followed by honing the bore to finish, provided the diameter:thickness ratio of the tube is less than 20:1. For higher diameter:thickness ratios it is generally impossible to hold very close tolerances because of the thinness of the walls.

Honing is a normal minimum finish for drawn cylinder tubes, not only to give a smooth and more uniform finish but to produce a more favourable surface pattern than that resulting from drawing. Surface finish measurements are no true guide in this respect, as a honed finish with the same nominal surface finish as given by drawing would normally be expected to give better sealing and longer seal life.

End Covers

End covers are usually – but not invariably – made from the same material as the cylinder tube. The main variation is in the method of fitting the end caps.

Screwed-on covers are commonly employed on smaller cylinders, particularly where there is ample cylinder tube wall thickness to justify threading without loss of strength. A leak-tight joint can be provided by the inclusion of a simple ring seal in the cover. One inherent disadvantage of this method is that it is difficult to ensure correct angular alignment of the two heads after assembly. It does also produce a relatively bulky head.

Where the tube wall is of sufficient thickness the cover may be attached directly by means of tapped holes in the cylinder tube and machine screws – Fig 9. This method, although neat, can be costly on a long cylinder, and is strictly limited in the diameter of threaded hole which can be accommodated without seriously weakening the cylinder tube (alternatively it may call for an excessive wall thickness for the pressure rating concerned); it is also troublesome should a corroded screw break off in its tapped hole or a thread strip during disassembly.

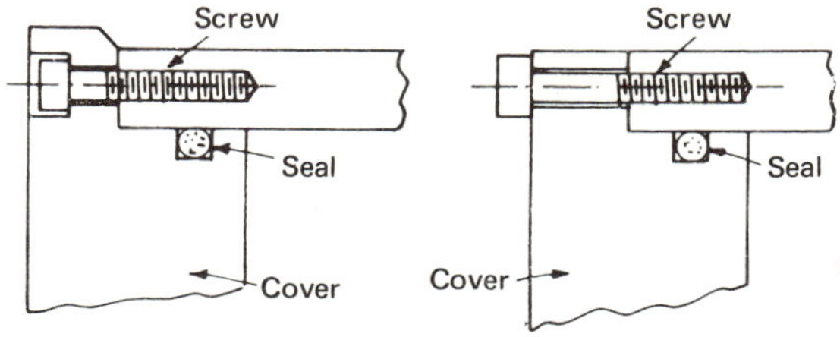

Fig 9

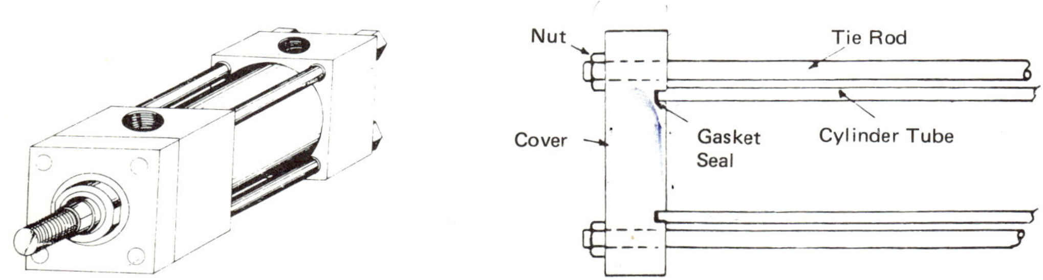

Fig 10 Tie-rod construction.

Tie-rod construction is shown in Fig 10. This method offers a simple and inexpensive solution and is often favoured for high pressure cylinders. A particular advantage is that there is no weakening or distortion of the cylinder walls involved. However, for high-pressure duties the possible elongation of the tie rods must be taken into account, and the type of seal employed selected accordingly. Tie-rods, however, increase the bulk and do not improve the 'streamlining' or appearance of cylinders. They may also be liable to damage when used on long cylinders.

With tie-rod assembly, cylinder ends are commonly made square so that each corner forms a convenient point for fitting the tie-rod parallel to the cylinder tube.

Welded assembly results in a minimum depth of cover and eliminates the need for seals – Fig 11. It does, however, have certain limitations, in particular the choice of material for covers which must be easily weldable – mild steel is normally used.

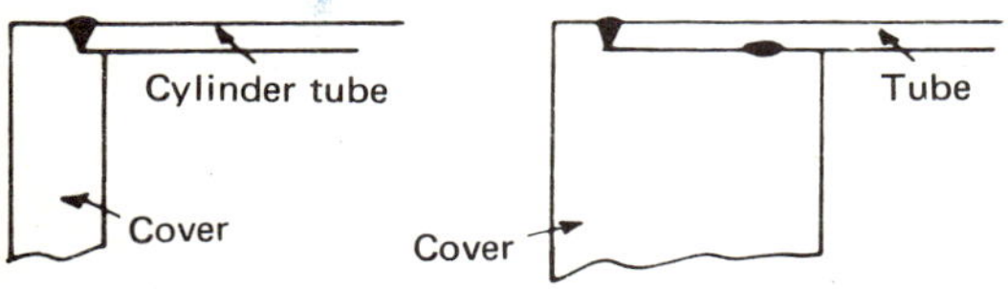

Fig 11

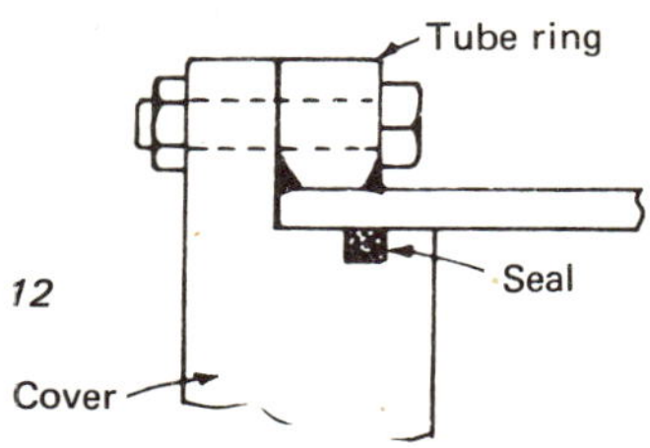

Fig 12

In the case of heavy duty cylinder tubes, tube-ring construction may be employed where a ring is welded on close to the end of the tube and the cover bolted to this ring, also seating on a suitable gasket, but leaving a circumferential gap between the cover and tube ring – see Fig 12. An advantage offered by this method is that in the event of the cover bolts and nuts becoming

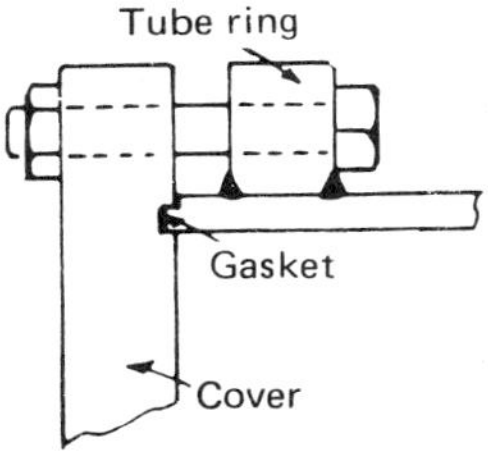

Fig 13

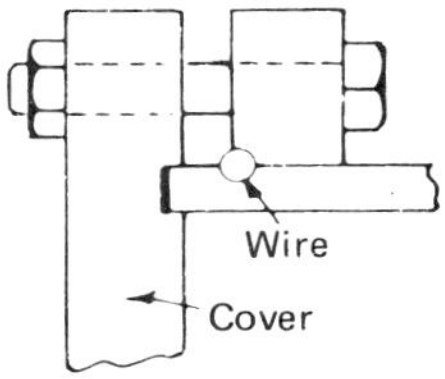

Fig 14

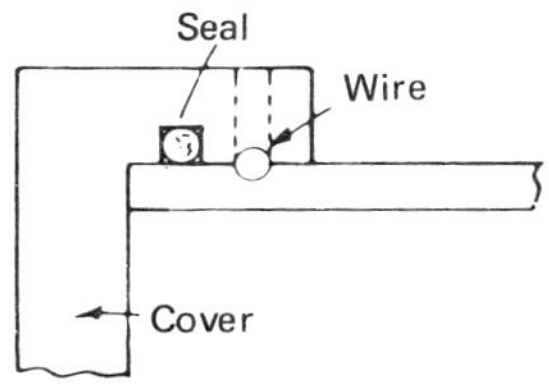

Fig 15

'frozen' through corrosion the cover can be removed without damage to cylinder or cover merely by cutting through the bolts. In some cases, however, the tube ring may be welded flush with the tube end, producing the equivalent of a flush flange fitting. (Fig 13).

A similar system which is more economic to produce employs a tube ring positioned by a wire or circlip situated in a groove cut in the cylinder wall (Fig 14). The cover is then bolted to the ring, sealing on to a gasket. Yet another method employs a wire which fits into matching semi-circular gooves machined in both the cylinder tube and cover. The wire is introduced through a hole, drilled tangential to the groove after the parts are assembled, the parts being rotated to bring them into the correct position and, if necessary, secured with a set screw. This type of fixing makes it difficult to remove the cover in the event of corrosion of the wire and to avoid this possibility either stainless steel wire or heavily plated wire is to be preferred. (Fig 15).

A variation on this method is the provision of a neat and positive end cover fixing by the use of snap rings, or similar key-type locking. A feature of the latter is often ease of stripping for assembling or dismantling on site.

In the case of cast cylinders, a widely employed method is to form a flange integral with the open end of the cylinder to which the cover is bolted, the size and number of bolts required being determined by the end load involved. If the cylinder is double-acting a gasket-type seal will also be required between the cover and cylinder flange, in addition to the rod seal or gland in the cover. Alternatively — and equally applicable to the fitting of covers to end tubes — tie-rods may be used.

Pistons

Materials used for pistons include cast iron or steel, brass, bronze and aluminium alloy, also sintered iron and steel. Ferrous metal pistons may be bronze faced, or hard chrome plated and polished. For use with water-based fluids bronze pistons are a common choice, although plated or surface-treated iron or steel may also be employed.

The form of the piston depends primarily on the type of piston seal used. The type of seal will normally dictate whether the piston has to be designed as a one-piece, two-piece or three-piece assembly in order to fit the seals. Solid (one-piece) construction is normally limited to pistons fitted with simple pressure-energized ring seals or automotive-type piston rings, the former normally providing the simplest and least expensive form of piston for smaller cylinders. Cup seals are generally preferred for larger sizes and heavier duties, and chevron seals or proprietary seal sets for the largest sizes of pistons and/or heaviest duties. In this case the piston needs to be of composite construction to allow the individual seals and headers to be mounted in position in matching grooves, bearing in mind that most seal sets have to be mounted in opposite pairs to provide sealing on a double-acting piston.

The most common method of attaching the piston to the piston rod is to machine a shoulder on the rod with a threaded end. The piston then locates on the shoulder and is held in place by a nut. For heavier duties the piston may be welded to the rod, and for lighter duties simply located by circlips.

Rods

Piston rods are normally made from steel bar, or flame-plated steel, turned to size, and preferably finished by hard chrome plating followed by fine grinding and/or polishing. A high surface finish is desirable in order to minimize wear on the rod seals. Plating is highly desirable to prevent corrosion, which again could result in mechanical damage to the rod seals and reduce seal life. However unless the rod is case-hardened there is the possibility of it being damaged (*ie* dented with the plating lifting and falling. Case-hardening avoids this chance of plating breakdown and the resultant damage to rod seals.

Polished stainless steel is sometimes used for rods to provide maximum resistance to corrosion, particularly for cylinders used with other than oil fluids.

Rod sizes for standard production are normally standardized at about one half the bore diameter, giving a ratio of areas of 4:3. With such geometry back-pressure effects are generally negligible for ordinary working. Standard cylinders are also usually offered with alternative sizes of rod — smaller in diameter for light duty or fast return double-acting cylinders, where the areas of ratios is usually of the order of 7:6, and larger in diameter for heavy-duty application where the area of ratios may range from about 7:4 to 2:1.

The cylinder rod has to be both supported and sealed by the rod side end cover (or both covers in the case of a through-rod cylinder). A simple plain bearing is adequate, but seal design ranges from the simple O-ring to lip-type seals and seal sets, according to the duty and the size of cylinder. A separate outer wiper seal is often included to clean the rod on the retracting stroke and so prevent abrasive dust, *etc,* collected by the 'wet' rod being drawn back into the main seals.

To simplify maintenance, the rod bearing and seals are sometimes combined in a single removable section. This is generally referred to as a cartridge gland or unit.

Standard production sizes of cylinders may offer different types of rod seals (fitted in alternative end covers), and even different bearing lengths, to suit different duties.

Mounting

Plain cylinders (often referred to as 'flush mounted') are commonly designed to accommodate a variety of alternative mounts. The exception is fixed centre-line mounts which are normally an integral part of the cylinder construction. The choice of mount is dictated by its suitability for a particular application, ease of adaptation to a machine, and whether the cylinder is to be rigidly or pivotally mounted, and the rod rigidly or loosely connected. Rigid mounting may be strictly necessary where the cylinder acts as a rigid structural member of the machine in which it is incorporated, otherwise a pinned mount may be preferred to simplify alignment and provide a certain tolerance to subsequent misalignment. Basic rigid mounts are:

- (i) Foot — usually fitted to the blank and rod ends of the cylinder.
- (ii) Centre-line — less common and confined to specific designs.
- (iii) Flange — either front (rod end) or rear (blank end).
- (iv) Nose — an alternative to front end flange mounting.
- (v) Pedestal — similar to a flange mount for fitting at either end or both ends of the cylinder, but incorporating a foot to bolt down to a surface parallel to the cylinder.

These may be used in combination, *eg* flange and centre-line, flange and foot, *etc*. Basic pinned mounts are:

(vi) Clevis – provoding a single pivot mounting point at one end.

(vii) Trunnion – providing a pivot mounting point at either end, or at the centre of the cylinder.

(viii) Tongue – providing a pivot mount in conjunction with a matching bracket.

(ix) Pivot – virtually the same as a trunnion mount except that the cylinder is fitted with a pair of pivot pins only for pivotal mounting, not necessarily in a trunnion.

(See also Figs 16 and 17).

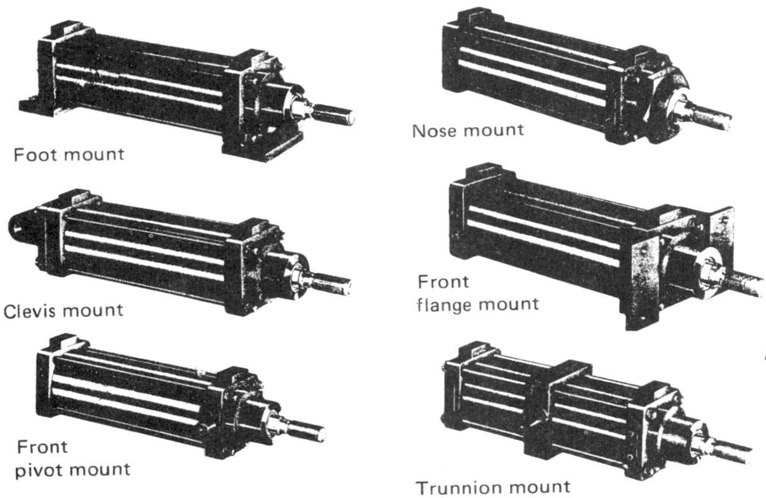

Fig 16 Basic types of cylinder mounts.

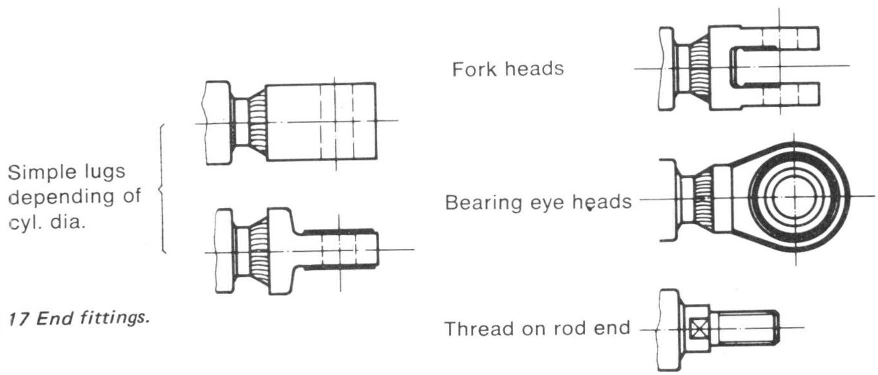

Fig 17 End fittings.

All pivoting mounts are normally, but not invariably, mounted on the centre-line of the cylinder.

Self-aligning or swivelling joints may be required at one, or both, ends of the cylinder, and the end fittings or bushes must be designed accordingly. Standard eye-type fittings can generally tolerate misalignment of up to about 3°, although this may need to be compensated for by flexible packings or washers. Cardan-type joints are generally used where the degree of misalignment is considerable and loads are heavy.

For rigid mounting of cylinders, the choice normally lies between centre-line, foot or pedestal mounting. The latter two are normally preferred, but the type of rigid mount available may be specific to a particular cylinder design. In the case of long cylinders, an additional foot mount may be provided at the centre to provide better support for bending loads. Also rigid mounts are not necessarily the same at each end of the cylinder, although they are usually so.

Rod ends are commonly screwed. The usual alternatives are plain, flanged and either clevis ended or tongued for pinned connections. The latter are capable of taking a small degree of misalignment without showing undue wear or imparting excessive side loads to the cylinder.

The majority of cylinder manufacturers aim to provide for as many alternative mountings as possible on standard production cylinders so that most likely applications can be catered for without resort to expensive special designs. They should be consulted as to the suitability of particular mounts for a specific application in any case of doubt.

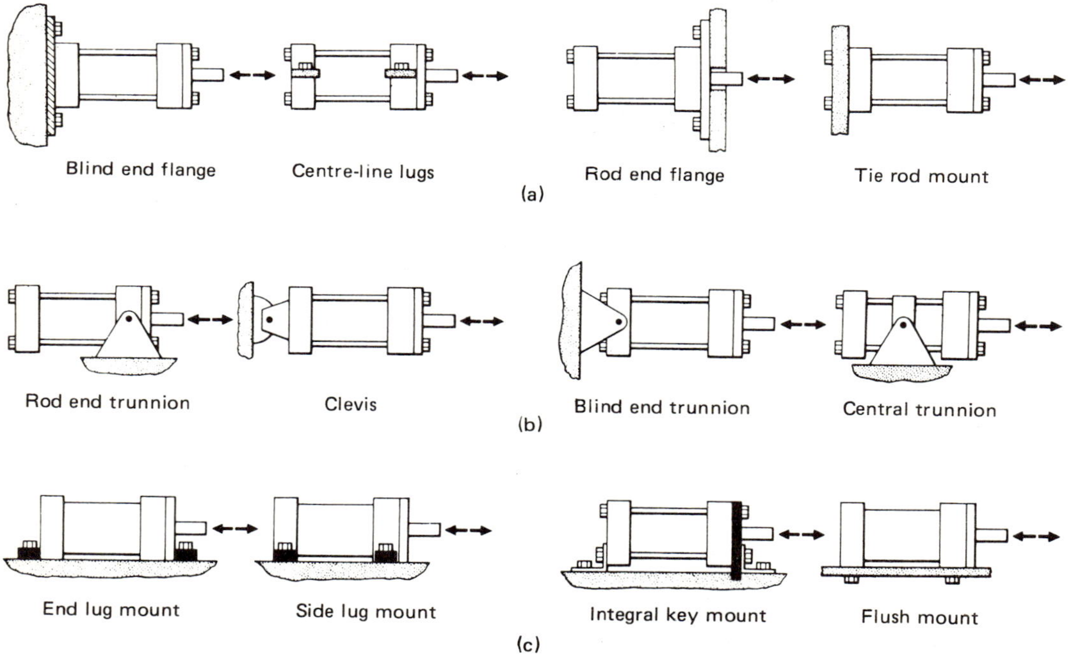

Fig 18 Cylinder mounts.
(a) fixed centre-line.
(b) pivoted centre-line.
(c) non-centre-line mounts.

Choice of Mount

In choosing a suitable mount the type of application is the major factor to be considered. The first consideration is whether the thrust line and the mounting are, or are not, coincidental. Fig 18 shows different types of mounting, (a) and (b) being for use when the thrust line and mounting are coincident. Listed are some of the factors affecting selection of suitable mountings:

1. Cylinders with mountings not on the centre-line tend to sway, probably resulting in extra wear and shorter life.
2. The frame holding the cylinder may not be strong enough. Certain mountings may well need a stronger anchorage to resist bending movements.
3. A pivot mounting is the answer when the object moved by the piston rod travels in a curvilinear path, but a fixed mounting type cylinder should normally be used when essentially linear motion is required. Pivot mountings can be used at either end of the cylinder or along its barrel.
4. Cylinder strength must be considered against stroke length. Long stroke, pivot-mounted, centre trunnion type cylinders can usually have smaller diameter rods without danger of bending compared with cylinders with fixed mountings which may require extra support to avoid sag.
5. The major force applied to the machine may result in tension or compression of the rod. The best mounting for thrust loads is the cap end flange type – with the rod in compression – while a head (rod) end flange mounting is better when the rod is stressed in tension.
6. If misalignment is possible between a cylinder and whatever it operates, it may be necessary to provide for compensation by selecting a suitable mounting. If the misalignment is primarily in one plane, for example, the simple pivoted centre-line mounting will give the necessary compensation. This type of mounting includes clevis and trunnion arrangements.

Installation

Correct installation can be controlled to a considerable extent while the layout of a machine is being arranged. These are some of the points to be considered at this early stage.

1. The inherent elasticity of a cylinder can mean the difference between success and recurring trouble. If high shock loads are expected the cylinder should be mounted to take advantage of this elasticity.
2. Fixed mounting cylinders should be keyed or pinned, provision being made at the design stage of the machine. If the appropriate member of a machine is thick enough to take key-ways, cylinders with integral key-mounts can be provided.
3. Separate keys to take shear loads are common. These should be at the correct end of the cylinder – at the head (rod) end if major shock loads are in thrust, and at the cap end if they are in tension. Only one end should be keyed to the machine to avoid losing the advantages of cylinder elasticity. Temperature and pressure effects should also be considered.
4. Locating pins can be used instead of keys to take shear loads and maintain alignment, again at one end or the other but not both. Whatever the problems of the machine designer there should not be pinning across corners.
5. In fixing a cylinder rod to, say, a machine member, the rod should not be rotated more than necessary to avoid danger of scoring the cylinder body. The danger is greater if there is misalignment when the rod is rotated.

6. Rod end knuckles, cam surfaces, *etc,* should not be permanently fastened to piston rods during installation. The joints would have to be broken during replacement of glands.
7. Installation engineers must guard against dirt, paint, *etc,* getting on the piston rod, and ensure that no damage is caused to exposed rods.

Critical Rod Lengths

Cylinder rods can be stressed as rigid rods, provided the length of the rod does not exceed ten times its diameter. The stress formula in this case is:

$$\text{material stress} = \frac{F}{A}$$

where F = compressive or tensile force or load
A = cross-sectional area of rod

For general working the maximum permissible material stress may be based on the ultimate tensile stress of the material and a suitable factor of safety. This will then give adequate rod strength either in tension or compression.

If the rod length exceeds ten times its diameter, then it may be subject to buckling under compressive loading. Adequate strength in tension can then no longer be taken as an indication of adequate strength in compression. The case of compressive loading must be analyzed separately, when the rod is considered as a column. The material stress then depends on the method of end fixing, *viz*

$$\text{material stress} = \frac{\pi E d^2}{4 L^2} \quad \text{for rod end fixed}$$

$$= \frac{\pi E d^2}{8 L^2} \quad \text{for rod end pinned}$$

$$= \frac{\pi E d^2}{64 L^2} \quad \text{for rod end free}$$

where d = rod diameter
E = modulus of elasticity of rod material
L = rod length
in consistent units

Equivalent strut lengths are summarized in Fig 19. Fig 20 then gives safe rod lengths for steel rods, assuming that there is no eccentric loading. In the case of horizontal cylinders, these may need supporting to counteract bending movements. Where bending loads are present in the rod compression, the critical length is reduced substantially.

In certain applications the bearing length with the rod of a long-stick cylinder fully extended may be increased by the fitting of a stop tube *eg* see Fig 21. The way in which the cylinder is mounted and the manner in which the rod is supported at the extremity of its movement will determine whether or not a stop tube can be advantageous.

CYLINDER MOUNTING	RIGID ROD END GUIDED	PIVOTED ROD END GUIDED	FREE ROD END
FRONT FLANGE	EQUIV L = $\frac{\text{STROKE}}{2}$	EQUIV L = $\frac{\text{STROKE}}{1\ 42}$	EQUIV L = STROKE x 2
REAR FLANGE	EQUIV L = $\frac{\text{STROKE}}{1\ 25}$	EQUIV L = STROKE x 1 12	EQUIV L = STROKE x 3 2
FOOT	EQUIV L = $\frac{\text{STROKE}}{2}$	EQUIV L = $\frac{\text{STROKE}}{1\ 42}$	EQUIV L = STROKE x 2
REAR EYE OR PIN	EQUIV L = STROKE x 1 12	EQUIV L = STROKE x 1 6	
TRUNNION	EQUIV L = $\frac{\text{STROKE} \cdot 6x}{1\ 42}$	EQUIV L = STROKE · 6x	

Fig 19 Equivalent strut length (L) of cylinder rods. (Sacol).

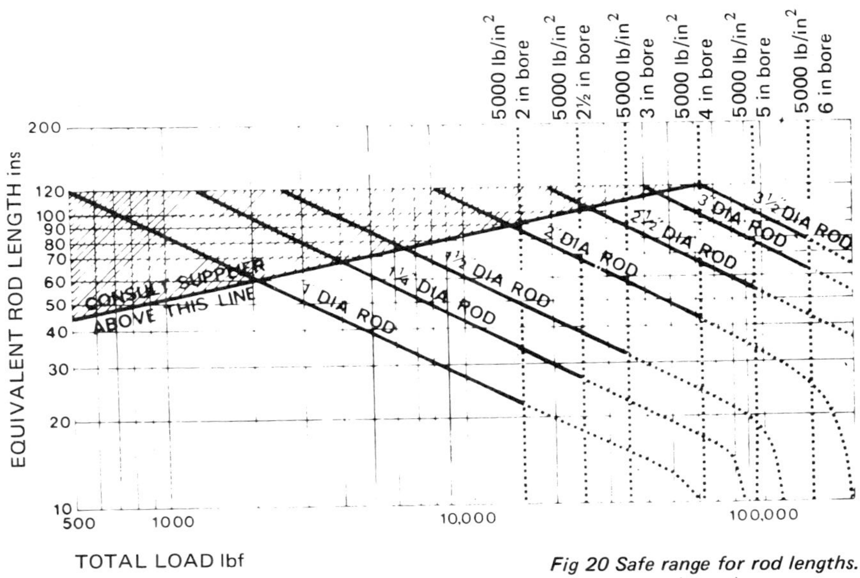

Fig 20 Safe range for rod lengths. (Sacol)

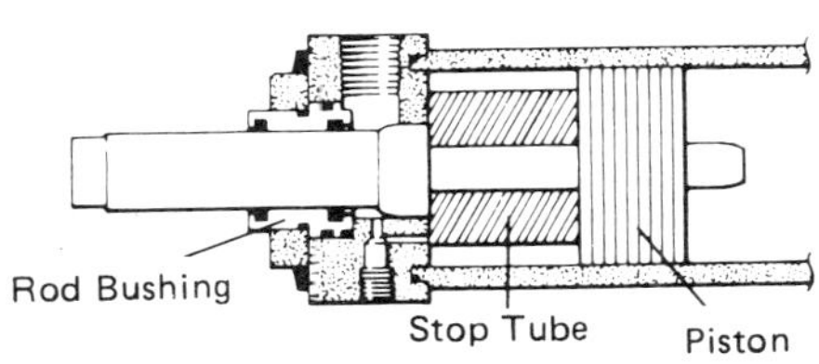

Fig 21 Cylinder with stop tube. (Double A Division, Brown & Sharpe Limited).

Stroke Adjusters

Stroke adjusters enable the working stroke of a cylinder to be varied between two extremes of the adjuster setting. These usually take the form of some sort of screwed attachment.

Position Indicators

Position indicators are comparatively rare on hydraulic cylinders and would normally only be required for specific applications where it is necessary to indicate that the piston has reached the extreme end of its stroke. The indicator could then be used to activate a micro-switch or similar device to make or break an electrical alarm, indicating a control circuit.

Line Connections

In relatively few applications is a hydraulic cylinder completely static (even with rigid mounting the cylinder itself may have motion as a unit). This means that the most common method of connection on inlet and outlet lines is by flexible hose. Provision is usually made for this by female tappings in the end covers at each port position capable of accepting standard couplings, or a male stub pipe accepting a proprietary sealing ring and nut. Obviously the same tappings can be used with either flexible or rigid lines.

The physical positions of the inlet and outlet ports are separated by the length of the cylinder. To make for a neater (and often less vulnerable) hook-up, both line connections are sometimes brought to the same end of the cylinder.

Logically the input connections to cylinders should be at the top, or uppermost. This may not always be possible in which case an *air bleed* or air bleeds may have to be incorporated to eliminate the possibility of air entrainment.

Rotary Actuators

ROTARY ACTUATORS fall into three main categories:

(i) *Torque actuators* or devices where the output motion is limited to less than one complete revolution.

(ii) *Semi-motors* or devices providing an output motion of one revolution or more, but not continuous rotation.

(iii) *Hybrids*.

With all types the output motion obtained is reversible. The particular advantage offered by ydraulic rotary actuators is direct rotary output from a (usually) compact hydraulic device with limination of linkages and lost motion associated with cylinder/crank rod arrangements. Also utput torque is substantially constant.

Torque Actuators

Torque actuators again divide into two main types — *all-hydraulic* and *hydro-mechanical.* All-ydraulic semi-rotary actuators are of vane type. Hydro-mechanical types are basically hydraulic ylinders with rotary output derived directly from the piston movement.

Vane-type actuators are usually of single-vane or double-vane configuration. Multiple-vane ctuators offer no significant improvement in performance. Construction takes the form of a ylindrical housing through which passes a central output shaft to which the vane is rigidly ttached. The housing itself has a second vane or abutment fixed to its internal diameter and xtending to the output shaft, dividing the interior space into two variable-volume chambers. In he case of two-vane actuators, there are two vanes fixed to the output shaft 180 degrees apart, nd two fixed vanes on the casing providing two separate variable-volume chambers — Fig 1.

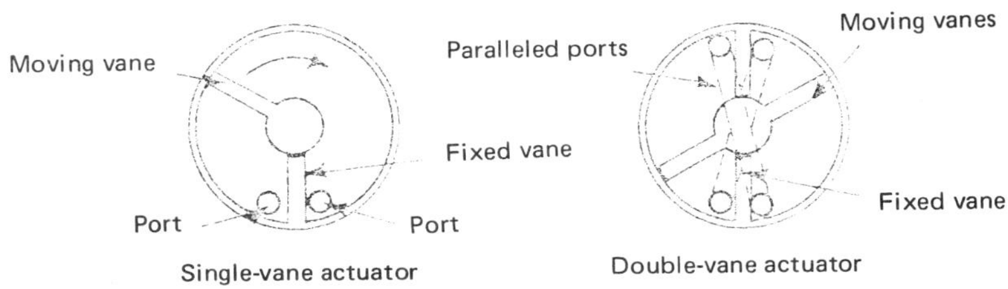

Fig 1

Design geometry normally limits the rotary movement of a single-vane actuator to about 280° naximum, and rather less than one half this figure with two-vane actuators. The torque is directly roportional to the effective vane area and the effective fluid pressure. The effective fluid pressure the applied (inlet) pressure less inlet losses and internal leakage.

In the case of a single-vane actuator a practical figure for torque output is given by:

$$\text{Torque} \approx 0.5\ (D^2 - d^2) \times L \times P$$

where L is the length of vane, in the same units as D and d,
P is the effective inlet pressure,
D is the internal diameter of the body,
d is the overall diameter of the spindle.

For a two-vane actuator:

$$\text{Torque} \approx 0.225\ (D^2 - d^2) \times L \times P$$

It is significant that this torque is less than one half that of a single-vane actuator of similar size, due to the doubled leakage path.

The displacement of a vane actuator is given by:

$$\text{Displacement} = \frac{(3D + d) \times L \times \theta}{1440}$$

where θ is the rotary movement in degrees

For any given design and style of vane actuator the torque rating is directly proportional to the fluid pressure; thus to increase the torque it is only necessary to increase the pressure. However, there are limits to this pressure increase and 70 bar (1 000 lb/in^2) is about the usual maximum pressure with standard designs. This is because of the limitations of the vane seal, higher leakage rates and consequent loss of efficiency developing rapidly with increasing pressure. In practice, too, torque output does not necessarily show an exact linear relationship to pressure. With conventional seals efficiency will usually increase slightly up to a pressure of 70 bar (1 000 lb/in^2), and then decrease again rapidly with increasing pressure. Nevertheless very high torques can be produced from quite small units.

In addition to the vane seal, further seals are required on the shaft where it emerges from the body of the actuator.

Internal leakage, and thus efficiency, can also be controlled by the fluid viscosity. If efficiency is important, *ie* the maximum output is required from a particular size of actuator, then the use of a fluid with a higher viscosity may be advantageous. A change from an oil viscosity of 30 centistokes to one of 100 centistokes, for example, could be expected to at least halve the leakage. However, it must not be overlooked that fluid temperature will also be significant and if the fluid itself is subjected to marked turbulence and heating, or the ambient temperature is high, the actual viscosity of the fluid in the actuator may be appreciably less than its nominal or room temperature viscosity.

The problem of leakage may also be present where the actuator has to be locked or held under load with a simple fluid-column lock. The lock will not be positive — *ie* the actuator will tend to retreat from the load due to internal leakage unless provision is made to supply extra fluid to compensate for the leakage.

The speed of a vane type actuator can be calculated directly from its displacement and the delivery. If necessary, speed can be controlled by throttling or by fitting orifice plugs controlling the inflow or outflow. Speed is seldom an important factor for such devices, which are normally used for high-torque outputs with an operating time of some 0.02 to 0.05 seconds per degree of rotation.

Friction is generated by the seals and also the spindle bearings. The latter may be a plain bearing or a roller bearing (usually needle roller type), depending on the application and the design bearing loads. Friction is responsible for the difference between starting torque and dynamic torque, this difference being minimized by using seals with low break-out forces and low friction spindle bearings. As a very rough guide the starting torque can be estimated as 80% of the dynamic torque in good designs, but can be improved upon if necessary. Efficiency of vane actuators is between 70% and 90%.

Piston-type Actuators

Three basic configurations for a piston-type semi-rotary actuator are shown in Fig 2. The internal crank provides a simple solution, although rotary output is normally limited to a maximum of about 90°. It has the advantage over an external crank mechanism that only a shaft seal is required. The rack-and-pinion layout offers the possibility of greater rotary motion, although both rod and shaft seals are necessary. The tandem rack-and-pinion configuration with two separate pistons is a 'tighter' design with more balanced reaction forces. Output motion is derived by pressurizing the two end-cylinder compartments to drive the pistons towards each other, or the centre portion to drive the pistons apart (with reversed output motion). In practice piston-lengths capable of generating up to six complete revolutions are quite possible, although standard rotation is usually 360° or less. Again only a shaft seal is required.

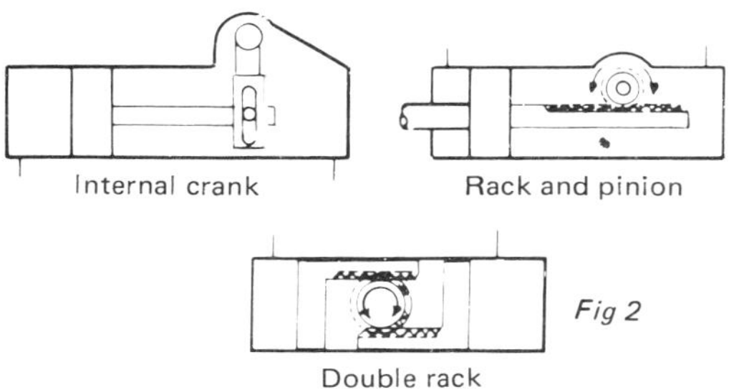

Fig 2

Cutaway illustration of 'Servotel' semi-rotary actuators. (Servotel Controls Ltd).

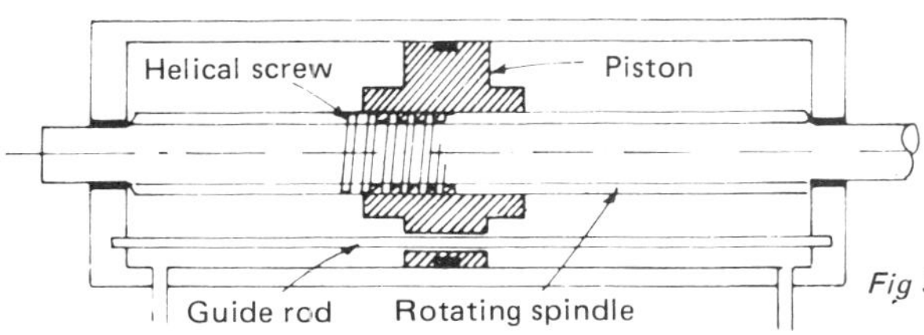

Fig 3

Rack-and-pinion actuators incorporating a double rack-and-pinion obviate side thrusts and are thus particularly useful for heavy-duty applications. There are many proprietary varations of such layouts, including other forms of internal pinion gearing. Rack-and-pinion drive motions are also adaptable for single-line working, return motion being accomplished by a spring (similar to the working of a spring-return single-acting cylinder). In addition, a spring-loaded reservoir or expansion chamber can be incorporated in the piston itself to provide automatic compensation for changes in fluid volume due to temperature changes.

In another design of piston actuator rotary output is provided by the piston rod itself. The piston rod is, in fact, a spindle on which the piston slides, the two units being inter-connected mechanically by a helical thread – Fig 3. The piston is prevented from rotating by a guide rod.

Thus application of pressurized fluid to either side of the piston will generate translational motion on the piston, with corresponding rotary motion of the spindle.

A variation on this layout is a double-piston design with a double-helical thread on the spindle and a third (central) pressure chamber. This has the advantage that the pistons are always working in opposition and thus thrust loads on the shafts are eliminated. All designs of helical thread movement tend to suffer from relatively low mechanical efficiency. However, there is the advantage that this type of mechanical linkage provides extremely good rigidity and irreversibility, using square threads. The length of the cylinder can also be increased, if necessary, to provide more than one complete revolution of output movement.

A variation on this configuration is where the cylinder is associated with a re-circulating ball and thread mechanism. The load which an actuator of this type can carry is limited by the capacity of the thread, and the number of revolutions by the length of cylinder.

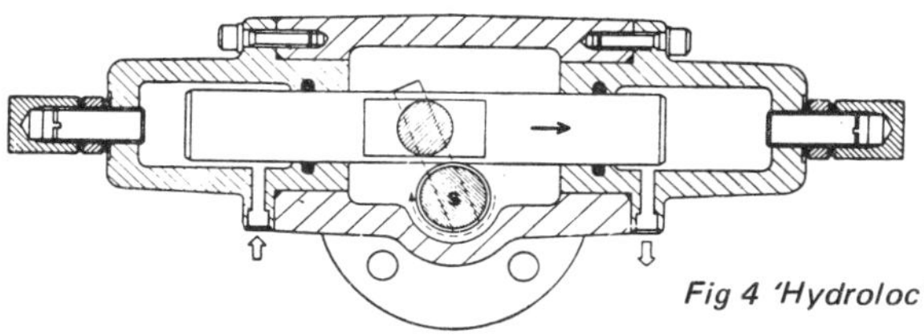

Fig 4 'Hydroloc' actuator.

A twin-cylinder (tandem opposed) semi-rotary actuator is shown in Fig 4. Here the common piston is in the form of a plunger, slotted at the centre and connected by internal linkage to a separate output shaft. The rotary output provides up to 90° angular movement, and these actuators are designed specifically to operate plug, butterfly and ball valves, under completely immersed conditions, if required. Position indication is also incorporated, either by a pointer attached to the top of the output shaft, or by integral hydraulics. In the latter case power for the indicating equipment is provided by a separate pump unit, or from one side of a double-ended main pump unit.

A variation on this design employs only a single working plunger sliding in both cylinders. The non-working cylinder then acts as a housing for a return spring. This provides for single-line operation, and also automatic valve return (*eg* automatic valve closure) in the event of a failure of the hydraulics.

A further example of a hydro-mechanical cylinder where all the mechanical system is contained within the cylinder is shown in Fig 5. Here piston motion is linked by a chain loop passing over sprocket wheels at each end of the cylinder body. The return run of the chain passes through a separate small-diameter cylinder containing a second piston, the purpose of which is to act as a seal for the chain return. When movement is initiated, both pistons are pressurized in the same direction but the larger area of the main piston ensures motion in the pressurized direction.

The same method of deriving rotary motion from a cylinder is used in *cable cylinders,* but here the cable and sprockets are mounted externally.

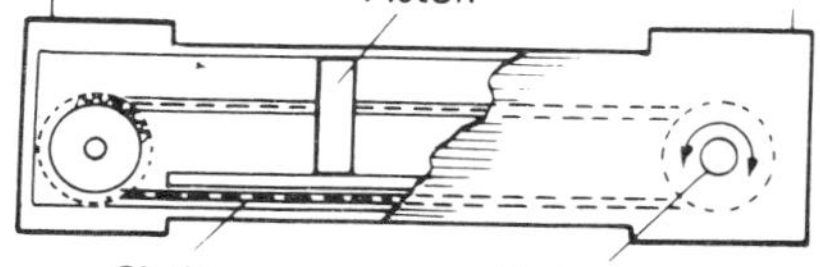

Fig 5

Sacol 'Torqmaster' semi-rotary actuator.

Sterling rotary actuator with end mount (left) and foot mount (right)

Hybrids

An example of a hybrid rotary actuator is shown in Fig 6. It has the same overall configuration as a vane actuator, but with the moving vane(s) replaced by an annular piston of square section. This piston is connected rigidly to the output spindle and has a rotary movement limited by its circumferential length and the presence of the sealing wall or partition. The clearance spaces between the piston ends and partition provide two separate pressure chambers for driving the piston in one direction or the other.

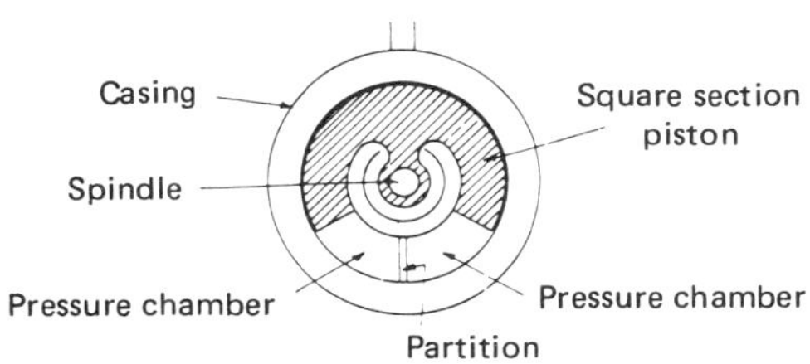

Fig 6

Sealing is a particular problem with this configuration, one solution being the fitting of floating or flexible seals to each end of the annular piston.

SECTION 2B

Hydraulic Valves and Selectors

Hydraulic valves group broadly into five categories.

(i) *Flow-control valves* — for controlling flow rates.

(ii) *Pressure-control valves* — for controlling system pressure.

(iii) *Directional-control valves* — for determining the routing of the fluid in a circuit; also known as *selectors.*

(iv) *Flow-modulating valves.*

(v) *Proportional-control valves.*

Hydraulic valves may also be classified by mode of operation — *eg* manual, mechanical, pilot-operated, electro-hydraulic, electro-modulated; particular form; specific duty; *etc.* Also by their general geometry and method of mounting, *eg*

(i) Pipe mounting; (in-line mounting);

(ii) Surface (gasket) mounting;

(iii) Sub-plate or manifold mounting;

(iv) Modular mounting.

Yet a further method of classification is by working elements *eg* poppet, ball, plug, spool, *etc.*

Flow-Control Valves

Simple flow-control valves work on the basis of restricting the flow with either a fixed or variable orifice. In the latter case flow-control valves fall into two categories.

(i) Non-compensated valves (*ie* simple throttle valves);

(ii) Compensated, when the valve is designed to compensate for the effects of varying pressures and temperatures in the circuit and thus maintain a constant performance.

An inherent disadvantage of simple restrictors is that the flow through them is strictly dependent on pressure drop across the valve and thus through-flow will vary with changes in load. As a consequence their application is virtually limited to those systems where the load is constant, or where variations in flow rate, and thus operating speeds, are permissible.

Pressure-compensated Valves

To provide a constant pressure drop across an orifice, and thus constant flow characteristics, a combination of two restrictors can be used, one fixed and the other automatically variable. These

Example of pressure-compensated flow control valve with fixed orifice. (A & D Fluid Power Ltd).

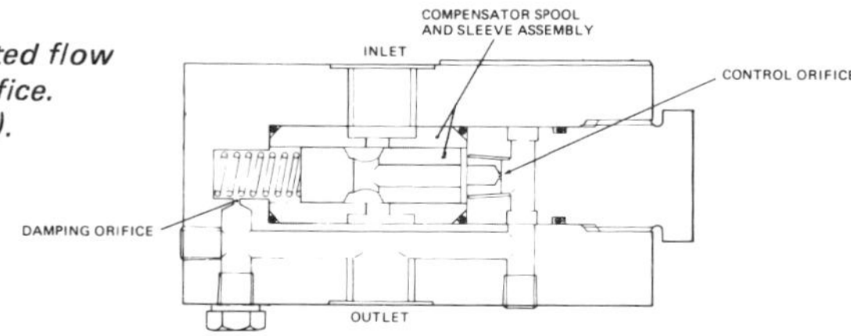

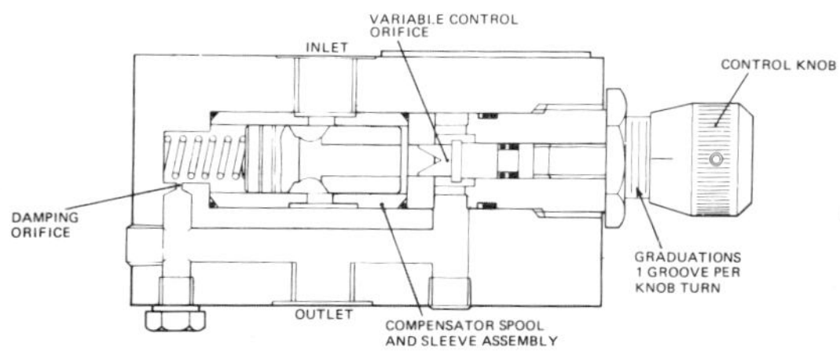

Example of pressure-compensated flow control valve with variable orifice. (A & D Fluid Power Ltd).

two elements are normally combined in a single unit to produce a *pressure-compensated flow restrictor.* A further refinement may be feedback or pressure from downstream to provide 'meter-in' control, when the variable throttle acts as a sensing orifice. One-way working can be provided by incorporating a non-return valve in the unit to provide free flow in one direction.

The main limitations of a pressure-compensated flow restrictor are that whilst constant flow characteristics are provided, independent of load, the controlled flow is throttled and surplus flow must be directed through another valve, resulting in the pump working at full relief setting continuously. To overcome this loss of efficiency, a pressure-compensated bypass regular valve can be used. Here the surplus flow is bypassed through the valve at working pressure, and at the same time the controlled flow is not subjected to marked throttling losses. The use of a pressure-compensated bypass regulator, therefore, enables much higher circuit efficiencies to be achieved. Its main limitation is that, unlike the pressure-compensated restrictor, it cannot be used in parallel configurations. Also it is unsuitable for constant-pressure systems, since it would bypass an unnecessarily high volume of fluid, resulting in lowered operating efficiency.

The difference in working of the two types can be seen in Figs 1 and 2, here drawn in the form of two-port and three-port valves respectively, each with a fixed restrictor inserted in the outlet. In proprietary units the actual geometry may vary appreciably, but spring-loaded spool or piston types are normally used.

Temperature compensation may also be incorporated in flow-control valves to adjust to change in fluid viscosity.

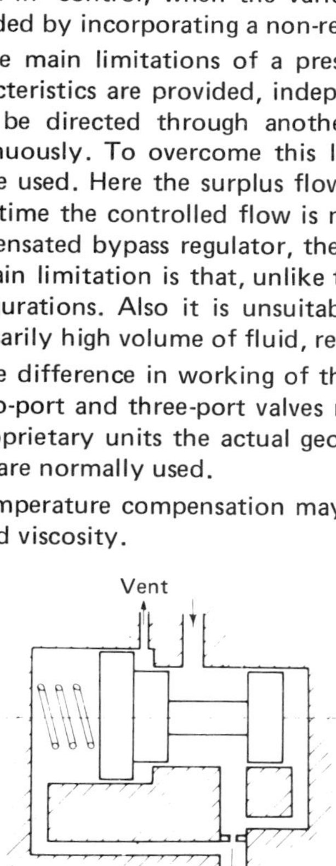

Fig 1

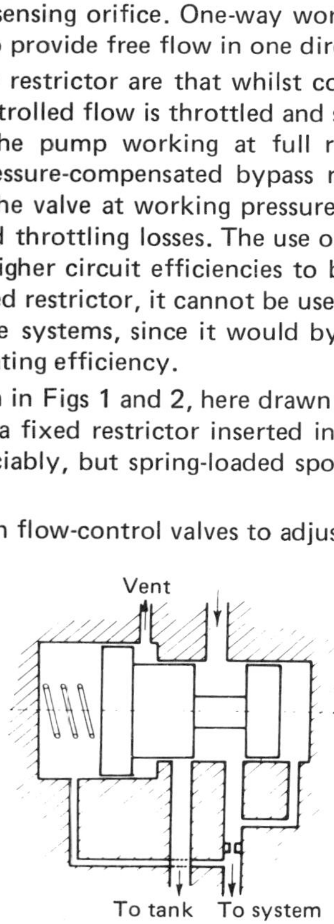

Fig 2

Table I provides a summary of various types of flow-control valves and their particular applications.

TABLE I – SUMMARY OF FLOW CONTROLLING VALVES

Type or Name	Sub-type(s)	Remarks
Fixed restrictor (Fixed throttling valve)	Two-way	Fixed orifice or capillary tube.
	One-way	Spring-loaded poppet with orifice through poppet; sharp-edged orifice and fixed end needle.
Adjustable restrictor (Adjustable throttling valve)	Orifice	Sharp edged orifice with adjustable needle.
	Screw-down	Needle or plunger screwed down into orifice or port.
	Screw	Two-way, with leakage past screw threads; or one-way with non-return valve and bypass.
Pressure-compensated flow restrictor	Two restrictors in series	Two-way; or one-way with non-return valve and bypass.
	Also with 'meter-in' characteristics	Can be used in parallel or multiple combinations.
	Bypass type	Gives maximum overall efficiency. Cannot be used in parallel.
Flow divider	Dual-orifice	To divide supply pressure equally between two outputs. Can be cascaded.
Priority	Orifice	To maintain pressure in a priority circuit on multiple services fed from a single pump.
Transfer	Suction-operated	To short-circuit a cylinder on part of a hydraulically powered stroke.
	Pressure (or mechanically) operated	To short-circuit a single-acting cylinder on a return stroke.
Non-return (check)	Ball poppet	Limited performance with metal-to-metal seal or resilient face seal.
	Chamfered plunger damped	With metal-to-metal seal or resilient face seal usually plunger type (used with resilient seals).
Sequence	Poppet	Can also act as a non-return valve for return flow.
	Piston	Tends to be less reliable due to spring friction.
	Drain-to-reservoir	Commonly applied to piston type valves to prevent leakage from the inlet side causing premature operation.
Locking	Single	Used for 'holding' actuator positions.
	Double	Normally employed with open-centre systems.
Shuttle	Piston	Three-port valves to provide duplicated line connections from alternative supplies.
Spill-off	Various	Relief valve with speed control.
Braking (Deceleration)	Tapered plunger	Automatic deceleration of cylinder movements (similar to that otherwise obtained by cylinder cushions).
Shut-off (fuse)	(i) plunger (ii) quantity measuring (iii) fluid sampling	To shut-off and isolate line in the event of excessive leakage developing.

Note: Flow-controlling valves may also be used for pressure control, or be combined with such valves – see also Table II.

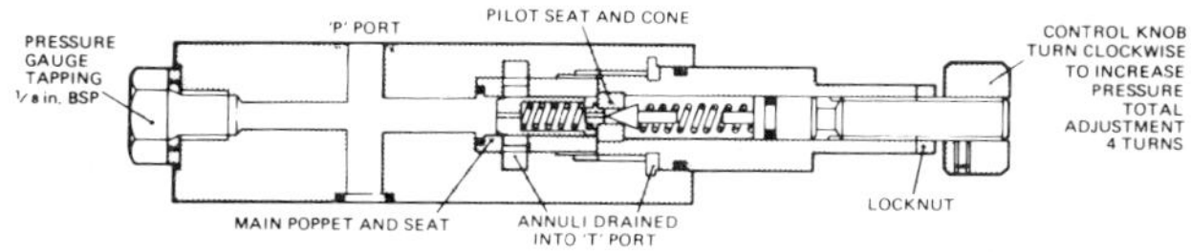

Example of pressure relief valve construction. (A & D Fluid Power Ltd).

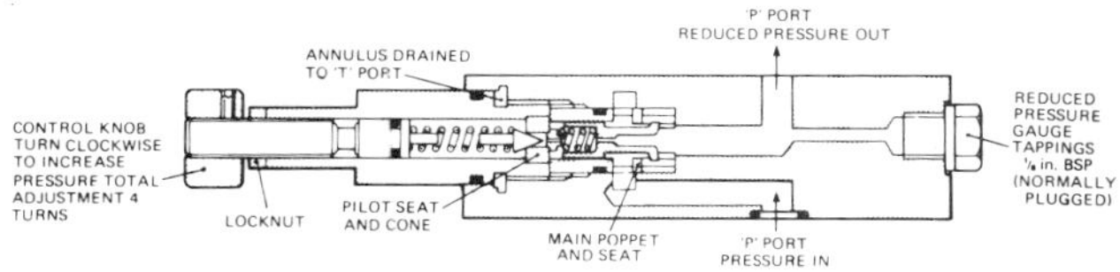

Example of pressure reducing valve. (A & D Fluid Power Ltd).

Pressure-Control Valves

The most common type of pressure-control valve is the *pressure-relief valve,* fitted to limit the fluid pressure which can be built up in a system. This is basically a two-way normally closed valve which opens at a pre-determined pressure. Activation may be direct, or by pilot, depending primarily on the circuit power control. Important parameters in the design of such valves are:

(i) pressure hysteresis — or the difference between 'cracking' or opening pressure and sealing pressure.

(ii) proportional band-width — or the change in inlet pressure in response to increase in flow rate as the valve opens.

The simplest type of pressure-relief valve is the spring-loaded ball valve. Its main limitation is a tendency to 'chatter'. Poppet valves provide much greater control in design and manufacture over both pressure and proportional band-width, with throttling characteristics proportional to the square root of the valve stroke. Sealing may, however, be a problem and an O-ring or similar seat seal is generally preferred to a metal-to-metal seat.

The other configuration used for simple pressure-relief valves is the plunger-type, incorporating a hardened steel plunger sliding in the valve body. The degree of lift is dependent on flow rate as well as pressure so that flow variations can again initiate chatter, unless suitable damping is present. One way of introducing self-damping is to use a two-diameter plunger to provide a bigger discharge area, with this end of the body sealed and discharge being directed through small holes in the body wall to restrict flow.

The characteristics of this type of pressure-relief valve can also be adjusted by contouring the plunger, rendering the throttling characteristics proportional to the reciprocal of the stroke.

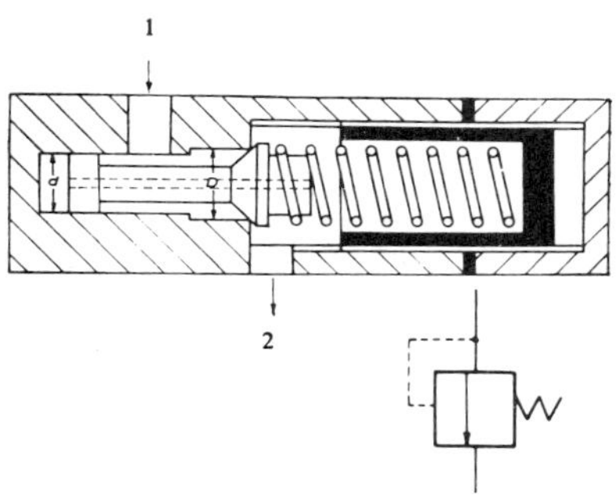

Fig 3 Differential relief valve. Effective force on spring depends on

$$\frac{\pi}{4}(D^2 - d^2).$$

Capacity depends on $\frac{\pi}{4}\ D^2$

Differential-Relief Valve

A two-diameter plunger is also used in the *differential-relief valve,* but retaining conventional packing — *eg* see Fig 3. The main attraction of this configuration is that it reduces the size of spring required since the cracking pressure is applied to the annulus rather than the full plunger diameter. Port areas, on the other hand, are proportioned on the larger diameter. The design may be further elaborated in the case of pilot-operated differential-relief valves to incorporate over-ride so that the valve will always open at a pre-determined maximum pressure regardless of adjustment of pilot pressure. In other words, it is impossible to set the valve beyond a specified maximum pressure by adjustment of pilot pressure.

Pilot-Relief Valves

Pilot-operation of relief valves is normally employed at higher power levels, but can show advantages at all power levels. They are normally of integrated design with line and pilot connections.

A typical design for a small pilot relief valve is shown in Fig 4. The main relieving valve is arranged so that it is hydraulically balanced and is kept closed by a light spring. Normally the pressure is equalized on either side of the spool by a small orifice. If the pilot valve opens it allows liquid to escape at a faster rate than it can be replenished through the orifice, so that the pressure falls on one side of the centre position. The excess pressure on the other side then lifts the valve and opens the exhaust port. Liquid then flows away to the tank in just sufficient quantity to keep the pressure constant.

In use the valve is set to the maximum working pressure, or to some higher pressure if the working pressure is determined elsewhere.

Valves of this type are very stable and their accuracy depends on that of the pilot valve and is not much affected by flow. They are, however, suitable for clean fluid only. They must not be used in a system where there is a chance of contamination building up in the liquid, as this can cause the valve to become inoperative and therefore dangerous.

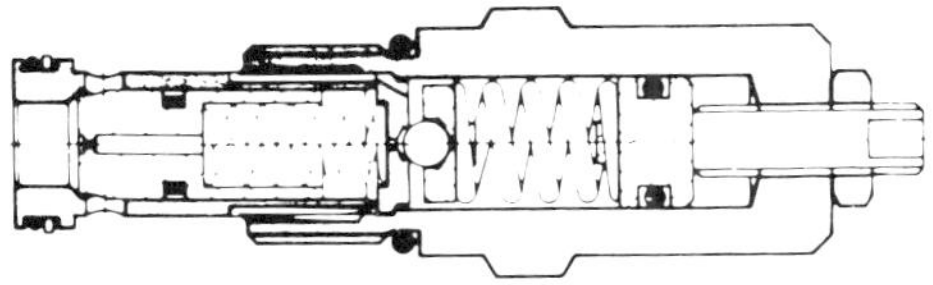

Fig 4 Pilot-operated relief valve. (Sterling Hydraulics)

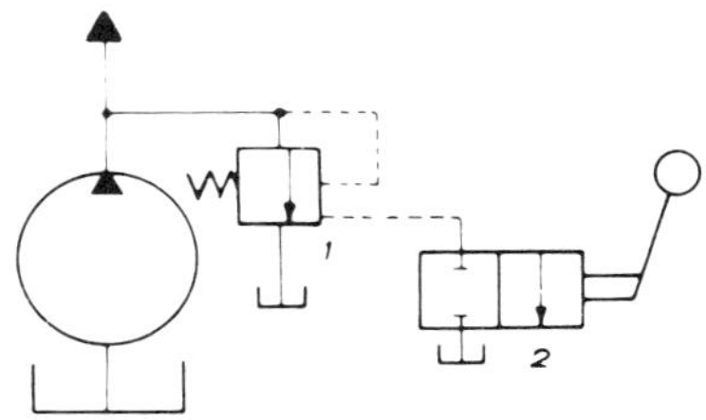

Fig 5 Pilot-operated relief valve. 1—With unloading by 2-way pilot valve (2).

There is no need for the pilot relief valve to be integral with the main valve and it is often convenient to connect it to another valve *eg* see Fig 5. A small two-way valve connected to the drain will, when opened, cause the pressure in the pilot system to fall to zero and the main valve will be open, bypassing the pump output at a low pressure. The actual pressure would depend on the strength of the spring and the pipe friction. This method is often used for unloading a pump, making a second valve unnecessary.

It is also possible to work at more than one pressure by inserting a second pilot relief valve between the two-way valve and drain, so that when the valve is open the maximum pressure setting is determined by this valve, providing the setting is less than the primary pilot valve.

Pressure-control valves and their application are summarized in Table II.

TABLE II – PRESSURE CONTROL VALVES

Type or Name	Sub-type(s)	Remarks
Relief	(i) direct loaded (ii) differential (iii) dual (iv) pilot (v) combined (unloading and relief) (vi) variable orifice	Ball, poppet or plunger types – see text.
Pressure control	(i) screw-down (ii) restrictors	Fixed pressure reduction. Variable with load.
Reducing	(i) fixed (ii) proportional	Pressure reduction. Output proportional to input.
Unloading	Various	Pressure-relief valves.
Cut-out	Direct acting Pilot operated	Used for unloading pump. Used for unloading pump.
Back pressure (Counterbalance)	(i) ball (ii) plunger	Suitable for low pressures. Maintains a minimum pressure in a particular line.
Priority	Plunger	Maintains pressure in a primary line at the expense of other services if necessary.
Thermal relief	Simple relief valve	Used on 'locked' section to relieve excess pressure which may be generated by thermal expansion of the fluid.

Directional-Control Valves

Directional control valves or selectors provide the means of changing the direction of flow in a circuit (*eg* from one end of a cylinder to the other). They are usually spool valves although poppet and rotary valves may be used for specific applications. The recently introduced cartridge-type directional-control valve is a direct seat poppet-type valve which has become highly competitive with the spool valve and can be designed for negligible leakage with very rapid response times.

Directional-control valves are normally classified by the number of positions and the number of ports or 'ways' in the valve body. Standard configurations and symbols are shown in Fig 6. Valve symbols comprise a number of squares, side by side, each square representing a possible *position* for that valve. Thus a two-position valve is represented by two squares, and a three-position valve by three squares. Ports are indicated in each square, but separately annotated to show the inter-connection (or absence of inter-connection) in each position. The valve symbol also (usually) includes an indication of the method of actuation of the valve and return to its normal position.

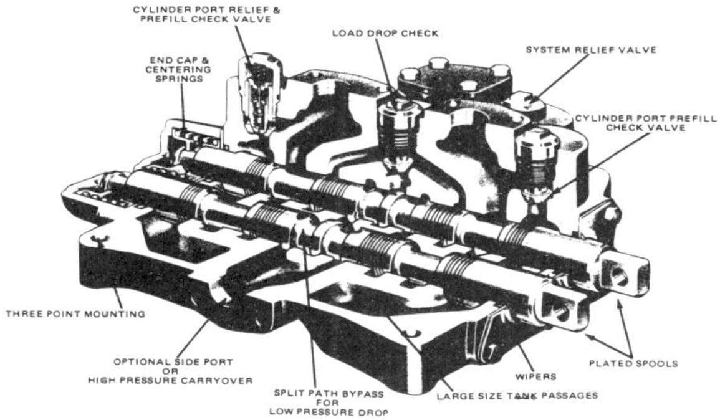

'Enbloc' directional valve. (Sperry Vickers)

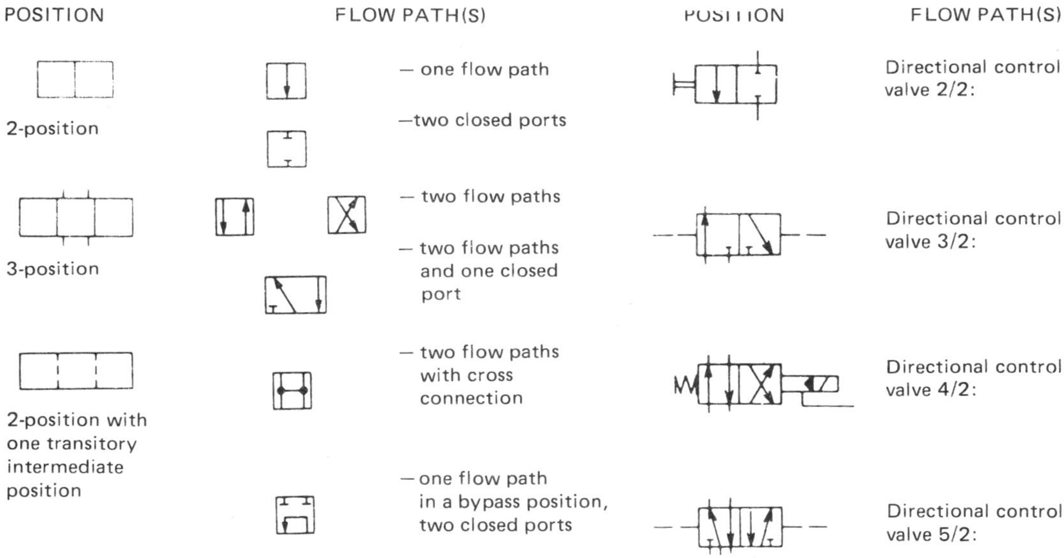

Fig 6a Basic ISO symbols for directional valves.

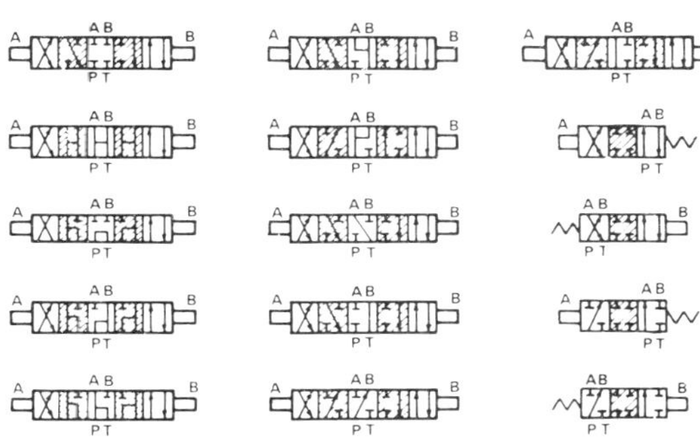

Note: Shaded portions of symbol indicate port connections whilst spool is shifting

Fig 6(b) Connections of 2-, 3- and 4-way selectors.

TABLE III – SELECTORS OR DIRECTIONAL-CONTROL VALVES

		Two-Way	Three-Way	Four-Way
Size range	litres/min gal/min	1–1 800 1–400	1–45 1–10†	1–1 800 1–400
Pressure range	bar lb/in^2	7–350 100–5 000	up to 420 up to 6 000	up to 420 up to 6 000
Actuating pressure range		*	*	*
Pilot pressure	bar lb/in^2	3.5–14 50–200	3.5–14 50–200	3.5–14 50–200
Response time (milliseconds) Δ		10–250	10–250	10–250

* 4 bar (60 lb/in^2) up to full system pressure
† 1–450 litres/min (1–100 gal/min) for mobile hydraulics
Δ 50 milliseconds typical for valves for mobile hydraulics

Essentially, therefore, selectors can be considered as two-way, three-way or four-way; a further comparison between these types is drawn in Table III – see also Fig 6(b).

Two-Position Two-Way Valves (2/2)

A two-way or two-port valve is basically an on-off switching element and can be simply described as a shut-off cock. Obviously a variety of designs can be used to provide this function – gate, globe, rotary, slide, poppet and spool valves. The first two are excluded, since their mode of operation is not suited to small sizes. Rotary or slide valves are practicable, with a capability of sealing in either direction. Poppet valve cocks are generally capable of sealing in one direction only. Spool valves are the most logical choice for precision applications.

The simplest type of 2/2 valve is normally operated with spring return. For a 'normally open' valve, the spring holds the valve in the open position; and for a 'normally closed' valve the spring holds the valve in the closed position.

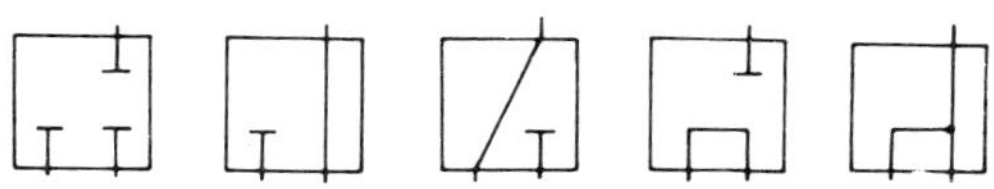

Fig 7(a) Connections for 3-port selectors.

Three-Way Valves

A three-way valve has three ports with five possible modes of inter-connection – Fig 7(a). It is most commonly used as a two-position valve for the control of a single-acting cylinder. It can, however, be used as a three-position three-way valve with open-centre characteristics (*ie* with a straight through pump-to-tank connection in mid-position) or as the operating mechanism for a two-way valve. Three-way selectors are also used for the control of double-acting differential cylinders.

Poppet valves are not generally used as three-way selectors, since an additional non-return valve would normally have to be included to prevent return flow passing through the inlet. An additional relief valve may also be necessary to accommodate thermal expansion of the fluid volume between the two valves.

It is possible to construct four- and five-position three-way valves, but they are rarely, if ever, used.

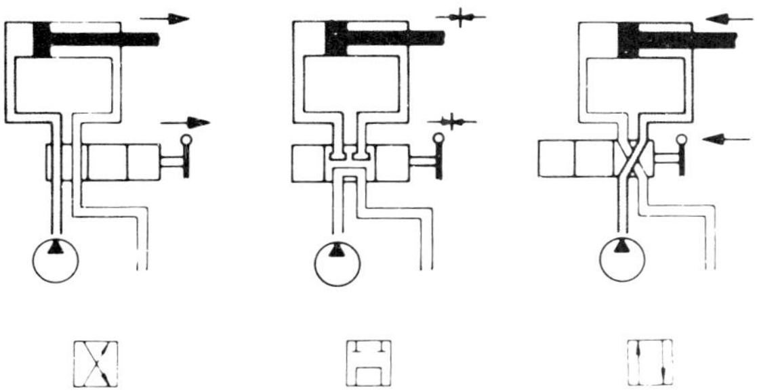

Fig 7(b) The three 'working' positions of a 3-position directional control valve.

Four-Way Valves

The four-way selector is the normal type of valve used for the control of reversing systems (cylinders, motors and pilot systems), and it is employed as a simple two-position reversing 'switch' to reverse the connections between two pairs of ports. The changeover conditions may provide either open-centre or closed-centre characteristics. (Fig 7(b)).

The four-way valve also offers numerous other useful configurations by alternative interconnection of the four ports. This may be further extended by making the valve a three-position type, which again has its particular applications. Four-position, four-way valves may also be used for a minority of special applications.

Five-Way Valves

A five-way selector is, basically, a modified four-way directional-control valve in which the two sets of return-to-tank ports in the valve bore are usually taken to separate external ports on the body. This provides independent speed control of exhaust flow, without the need for separate non-return valves, or can provide for dual pressure operation of an actuator. They are also used for interlocking circuits or circuit unloading.

Six-Way Valves

A six-way selector is, basically, a four-way directional-control valve providing pump unloading in the centre position. It is a type which has been developed specifically for mobile equipment applications, the flow control provided in the centre position saving unnecessary waste of power in the case of a continuously running pump by providing simple unloading and reducing the heat generated in the system.

TABLE IV – COMPARATIVE PERFORMANCES FOR OPEN- AND CLOSED-CENTRE WORKING

Performance	Open-Centre	Closed-Centre
Pump	Fixed displacement	Variable displacement
System idling time	High	Low
Actuator(s)	Single, independently positioned	Can operate more than one actuator from independent control valves
Actuator response time	Not critical	Critical
Pilot pressure	Generally too low to operate control	Adequate pressure system

TABLE VII – VALVE DESIGN ELEMENTS

Element	Advantages	Limitations	Remarks
Spool valve	(i) Very flexible as regards porting configuration. (ii) Relatively straight-forward manufacture up to 105 bar (1 500 lb/in^2) rating. (iii) Low inertia giving good acceleration and deceleration. (iv) Low operating load.	Necessity for precision manufacture increases with increasing pressure.	Most common type of valve.
Poppet valve	(i) High response. (ii) Low leakage. (iii) Easy to manufacture with higher pressure rating. (iv) Relatively insensitive to contaminants.	(i) Strictly limited design flexibility – each valve is only capable of two-way function. (ii) Lacks modulation characteristics.	Fairly common. Can be used with resilient seals.
Sliding plate	(i) Very flexible as regards porting configurations.	(i) Demands precision manufacture to control internal leakage. (ii) More expensive to produce.	Main application is for servo-valves.
Rotary plate	(i) Good flexibility as regards porting configurations. (ii) Generally suitable for low-cost production.	(i) Requires high operating forces unless pressure balanced. (ii) Elimination of internal leakage difficult.	Low-cost selector for pressures up to 35 bar (500 lb/in^2). Suitable for higher pressures with special seal designs.
Rotary spool	(i) Good flexibility as regards porting configurations. (ii) Generally suitable for low-cost production with simple porting. (iii) High flow rate for given size.	(i) Relatively limited in pressure rating, *eg* up to 35 bar (500 lb/in^2). (ii) Cost increases with increasing complexity of porting.	
Rotary ball	(i) Low-cost production for low pressure ratings. (ii) High flow rate for given size.	Limited flexibility as regards porting configuration.	Can be used with resilient seals.
Ball-poppet	(i) Low leakage. (ii) High response. (iii) Low cost. (iv) Generally insensitive to contaminants.	Strictly limited as regards porting configurations.	Suitable for pressures up to 210 bar (3 000 lb/in^2), but mainly used for check valves.

Open- and Closed-Centre

An *open-centre* valve connects the supply to reservoir ports in the centre position, thus reducing power consumption when the system is idling. A *closed-centre* valve blocks off both the supply and system in the centre position, thus holding the system pressurized. Both have their specific advantages and applications, some comparative data being summarized in Table IV.

Particular advantages are obtainable from pressure-compensated closed-centre valves in having a wide potential 'meter-in' band independent of load pressure. With open-centre valves the metering characteristic and 'meter-in' flow forces are dependent on the flow rate and service-load pressure.

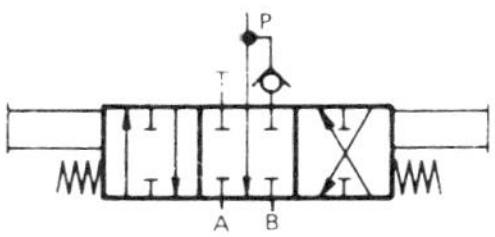

"D" spool — normally used to direct pump flow to either end of a double-acting cylinder. Both cylinder ports blocked in neutral position.

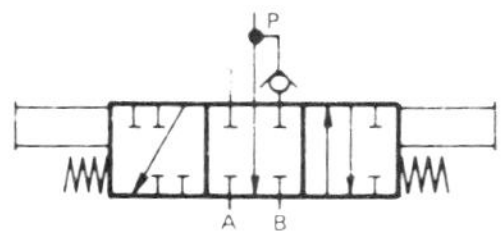

"W" spool — reverse operation to "T" spool.

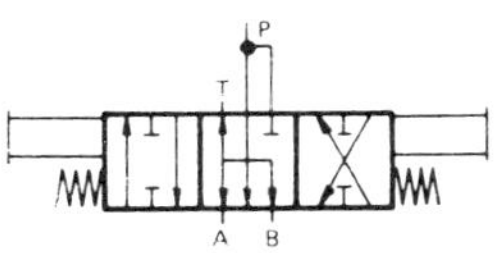

"B" spool — to direct flow to a bi-directional hydraulic motor. Ports partially open in neutral.

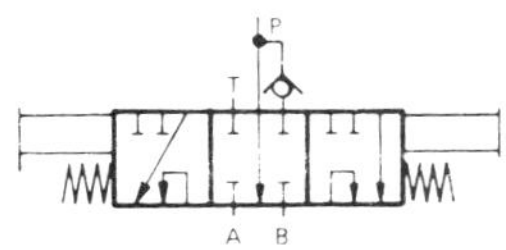

"S" spool — double-acting spool which checks oil back down bypass for additional functions.

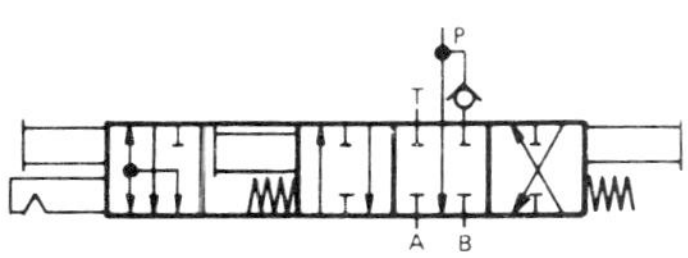

"C" spool — double-acting spool with fourth 'float' position which can provide free movement of a double-acting cylinder.

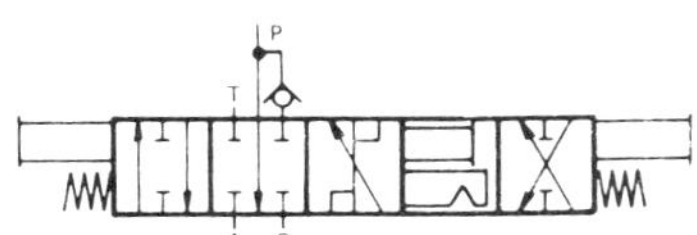

"R" spool — four-position regenerative spool allows cylinder exhaust at one end to combine with pump flow at the other for rapid rod extension.

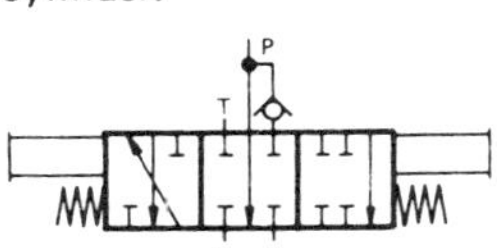

"T" spool — directs flow to one end only of a double-acting cylinder (*eg* for fork lift truck operation).

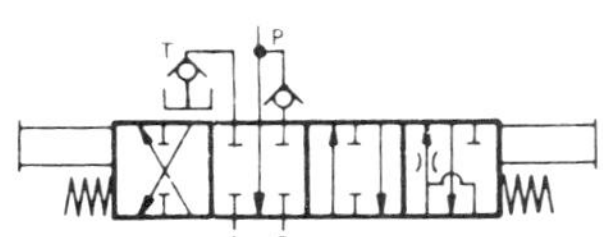

"G" spool — four-position regenerative float type action, normally used in conjunction with a back-pressure valve.

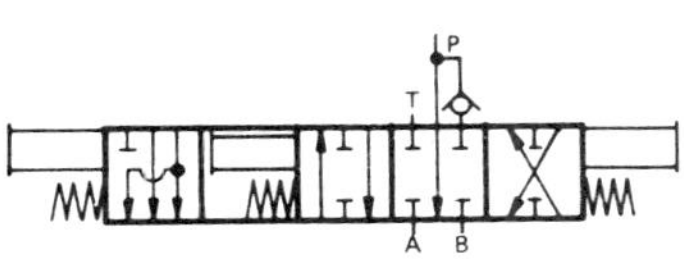

"F" spool — normally used to control dump cylinders.

Fig 8 Examples of spool configuration and applications. (Sperry Vickers).

Valve-Working Elements

A summary of valve-actuating elements is given in Table V. *Spool valves* are the most favoured type of valve for selectors, particularly for industrial hydraulic applications. Spools themselves are generally classified by the porting conditions given when the spools are in the centre position. Possible variations are numerous (*eg* see Fig 8) but the most common spool types are:

(i) *Fully locking spool* – normally used on spring-centred valves or on spring-return two-way valves. Locking is not necessarily completely positive.

(ii) *Open-centre spool* – this provides locking of the two service ports with pump-to-tank flow. As well as permitting pump unloading on an open-centre circuit, this type of spool can also be used as a selector on a circuit having several actuators in series operating from a single pressure source.

(iii) *Partial open-centre spool* – in this case partial locking is provided on one service port, with pump-to-tank flow for unloading.

(iv) *Fully bypass spool* – this again provides unloading, bypassing the pump and service ports-to-tank in the centre position.

(v) *Partial bypass spool* – with this configuration the service ports are bypassed to tank in the centre position but the pump remains loaded.

Spool control may be by means of a spring (either returning the spool to centre or to one extreme position), mechanical detent or hydraulic. In the latter case, the pilot pressure may be used to provide offset control or to operate mechanical detents.

A more comprehensive listing of spool types and spring arrangements is given in Table VI.

TABLE VI – SPOOL TYPES AND SPRING ARRANGEMENTS

Closed centre
Open centre
Open centre over tapers
Tandem centre, closed crossover
Tandem centre, open centre crossover
Blocked actuator connection(s)
Centre blocked actuator and pressure port
Centre blocked tank port
Centre blocked pressure port, actuator port(s) bleed
Centre blocked pressure port
Spring offset (return to end)
Spring centred
Hydraulically biased

Since spool valves operate on a sliding principle, design normally follows the basic requirements of all slide valves, *ie*

(i) Pressure-balanced ports are required, so that there is no net pressure force acting axially on the spool.

(ii) Valve diameter should be a minimum consistent with suitable stiffness.

(iii) The valve body or sleeve must have adequate rigidity.

(iv) Friction forces must be minimized, and are largely controlled by material selection for rubbing/sliding parts.

(v) Annular flow should be symmetrical, in order to avoid radial unbalance forces, which could increase friction.

(vi) Bernoulli forces, arising from changes in fluid momentum, must be minimized.

Parameters (v) and (vi) are largely controlled by the detail design of the spool.

Sliding plate valves may be preferred to spool valves for particular applications, although this type tends to be more expensive to produce and is prone to higher leakage.

Poppet Valves

Poppet valves have the advantage of high response and relative insensitivity to contamination. They are also well suited to high-pressure duties and so may be preferred for specific applications. They have low leakage, can be made to seal properly for long periods of time and are relatively cheap to manufacture. They are less suitable for large valve sizes, however, since the opening load becomes excessive and calls for the use of a pilot-valve system. Once the limiting size of orifice through the selector has been established, a poppet valve of the smallest possible diameter consistent with this throat area is chosen, with other passages slightly greater than the throat area.

The lift required is dictated by the poppet angle. With a 45° valve, a lift of about one half the valve throat diameter is required for full opening, with no valve stem in the throat. Pushrod (rather than stem) operation is generally to be preferred since the stem size and thus the restriction in the throat can then be reduced to a minimum. Pilot systems are widely used for reducing the valve opening force required, flow-operated valve mechanisms generally offering the most benefits.

Rotary Valves

Rotary valves rely on close contact being maintained between a rotating port plate and a back-up member and are rather more difficult to produce with adequate sealing for high pressures. Also high operating forces may be required unless the elements are pressure balanced. They are, therefore, more usually applicable to lower-pressure systems, where the relative simplicity of the configuration, and its flexibility as regards porting configuration, can be used to advantage. Some designs are produced for higher pressures, although the provision of a suitable rotary seal on the spindle remains a problem.

Alternative valves include the rotary spool or plug valve and rotary ball valves. Porting configurations are somewhat limited with the latter.

Cartridge Valves

A cartridge valve is simply a 2-port valve which will either block flow, allow free flow, or modulate flow as a function of a pressure signal. Adapting this concept to a specific machine function can often result in a more efficient, faster responding, smoother and reduced size hydraulic system.

An example of a cartridge valve is shown in Fig 9, comprising a sleeve, poppet and spring mounted on a manifold. The poppet moves within the sleeve to control flow through the valve, with the pilot-flow passage contained in the cover. The two main ports 'A' and 'B' each have an active area to open the valve. The 'A' area can be seen as the circular area within the valve seat. The 'B' area is the annular area between the seat and the o.d. of the poppet.

The pilot port 'X' is connected through the cover to the spring chamber of the valve. The pressure in the pilot port 'X' acts over the entire area of the poppet, which is equal to the 'A' area plus the 'B' area.

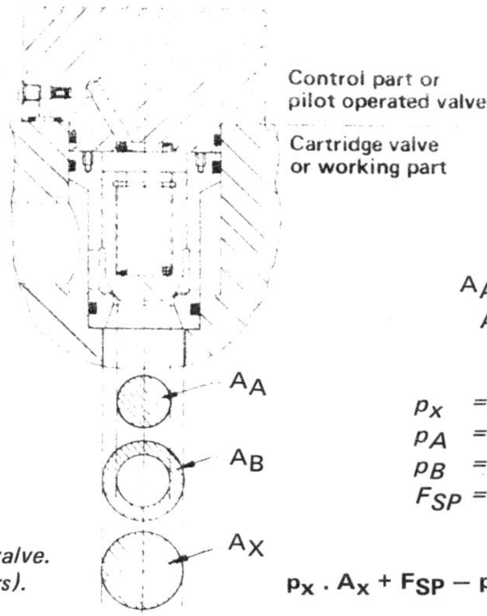

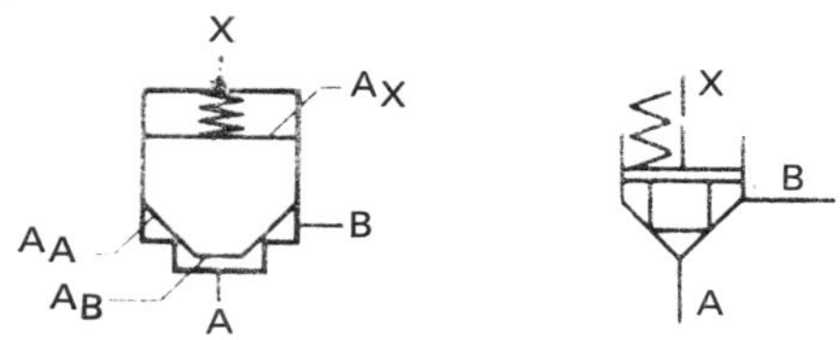

p_X = *pressure X*		F_{FL} = *flow force*	
p_A = *pressure A*		A_X = *area X*	
p_B = *pressure B*		A_A = *area A*	
F_{SP} = *spring force*		A_B = *area B*	

Fig 9 Cartridge valve. (Sperry Vickers).

$$p_X \cdot A_X + F_{SP} - p_A \cdot A_A - p_B \cdot A_B - F_{FL} = \begin{matrix} < 0 \rightarrow \textbf{open} \\ > 0 \rightarrow \textbf{closed} \end{matrix}$$

The valve will be open or closed as a function of the force balance equation shown, where the pressure in 'A' plus the pressure in 'B' acting over their respective area, plus the flow force tend to open the valve. The pilot pressure acting over the 'X' area and the spring force tend to close the valve.

Proportional-Control Valves

Proportional valves are the modern choice for applications calling for pressure, fast control of velocity, acceleration and deceleration of actuators. Basically these are solenoid-operated valves with electrical position-feedback control (proportional solenoid) – Fig 10. Spool deflection is measured by an inductive position sensor and fed back to a proportional amplifier. The pilot stage is thus an electric position-control system unaffected by load disturbances, whilst responding rapidly and precisely to input signals. Such valves also normally have built-in safety features so that in the event of supply failure, signal failure or loss of spool position-feedback, the valve will automatically go to the centre or safe position.

Proportional valves are most commonly applied to open-loop applications, where in effect they give 'closed-loop' performance normally associated with a servo-system. They are not, however, suitable for *positional* control applications, *ie* to replace a servo-valve with actuator feedback. Basic functional types are proportional throttles and proportional selectors (directional valves). Normally such valves are not pressure/load-compensated or temperature-compensated, thus the volume of fluid passing through the valve will be affected by changes in pressure drop and/or fluid

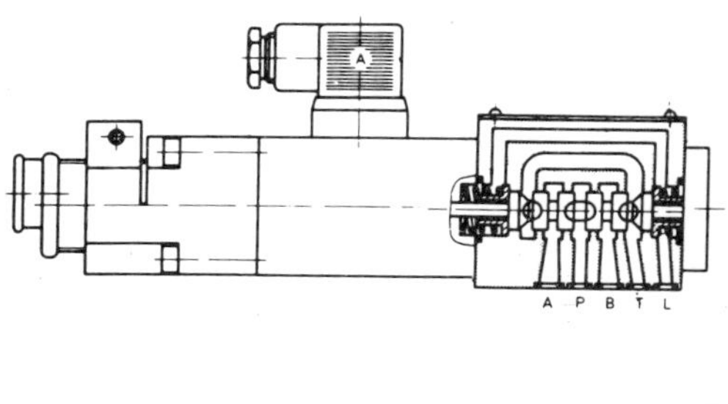

Fig 10 Proportional throttle valve.

viscosity. For applications where consistent flow is required independent of changes in supply pressure or load pressure they can be associated with a further device to maintain a constant pressure drop across the valve.

A basic requirement with such valves is that they should be used only with the electronic controllers specified by the manufacturers for each valve.

Valve Actuation

The operation of selectors may be manual, mechanical or hydraulic, although it is more usual to classify operation by the actuation method employed. Manual control is usually by a lever moving the valve against spring bias. Detents may be provided to give intermediate 'hold' positions, if necessary, so that such positions can be held with no load on the lever. The general description 'mechanically controlled' is usually given to valves operated by cams or similar mechanical (rather than manual) systems. Pilot-operated valves are normally described as hydraulically controlled.

Hydraulic pilot-operated mechanisms are the most popular, and are particularly adaptable for spool valves, since the spool can readily be driven, like the piston in a cylinder, by an applied hydraulic force or pilot pressure applied to one end. Numerous variations on this theme are possible, with spring centring or spring biasing, and with the incorporation of detents if required. Operating speeds can be at least as high as (and can be made higher than) any mechanical operating system; and speed control can readily be provided by incorporating a restrictor or throttling valve in an appropriate part of the circuit. This has the advantage that high operating forces are available to overcome friction, although such a method of biasing will not have 'fail safe' characteristics.

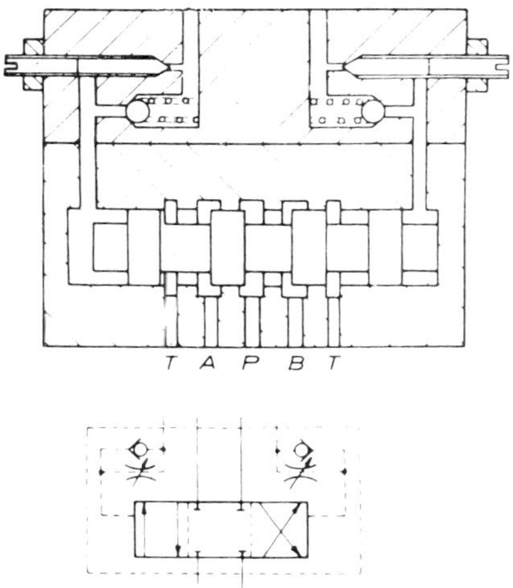

Fig 11 Pilot-operated 4-way valve with restrictors to limit speed of change-over.

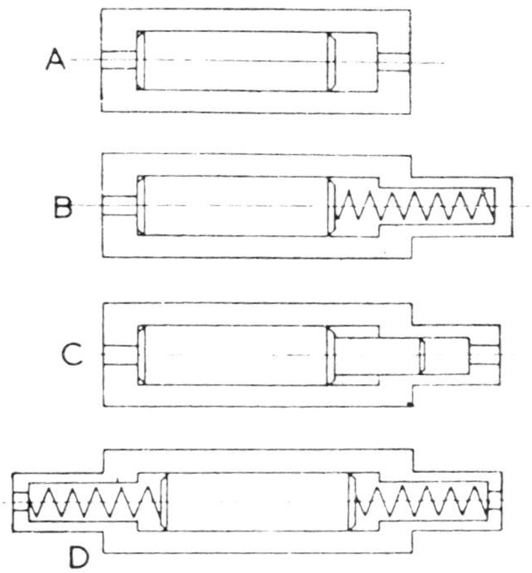

Fig 12 Some arrangements of pilot-operated valves.
A—Two-position free spool or 'flip-flop'.
B—Two-position spring biased.
C—Two-position pressure biased.
D—Three-position spring centred.

Pilot-Operated Valves

Pilot-operated valves are used both as units and integral with a solenoid-operated pilot valve. Fig 11 shows a four-way two-position (springless) valve on which is super-imposed a block incorporating restriction and check valves in the pilot lines. By using these the rate of shift can be controlled and shocks in the system due to sudden pressure changes minimized.

The more common spool and spring arrangements are shown in Fig 12. The two-piston type must be mounted with its axis horizontal and is suitable for pulse operation so that pilot pressure need not be maintained for longer than is needed to shift the spool.

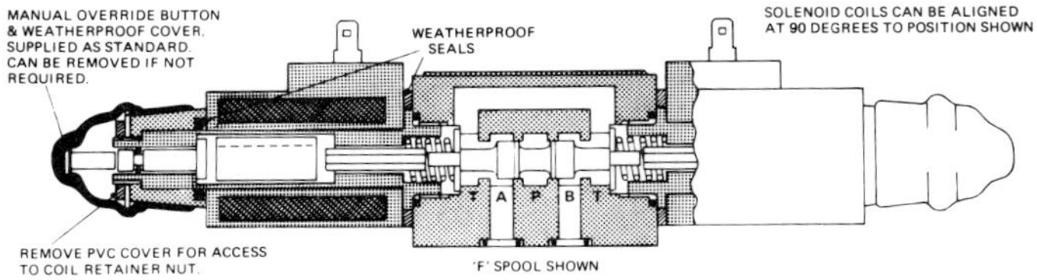

Double solenoid-operated valve.
(A & D Fluid Power Ltd).

Solenoid-Operated Valves

In a solenoid-operated spool valve an electric solenoid and spring are used to move the spool. Generally it is best to use the solenoid for a 'push' operation, utilizing spring action for 'pull' motions. The solenoid must be powerful enough to over-ride inertia and friction and also the spring and hydraulic forces. The latter may be extremely variable and not completely predictable, calling for a generous margin in the power of the solenoid and springs.

Solenoids may be of the 'dry' or 'wet' type. In general, 'wet' solenoids can be smaller for the same duty because of their lower static and dynamic friction. They also have the advantage that all moving parts are enclosed and lubricated, and seals between the solenoid and valve body are eliminated. They are also described as glandless valves.

The size of directly operated solenoid valves is generally restricted to flow rates up to about 45 lit/min (10 gal/min) – *ie* 3 mm (1/8 in) and 6 mm (¼ in) nominal valve sizes. Many of these valves can be switched directly from static systems, the outputs usually being 24 V d.c. and 20–65 watts, depending on the system.

For higher flow rates demanding larger port sizes and larger spool diameters, directly operated solenoid valves become excessively bulky and costly. This can be overcome by pilot operation where a small solenoid valve is used as a piloting first stage in a large body valve whose spool is moved by differential pressure applied to its terminal surfaces.

The pilot valve controls pressure and exhaust flows in the piloting chamber and shifts the spool – a critical factor in design being the waterborne effects commonly associated with large, high-pressure flows. Reaction time of the piloted valve is the sum of the pilot reaction time and the shifting time. It is usual to aim for a slow, controlled movement of the main spool, which can be achieved by incorporating restrictors between the pilot valve and piloting chamber. Where pilot pressure is derived from the pressure port of the valve itself it may be necessary to control the pilot pressure reducing valve in order to ensure constant operating times.

The sizing of piloted valves requires careful consideration as regards pressure drop, especially where large flow rates are involved.

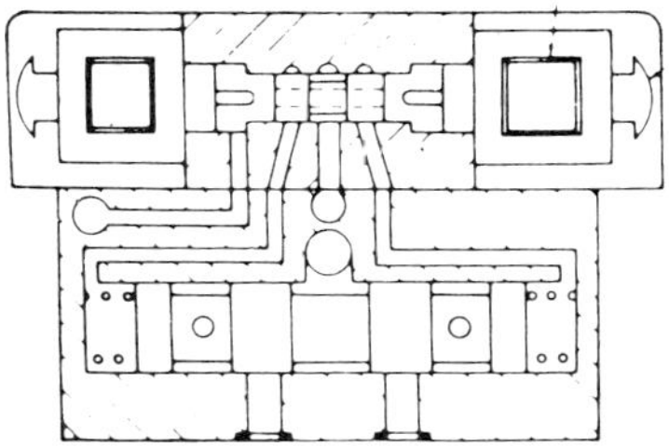

Fig 13 Sub-plate or gasket mounted solenoid-operated pilot-actuated valve. The upper valve is separated and can be used alone.

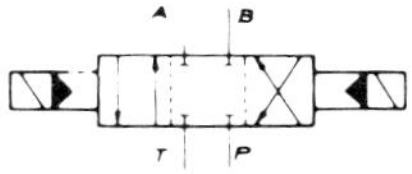

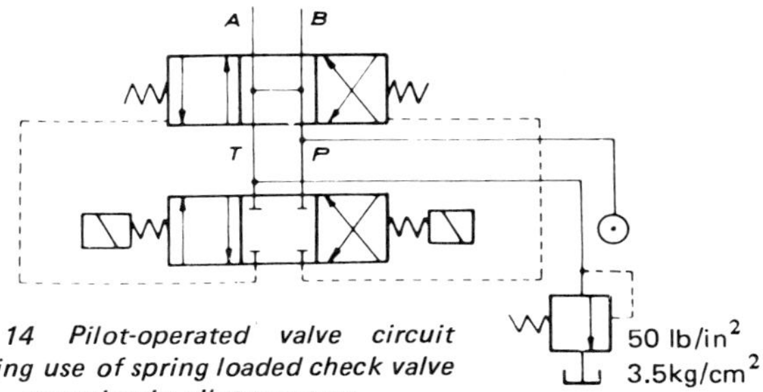

Fig 14 Pilot-operated valve circuit showing use of spring loaded check valve to maintain pilot pressure.

A typical layout of a miniature valve mounted pick-a-back on a pilot-operated valve is shown in Fig 13. The same spool arrangements are available as for pilot valves and the solenoid and pilot valve spools are matched. One point which must be watched is that pilot pressure is available if an open-centre valve is used to unload the pump (this can be ensured by inserting a back-pressure valve — a spring-loaded check valve is often used — as shown in Fig 14).

D.C. or A.C. Solenoids

D.C. solenoids are generally preferred to a.c. since d.c. operation is not subject to peak initial currents which can cause over-heating and coil damage with frequency cycling or accidental spool seizure. A.C. solenoids are preferred, however, where fast response is required, or where relay-type electric controls are used. Response time with a.c. solenoid-operated valves is of the order of 8–15 milliseconds, compared with the 30–40 milliseconds typical for d.c. solenoid operation.

Glandless Solenoid Valves

By arranging the solenoid armature to work in a sealed tube with the solenoid coil enveloping it, the sealing glands can be dispensed with, so simplifying the construction and eliminating one possible point of leakage.

This principle has been applied extensively to the smaller valves. A typical type is shown in Fig 15.

This valve is tee shaped with two ports opposite each other, whilst the third is at right angles to them. The plunger, usually of a corrosion-resistant ferrous material, is spring-biased so that when unenergized it closes the lower orifice, whilst leaving the other open. When energized, the plunger is pulled up so that the lower orifice is opened and the upper closed. If desired, the spring can be arranged to bias the plunger in the other direction.

The plunger is provided with plastic valve discs, *eg* nylon, or of synthetic rubber. Because the plunger is unbalanced, the force due to the pressure must be limited and the size of orifice, and therefore the flow and pressure drop, is usually related to the pressure.

The maximum pressure is also related to the type of valve and may be as high as 210 bar (3 000 lb/in^2) with a 0.85 mm (1/32 in) orifice. Flow depends on the allowable pressure drop and this in turn on the orifice size and fluid.

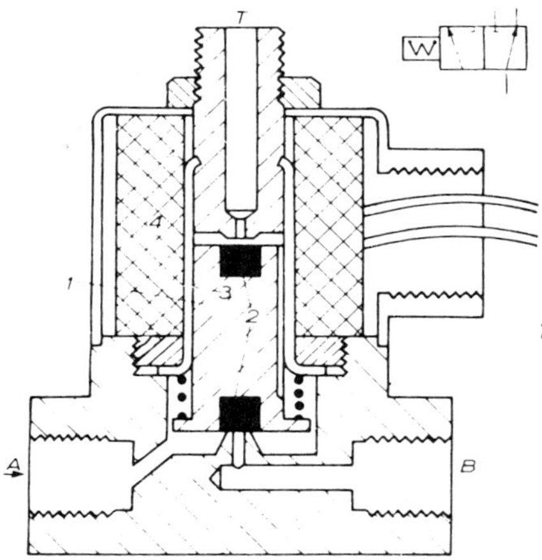

Fig 15 Glandless solenoid valve.
1—Plunger. 2—Synthetic seats. 3—Sleeve.
4—Coil. A—Cylinder. B—Pressure.
T—Exhaust.

Sealing is normally 'bubble-tight' but this is to some extent dependent on the cleanliness of the fluid. Lubrication is not essential but if used with air the valve life is increased by air-line lubrication.

Glandless valves can be installed in any position and will withstand appreciable shock loads. Response time is extremely short, 5 milliseconds on a.c. and 10–15 milliseconds on d.c. and it is said that speeds of up to several hundred cycles per minute are possible.

For hazardous atmospheres most makers supply explosion-proof materials which are slightly heavier and bulkier than the standard type.

Although these valves were originally developed for aircraft and missile application, there is no doubt that they have many uses in hydraulics, both as main valves for low-power jobs and as pilot valves.

Where pressures do not exceed 17.5 bar (250 lb/in^2) a 1.6 mm (1/16 in) diameter orifice is suitable and this gives a flow of 100 in^3/min (0.50 gal/min) for a 3.5 bar (50 lb/in^2) pressure drop. On a 50.8 mm (2 in) diameter cylinder, this would give a piston speed of 760 mm (30 in) per minute, whilst when used as a pilot valve for 25.4 mm (1 in) diameter spool with 12.7 mm (½ in) movement, the operating time is about half a second.

When acting as a pilot valve the actual flow would almost certainly be greater than that for a 3.5 bar (50 lb/in^2) pressure drop, as for a large part of the time the pressure drop will be nearer 14 bar (200 lb/in^2), until the resistance to piston or spool builds up.

Most of the makers of these valves also supply pilot-operated valves incorporating the basic glandless valve; four-way valves and two- and three-way valves for larger flows are made in this way.

These valves may prove useful in acting as pilots to larger valves when the 'pressure release' principle is adopted. Fig 16 shows this applied to a differential area spool valve. Normally the larger area keeps the spool to the right, with the solenoid valve SV closed. Opening the solenoid valve causes the pressure to drop in the left hand chamber and the spring, with the excess pressure, causes the spool to move to the left. A two-position springless valve could also be operated in the same way, with jets and solenoid valves at each end.

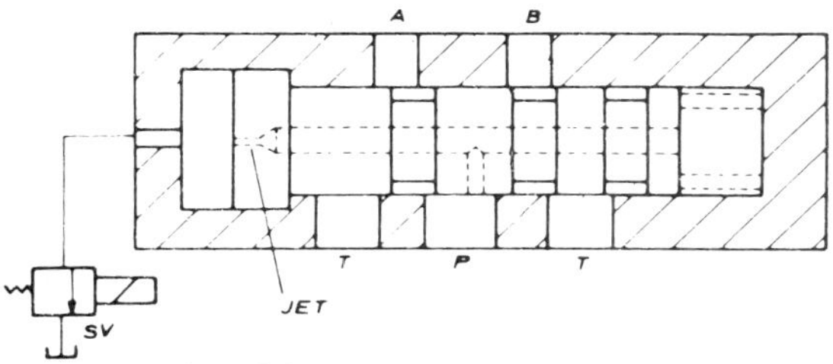

Fig 16 Small solenoid valve used as pilot, employing pressure release principle.

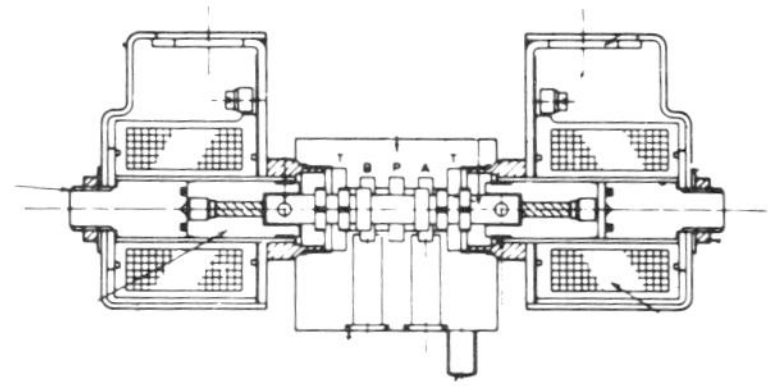

Fig 17 Glandless solenoid valve. (Flui-Trol).

Glandless Solenoid Valves – Spool Type

The construction of a four-way spring-centred closed-centre double solenoid valve is shown in Fig 17. It has push solenoids with a spool movement of 1 mm (0.040 in) either side of centre. It is suitable for pressures up to 140 bar (2 000 lb/in^2) and has a flow of 9 lit/min (2 gal/min) for a pressure drop of 2 bar (30 lb/in^2) on light hydraulic oil at 27–38°C (80–100°F). It is gasket-mounted and when used as a pilot valve is bolted on top of the main valve.

Miscellaneous Valves in Use

Non-return valves or *check valves* are used in circuits, or combined in the body of other valves, to provide flow in one direction only. The simplest type is the spring-loaded ball valve, although this has limited suitability for hydraulic services and rather more sophisticated designs are normally employed. In high-pressure services, specially good sealing is essential, and it may be necessary to design the valve with a resilient seating seal. Damping may also be provided to prevent the seal being damaged by the impact of sudden flow reversal. A further consideration in a design involving high flow rates is that it must be impossible for the sealing member to travel to a position where it could obstruct the through-flow in the valve-open position. Also, of course, in high-pressure valves with resilient seals, the seal inserts must be designed to prevent extrusion or displacement of the seal.

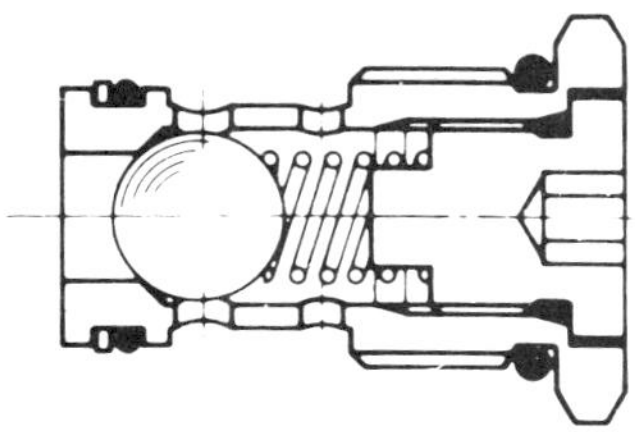

Fig 18 Ball type check valves.

A typical design of a simple non-return valve is shown in Fig 18. Spring pressure would normally be such that a back-pressure of about 0.3 bar (5 lb/in^2) is present when the valve is opened, although a stronger spring may deliberately be used to provide higher back-pressures for specific applications, *eg* to give the valve 'restrictor' characteristics. With a straight-through configuration the flow path is usually arranged through the centre of the spring in order to minimize restriction.

Check valves are used in a circuit to eliminate actuator movement (*eg* cylinder movement) and to maintain it in a 'hold' position without creeping, as might otherwise occur due to directional valve spool leakage. The most usual type for this particular duty is the pilot-operated check valve.

Most pilot-operated check valves are supplied with a pilot ratio of at least 2:1 with the main piston seat area. This means that to open the valve a pilot pressure is required equal to half the pressure being applied to the primary side of the valve. If the pilot-operated check valve is being used on the annular side of the cylinder, then, with a 2:1 ratio pilot, the cylinder area ratios must be less than 2:1, otherwise it will be impossible to open the valve.

When there is no pressure on the primary side of the cylinder, the pilot-operated check closes and allows no flow from the annulus side of the cylinder, therefore retaining the load in a fixed position. When pressure is introduced to lower the load, the check valve is opened by pilot pressure to allow free passage of the oil from the bore side of the cylinder. In the reverse condition, to lift the load, the pilot-operated check operates as a normal check valve.

Spill-Off Valves

A spill-off flow-control valve is shown in Fig 19. It is of the balanced-piston type and is based on the relief valve which embodies this principle. The main oil supply passes through the lower balance chamber, past the speed-control orifice to the actuator. The upper balance chamber is connected to the actuator line. If there is no load on the actuator there will be no back-pressure in the upper balance chamber. The relief valve will therefore open at the spring setting – about 1.4 bar (20 lb/in^2). If a load is applied to the actuator, a back-pressure will be created in the line and this will act on the valve plunger to increase the blow-off pressure in direct proportion. The pressure across the speed-control orifice is, however, always maintained at the spring setting and the small ball valve set to blow off at the maximum safe pressure enables the valve to act as a safety valve also. This circuit has no control over negative loads and the movement under varying positive loads may be irregular, due to the initial compressibility of the oil. This can be overcome, for many practical purposes, by placing a spring-loaded check valve in the exhaust line, which ensures a back-pressure of 3.5 bar (50 lb/in^2) or so.

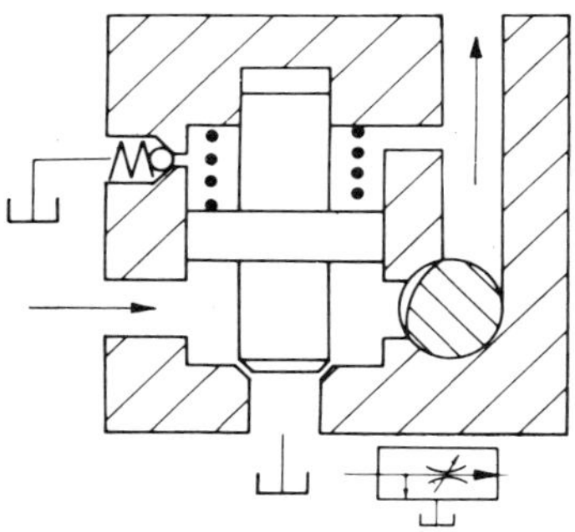

Fig 19 Spill-off flow control valve.

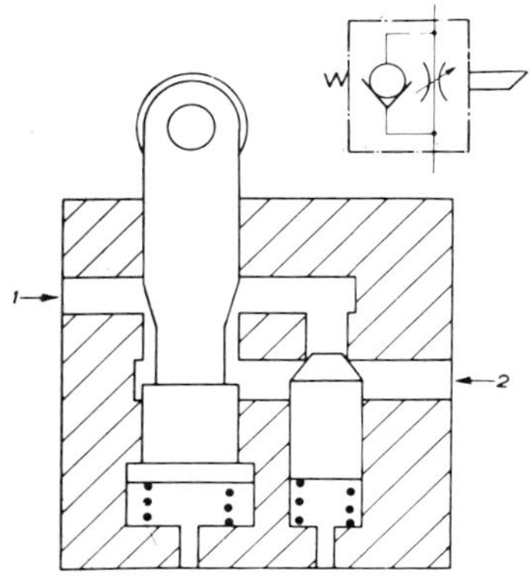

Fig 20 Deceleration valve with integral check valve to give free flow on return stroke.
1–Free flow. 2–Controlled flow.

Deceleration Valve

When working a ram at high speeds, it may be necessary to provide some means of slowing it down before the end of the stroke, so as to prevent shock. If the load is not too great, this can be done effectively by a simple deceleration valve such as that shown in Fig 20. The tapered plunger is depressed by a cam on the moving part so that the amount of throttling can be controlled. With another type, the valve is mounted parallel to the ram and comes into operation during the last half-inch or so of travel. The same effect can also be obtained with a cushioned cylinder.

If an attempt is made to decelerate too great a load with this type of valve, the pressure may build up dangerously and it is then necessary to employ a braking valve. The simplest form of braking valve is merely a relief valve which comes into operation at the appropriate point of the stroke. The energy is then dissipated by blowing through the relief valve.

Flow Dividers

A flow-divider valve is used where the input needs to be split into two equal outputs, *eg* for the operation of two actuators at the same speed. A simple form is shown diagrammatically in Fig 21 where the input is fed to two orifices of equal size feeding two output ports. This has the same characteristics and limitations as a fixed restrictor in each outlet line, with equal division of the inlet flow.

A more satisfactory form is the pressure-compensated flow divider where two spools are used, mechanically connected, with the total flow passing across metering orifices in each of the spools. If the flow to one inlet increases, pressure drop across the spool is increased, causing it to move against its spring and so provide more throttling effect. At the same time the mechanical connection transmits movement to the other spool, the result being that the two outputs are substantially self-adjusting to the output loads. A further method is to provide variable orifices during reverse flow, so that the flow divider can also be used for combining two input flows in equal volumes, thus controlling speed on the return stroke of the actuators.

Flow dividers have certain inherent limitations, notably a tendency to become slightly out of phase, which effect can be cumulative unless they are re-phased at the end of each actuator stroke – *ie* loss of synchronization could 'grow' in a series of part-stroke movements. Synchronization can also be affected by the number and manner in which further flow dividers may be connected in the circuit, *eg* to produce synchronization of more than two actuators.

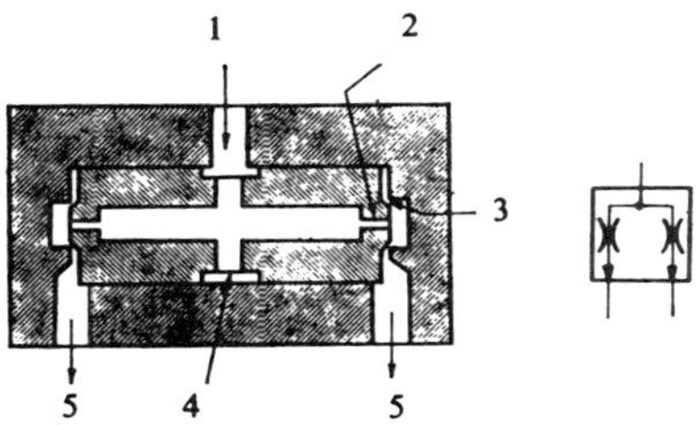

Fig 21 Flow dividing valve.
1–In. 2–Fixed jets. 3–Variable gaps. 4–Plunger. 5–Out.

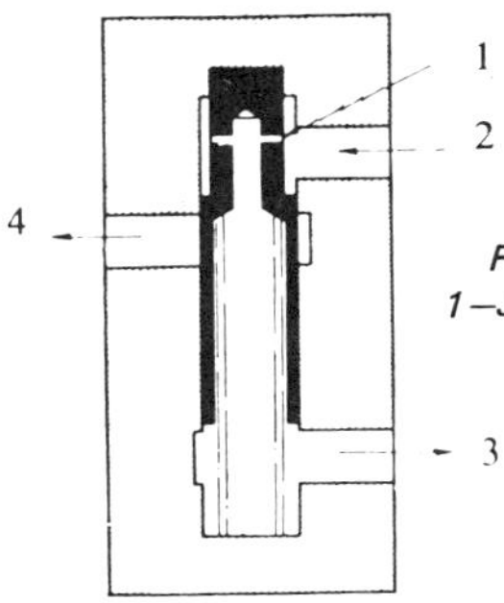

Fig 22 Priority valve.
1–Jets. 2–In. 3–Priority. 4–Secondary.

Priority Valves

A priority valve may be used in a circuit where it is essential that one service in a multi-service system, fed by a single pump, must always have priority irrespective of other demands. A simple form of such a valve is shown in Fig 22. The hollow plunger acts as a flow regulator by keeping the pressure drop across the orifice constant, the surplus capacity being bypassed to a subsidiary output. If the subsidiary pressure becomes higher than the priority service pressure, the plunger is forced down to the lower seat and a metering annulus is then formed by the lower end of the plunger.

The valve will only function when there is a constant flow through the main outlet, which presupposes an open centre directional-control valve being used. If the flow is completely stopped, there is no pressure difference to force the plunger down. The back-pressure due to the load is the highest pressure (main or subsidiary), plus the pressure required to overcome the spring.

Transfer Valves

Transfer valves are used to short circuit an actuator, to allow for rapid mechanical movement without the 'braking' effect which might otherwise be caused by cavitation of the fluid. This could apply, for example, when initial movement of an actuator load was under gravity or mechanical force, with the final movement accomplished by hydraulic pressure. To increase the speed of movement above that obtained by hydraulic flow alone, the suction caused by over-speeding opens a valve to short-circuit fluid from the other side of the actuator, making for sufficient volume to prevent cavitation.

A 'dump' valve performs a similar function, but in this case is used with a single-acting actuator. To speed up the return stroke, the throttling effect of the fluid being forced out through the return line is relieved by a separate valve opening to 'dump' fluid at a higher rate into an appropriate part of the return circuit. In this case the valve is pressure-operated rather than suction-operated, and can be similar in form to a pressure-relief valve with suitable locking characteristics, or even be a simple cock with mechanical actuation derived from the return movement.

Sequence Valves

A sequence valve can be used where it is necessary to ensure that sufficient pressure has built up in one circuit before fluid is admitted to another circuit. In its simplest form it comprises a spring-loaded spool valve with primary and secondary ports, the spool being normally positioned so as to shut off the secondary port. The primary pressure acts on the end of the spool, against spring pressure. When sufficient pressure is present, the spool is moved against the spring, opening a connection through to the secondary port — Fig 23. The valve also has a throttling action, which prevents the primary pressure falling suddenly, but opens fully when the working pressure is reached. A non-return valve may be incorporated in the same body to provide free return flow of fluid in the non-controlled direction.

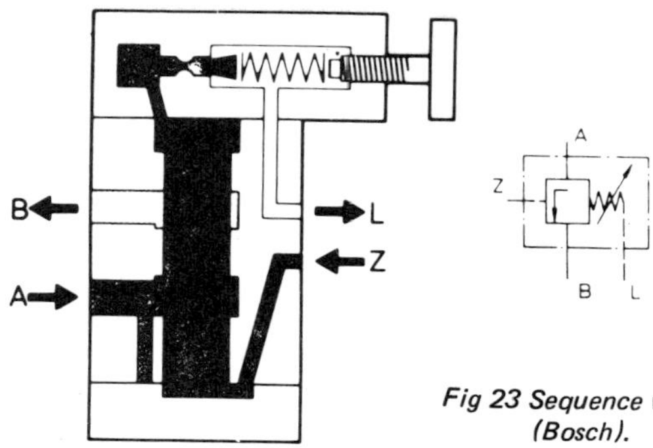

Fig 23 Sequence valve. (Bosch).

Basically, a simple sequence valve is similar to a pressure-relief valve, and a pilot-operated pressure-relief valve may, in fact, be adapted for this purpose merely by connecting the drain from the low-pressure side of the pilot section to the tank. Spool valves are, however, more normally used.

Sequencing valves have the particular limitation that they are reliant on system conditions remaining stable. Pressure variations can cause operation to be premature or delayed and also troublesome. In general, their use is best avoided if possible.

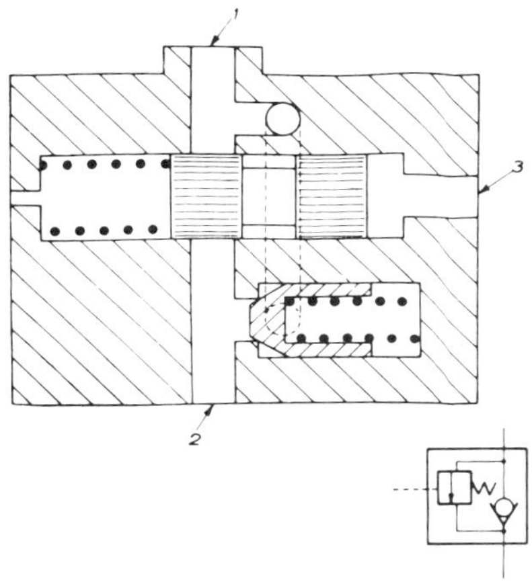

Fig 24 Cylinder locking valve.
1—Cylinder connection.
2—Directional control valve connection.
3—Pilot connection.

Locking Valves

Locking valves are also designed for use with cylinders, working as check valves but 'locking' in position hydraulically. The pressure needed to unlock the valve is independent of the pressure in the system, as can be seen by reference to Fig 24.

On the forward stroke, fluid passes the non-return valve. On the return stroke, pressure from the other side of the cylinder acts on the spool valve and opens it. The advantage of this valve is that the piston cannot move until there is definite pressure. If there is negative load on the piston causing it to run away, it can only move as fluid is admitted to the cylinder, the valve opening just sufficiently to allow the piston to move at the desired rate. Both single and double valves of this type are made.

Such valves are generally unsuitable for use in systems where the loads are high and unidirectional, since the resulting movements can become uncontrolled unless restrictors are also added to the circuit.

Hydraulic Fuses

Hydraulic fuses are intended to shut off automatically any line in which failure has occurred. The simplest type works as a flow-sensitive valve, closing and sealing the line should the flow rate exceed a pre-determined amount, *eg* as it would in the case of a burst or massive leakage. Such fuses are generally characterized by low sensitivity, so that they will not respond to smaller leaks; and they are also set to a value above the maximum flow rate for the system, which is not necessarily representative of average working conditions (further reducing their sensitivity to leaks).

The usual form of such a fuse is a hollow piston, with orifice, spring-loaded in a close-fitting cylinder. The piston is extended in the form of a stem, with the end shaped to provide a seal in the outlet end of the cylinder at full piston travel. At fluid pressures up to the maximum setting, the piston remains in the 'open' position under spring pressure. Higher pressures result in a differential pressure across the piston orifice greater than the spring pressure, causing the piston to move until the stem engages in the valve seat and shuts off the flow.

Alternative types of fuse are used for:

(i) Quantity measuring and

(ii) Fluid sampling.

With a quantity measuring device, the body capacity is arranged to be slightly in excess of the demand of the system protected, and the fuse automatically empties on the return stroke. Any greater volume entering the fuse operates a valve inside it to shut off the flow, this additional volume being detected as a leakage.

The fluid-sampling fuse works as a two-stage unit. First, pressure is applied to the protected circuit through the fuse with the circuit inoperative. Any flow resulting must then be due to leakage, in which case the fuse is 'triggered' to shut off the supply. Only if there is no leakage does the fuse remain 'open' to accept normal full flow through the second stage to the circuit.

A particular advantage of hydraulic fuses, apart from shutting off a damaged circuit and minimizing fluid loss, is that they enable other circuits on the same supply to continue to be operated with the faulty circuit isolated by its fuse.

Valve Construction

There are two basic forms of valve construction, one being made of sections each containing its own spool and service ports. When a multi-spool valve is required, these sections are bolted together. The second is known as a monobloc construction in which the valve body is cast with all passages, spool bores and service ports in one piece.

Sectional construction has the advantage that various standard sections can be stocked to provide the valve configuration required for a specific application. Generally valves are purchased ready assembled together with spare sections. The user can then add (or subtract) sections to make up a valve to suit his requirements. Equally, faulty sections can easily be replaced.

There are, however, disadvantages with sectional construction. Thus the faces of individual sections, which must be absolutely parallel and flat, can easily be damaged when handled or placed in store. Any scratch makes it difficult to obtain perfect sealing between the sections when they are bolted together.

Another disadvantage is the critical torque range required when tightening up the assembly bolts. If they are not torqued tight enough, leakage can occur between the sections. If overtightened, problems can arise with the spools. Sticking spools can also cause trouble when the valve stack is mounted on a machine. This type of construction is generally less rigid than a monoblock design, and if the mounting face is not flat and rigid, or if the mounting blocks are not tightened properly or torqued evenly, it is possible to create sufficient twisting in the stack to cause the spools to stick. In addition, because of the problems just mentioned, even greater difficulty would be experienced in servicing the valve under field conditions.

To minimize these problems, the sections are usually designed thicker, and thus have greater overall width compared with the monobloc design, but the small size of an individual section is much easier to handle through the manufacturing process, and in the event of faulty manufacture or material there is less scrap cost.

Valve Manufacture

The flow rate of a valve is approximately proportional to the port area uncovered by the movement of the spool and since most ports are of circular section and the spool travel is small compared with the port diameter, the flow rate is then approximately proportional to the valve travel raised to the 3/2 power. However, in systems where linearity is of prime importance rectangular

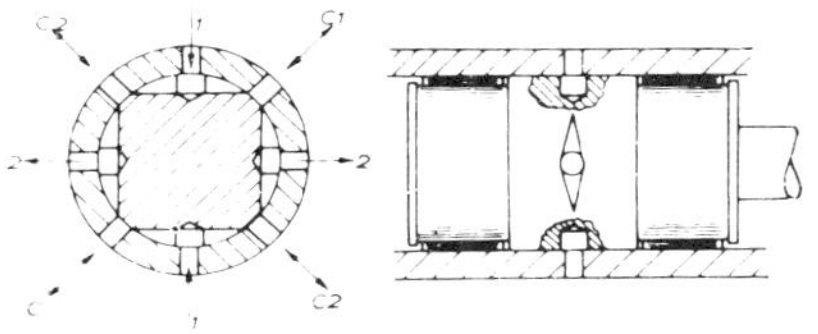

Fig 25 Rotary valve with needle bearings. (S.A.M.M.) 1—Supply. 2—Exhaust C1 and C2 to load.

sectioned ports are often used. The latter are generally more difficult to manufacture and various methods are in current use. These include broaching, spark erosion and fabricating the sleeve from a series of rings, the latter often being brazed together to form the valve sleeve. Valves having flow rate against spool travel characteristics other than those described previously can be made by using ports having the necessary geometry and this is occasionally used in certain system designs.

Rotary spool valves are generally made such that each port is duplicated and these two openings are diametrically opposite. This eliminates any radial, out of balance, hydraulic forces, but nevertheless the rotational resistance of the spool is generally an important problem with this type of design. One solution favoured in both France and Britain involves the use of needle bearings and is shown in Fig 25.

Contamination

In the earlier designs of servo-valves, which were mostly for missile applications, the object was to produce a valve of light-weight design with a good frequency response (above 100 Hz/sec if possible) and a high sensitivity. This was partly achieved by the use of light-weight first stages and small torque motors. Considerable trouble arose from the effects of working-fluid contamination, and this resulted in the modification of some basic valve designs to reduce the first stage susceptibility to contamination, and also in the development of new designs with a higher level of contamination acceptance.

Contamination has two main effects: one is to reduce the reliability of the valve by clogging or silting and the other is to reduce its life by increasing the wear rate of the spool. Although an improvement in the valve's ability to operate in relatively dirty oil can often be obtained by suitable re-design, (resulting in an increase in reliability over the longer periods necessary for aircraft and industrial use) adequate filtration is still essential to reduce wear. Wear occurs mainly in the second-stage metering orifices where high fluid velocities and accelerations occur. This wear results in increased leakage flows, increased threshold and a general deterioration in performance.

Torque Motors

The smallest clearance in a valve is generally between the spool and sleeve when the former is in its null position. To overcome the increased stiction due to the silting up of this stage the trend has been to increase the power capacity of both the torque motor and first stage whenever the resulting increase in weight is permissible. A further modification has been the introduction of 'stagnant' or 'dry' torque motors to eliminate the problem of magnetic contamination in the first stage. It is of interest to note that for one type of double nozzle-flapper valve a user returned, due to failure, 9.5% of those valves having wet torque motors and only 3.2% of those having dry torque motors. Of the former, 38% of the failures were due to torque motor contamination while there were no failures due to this cause with the dry torque motor type. However, only 0.5% of the failures of the wet torque motor type were attributed to a damaged flapper, while 10.4% of the dry type had failed from this cause. It is possible that this increase in flapper damage was caused by the presence of the sealing diaphragm isolating the torque motor from the working fluid.

Filtration

Although in nozzle-flapper valves the diameter of the nozzle is about 500 μm (0.02 in) and the diameter of the fixed orifice is about 140 μm (0.006 in), the clearance between the nozzle and the flapper is only 25–50 μm (0.001–0.002 in) and particles larger than this will cause jamming of the flapper, silting and consequent clogging of the nozzle, or a transient movement of the load until the particle is cleared. To prevent this, internal filters are usually fitted immediately upstream of each orifice and nozzle. In double nozzle-flapper valves unbalance can occur due to filter blockage at one of the nozzles, and to prevent this one valve type uses a common filter for both nozzles. This filter, however, cannot be placed immediately upstream of the nozzles and thus does not remove contamination which might be left in the passageways during assembly. Other filtration techniques have been used such as flushing the outside of the filter with the second-stage flow, and the use of filter inlets arranged at 90° to the main flow direction so that the momentum of the larger particles helps to prevent them from entering the filter.

Chip Shearing

Some manufacturers include a 'chip shearing' action in the valve designs so that if they become jammed by a chip a large force capable of shearing the chip is applied to the spool or sleeve. It can be argued through that this self-clearing action is very seldom a desirable feature since it has two over-riding disadvantages, the first being that the resulting transient disturbance of the system whilst the chip is being sheared, followed by the restoring action, could result in serious damage, for example, in machine-tool applications. The other disadvantage is that if it were a steel chip, it is probable that the resulting damage and chipping of the hardened steel metering edges being thus used as shearing blades, would ruin the valve for critical applications.

Hydraulic Forces in Valves

An often simple but relatively important modification to spool-valve design concerns methods of eliminating hydraulic lock. It can be shown that the locking force increases with operating pressure to a maximum value and then decreases to zero as this pressure is exceeded. This effect can be attributed to the presence of taper on the spool glands producing adverse pressure distributions which in turn create significant radial locking forces. The release of hydraulic lock is eventually due to expansion of the sleeve and the compression of the spool as higher operating pressures are applied.

The locking of the valve spool is also sometimes attributed to oil contamination but the latter is more generally recognized for its eventual reduction of the flow rate at small valve openings.

Methods of eliminating or reducing the locking force include the grinding of a small taper or series of steps on the glands, the machining of a series of small circumferential grooves in each gland such that the possibility of circumferential pressure gradients around the glands is virtually eliminated, the use of a small and relatively high frequency input signal such that a small amount of spool 'dither' occurs, and the addition of a spool-rotating device to the valve design. Whilst the first three methods are quite common in practice, the last is seldom encountered.

Valve instability may be due to many different causes both in the design of the valve and the system operating characteristics. Two factors in particular, however, may relate to valve design. The first concerns the 'damping'length of the spool stem, and to avoid the effects of spool instability it is generally essential that the damping length is positive. The second is due to the fluctuations in the reactor forces as the pressure drop across the system varies and can be eliminated by the reduction of these forces to an acceptable level. (See Fig 26).

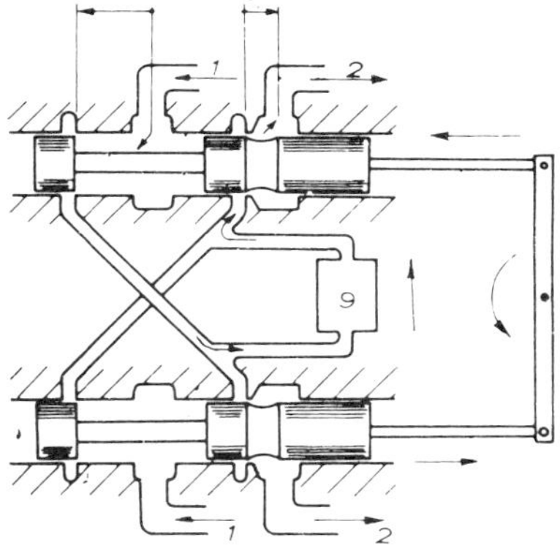

Fig 26 Split-spool valve with positive damping length.
1—Supply. 2—Exhaust. 9—Load.

In addition to friction, forces may also be set up due to the changes in fluid momentum through the valve, generally described as Bernoulli effects or Bernoulli forces. Thus, typically, there may be a reduction in pressure on the valve plunger at the controlling edge, leading to a force being generated producing unbalance or tending to close the valve. At the same time, if backlash is also present in the system, Bernoulli forces may produce high frequency 'chatter' of the valve plunger.

Bernoulli forces are most likely to be marked when flow rates are high. However, various methods to offset Bernoulli effects can be incorporated in the design of the valve itself to provide better balance when the valve is pressurized with flow. Two such examples are shown in Fig 27a and Fig 27b, the latter being the simpler to accommodate from the production point of view, although not quite as effective as the former.

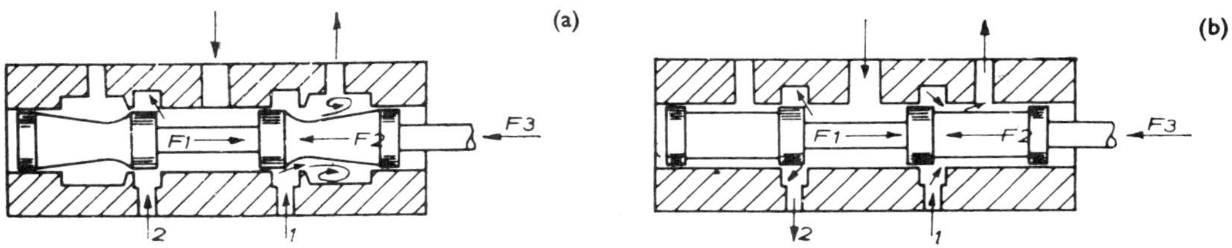

Fig 27(a) and (b) Reaction force compensation.
1—Supply. 2—Exhaust. 3—Input force. (a) Spool and sleeve shaping. (b) Thick stem valve.
F1—Outflow reaction force opposing valve opening.
F2—Inflow reaction force reversed in sign owing to shaping the thick stem such that it opposes F1 and produces force compensation. F3—Input force.

Empirical formulas sometimes used for calculation of Bernoulli force are:

(i) $F = \dfrac{QP}{K_1}$ (ii) $F = K_2 \dfrac{Q^2}{A}$

where F = Bernoulli force tending to close the valve
P = pressure
Q = flow rate
A = area of one metering point

Approximate values of the factors are $K_1 = 50$ and $K_2 = 10^{-3}$ for the F in lbf, P in bars and A in in^2.

SECTION 2C

Reservoirs and Tanks

THE DESCRIPTIONS *reservoir* and *tank* are used synonymously to describe the vessel used in a hydraulic circuit to store fluid and also to accommodate changes in volume in the working part of the system. Fluid volume changes are produced by thermal expansion/contraction of the fluid, and also any unbalanced working volumes in the system — *eg* a hydraulic cylinder works with unequal (unbalanced) volumes unless it is a through-rod type.

Other functions which are performed by a reservoir or tank are:

(i) It is the filling or topping up point for the system.

(ii) Removal of entrained air from the fluid.

(iii) Removal of solid contaminants in the fluid by settlement.

(iv) Fluid cooling (depending on circumstances).

Basic Types

Reservoirs can be divided broadly into two types — vented or sealed. The former act purely as storage tanks, open to atmospheric pressure. Sealed tanks may be pressurized or unpressurized. In the former case they have a certain ability to act as accumulators in addition to 'tank' duties and may in fact combine the duties of both reservoir and accumulator in a single unit where bulk and weight saving is important and 'accumulator performance' demand is relatively light. There are, however, many other reasons for selecting a sealed (normally) unpressurized tank, such as:

(i) On mobile equipment, aircraft, *etc*, where movement could cause spillage from a vented tank.

(ii) To exclude contamination when working in a heavily contaminated environment beyond the ability of a breather/filter to cope with.

(iii) To provide an accumulator effect.

Item (ii) does not necessarily apply in ordinary atmospheres. A vented reservoir is almost invariably fitted with a dust-tight cover to exclude foreign matter, when the fluid filler should be sealed and the vent opening protected with a breather/filter. It has the particular advantage over a sealed reservoir in that it can be made smaller for the same fluid volume. This is because a sealed reservoir requires a fairly generous air volume in order to minimize changes in pressure as the fluid level rises and falls, either due to volumetric displacement or heating or cooling.

Sealed Reservoirs

Sealed reservoirs may work at atmospheric pressure, or be sealed under a normal air pressure of the order of 0.7–1.4 bar (10–20 lb/in^2). A relief valve may be fitted above fluid level to blow off excess pressure above the nominal filling pressure and a vacuum relief valve may also be fitted to protect the reservoir should the pressure fall too low. Considerable care must also be taken to avoid overfilling, since this will reduce the air volume and produce wider changes of pressure during working. The sealed type must, therefore, be considered as a more specialized design. A vented type would normally be used for most applications.

Filler and Venting

All types of reservoir require a filler and a vent to prevent pressure build-up during filling. In some designs this may be replaced by a single filler/breather; or in the case of a sealed reservoir, by incorporating a suitable vent opening in the filler hole. The filler opening itself should be protected for filling. This is merely an additional — and necessary — precaution to prevent any contamination being introduced with the fluid.

Where intake filters are employed these are commonly fitted inside the reservoir. An access opening must then be provided so that filter strainers or elements can be removed for cleaning without draining oil from the tank.

Unsealed tanks are also vented to atmospheric pressure, the vent normally being located on top of the tank. Again the vent is normally covered by a strainer or filter element in the form of an air breather to prevent induction of atmospheric contaminants. The size of this breather is important, for it must be adequate to allow for the discharge of air from the tank at the maximum rate of fluid level rise, and induction of air at maximum fluid level fall. In addition it must allow the escape of entrained air which is released from the fluid surface in the tank. The size of the breather must be sufficient to accommodate all such air flows whilst maintaining atmospheric pressure in the tank. The geometry of the breather must allow ready escape of air, this being particularly important in the case of a sealed reservoir. Inadequate venting during filling in this case can lead to air entrainment in the fluid itself.

Line Connections

Line connections to the reservoir are particularly important. Return pipes must always discharge below the lowest possible fluid level in order to prevent aeration. It may also be possible to direct the return flow to advantage so as to promote circulation and better cooling, *eg* by cutting the end of the pipe at 45° and 'aiming' the flow towards the wall of the reservoir.

The suction pipe is also located as low as possible, but not so low that it can pick up contaminants which may have settled. Further protection in this respect can be given by fitting the suction pipe with a strainer. The actual suction entry position must also be chosen to minimize any vortex action which may be generated in the fluid, and any possibility of drawing air in with the fluid.

It should be noted that the return and suction pipes themselves normally emerge through the top of the tank and are, in fact, top mounted, both pipes running down inside to almost the full depth of the tank. In some designs one (or both) pipes may be bottom mounted, in which case the same considerations for the pipe ends apply — they must be near the bottom of the tank, and never be exposed above or near the surface of the fluid.

A baffle separating the return and suction lines is helpful in restraining agitation and preventing the incoming flow from impinging on the suction filter. Baffles also help stop solid contaminants which may enter with the return flow, causing them to settle on the bottom of the tank rather than circulate from return to suction.

Design

Reservoir design is far from standardized, and even size (volume) is selected largely on arbitrary lines. In general, design tends to follow accepted practice for the application, ranging from making it part of a machine structure to using separate free standing or pump-mounted tanks. Mounting the pump on the reservoir is quite common and in some cases the pump may be submerged in the fluid. JIC (Joint Industry Conference) Hydraulic recommendations are largely followed for vented reservoirs for industrial applications, whilst aircraft reservoirs and other special types may have their own rigid specifications. A reservoir to JIC recommendation is shown in Fig 1 and can be used as a general guide to good practice. At the same time it should be noted that it is generally preferable to design the reservoir as a separate rather than an integral tank since this will improve both accessibility and the ability of the reservoir to act as a cooling radiator. A further point that may be considered is that where the pump is mounted on the reservoir and positioned close to a sensitive machine, the unit may have to be isolated from the machine to prevent pump vibration being transferred to the machine.

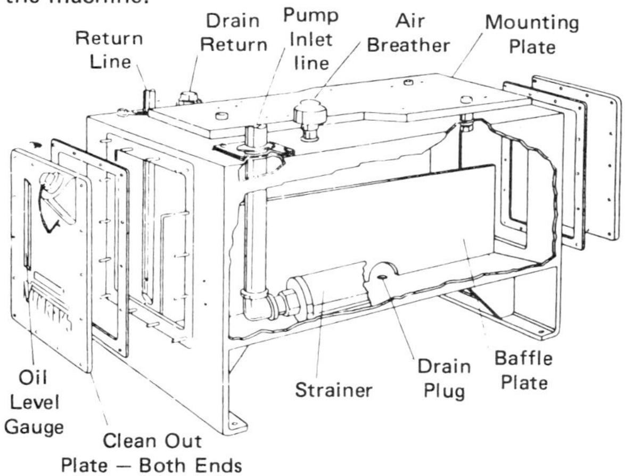

Fig 1

A primary requirement of any reservoir design is that it should be easily cleaned. A drain plug is normally fitted to the bottom for draining the reservoir completely. Removable end covers or side plates are also desirable for access to the interior to clean out sludge and attend to the suction line filter or strainer periodically. In some cases provision is made to remove the suction line completely through the top of the tank so that the strainer can be attended to without having to drain the tank and remove any clean-out covers.

Baffles

In addition to separating the incoming and outgoing flows, baffles can serve a number of other useful purposes. Simple reservoirs are normally designed with a single baffle extending the length of the tank and approximately three-quarters of its height (as in Fig 1). Many variations are, however, possible in both the number and design of baffles used.

A series of baffles may be preferred to minimize surging of the fluid when the reservoir is used with mobile equipment. In this case the baffles would normally be arranged at right angles to the direction of the 'surge' motion, virtually dividing the tank into a number of smaller volumes with small interflow. An alternative is to arrange the baffles in 'maze' configuration so that end-to-end (or side-to-side) flow is tortuous and thus heavily damped during surge motions.

A somewhat similar baffle treatment may be adopted, with return and suction lines separated as far as possible, to provide a long crossflow path. This will generally improve cooling (particularly if the flow can be directed over the tank surface to the fullest extent) and also assist in the release of entrained air. Such a solution may therefore be adopted where maximum cooling effect is required from the reservoir or air entrainment is likely to be a problem. Note, however, that in the former case maximum cooling will only be achieved if the tank's sides and bottom are free standing and can radiate heat readily; in the latter case problems of air entrainment may be due to bad reservoir design and pipe placement in the first place.

The use of horizontal baffles can also be helpful in removing entrained air or air bubbles. In this case the baffle normally takes the form of a fine mesh wire screen mounted at an acute angle, but with its highest point below the lowest fluid level. Such a screen can be effective in constraining bubbles on the inlet side of the tank; the air thus trapped subsequently escapes *via* the surface of the fluid. Horizontal wire mesh baffles are also effective in breaking up droplets in water-emulsion type fluids and avoiding separation of water in the bottom of the reservoir. Conventional baffles may not be desirable in reservoirs where such fluids are used because they tend to collect water droplets. With water-emulsions, strong circulation in the reservoir is usually more important than damped or controlled internal flow, since this will inhibit separating of the water content.

Construction

The majority of industrial type (vented) reservoirs are of mild steel with welded joints. Material thickness chosen is often arbitrary, but 3 mm (1/8 in) thickness is a recommended minimum. After fabrication the interior should be cleaned by shot blasting to remove scale and surface corrosion. Both the inside and outside of the tank should then be painted with a suitable resistant finish. The choice of a suitable paint may be strictly limited where the reservoir is for use with synthetic fluids, and one finished with an oil-resistant paint would be quite unsuitable for such services.

A fluid level indicator should be part of every reservoir design. This may simply take the form of a transparent window to act as a fluid level sight or an external sight tube. More sophisticated level indicators may be fitted to some reservoirs, or a simple dipstick marked with high and low levels. If a dipstick is fitted, this should be of screw-type, sealing on a gasket. Dipsticks are the most common type of level indicators fitted to sealed reservoirs where maintenance of correct fluid level is more critical, although windows are equally suitable.

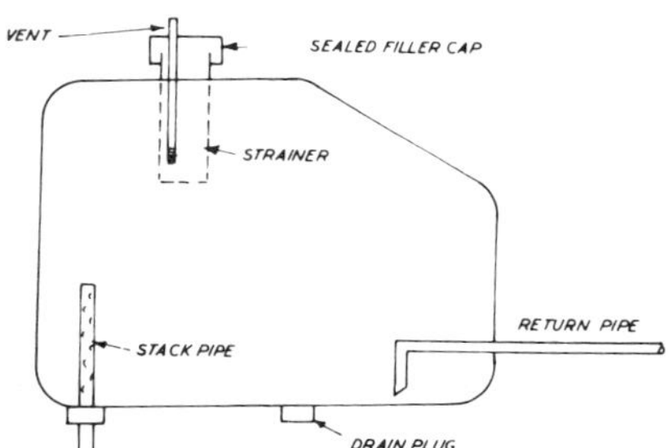

Fig 2

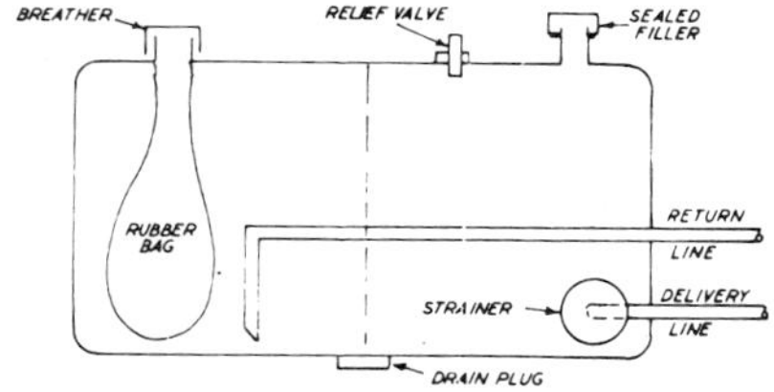

Fig 3

Problems with Sealed Reservoirs

With a pressurized sealed reservoir, direct pressurization of the air above the fluid in a simple tank, as in Fig 2, can lead to super-saturation of the fluid with dissolved air, which is subsequently released in some other part of the system. Nevertheless, such simple pressurized tanks have proved suitable for a wide variety of services, including aircraft installations. To reduce the size of the tank and still keep sufficient volume of air to produce a satisfactory compression ratio for minimizing pressure difference between high and low levels, the top of the tank may be connected to a separate air chamber, or even to an external air supply with a reservoir relief valve.

The problem of deciding on a suitable compression ratio and air:fluid volume is exactly the same as that for accumulators – see chapter on *Accumulators.*

An alternative solution, now generally preferred, is to fit a sealed reservoir with a 'breathing bag' to accommodate changes in fluid volume. The design of the reservoir is then essentially similar to that of a bladder-type accumulator, although nothing like as strong a construction is required. The air volume will be only lightly pressurized (0.7 to 1.4 bar, 10 to 20 lb/in^2), or the bag open to atmospheric pressure. The use of such a design with physical separation of air and fluid also eliminates the question of super-saturation of the fluid under pressurization. The elements of such a design are shown in Fig 3 where it will be seen that the fluid side of the reservoir is still capable of venting off entrained air, and collecting sediment in the bottom of the tank.

Aircraft Reservoirs

Reservoirs for aircraft systems pose special problems – a major one being the loss of atmospheric pressure with altitude. The pump cannot produce a suction pressure greater than the barometric pressure in an open tank, and thus a sealed tank is virtually essential for proper working at altitude (*ie* at an altitude where the absolute pressure required at the pump inlet, plus the pressure drop between reservoir and pump, exceeds the atmospheric pressure). The pressure drop figure can be a critical factor, for fluid temperature may be very low at altitude. A pressurized reservoir can thus become essential to avoid pump cavitation.

In the case of aerobatic aircraft feed must be provided by the reservoir at all attitudes of the aircraft. The simplest solution in such cases is the use of a 'klunk' tank where the delivery pipe enters roughly at the middle of the tank and terminates in a flexible pipe with a weighted end. Gravitational and/or acceleration forces displacing the oil content of the tank will similarly displace the weighted end of the pipe, so that the end always remains in the fluid – Fig 4. A particular advantage of this type of tank is that it usually permits a smaller size of pressurized reservoir to be used on aircraft, with lower compression ratio.

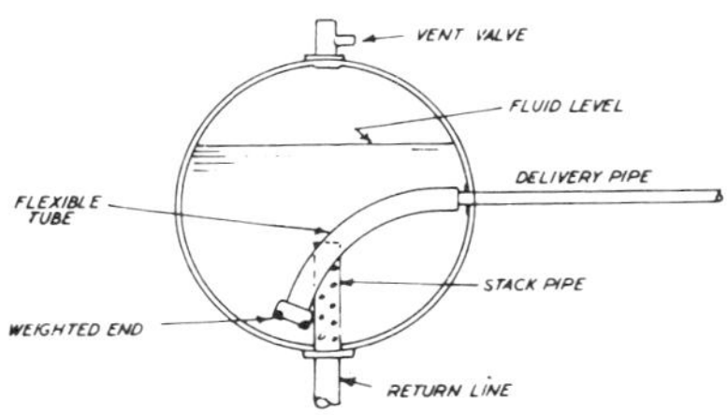

Fig 4

The same principle can be adopted for hydraulic circuit reservoirs on industrial machines which have extensive two- or three-phase motions. In this case the klunk tank principle can be used with both open-to-air (unpressurized) and pressurized reservoirs.

A further alternative, both for aircraft and industrial hydraulics, is the use of a spherical tank with stack pipe suction and return, and sufficient capacity to ensure that the reservoir is always more than half full of fluid under all conditions in order to give uniform coverage of the pipe entry.

Accumulators

A HYDRAULIC accumulator is a device designed specifically for the storage of liquid under pressure. At the same time it can be effectively used for pump control (usually in special designs), shock absorption, surge damping, volume compensation, *etc.* The various types in use are described under separate headings.

Non-Separated Gas-Loaded Accumulators

Non-separated gas-loaded accumulators take the form of a cylindrical pressure vessel or air bottle mounted vertically, with a fluid port at one end and a gas port at the other. Fluid is first introduced into the vessel and then pressurized by a precharge of gas to the required minimum pressure. Further fluid can then be pumped in, compressing the gas and increasing the pressure. Because of the high compressibility of gases, a large volume of fluid can be accommodated, capable of being delivered from the accumulator above the minimum pressure specified (Fig 1).

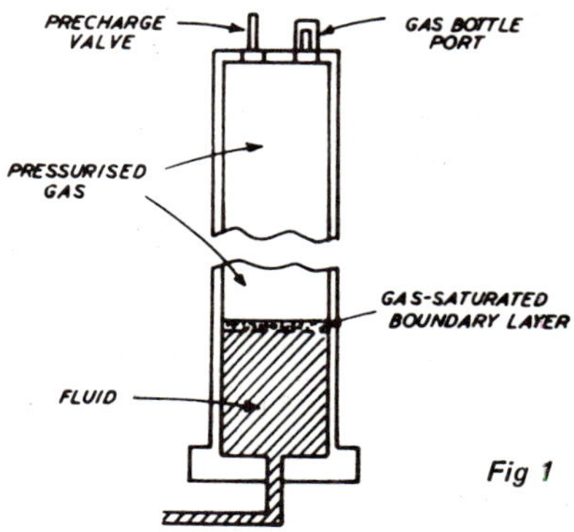

Fig 1

The particular disadvantage offered by such a simple system is the mixture of fluid and gas at the interface. This limits the amount of fluid which can be drawn off without incurring the danger of drawing off gas into the hydraulic system. Also the cylinder must be mounted vertically. On the other hand such accumulators are simple and comparatively inexpensive and well suited to handling large volumes of fluid; and thus find widespread application. Storage volume can be readily increased by multiple installations.

The shape of the pressure vessel is invariably tall and narrow so that the contact area between gas and fluid is small. Even so, probably not more than two thirds of the fluid volume can be used without the danger of gas being drawn out into the hydraulic circuit. The pressure vessel may be

Acclaimed Accumulators

Acclaimed worldwide for our hydraulic accumulators, which have an international distributorship.

Acclaimed for our guaranteed quick response to your every need, whether off-the-shelf or tailor-made.

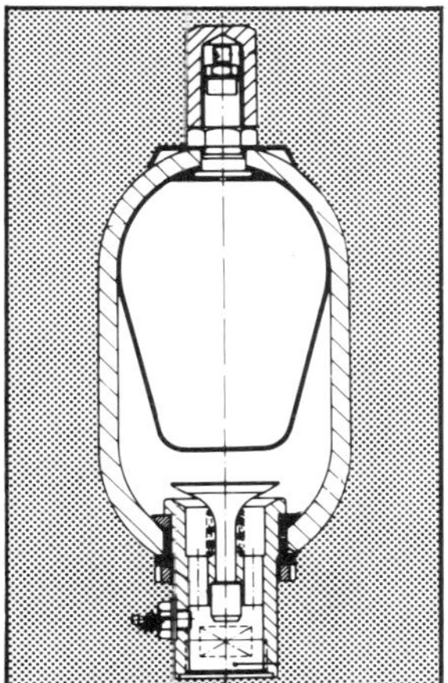

Christie diaphragm long-life accumulators
0.075-5 litres, up to 330 bar

Christie piston minimum leakage accumulators
1-120 litres, up to 690 bar

Christie shock alleviators
Low pressure (30 bar), large volume (10 - 216 litres)

Christie bladder accumulators
1-54 litres, 207 and 345 bar

Christie 101
Simple, inexpensive, non-repairable design

Christie Unitest digital hydraulic test & monitoring unit
Monitors pressure, flow or temperature simultaneously

Christie high frequency pulse attenuator
Reduces hydraulic ripple and fluid borne noise

Accumulator safety block
For extra safety. 3 sizes.

Pronal flexible separators
Protect hydraulic system from containment

Dunlop nitrogen booster set
Reduces nitrogen cylinders required for pre-charging hydro-pneumatic accumulators to high pressures.

Stainless steel accumulators
Available in most ranges

Accumulator station
A group of accumulators piped up on a stand

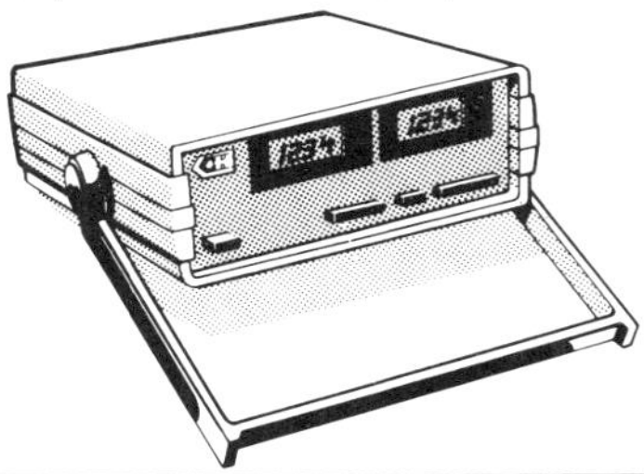

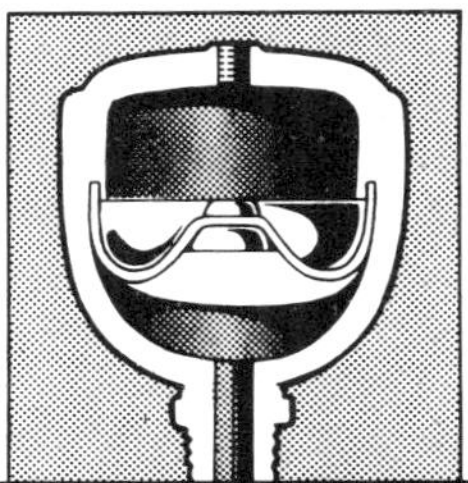

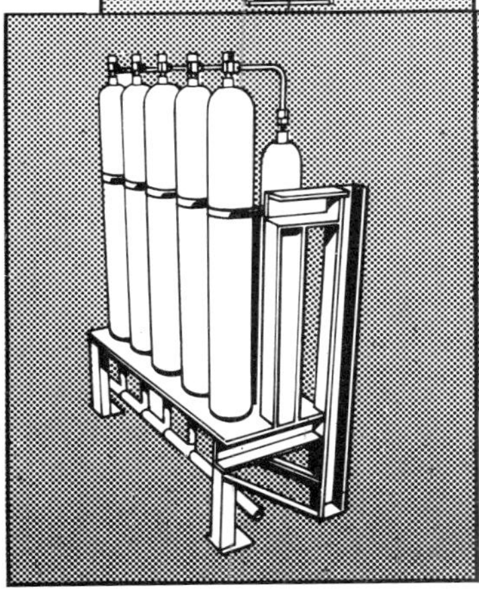

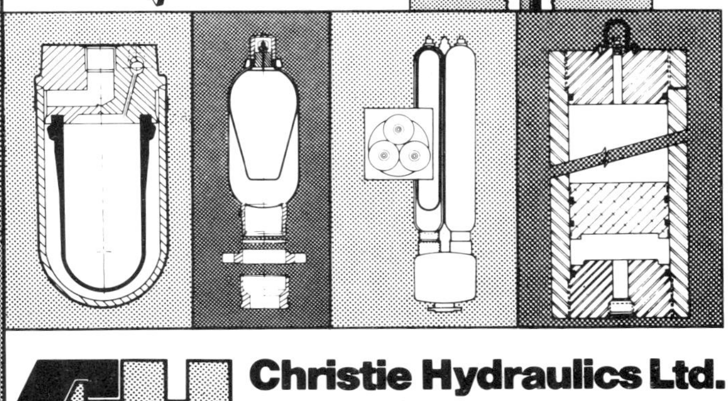

Christie Hydraulics Ltd.
Sandycroft Industrial Estate, Chester Road,
Sandycroft, Deeside, Clwyd CH5 2QP
Tel: 0244 535515 Telex: 617026

connected directly to a high-pressure air compressor operating against a level regulator, and with pressure regulator valves and controls.

Initially the accumulator is partially filled with fluid and charged with compressed gas at a pressure corresponding to the nominal hydraulic pressure required. If fluid is withdrawn on demand the gas expands, causing a slight reduction in pressure. If further fluid is introduced the precharge of gas is compressed, increasing the fluid pressure. Thus the degree of pressure variation is controlled by the ratio of the gas volume to the fluid volume; if necessary the gas volume can be increased (and pressure variation decreased) by introducing a further supply of gas. Additionally, capacity can be increased by adding further accumulators in parallel. The system is particularly flexible, both as regards control of pressure variation and capacity available.

In normal practice a pressure variation of 10% maximum is generally considered satisfactory. Volume ratios are normally calculated on the basis of isothermal conditions, when a 100 to 90 variation in pressure between high and low fluid levels is equivalent to a gas volume increase of from 9 to 10, or a gas:liquid ratio of 9:1. A rather higher ratio (10:1 or 11:1) would normally be adopted, however, to allow for a reserve of fluid, *ie* a sufficient fluid volume in the bottom of the liquid bottle to ensure that the gas-rich boundary layer is maintained at a minimum working height well above the fluid port. An equally important factor is the optimum positioning of the fluid port to avoid any possibility of vortex formation at low fluid levels which could allow gas to be drawn out through the liquid port.

To avoid the bulky design necessary to accommodate a 10:1 or higher gas:fluid volume, a more suitable arrangement is to employ a separate liquid bottle with a high fluid content connected to additional gas bottles – Fig 2. This diagram also shows the controls necessary in diagrammatic form. The low level control on the accumulator is made to cut in whilst there is still an adequate volume of fluid remaining, and normally before the design low level is reached. This allows a time safety margin for the switchgear to shut down the accumulator or bring the pump to maximum output before the actual design low level is reached. A stop valve is also needed to prevent complete emptying in the event of exceptionally heavy demand or failure of the fluid pump.

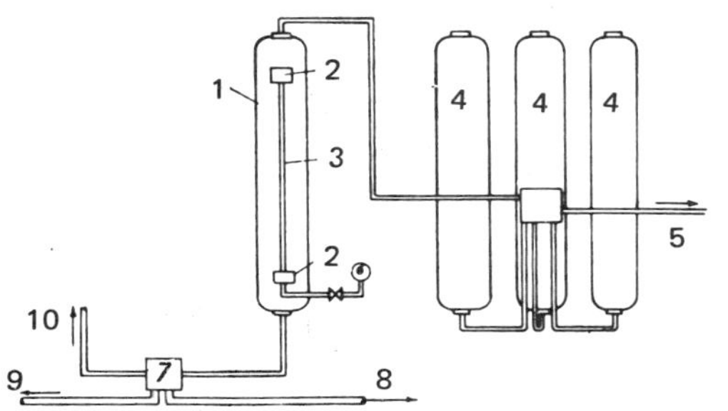

Fig 2 Typical circuit for non-separated gas-loaded accumulator. 1–Liquid bottle. 2–Isolating valve. 3–Control tube. 4–Gas bottles. 5–To compressor. 6–Pressure gauge. 7–Automatic stop valve. 8–To working circuit. 9–To pump(s). 10–To isolating valve and servo-valve(s).

Controls are normally automatic, and could include monitoring of the actual fluid level inside the liquid bottle. This could be by continuous pressure measurement; or by floats coupled indirectly to external indicators; or electronically.

Once precharged, no gas is actually used up during cycling, other than small amounts that may be lost by absorption. Thus no periodic topping up of the precharge is necessary, unless there is a definite discharge of the gas into the fluid circuit through operational error or malfunction of the accumulator.

Non-separated accumulators of this type for general industrial applications cover pressure ranges from 35 bar (500 lb/in^2) to 420 bar (6 000 lb/in^2). A typical installation would comprise a single liquid bottle with three auxiliary gas bottles of similar size and the control system already shown in Fig 2.

Float-Type Accumulators

Float-type accumulators were developed to overcome the interface mixing of gas and fluid inherent in non-separated accumulators by reducing the area of fluid in contact with the gas. Construction is considerably more complex (see Fig 3), but such units find application in very large installations where the cost of making a mould tool for the manufacture of a large membrane to separate the gas from the liquid is prohibitive, and where the cost of honing the bore of a large pressure vessel in order to run a seal piston separator in it is equally prohibitive. They are also still used in installations which cannot easily obtain spare parts, *eg* for blow-out preventer standby hydraulic energy on oil rigs at sea; and systems in less developed semi-industrial countries where seals and rubber technology are not readily available.

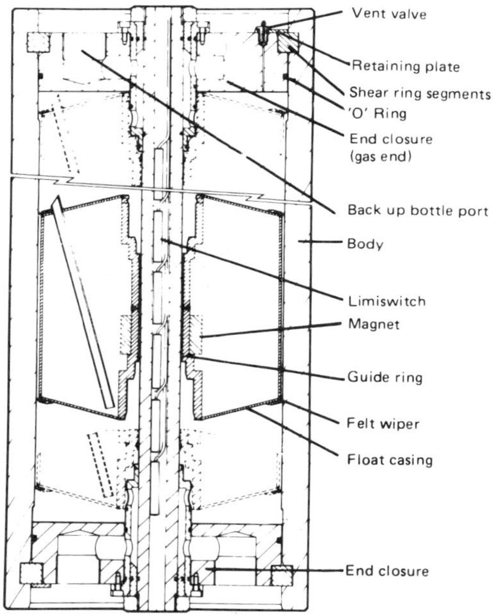

Fig 3 Float type accumulator.

Separator-Type Accumulator

Separator-type accumulators employ a similar principle of accommodating a pressurized (compressible) gas and (incompressible) hydraulic fluid in a single pressure vessel, but in this case gas and fluid are separated by a physical barrier. Thus no gas-saturated boundary layer is involved and the capacity of such an accumulator is virtually its full fluid volume. They are the most common types of accumulator in use today, particularly as they can be rendered in compact form.

Separator type accumulators can be broadly classified as having either flexible or rigid separators. Flexible separators include, bags, diaphragms, bladders and bellows constraining the gas charges and mounted within a suitable pressure vessel. Rigid separators take the form of a piston mounted in a cylinder.

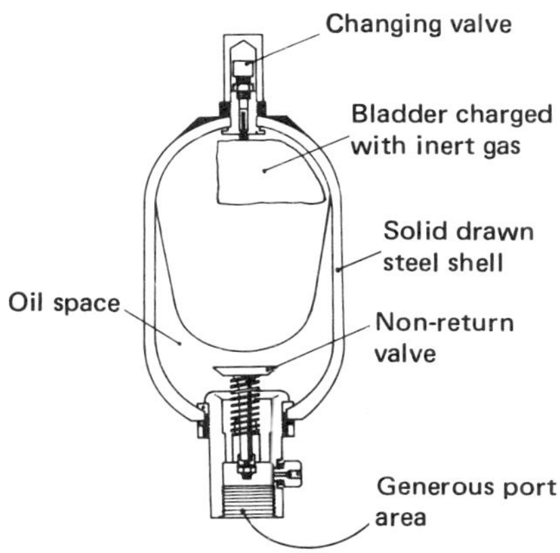

Fig 4(a) Bladder-type accumulator.

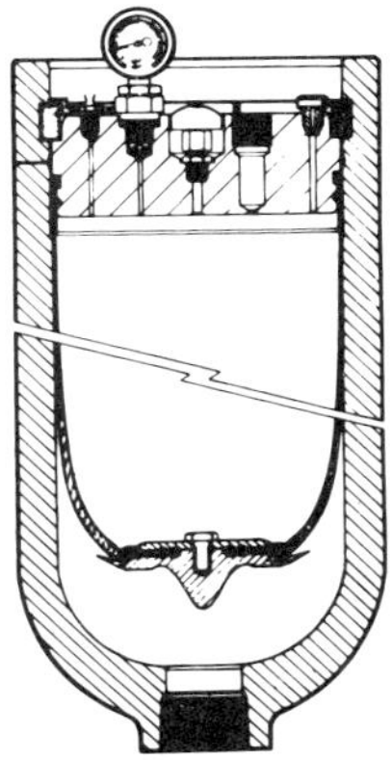

Fig 4(b) 'Hydrostor' open bag type accumulator.

Bag-Type Accumulators

Improved techniques in fabricating gas-containing bags by bonding a number of moulded components together have resulted in bag-type accumulators becoming the most popular type. A typical design is shown in Fig 4. It consists of an outer pressure vessel, normally cylindrical in shape with hemispherical ends, (but sometimes a plain cylinder), and with drawn or forged construction to eliminate welding or mechanical joints. Inserted into this pressure vessel is a bag of elastomeric material, chosen for compatibility with the hydraulic fluid being used (normally nitrile rubber for general use). The bag is drawn into position during assembly through the wide opening which accommodates the poppet valve. The bag and shell are assembled by means of a high-pressure valve moulded integral with the bag and clamped in place on the top of the shell with an external nut.

The oil port is assembled in the other end of the shell, the joint commonly being closed with an O-ring and the design adjusted so that the lower mouth of the shell will spread at a pressure below the design pressure of the shell as a safety precaution. The valve opening is large to allow an unrestricted flow of oil. The large opening also allows the bag to be removed for inspection or replacement should this be necessary. The poppet type valve provides high volumetric efficiency and also serves to prevent extrusion of the inflated bag when the fluid side is depressurized, or should all the fluid be drawn off. Other safety factors normally include some safeguard to prevent removal of the fluid discharge plug while there is any pressure remaining inside the bag.

In use, the bag is pressurized (with air or nitrogen) to the specified precharge pressure and fluid is pumped into the main chamber to compress the bag. The gas precharge pressure is invariably much lower than the fluid pressure and compression ratios of up to 5:1 may be achieved, according to the particular requirements of the installations. Nominal maximum working pressure with this type of accumulator is commonly 210 bar (3 000 lb/in^2), although the same type can also be designed for higher rated working pressures, for example, 350 bar or even 420 bar (5 000 lb/in^2 or 6 000 lb/in^2).

The flexible bag is normally pear-shaped or similarly tapered, this form giving optimum pressure distribution. Some modification of optimum shape may, however, be dictated by the material used and the method of construction. Bag failure is unusual with modern designs, although this can happen at a dangerously low level. This type of failure can occur where an external gas bottle is used and the total gas volume falls appreciably due to a considerable drop in temperature (as may happen over-night).

Bag type accumulators are particularly suitable for use with oil fluids, but can also accommodate other types of fluid provided that the bag material is compatible. Used with water or water-fluids it is generally necessary to pre-treat the steel shell to prevent corrosion. Stove-enamelled interior finish for the shell is a typical treatment.

In the use of a bag type accumulator it must always be remembered that if the bag is punctured, the gas loss is likely to be sudden and total. For this reason, in order to prevent the bag being damaged by contacting the bag anti-extrusion mechanism, the flow rate capabilities of this design are severely limited. If used with gas bottles to increase capacity, it must be remembered that in order to prevent the bag from being forced into the piping system which runs to the bottles, it is necessary to reduce the ratio between the volume of the accumulator and the total volume of the gas bottles. Additionally, to prevent bag damage caused by violent changes of gas temperature entering the bag through small apertures or sintered plugs, the flow rate of bag type accumulators connected to gas bottles must be kept to the minimum. It may even be necessary to increase the number of bag type accumulators connected to the gas bottles. As a result gas-bag type accumulators of the closed-bag type are often used very largely to house gas rather than to accumulate liquid.

Bag-type accumulators are the most versatile of all types and are equally effective for energy storage, shock absorption and 'holding', and 'reservoir' duties (fluid leakage make-up and fluid volume compensation due to temperature changes). They also provide effective damping of pump pulsations and are widely used for such other duties as hydro-pneumatic springs, pressurized fluid dispensers and transfer barriers between two fluids.

Such devices are frequently called open mouth units. Apart from being manufacturable in bag materials which are glueable, they also have servicing and inspection benefits as the complete cross section of the tube is laid open by the removal of a header plug. This header plug is also used to seal the open mouth of the bag and because the bag has an open mouth the internal shape mould former is easily removed from the moulding, thus allowing the bag to be produced in a single operation. The header plug of the assembly therefore provides a surface through which charging valves, venting valves, gauge installation with isolation valves and gas back-up bottle connections can easily be introduced in conjunction with safety over-load vent gas depressurization rupture discs. In general, these units are more versatile both in temperature, liquid, and ancillary equipment terms.

Diaphragm Accumulators

The original form of diaphragm accumulator was a spherical vessel accommodating a moulded flexible membrane separating the chamber into two numerically equal volumes – Fig 5. Today diaphragm designs and materials may vary considerably. The favoured construction is a convoluted form in synthetic rubber, the convolutions providing minimum creasing with maximum flexibility of movement, and, in particular, maintenance of flexibility at lower temperatures where elastomeric materials tend to harden. A rather simpler form of diaphragm is shown in Fig 6.

The gas volume is charged at a lower pressure than the fluid volume. When the accumulator is initially charged with gas to the required pressure the diaphragm is fully flexed with the gas occupying the full volume of the accumulator. Fluid is then pumped into the high-pressure side,

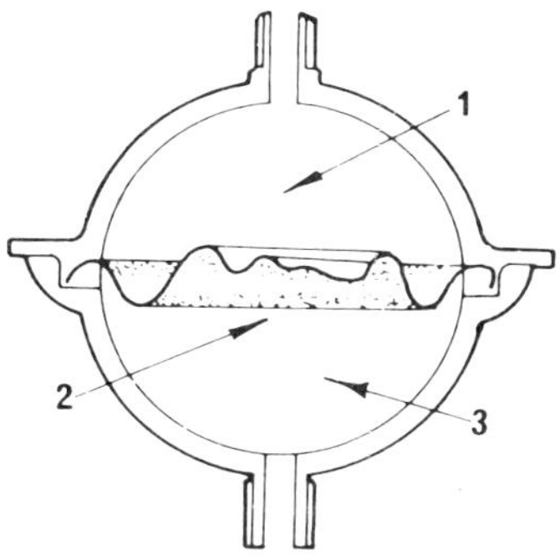

Fig 5 Diaphragm type accumulator. 1—Oil chamber. 2—Diaphragm. 3—Air chamber.

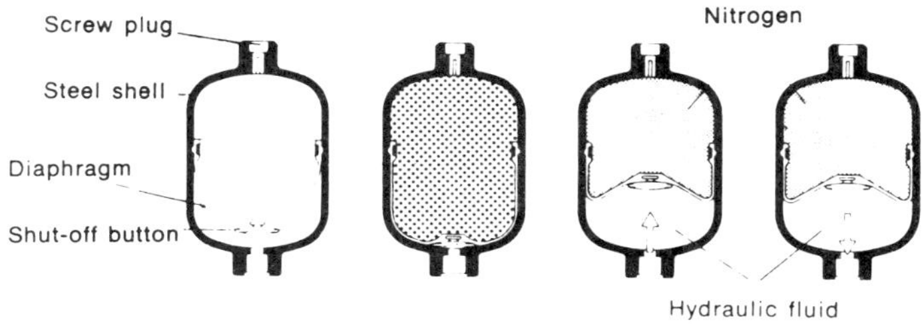

Fig 6 Bosch diaphragm-type accumulator

compressing the gas and establishing a balance with equal pressure on each side of the diaphragm, and therefore no stress (other than pure compression) is placed on the diaphragm material.

Under normal operation the pump will build up gas pressure in the accumulator from discharging to the reservoir. A safety device would also be incorporated to ensure that should the fluid chamber become completely exhausted the diaphragm cannot extrude through the fluid port under gas pressure.

Accumulators of this type are compact and light, but best suited to systems where demand is intermittent and the fluid volumes required from the accumulator flow are not very large, as this type of accumulator is not normally made in large sizes.

Piston-Type Accumulators

A basic design of piston-type accumulator is shown in Fig 7. Normally the piston is 'free', but in some cases it may be connected to a conventional piston rod. Design and construction are relatively straightforward and the type can be made in a wide range of sizes. Cost is, however, relatively high. It is particularly suited for high-pressure systems since cylinder stresses are readily determined and standard hydraulic quality cylinder tubes can be employed for the barrels. Construction can follow that of hydraulic cylinders, with tie-rod assembly for higher pressures or heavy duties. However, it is necessary to make provision to prevent disassembly of the end covers when either the fluid or gas side of the accumulator is under pressure.

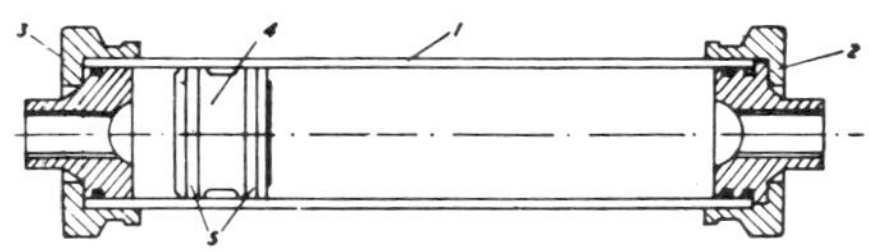

Fig 7 free piston type accumulator. 1—Cylinder. 2—Fluid head. 3—Gas head. 4—Piston. 5—Piston seals.

The main disadvantages of piston-type accumulators are associated with the piston seal, the most difficult task being that of maintaining an adequate seal when the fluid system is shut down with the gas end still under pressure. A number of designs include a liquid seal for additional protection.

With a liquid seal the fluid port is closed by a probe attached to the piston entering the end cover at the end of its stroke and trapping a certain amount of fluid in the liquid end of the cylinder. This amount of fluid is pressurized by gas pressure on the piston, but the area on which it acts is less than the piston area. As a consequence the pressure generated on the trapped fluid is greater than the gas pressure, preventing gas leakage into the liquid end. Liquid seals also prevent hydraulic shock in the event of the fluid content of the accumulator being fully discharged.

An alternative method of 'cushioning' the end travel of the piston is shown in Fig 8. This is a dashpot-type piston-accumulator, where the fluid side of the piston carries an extension of reduced diameter entering a cushion chamber in the cylinder end. Should the fluid level fall to such an extent that the piston nose enters the cushion chamber, dashpot damping is provided over the remainder of the stroke.

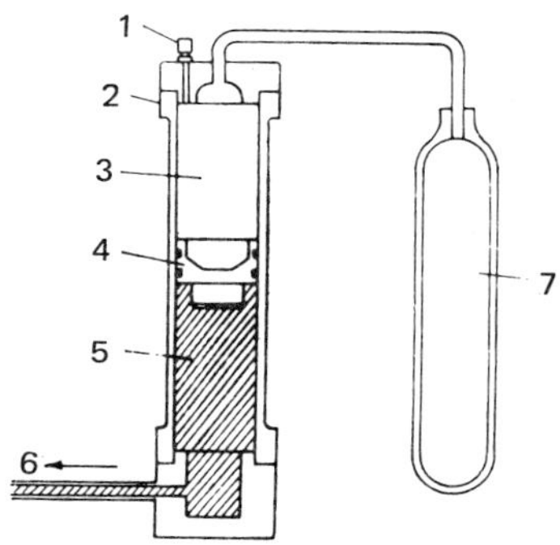

Fig 8 Dashpot piston accumulator with auxiliary gas bottle.
1—Gas precharge valve. 2—Cylinder. 3—Gas under pressure. 4—Piston. 5—Fluid. 6—Fluid discharge to system. 7—Auxiliary gas bottle.

Solid piston accumulators are used where any sudden gas loss would be catastrophic. They are nearly always used in conjunction with back-up bottles. The response time of such devices is frequently as fast as 3 milliseconds. Any gas loss from these units will be relatively easy to determine and it is therefore usually possible to plan their maintenance and the replacement of seals to coincide with machine maintenance intervals.

Virtually all piston-type accumulators suffer to some extent from gas leakage which may develop in use since no effective seals can be entirely free from wear. Such conditions are aggravated by contamination of the fluid, or corrosion, which could affect the bore finish or seal material. Periodic topping up of the gas charge is therefore normally necessary in order to maintain a minimum working pressure.

Fluid capacity (and thus size of cylinder) can be determined from the following:

$$V_2 = \frac{P_1 V_1}{P_2} - \frac{P_1 V_1}{P_3}$$

where V_1 = Total gas volume
V_2 = Fluid volume
P_1 = Initial precharge pressure (gas)
P_2 = Minimum working pressure
P_3 = Maximum working pressure
$(P_3 - P_2)$ = Maximum permissible pressure variation

For continuous operation, P_1 should be taken as near as possible to P_2. The actual pressure difference over the working range $(P_3 - P_2)$ can be adjusted by varying the compression ratio.

Where this would require the use of an excessively large cylinder a separate gas bottle can be used to provide the additional gas volume. In this case the gas end of the cylinder would have to be designed with a large-size port in addition to the charging valve for the precharge.

Piston-type accumulators are useful for handling special duty fluids which may attack conventional low-cost elastomers used with flexible separator accumulators. They would normally be vertically-mounted, but horizontal or angular mounting is not necessarily excluded. They are not so suitable as other types for shock-absorbing duties, due to the inertia of the piston and the friction of the piston seals.

Some solid-piston accumulators are designed with a tail rod which indicates the fluid volume stored at any time. These are known as *indicating piston accumulators.*

This is an advantage which can rarely be obtained from a membrane type unit, and even then, not accurately. In addition, by knowledge of the original precharge pressure and the use of a gauge connected to the gas side of the unit, it is possible by correlating the position of the indicator rod to the pressure shown on the precharge gauge to ascertain that the unit has not lost any gas precharge. This may be done at any time *ie* it is not necessary totally to discharge the unit of liquid before checking the precharge pressure to ascertain whether any of it has been lost.

The tail rod is frequently used to actually cut in or out of circuit one or more pumps and this can be done either non-electrically by directly tripping the unloading valve, or by tripping switches. The result is that it is not necessary to have more than about 6.3 bar (5 lb/in^2) between one pump cut-in and the next or between pump cut-in and cut-out, even when working at 350 bar (5 000 lb/in^2) when these indicating-type units are connected to gas bottles. This amount of control is extremely difficult to obtain even with the use of the most sensitive pressure switches, and it is this degree of controllability which has caused these units to be classified as 'pump controllers' rather than accumulators.

A tandem type piston accumulator is shown in Fig 9 and is used for special duties. It is also called a self-displacing accumulator and comprises an accumulator combined with a pressurized reservoir; it is thus capable of maintaining a constant volume of active fluid in the hydraulic circuit. The gas precharge displaces the tandem piston to fill the low pressure cylinder with fluid. When the system is pressurized, the high pressure (hydraulic) side of the accumulator is filled with fluid and the gas compressed. The fluid to fill the high-pressure side is drawn from the low-pressure side. With the system working, any fluid withdrawn from the high-pressure side is simultaneously replaced on the low-pressure side, thus maintaining a constant volume of fluid both in the system and the accumulator.

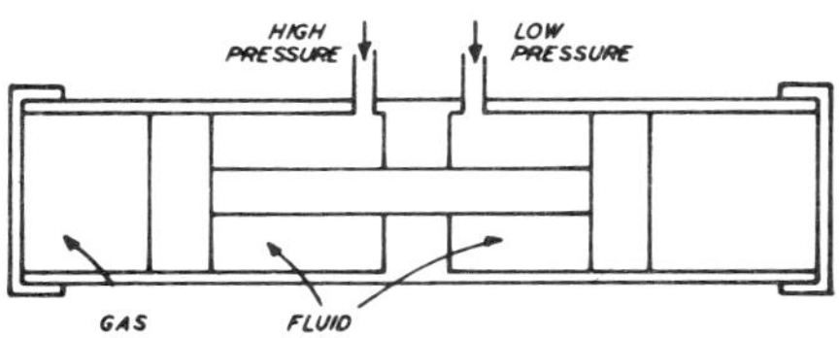

Fig 9

Annular-Piston Accumulator

An annular-piston accumulator is of conventional construction embodying a cylinder and liner, the piston moving within the liner and the outer annular volume providing additional gas volume – Fig 10. This has the advantage of providing a substantially larger gas volume without increasing the length of the cylinder or employing a separate gas bottle. Constructionally it has the advantage that the liner can be relatively thin since the gas pressure on the outside is equal to the fluid pressure on the inside, and they thus offset each other.

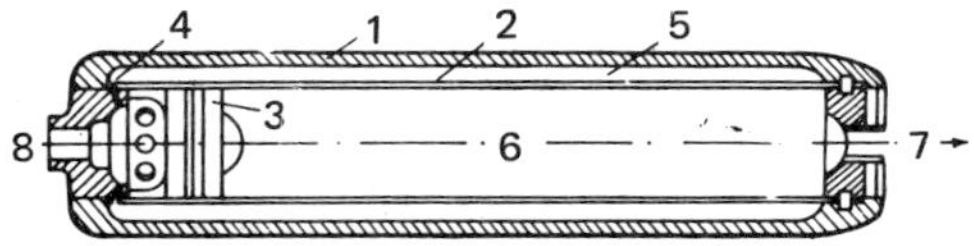

Fig 10 Annular type piston accumulator. 1—Cylinder. 2—Cylinder liner. 3—Piston. 4—Gas ports. 5—Gas chamber. 6—Fluid chamber. 7—Fluid discharge. 8—Gas precharge.

Fig 11 Membrane-cum-piston type accumulator.

Membrane-Piston Accumulators

A membrane-piston accumulator incorporates a hollow piston constructed like a diaphragm — Fig 11. This is a stronger form of separator than a diaphragm or membrane and so such units can be designed to accommodate flow rates up to four times as great as bag-type accumulators. The diaphragm within the piston also gives an advantage over pure solid piston type units because the dynamic seals on the piston do not have to move for every minute pulse or fluctuation in the system caused by pump- or servo-oscillation. This advantage enables its dynamic seals to outlast those of a pure piston type accumulator many times over.

The arrangement is such that the ports to the chamber are on a pitch circle diameter and the ports to the diaphragm inside the piston are on the centre line. This makes it impossible for the diaphragm to escape down the ports of the system and as the piston moves at a much lower pressure differential than the pressure required to extrude the diaphragm the piston itself acts like the anti-extrusion valve at either end of stroke. As a result of this arrangement there is no anti-extrusion valve and in consequence the diaphragm's sensitivity, unhindered by an anti-extrusion valve, is capable of dealing with high frequencies from 10 Hz to 1 000 Hz. This combination piston-cum-diaphragm unit is also used where the whole of any additional gas volume is required to be stored in additional gas bottles.

Weight-Loaded Accumulators

The basic design of a weight-loaded accumulator is shown in Fig 12. A heavy walled cylinder is mounted vertically on a substantial base and carries a ram. A cross-head is attached to the top of the ram, from which is slung a weight box. This is filled with any high-density waste, such as ballast, iron scrap, concrete, *etc.* Alternatively, in the case of smaller units, specially made weights may be slung from the ends of the crosshead.

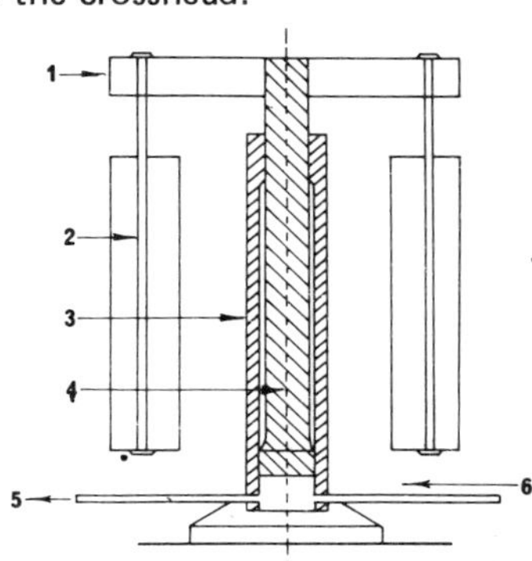

Fig 12 Weight-loaded accumulator (diagrammatic). 1—Crosshead. 2—Weights or weight box. 3—Cylinder. 4—Ram. 5—Service line. 6—Pump.

There are two main types, depending on the method of constraining the weights or weight box. On a self-guided design the weight case is provided with internal guides. On externally-guided designs the weight case is constrained against radial movement by external guides or channels, usually mounted on a steel structure. The latter type is normally preferred to large high-pressure accumulators to minimize bending stresses.

The ram is raised by pumping fluid into the cylinder. Once raised, the fluid in the cylinder is pressurized by the combination of the weights and ram acting on the cross-sectional area of the fluid column.

Full theoretical hydrostatic pressure should be available from a weight-loaded accumulator, less a nominal allowance for seal friction. However pressure variations are likely to occur with differences or variations in falling speed. Thus a pressure variation of 5% is likely to be experienced with a maximum falling speed of 0.3 metres per second (1 foot per second), but may be higher with higher falling speeds. Momentary peak pressures may also be higher or lower than the nominal pressure by an appreciable amount, depending on the rate of deceleration or acceleration of the ram, respectively.

Falling speed can be controlled by the stroke/bore ratio of the ram. A stroke/bore ratio of between 10 and 15 is commonly adopted for accumulators working up to 105 bar (1 500 lb/in^2), although higher ratios are generally to be preferred for higher pressures. This, however, increases the problem of obtaining mechanical rigidity and also increases the overall height of the accumulator. This could make it unsuitable for indoor installation. As a rough guide, the overall height of a weight-loaded accumulator is at least twice the stroke.

Cast iron cylinders are commonly employed for accumulators working up to 105 bar (1 500 lb/in^2). Cast steel or forged steel cylinders are used for higher pressures. Honed bores are required, although satisfactory performance may be obtained with rougher bores using leather seals. Rams may be made from cast iron (the original choice and still widely employed), but preferably chrome-plated. Stainless steel or alloy steel rams are more usual on smaller sizes of modern weight-loaded accumulators.

Weight-loaded accumulators continue to be used to meet heavy industrial requirements and large units usually employ water as the fluid. The large weight-loaded accumulator offers the advantage of extremely high capacity at relatively low cost per unit volume. Construction is rugged and durable, and the units are capable of accommodating shock loads. Only simple control gear is necessary.

The disadvantages of a weight-loaded accumulator are:

(i) The accumulator is extremely bulky and heavy and thus could not be considered where space- or weight-saving is an important factor.

(ii) Pressure output is not constant, largely due to the effects of seal friction and inertia.

(iii) Certain restrictions are imposed on delivery, largely due to limitations on falling speed to minimize hydraulic shock.

(iv) The seals themselves may pose problems, both in providing adequate sealing with low friction when they are used with such a low-viscosity fluid as water, and when expected to give long seal life. Where such an accumulator is used as a central source, failure of the seals would result in loss of supply to all the hydraulic machines on the circuit.

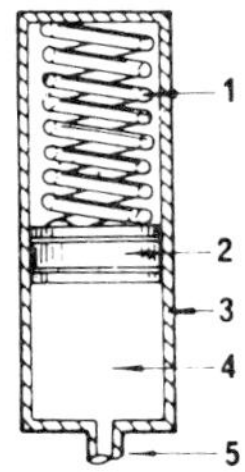

Fig 13 Spring-loaded accumulator. (diagrammatic).
1—Spring. 2—Piston. 3—Cylinder. 4—Oil reservoir. 5—Oil port.

Spring-Loaded Accumulators

Spring-loaded accumulators can perform the same duties as a weight-loaded accumulator where volume demand is small. A basic design is shown in Fig 13. This comprises a free piston sliding in a cylinder, with a compression spring (or springs) mounted in the blind end of the cylinder. The accumulator is charged by pumping fluid into the cylinder, displacing the piston towards the blind end and compressing the spring. The fluid is thus pressurized by the spring force.

The pressure generated by a spring-loaded accumulator depends on the initial loading and spring rate of the spring, and thus is not constant throughout the stroke of the piston. Pressure will vary from a maximum when the spring is fully compressed, to a minimum when the spring is fully extended, unless constant-rate springs are employed.

The particular advantage of a spring-loaded accumulator is that it is compact in design, light in weight and can be used in any attitude. Basically, however, it is only suitable for relatively small capacities and low pressures. It is also not generally suitable for high cycling rates because of the limited cycling life of mechanical springs.

See also chapters on *Accumulator Performance and Duties, Accumulator-Type Devices.*

Accumulator Performance and Duties

GAS-LOADED ACCUMULATOR performance can be calculated directly from the general gas law PV^n = a constant. For fully adiabatic conditions, n = 1.4. For applications where the accumulator has sufficient time to return to normal temperature before a second discharge is required, n = 1. Where fairly rapid cycling is required practical values of n usually lie between 1.1 and 1.3, depending on the amount of heat produced and the actual time of cycling. For example, 1.1 for low to moderate rates of cycling and some heating effect; 1.3 for rapid cycling where heating effects are very apparent.

$$\text{Nominal or average pressure} = \frac{P_H + P_L}{2}$$

$$\text{Pressure variation (\%)} = \frac{P_H + P_L}{P_H} \times 100$$

where P_H = pressure at high fluid level
P_L = pressure at low fluid level

Accumulator Sizing

The fluid volume or effective fluid capacity (V_f) required for a given *working pressure range* from minimum (P_2) to maximum (P_3) is given by:

$$V_f = P_1 V_1 \left(\frac{1}{P_2} - \frac{1}{P_3} \right)$$

where P_1 is the initial precharge pressure (gas pressure)

The graph (Fig 1) can be used for sizing all gas-loaded accumulators, using the pressure ratio required (P_3/P_2) and the fact that for true adiabatic conditions (n = 1.4), then $P_l = 0.9\ P_2$. In the example shown, if P_3 = 3 000 lb/in^2 and P_2 = 2 500 lb/in^2 then the gas precharge will be 0.9 x 2 500 = 2 250 lb/in^2, and the pressure ratio will be 1.2. Using the graph, locate 1.2 on the vertical axis and take a horizontal line across to the curve. Project this line vertically downwards and it will be seen to cut the horizontal axis at 10.5%. Check the manufacturer's literature to find the column headed 'Actual gas volume'. This is the volume of the accumulator at condition V_l. If this column is not given take the nominal volume of an accumulator, *eg* 20 litres. Multiply this figure by 0.105, to find the actual volume of oil stored. *ie* 0.105 x 20 – 2.1 litres of oil stored between the pressures of 3 000 lb/in^2 and 2 500 lb/in^2.

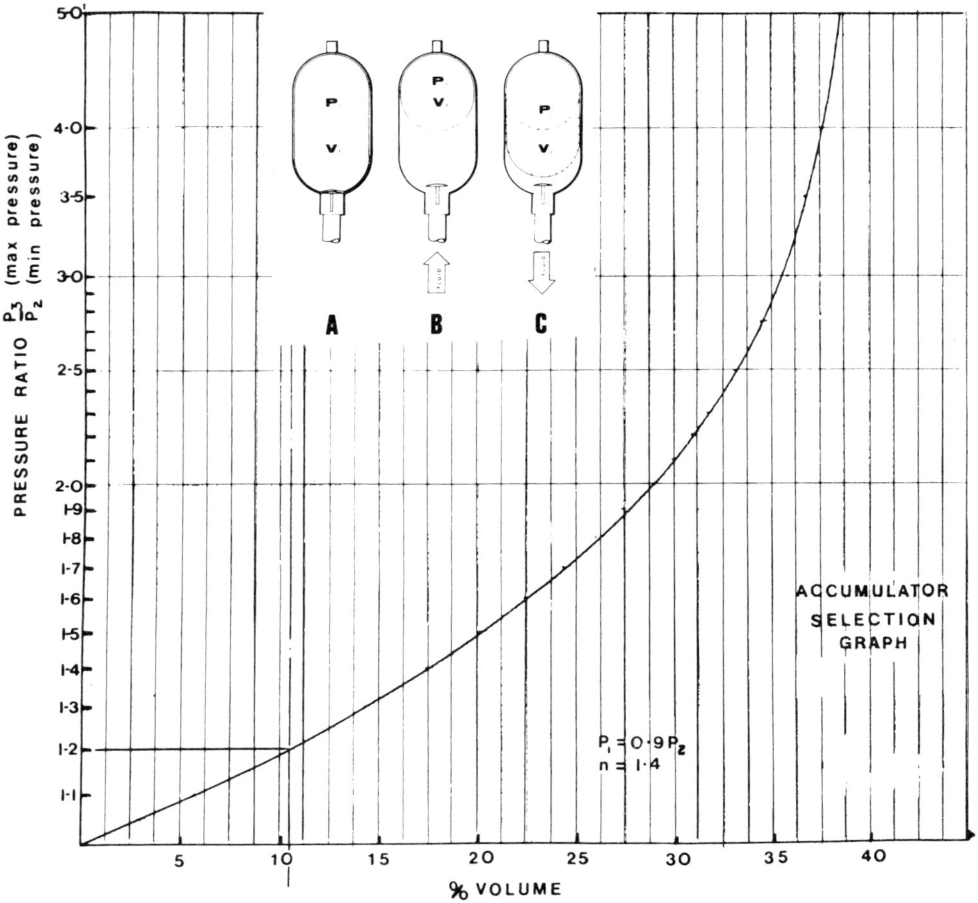

Fig 1
(Christie Hydraulics Ltd).

Precharge Pressure

$$P_1 = \frac{V_f \, P_2 \, P_3}{V_1 \, (P_3 - P_2)}$$

For a practical solution, P_1 must not exceed P_2, V_f must not exceed the capacity of the accumulator, and P_3 must not exceed the maximum pressure rating of the accumulator. If necessary the critical volume (V_1) can be increased by adding a gas bottle, contributing an extra gas volume V_4. V_2 and V_3 are now the fluid volumes at P_2 and P_3, respectively.

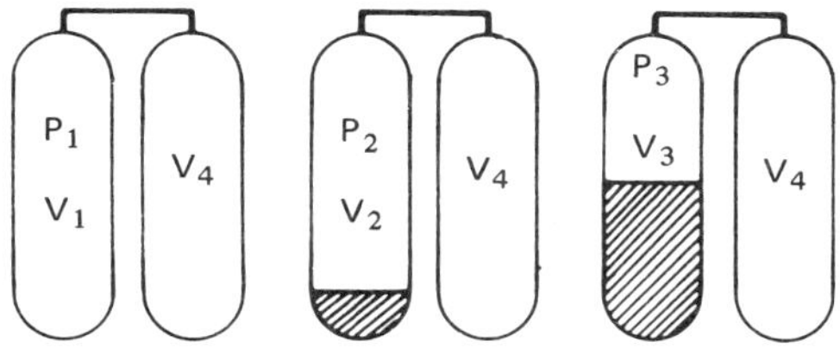

Fig 2

Gas-Loaded Accumulators with Auxiliary Bottle (Fig 2)

Relevant formulas are:

Fluid volume, $V_f = V_2 - V_3$

Precharge pressure, $P_1 = \dfrac{V_f\, P_2\, P_3}{(V_1 \times V_4)\,(P_3 - P_2)}$

For working between given pressure limits P_3 to P_2

$$P_2 = P_3 \quad \frac{(V_2 - V_3)}{(V_2 + V_4)}$$

Volume Relationship for Non-Separated Accumulators

For any given pressure difference $(P_2 - P_1)$

$$\frac{V_f}{V_g} = \frac{P_2 - P_1}{P_2}$$

where V_g = gas volume

Thus for an x% pressure difference:

$$\frac{V_f}{V_1} = \frac{x}{100}$$

or for a 10% pressure difference:

$$\frac{V_f}{V_2} = \frac{10}{90} = \frac{1}{9}$$

Compression Ratio (See Fig 3)

Compression ratio $= \dfrac{V_2 + V_4}{V_3 + V_4}$

With no gas bottle, $V_4 = 0$

Thus

Compression ratio $= \dfrac{V_2}{V_3}$

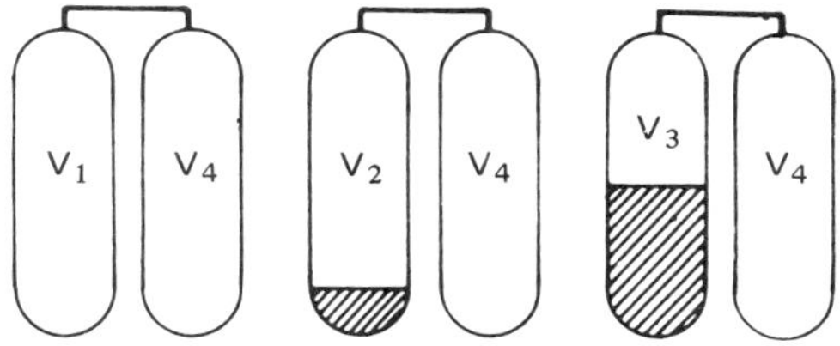

Fig 3

Compression ratios are generally in the range 1.5:1 to 3:1, depending on the application, with 2:1 a typical average figure. This may be further modified, and the pressure difference over the working range reduced, by coupling the accumulator to an auxiliary gas bottle.

The choice of size and type of piston accumulator is largely dictated by the particular application. Thus relatively large capacities are required to cope with continuous operation and high demand. A much smaller size could be used where the accumulator has only to supply peak demand or is worked only intermittently, or is mainly employed as a shock absorber. Where the accumulator is employed solely as a source of emergency power the size can be calculated on the demand.

For continuous operation with piston type assemblies the inflation pressure (P_1) should ideally be equal to the lower or cut-in value of the system working pressure (P_2) as this will give the greatest swept volume over the working pressure range and thus minimize the number of pressure cycles. For intermittent use, or where the accumulator is used as a source of emergency power, lower inflation pressures and consequently higher compression ratios can be used.

Other Accumulator Duties

Pressure surges caused by pump pulsations are a common cause of shock wave generation in hydraulic systems. The severity of the pressure pulsations and their frequency depends on the type of pump and its speed. All positive-displacement pumps generate pulsating flow. Piston pumps are the worst in this respect, and the fact that they are the type most used for generating higher pressures aggravates the problem. The use of several piston pumps discharging simultaneously into a pipeline may create very severe pressure surges liable to cause damage or failure. Pulsations and pressure surges such as these can be minimized or eliminated only by the use of a bag type accumulator whose absence of inertia or friction permits the extremely rapid response essential for effective pulsation damping.

A formula which can be used to determine a suitable size of accumulator to eliminate the effect of pump pulsation is:

$$V = \frac{K \times Q}{\text{pump rev/min}}$$

where Q = pump discharge

K = a factor dependent on the type of pump, *eg*

Simplex pump (single-cylinder, single-acting) — K = 5

Simplex pump (single-cylinder, double-acting) — K = 2.5

Duplex pump (two cylinders, double-acting) — K = 1.3

Triplex pump (three cylinders, single- and double-acting) — K = 0.45

Note: when the pump rev/min exceeds 100, then the denominator in the equation should be taken as 100.

Pressure Holding and Leakage Compensation

In a closed system where pressure must be held against the work by a holding ram for long periods while further duties in the operating cycle call for pump capacity, the use of an accumulator to replenish oil lost through leakage is advantageous.

The accumulator in a blocked circuit eliminates the problems of holding-pressure variations created by the varying demands of branch circuits on the pump in open-centre systems. In addition, system leakages, which are normally present or which develop over a period of time are automatically taken care of.

When lengthy holding times are required, two or more hydraulically-operated presses can be run economically with the use of accumulators. External or internal leakage through ram packings, valves or seals, results in piston creep and variation of the load on the work. The accumulator compensates for such leakage, maintaining the correct loading for the required period of time. Providing each press and accumulator is isolated during the holding cycle, the system pump is free to meet the volumetric requirements of the other presses.

Volume Compensation

Accumulators are now making possible many uses of hydraulic mechanisms which hitherto have not been feasible. This is especially true where the danger of increased pressure due to thermal expansion of the fluid in closed systems would cause rupture of the lines. The installation of an accumulator, pre-charged to the normal working pressure in the line, readily takes up the expanded volume and, more important, feeds it back into the line as thermal contraction takes place.

Hydraulic push-pull control mechanisms have been greatly complicated and limited in use due to thermal expansion problems. The application of accumulators with a high precharge simplifies them and extends their use.

Transfer Barrier Between Different Fluids

In some systems it is necessary to develop pressure in one side of the circuit, and transfer the pressure developed into another fluid without the fluids intermixing. In this type of application the flexible diaphragm in the accumulator acts as a barrier between the fluids, allowing some movement without diminishing the pressure.

Fig 4 shows two accumulators employed to ensure an emergency gland-sealing oil supply to a compressor or fan. In the event of pump failure, gas pressure forces contaminated oil to expand the separator bladder, thus expelling clean oil into the system to seal the glands and prevent gas escaping into the plant.

In some cases an additional gas bottle is used to back up the accumulator and thus increase the available capacity. This course is frequently recommended, the object being to allow delivery of a given quantity of fluid at a smaller pressure drop than would occur with the accumulator alone. In a correctly designed system, exact pressures are known and the volume of the accumulator bladder can be controlled so that it need never decrease to less than one-quarter of its original expanded size.

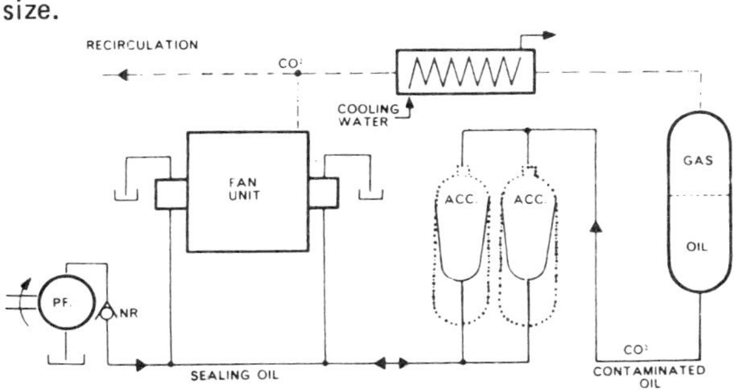

Fig 4 Accumulators used as transfer barriers.

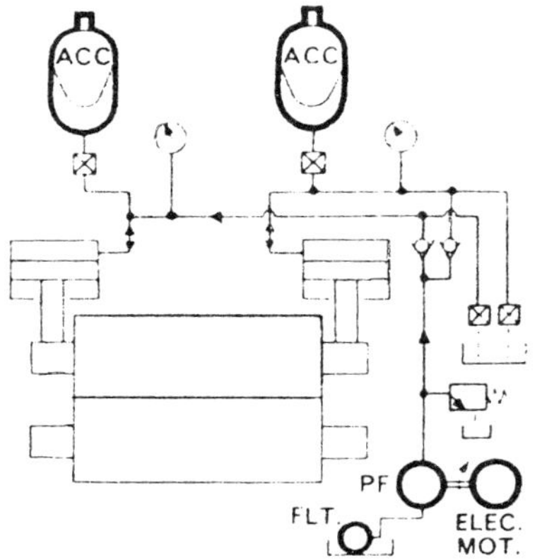

Fig 5 Accumulators installed in a constant-pressure hydraulically loaded mill roll.

Hydro-Pneumatic Spring

The installation of an accumulator in a rigid hydraulic system introduces hydro-pneumatic springing which can be used to advantage in many applications.

For instance, steel mill rolls or sugar mill cylinders (Fig 5) are required to exert a constant pressure as material passes between them. If foreign matter or over-size material is introduced the rolls must move apart to prevent damage and automatically resume their normal positions at the required pressure. This springing action is accomplished by an accumulator, or a series of accumulators if necessary, of sufficient capacity to absorb and release displaced fluid at almost constant pressure.

In all springing applications, the rapid action of bag type accumulators due to lack of inertia, friction and 'stiction' is advantageous, particularly where movements are small and, even more important, where pressures are low.

Fluid Make-Up

Unequal volumes of fluid will be displaced in the system by double-acting single-rod cylinders and similar devices. An accumulator in the circuit provides a means of accommodating such volume changes while offering other advantages which a simple tank or reservoir cannot. An accumulator, for example, will return excess fluid at substantially constant pressure.

Control

Two (or more) accumulators in a circuit can be used to provide synchronization of movement of hydraulic cylinders, constant velocity or constant pressure operation, as required. This is merely a case of using the 'stored energy' and 'constant pressure' characteristics of an accumulator, which are maintained independent of demand or pump output, provided the accumulators are properly sized.

Holding Devices

An accumulator is a convenient source of pressure to operate a holding device at constant pressure, regardless of demand from other sections of the system, or for maintaining a high working pressure on a workpiece during a long standby period. A particular advantage is that this is accomplished without power being absorbed or heat generated, as would be the case were the 'hold' maintained by a continuously running pump.

Dual-Pressure Circuits

An accumulator is a ready source of high pressure or increased capacity on a dual-pressure circuit and in such cases may be charged by a separate pump. Thus a large-volume low-pressure system demand could be met by a low-pressure pump (with its own accumulator, if necessary) and the second high-pressure service met by an accumulator charged by a small, high-pressure pump.

Emergency Power

An accumulator can be used as an emergency source of power in the event of failure of the primary power source (for example, failure of the pump driver). They are widely employed in this way in aircraft systems for maintaining operation of emergency services such as undercarriage operation, and also on large road vehicles utilizing hydraulic brakes and steering systems, but they also have industrial applications. The main requirement is that the accumulator should be of a suitable size to meet the emergency demand. Reversion to emergency operation can be fully automatic or manually selected, as required.

Fluid Dispenser

Liquids and lubricants can be stored in an accumulator to be dispensed at will under controlled pressures. In cases where constant but extremely slow rates of lubrication are sufficient, an accumulator can provide an automatic supply requiring checking only at long intervals.

In other cases where actuation is necessary for a short period in the event of pump failure or incapacitation of the primary hydraulic lines, the accumulator serves as a source of energy to provide emergency lubrication and prevent damage.

See also chapter on *Accumulator-Type Devices.*

Safety Design Codes

Most countries have stringent requirements for hydraulic accumulator manufacture, but these are not necessarily applicable to all such vessels. Examples of such requirements are:

1. Accumulators must be capable of losing the energy they store by virtue of the gas within them prior to reaching destruct pressure.
2. The units must be manufactured in such a way that both tactile and close visual inspection can be made without difficulty at any place within the units.
3. The units shall not have less than 3:1 gas vent pressure to design pressure ratio.
4. Units shall not have less than a 4:1 safety factor, *ie* burst pressure to design pressure.
5. Design pressure shall not be less than 10% above maximum working pressure.
6. There shall be an allowance for corrosion to 10% of wall thickness.
7. Allowance shall be made in the design calculation for any stress raisers that could be caused by scratching of the bore of the unit, possibly resulting from malfunction of sliding parts.
8. Units shall be tested to destruction 1% or one off per batch, made from the same material.
9. The works order number shall be recorded on the material analysis and physical properties certificates of the material involved in the construction.
10. The units shall be designed in such a way that it is impossible to begin to disassemble them until the pressure has been vented from within them.

11. The complete accumulator assembly shall be pressurized to 1½ times working pressure before despatch.
12. They shall be made from sufficiently ductile material to enable the result of any over-pressurization to a dangerous level to result in a measurable retained deformation after they have been depressurized.
13. Any unit of a welded construction shall be pressurized to twice the working pressure and shall be de-rated by 50% of its calculated maximum working pressure for any cyclic duty exceeding 1 000 Pa. It shall not be sold at all unless it has been totally normalized after welding. This normalization procedure shall also apply to any swaged units. Any heat treatment of the vessel shall only be carried out under constant inspection and be signed for by the inspector.
14. Accumulators must be designed to retain above safety factors in the event of gas over-pressurization, *eg* due to a local fire.
15. Every accumulator must be used with a pressure-relief valve so positioned as to make it impossible to isolate the accumulator from the said relief valve.
16. A bleed valve shall be so positioned in relation to the accumulator, its non-isolatable relief valve, and its isolation valve, that it is impossible for the bleed valve to be isolated from any pressure remaining within the accumulator. Thus, in the event of the bleed valve being opened it cannot fail to empty any remaining pressurized fluid from the accumulator.

Accumulator-Type Devices

SPECIAL DESIGNS of gas-loaded hydraulic accumulators may be used to perform specific functions, *eg* as Pulsation Dampers, Liquid-borne Noise Silencers, Line Surge Alleviators, Pump Controllers, Transfer Barriers, *etc.*

There is yet another area in which the original accumulator was used but where specialist manufacturers now find it necessary to supply purpose built products. These products are shock absorbers for liquids and they fall into three basic categories:

1. Surge Collectors — these devices allow large amounts of kinetic energy contained by a pipe with a large mass of fluid flowing at a substantial velocity, to be decelerated over a substantial period of time by the increasing resistance caused by the compression of a pre-charge gas. These units are generally designed to have very large entry orifices which will in many cases allow the volume accepted by the unit to flow back into the system only slowly instead of surging back in the opposite direction.
2. Shock Preventers — specifically preventers or 'alleviators'. These are installed in a system so that their capacitance and impedance prevent a shock from occurring by allowing a deceleration time for the fluid, but in nearly every case these units have entries which are substantially as great as the diameter of the vessel itself.
3. Shock Absorbers — these units are capable of absorbing a shock, *ie* absorbing a standing wave which has already been created and is travelling down a pipe. They do so by opening instantaneously upon sensing the beginning of a shock, allowing the shock in against their membrane and closing behind it. (Fig 1).

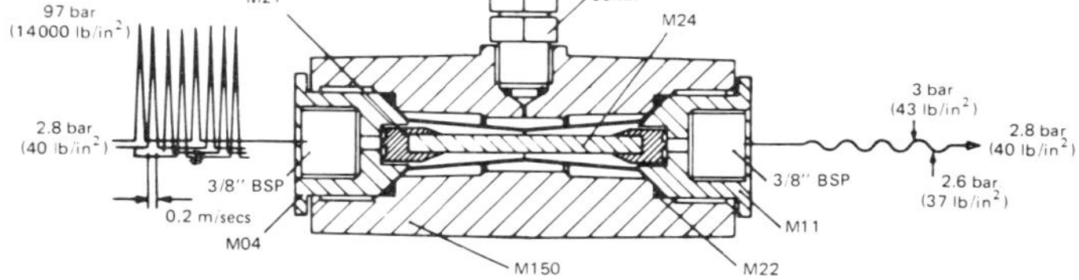

Fig 1 Example of shock absorber in section showing working characteristics.

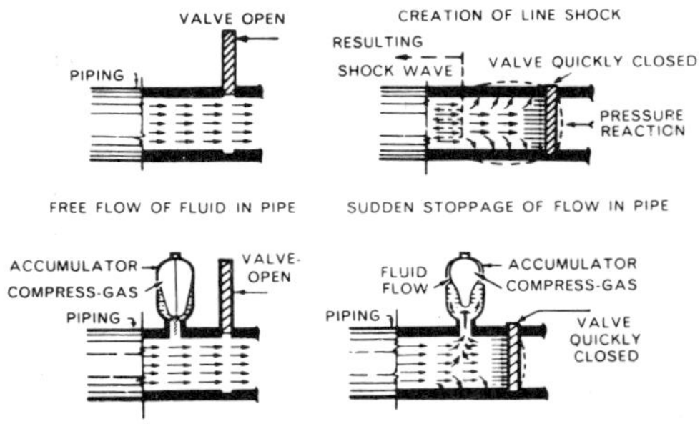

Fig 2 Line shock (water hammer) and its absorption.

Shock Absorption

Hydraulic line shock is the result of compression waves travelling from a source (*eg* a rapidly closed valve) up-stream to the end of the pipe and back again causing an increase in pressure and noise. This cycle is repeated at a regular frequency (depending on the length and the elastic modulus of the pipe) until the waves are dissipated by friction (Fig 2).

The size of accumulator necessary to alleviate shock waves can be calculated from:

$$V_C = \frac{0.004\,Q\,.\,P_2\,(0.005L - T)}{P_2 - P_1}$$

where

V_C = accumulator capacity required in gallons

Q = rate of flow in the pipe in gal/min

T = normal closing time of the valve in sec (if the valve closure is instantaneous T is equal to zero)

L = length of pipe in feet

P_1 = the flow pressure at the valve in lb/in^2 (this is the static pressure at the valve when the valve is open and the fluid is at its full flow rate)

P_2 = the pressure, the upper limit of which the surge should be permitted to attain in absorbing the decelerating flow on valve closure. This valve should be set at 1½ times the static pressure in the line when the valve is closed and the fluid is at rest

For Q in litres/min, L in metres and P_1 and P_2 in bar:

$$V_C\ (\text{litres}) = \frac{0.004\,Q\,.\,P_2\,(0.016L - T)}{P_2 - P_1}$$

Surge Damping

An accumulator has the inherent property of eliminating pulsations of a frequency greater than the cut-off frequency of the device. The cut-off frequency of an accumulator can be calculated from the geometry, fluid density and pressure.

$$\text{cut-off frequency (radians per second)} = \frac{2}{\sqrt{\dfrac{\rho L}{A} \cdot \dfrac{dV}{dP}}}$$

where ρ = mass density of the fluid
L = length of fluid chamber
A = cross sectional area of fluid chamber
V = volume of fluid chamber
P = pressure

All gas-loaded accumulators will have a low cut-off frequency. Thus they will pass all moderate to high frequency pulsations but tend to absorb all low frequency pulsations below the cut-off frequency.

This characteristic is put to advantage to eliminate pump pulsations or pump ripple. In general, however, for satisfactory performance in this respect it is necessary to 'size' the accumulator to provide a cut-off frequency higher than the known frequency of the pump. If necessary, two accumulators may be used, one on the inlet side and the other on the outlet side of the pump.

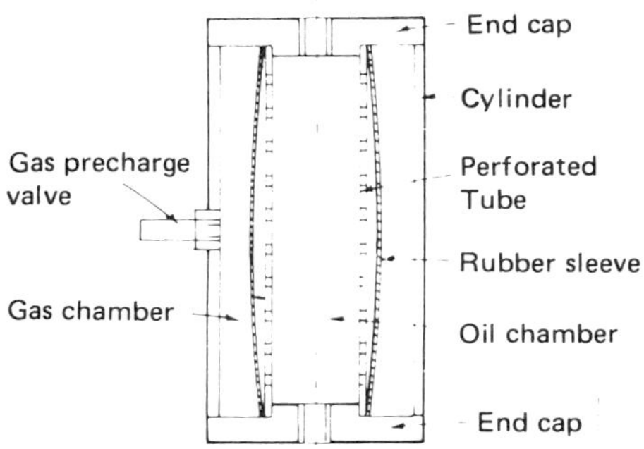

Fig 3

Tubular Accumulators

Tubular accumulators are intended primarily to act as pump pulsation dampers or shock absorbers, rather than energy storers. They comprise inner and outer cylinders, the inner cylinder tube being perforated and covered with a rubber sleeve — Fig 3. The unit is pre-charged with gas in the annular space between the two cylinders, fluid flow being through the inner cylinder, but with access to the inner side of the rubber tube through the perforations.

Bellows Type Accumulators

The bellows type accumulator is another form of tubular accumulator in which the rubber sleeve is replaced by a stainless steel convoluted sleeve. Again, it is intended to be used as a pulsation damper or shock absorber with fluids incompatible with elastomeric sleeves, *eg* in high temperature hydraulic systems. The operating principle is essentially the same as that of a tubular accumulator.

SECTION 2D

Pipes and Piping

PIPE SIZES commonly used for hydraulic lines range from 6 mm (¼ in) to 20 mm (¾ in), with smaller sizes down to 3 mm (1/8 in) in certain applications. Material choice for rigid lines is normally seamless drawn low-carbon steel of hydraulic quality. Other materials used are copper, aluminium alloy, special brass alloys, stainless steel and titanium.

Standard sizes of tubing produced for hydraulic lines are normally produced in two wall thicknesses, nominally rated as *standard* and *heavy gauge.* D/t ratios range from 6 to 10:1 in the former case; and 4 to 7:1 in the latter case, increasing with increasing diameter size. Corresponding pressure ratings for standard tubing range from about 420 bar (6 000 lb/in^2) to 295 bar (4 200 lb/in^2); and for heavy gauge tubing from 540 bar (7 700 lb/in^2) to 455 bar (6 500 lb/in^2) – see Tables 1A to IC.

TABLE IA – IMPERIAL TUBING TO BS 3602 CDS23

	Standard Tube					Heavy Gauge Tube				
Tube o.d. ins	Wall thickness		Bore in inches	Maximum recommended working pressure		Wall thickness		Bore in inches	Maximum recommended working pressure	
	swg	inches		lb/in^2	bar	swg	inches		lb/in^2	bar
1/4	20	0.036	0.178	5750	396	18	0.048	0.154	7700	530
5/16	20	0.036	0.240	4300	296	18	0.048	0.217	6500	448
3/8	18	0.048	0.279	5000	345	16	0.064	0.251	6500	448
1/2	18	0.048	0.404	3600	248	14	0.080	0.340	6500	448
5/8	16	0.064	0.497	3800	262	12	0.104	0.417	6500	448
3/4	14	0.080	0.590	4200	290	10	0.128	0.494	6500	448
7/8	14	0.080	0.715	3500	241	10	0.128	0.619	6500	448
1	12	0.104	0.792	4000	276	8	0.160	0.680	6500	448
1¼	12	0.104	1.042	3200	220	3/16in	0.187	0.876	6500	448
1½	12	0.104	1.292	2700	186	4	0.232	1.036	6500	448
2	10	0.128	1.744	2250	155	5/16in	0.313	1.374	6500	448

TABLE IB – METRIC TUBING TO BS 3603 CDS23

	Standard Tube				Heavy Gauge Tube			
Tube o.d. mm	Wall thickness mm	Bore in mm	Maximum recommended working pressure		Wall thickness mm	Bore in mm	Maximum recommended working pressure	
			lb/in²	bar			lb/in²	bar
6	1.0	4.0	6500	448	1.2	3.6	7500	517
8	1.0	6.0	4500	310	1.5	5	7500	517
10	1.2	7.6	5000	345	1.8	6.4	7500	517
12	1.2	9.6	4400	303	2	8	7500	517
16	2.0	12.0	4400	303	3	10	7500	517
20	2.0	16.0	3500	241	3	14	7500	517
22	2.0	18.0	3200	220	3.5	15	7500	517
25	2.8	19.4	4300	297	4	17	7500	517
30	2.8	24.4	3500	241	4.5	21	6500	448
38	2.6	32.8	2700	186	5.9	26.2	6500	448
50	3.5	43.0	2700	186	8	34	6000	414

TABLE IC – AMERICAN HYDRAULIC PIPE SCHEDULES

Nominal Pipe Dia. in	Pipe o.d. in	Standard (Schedule 40)			Extra Heavy (Schedule 80)			Schedule 160			Double Extra Heavy		
		Pipe i.d. in	Burst Pressure lb/in²	bar	Pipe i.d. in	Burst Pressure lb/in²	bar	Pipe i.d. in	Burst Pressure lb/in²	bar	Pipe i.d. in	Burst Pressure lb/in²	bar
1/8	0.405	–	–	–	–	–	–	–	–	–	–	–	–
1/4	0.540	0.364	16,000	1120	0.302	22,000	1540	–	–	–	–	–	–
3/8	0.675	0.493	13,500	945	0.423	19,000	1330	–	–	–	–	–	–
1/2	0.840	0.622	13,200	925	0.546	17,500	1225	0.466	21,000	1470	0.252	35,000	2450
3/4	1.050	0.824	11,000	770	0.742	15,000	1050	0.614	21,000	1400	0.434	30,000	2100
1	1.315	1.049	10,000	700	0.957	13,600	950	0.815	19,000	1330	0.599	27,000	1900

In general, tubes to specific overall diameter size are preferred since this standardizes the critical dimensions for fittings and couplings and minimizes the number of fittings required. All fittings of a particular size will fit all tubing of that o.d. regardless of the different pressure ratings, as governed by the wall thicknesses available. Overall diameter sizing is also standardized in tubes for threading.

Bore size tubing is, theoretically at least, the more useful in critical applications since it enables the tube size to be chosen on the flow requirements, which are determined by the actual bore size. In other words, it enables closer selection of an optimum hydraulic size. It is also more realistic for rating high-pressure tubes due to the greater variation in wall thicknesses possible. The equivalent hydraulic size of o.d. tubing can only be determined from a knowledge of the actual wall thickness used.

For lower pressures, 'nominal' pressure ratings are sometimes adopted for pipes and tubes, expressed in terms of a given maximum working pressure. This is usually substantially below the theoretical value possible. Sometimes these tubes may be produced in three different wall thicknesses, specified as light, normal and heavy gauge. These ratings have no precise meaning, unless related to definite pressure ratings or D/t ratios.

For very high pressure working, the small bore sizes necessarily associated with a low D/t ratio normally call for the use of special high strength alloys or composite construction in order to minimize wall thickness. The manufacture of such very high pressure tubing is critical, since all flaws must be eliminated. It is normally necessary to machine finish the bore and inspect internally, externally and through the section for cracks and other defects. An alternative method of construction for thick-walled, very high pressure tubing is to fabricate the section from two or more co-axial tubes in carbon steel and alloy steel, assembled as an integral tube. Tubes of this type are capable of accommodating pressures up to 14 000 bar (2 000 000 lb/in^2).

Cast Iron Pipes

Cast iron pipes continue to have a limited application for large-bore hydraulics — *eg* water-hydraulic systems. In the case of cast iron pipes and butt-welded and seamless steel pipes for water-hydraulics, rounded off pressure ratings are normally given, D/t ratios being selected to give a range of different pipe sizes with the same working pressure rating. Such pipes are normally specified in terms of nominal bore size, when actual bore size and actual overall diameter may differ appreciably from nominal values, especially in the case of cast pipes.

TABLE II — MAXIMUM PERMISSIBLE STRESS FOR TUBE CALCULATIONS
(Minimum UTS divided by 3)

Material	Condition	S_{max}	
		bar	lb/in^2
Low carbon steel	As drawn	1 280	18 300
	Drawn and polished*	1 950	28 000
20-ton steel	As drawn	1 000	15 000
Stainless (304) steel	Annealed	2 350	33 300
	Half-hard	2 800	40 000
	Hard	3 500	50 000
Light alloy 61S-T6	As drawn	1 000	15 000
Copper	Annealed	480	6 800†
	Half-hard	630	9 000†
	Hard	800	11 300†
Tungum	Annealed	1 550	22 000
	Precipitation hardened	1 550	22 000
Titanium 115/125		1 300	18 500
Titanium 150/160		2 100	30 000

* Cylinder tubes † Up to 65°C (150°F) only

Pressure Rating

Pressure ratings for pipes and tubes are normally taken from manufacturers' figures, but can also be calculated from first principles. In the case of homogeneous metal tubes the following simple formula can be used:

$$P_W = 2S_{max} \cdot \frac{t}{D}$$

where P_W = maximum permissible working pressure
S_{max} = maximum permissible material stress
t = tube wall thickness
D = tube overall diameter

Maximum permissible material stress is normally taken as one third (or 0.3) of the ultimate tensile stress of the material — see Table II.

Provided the maximum material stress figure is taken within the limit of proportionality of the material, this simple formula is valid. It does not hold true for higher stress values, and thus will not accurately predict bursting pressures, *eg* using S_{ult} in place of S_{max}. It is also not valid where the ratio D/t is 16:1 or less, as stress is then no longer uniformly distributed through the wall thicknesses, but ranges from a maximum at the inner surface to a minimum at the outer surface. The simple formula is thus restricted to thin-walled tubes (D/t greater than 16:1). It will overestimate the pressure rating for thick-walled tubes (D/t 16:1 or less), and in such cases an alternative formula must be used, *viz*

$$S_{max} = P \frac{R_i^2}{R_o^2 - R_i^2} \left(\frac{R_o^2}{R_i^2} + 1\right)$$

where R_i = inner radius of tube
R_o = outer radius of tube
P = internal pressure

Alternative formulas, written as a solution for wall thickness required are:

$$t = \frac{D}{2}\left(\sqrt{\frac{S+P}{S-P}} - 1\right)$$

$$t = \frac{D}{2}\left(\sqrt{\frac{3S+P}{3S-4P}} - 1\right)$$

For $S = S_{max}$ the corresponding value of P is P_W.
For $S = S_{ult}$ the corresponding value of P is the bursting pressure.

Modified formulas are used in the case of non-homogeneous tubes; and also for non-metallic tubes.

(i) In the case of welded tubes, a correction factor may be introduced; or alternatively, a larger safety factor may be used to establish P_{max} from the ultimate tensile strength of the material. This is not necessarily the invariable rule, as welded tubes can have the same working strength as drawn tubes. Corrections may be applied to tubes with welded connections on a similar basis, however.

(ii) In the case of cast tubes, a nominal (and substantially lower) value may be adopted for P_{max}. Cast tubes are associated with older hydraulic systems and large pipe sizes, where pressure rating is established on empirical lines, permitting fairly large tolerances in wall thickness.

(iii) In the case of copper pipes and tubes intended for brazed or soldered connections, the standard thin-walled formula is derated:

$$P_w = \frac{2S_{max}t}{D - 0.8t}$$

(iv) In the case of metallic tubes intended for threaded connections, an allowance is made for the reduction in tube strength due to threading. The following formula can then be used for thin-walled tubes:

$$P_w = \frac{2S_{max}(t - C)}{D - 0.8(t - C)}$$

where C is taken as equal to the depth of thread cut, with a minimum value of 1.25 mm (0.05 in)

(v) In the case of plastic tubes, an allowance is made for the higher elastic moduli of such materials, when a suitable formula is:

$$P_w = \frac{2S_{max}t}{D - t}$$

Brass and Copper Pipes

Copper tubing is attractive for small-bore, limited-input services, because of the ease of manipulation, as well as being resistant to general corrosion (*eg* for marine services). It is not suitable for high pressure lines, due to the limited strength of the material, and the fact that it is susceptible to work-hardening and early fatigue failure if subject to vibration. Nevertheless, copper tubes may be rated for pressures up to 210 bar (3 000 lb/in^2) in standard production sizes.

For higher pressure working with copper tubes, it is necessary to bear in mind that the strength of copper decreases with increasing temperature, so 65°C (150°F) is about a maximum for working at 70 bar (1 000 lb/in^2) or above. The strength of copper is halved at a temperature of 200°C (400°F).

A preferred alternative to copper for higher pressure working is alpha brass, offering far better strength and fatigue properties.

Al-Ni-Si-Brass (Tungum)

Tungum alloy is an aluminium-nickel-silicon-brass alloy with an alpha base structure. It combines high strength with good ductility, good fatigue properties and excellent corrosion resistance. It has been widely used on British aircraft hydraulic systems for a number of years because of its high tensile strength, ease of manipulation and freedom from season cracking and brittle fracture.

It has similar applications for high pressure industrial feed lines, where high working pressures and great reliability are required, as it has such excellent resistance – particularly in salt water or salt spray or similar corrosive conditions, which could attack steel tubes. Standard production sizes are available for pressure ratings up to 700 bar (100 000 lb/in^2).

Because of its high material strength and corrosion resistance, tungum alloy can often be regarded as an economic alternative to stainless steel.

Aluminium Tubes

Aluminium alloy tubes have been used in aircraft hydraulic systems in America, and in other applications where weight saving is important. The use of 61S-T6 alloy yields a material strength almost directly comparable with that of low-carbon steel. In general, however, aluminium is not regarded as suitable for pressure lines subject to vibration or pulsating pressures because of the relatively poor fatigue characteristics of the material. It may, however, be selected for low-pressure or return lines on a system where weight saving is important, or as a cheaper alternative to copper.

Stainless Steel Tubes

Because of their high cost, stainless steel pipes are normally only used for specialized applications, such as where resistance to a corrosive atmosphere is required, or mechanical strength has to be maintained up to very high temperatures. The strength/weight of stainless steel tubing is superior to that of any other material, with the exception of titanium.

Alloy type 321 is normally preferred for high-duty pressure lines, with a maximum permissible material stress of 2 320 bar (33 000 lb/in^2), rising to 3 500–4 600 bar (50 000–66 000 lb/in^2) in the fully hard condition. Stainless steel tubes can, therefore, be anything from two to four times as strong as low-carbon-steel tubes, permitting substantial reductions in wall thickness and weight for a given bore size and pressure rating.

Titanium Tubing

Titanium offers an exceptional strength/weight ratio, excellent resistance to corrosion, and a maximum service temperature of the order of 500°C (930°F). Tensile strength is of the order of 3500–8 600 bar (22–55 tons/in^2), which can be raised to 11 000–14 000 bar (70–90 tons/in^2) by alloying.

Pipeline Runs

Basic requirements in planning the layout of pipeline runs are:

(i) Pipe lengths should be kept to a minimum to reduce friction losses.

(ii) Straight runs between fixed points should incorporate at least one easy bend to accommodate thermal expansion/contraction. Dead tight straight runs should be avoided at all costs. These can set up severe tensile or compressive forces in the line and also make it difficult to couple to fittings where the tube has to be sprung back from a nipple to assemble or dismantle.

(iii) The number of bends should be reduced to a minimum consistent with the basic geometry of the layout, employing generous radii as far as possible. There is no advantage in employing an excessively generous radius at pipe or tube ends, however, as this may produce 'out-of-alignment' with unions or couplings. The pipe or tube should always approach end fittings as a straight length, adjusting the radius of any adjacent bends to terminate accordingly.

(iv) All pipes and tubes, and particularly pressure lines, should be properly supported, especially before and after bends on high-pressure systems. Any sudden interruption of flow will tend to produce a straightening out of the line at bends, leading to 'whipping' if the line is not supported. Supporting clips should not, however, clamp the pipe or tube rigidly but allow sufficient freedom of movement for thermal contraction or expansion, unless this is accommodated in an intervening bend.

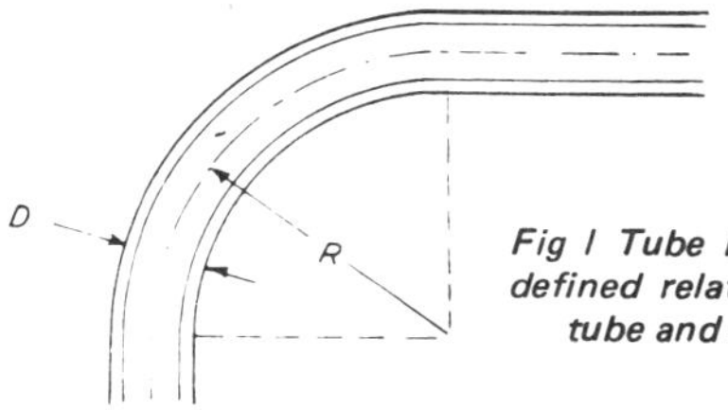

Fig 1 Tube bend radius (R) is normally defined relative to the centreline of the tube and in terms of tube o.d. (D).

ipe Bends

The bend radius (R) of a pipe or tube is normally defined with reference to the tube centre line nd the tube overall diameter (D) — Fig 1. As a general rule an *absolute minimum* bend radius is 5D. There are two reasons for this. One is that the more severe the bend (the smaller the bend adius relative to D) the greater the likelihood of the pipe wall buckling on the inside, or even ecoming overstressed on the outside. The other is that the smaller the bend radius (relative to D) the greater the fluid friction loss induced by the bend.

In the case of 180° bends (U-bends) where there is the likelihood of appreciable reduction in he wall thickness on the outside of the bend, the following formula is given by ASTM A-566-65 o specify the minimum allowable wall thickness after bending:

$$t_{(min)} = \frac{2\,R\,t}{2\,R + D}$$

where t = specified (minimum) wall thickness of tube before bending
$t_{(min)}$ = minimum allowable wall thickness after bending
R = bend radius (to centre-line of tube)
D = outside diameter of tube

Any bend will automatically tend to reduce the fatigue strength of the line at that point; and ny mechanical or geometric faults in the bend will affect the normal line strength. Early failure t bends can often be traced to ovality and the actual geometry of finished bends is worth hecking for this possible fault.

upporting Clips

ecommended spacing for supporting clips of high-pressure lines is summarized in Table III. This intended mainly as a general guide to good practice and may be over-ruled by local geometry on ny particular system. Closer spacing may be required at bends, for example, or where a number of dividual lines feed out from a common group. The main thing is to provide *adequate* support r all lines and it is generally better to use too many rather than too few pipe clips.

Various proprietary pipe clips are available, most of which have an elastomeric lining bonded to e inside to grip the pipe without locking it rigidly in place. Others take the form of simple clips,

TABLE III – PIPE SUPPORT SPACING – HIGH-PRESSURE LINES

Tube O/Diameter ins	Maximum Distance Between Clips ins
1/8	10
3/16	12
¼	15
5/16	18
3/8	18
7/16	24
½	24
9/16	24
5/8	24
¾	30
7/8	30
1	30
Over 1	30 x diameter

Tube O/Diameter mm	Maximum Distance Between Clips mm
3	250
4	275
5	300
6	350
7	400
8	450
9	500
10	550
15	600
20	700
25	750
Over 25	30 x diameter

requiring that the line be sleeved (*eg* with a short length of split rubber tube) before being clamped in place. If the clip is in a position where it is likely to be splashed with fluid or oil, the rubber used should be of oil-resistant type, otherwise it is liable to deteriorate rapidly and the clip lose its clamping action.

Besides providing normal support for the lines, pipe clips also assist in damping vibration which might set up oscillations in the pipe or tube run, again emphasizing the importance of an elastomeric or vibration-insulating layer between the line itself and the fixed rigid clip. In certain systems – *eg* aircraft systems – all piping is required to be electrically bonded and provision is often made for this in the design of the clip. A small tongue or similar shape is positioned so that it will contact the tube wall and effectively 'earth' the pipe through the clip. With a plain type of clip bonding can be introduced by inserting a layer of metal gauze between the pipe and clip and suitably connecting the end of the gauze to 'earth' if the clip is non-metallic (*eg* on the fixing screws or bolts).

Installation of Hoses

The following points are of particular importance in the installation of hydraulic hoses:

(i) Maintain proper bend radii. Smaller bend radii than those specified for the type and size of hose can materially reduce hose life.

(ii) Avoid twisting of the hose. The installation should be oriented so that when relative motion is present the hose will bend rather than twist.

(iii) Carefully plan routeing from port to port – again to avoid twisting or torquing the hose during and/or after installation; but also to keep the hose away from potentially damaging areas (*eg* near a hot exhaust manifold).

(iv) Add protection against external hose damage where necessary. This may apply to both hose and fittings.

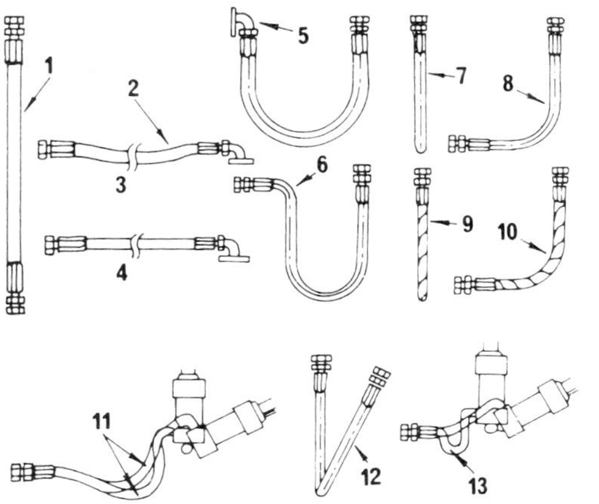

1–vertical runs preferred to (2) horizontal
3–slack is correct.
4–taut is incorrect.
5–correct.
6–incorrect.
7 and 8–correct – no twist
9 and 10–incorrect.
11–correct – large bend radius for flexing.
12 and 13–incorrect.

Fig 2 Correct and incorrect installation.

Examples of right and wrong hose line installation are shown in Fig 2.

(i) Problems of abrasion and minimum bend radius are especially critical in flexing applications. Hose end fittings are not flexible.

(ii) Be sure hose is not twisted.

(iii) Avoid criss-crossing of hose with resulting sawing action.

(iv) The hose should be bent in the same plane as the motion of the boss.

(v) Compensate for shortening effects by bending the hose or leaving slack in straight runs.

(vi) Use clamps of the correct size and avoid abrasion failures.

Bend Radii

Minimum Bend Radii – Bends are very common in hose installation and it is important that manufacturer's recommendations on this matter are followed closely especially on lines experiencing high surge pressures, constant flexing or vacuum conditions. Too tight a bend also tends to cause wear at the joint between the hose and any fittings, which may cause a fitting to blow off. The simple remedy is, upon installation, to allow sufficient hose to cope with flexing movements.

Other Possible Problems

Abrasion – Usually caused by contact with moving parts; sharp edges; criss-crossing of hose lines or improper use of clamps and poor assembly of elbow fittings adaptors. All of these can cause a severe wearing away of the outer cover of the hose, also weakening its reinforcement.

Criss-crossing is one of the most common of abrasion problems, any vibration, no matter how small, produces a sawing action, eventually wearing away the protective cover of both hoses. This can easily be avoided by correctly applying a clamp at the point of cross-over, thus efficiently separating the two lines.

Ambient Heat – Exposure to high ambient temperature can drastically shorten hose life, this excessive heat being transmitted through the outer cover to the inner tube, causing breakdown of the material.

Hose lines should therefore be routed away from ambient heat sources or protective measures taken to reduce the effect on the hose line, *ie* baffle or protective sleeve where re-routeing is impractical.

Correct Line of Hose — Just as important as minimum bend radii. Although the quickest route between two points is a straight line, no hose installation should contain straight taut hose. If it does, high operating or surge pressures can cause wear and eventually blow-offs.

Fluid Leakage — Is an exasperating costly problem which can be unsightly, troublesome and hazardous, if not controlled. Every point is a potential leakage point, so it makes good sense to keep connections to a minimum. Straight 'O' thread ring adaptors or split flange fittings provide superior seals and perform well in controlling leaks.

'On-Line'

Once the hose is 'on-line', and is part of a plant function, the need for additional safeguards can be established. Spring guards are required if there is even the slightest evidence of flexing at the couplings. Folding back or kinking of a soft hose must be avoided since it could rupture a hose carcass. The design of the hydraulic system should also provide sufficient restraining and shut-off valves for isolating hoses to prevent them from being subjected to pressure when they are non-operational over extended periods. Hoses are then placed in operation only at or below their specified working pressure.

See also chapters on *Pipe Couplings and Fittings* and *Hydraulic Hose.*

Pipe Couplings and Fittings

MOST MEDIUM and virtually all high-pressure pipework is small-bore, seldom exceeding 19 mm (¾ in) diameter. As a general rule the higher the pressure the smaller the bore size selected, provided this does not result in an excessive pressure drop.

Fittings used with small-bore pipes are usually of compression type, either flared or flareless. Normally fittings are made of the same material as the pipe, *eg* steel fittings for steel pipes, although aluminium, brass or gunmetal fittings may also be used with steel pipes.

Compression fittings are made for pipe sizes up to about 100 mm (4 in) with pressure ratings up to 700 bar (100 000 lb/in^2) depending on size. They are also available in a very wide variety of shapes, offering a convenient alternative to bent pipework in laying out the geometry of the system (although a fitting will normally have a higher pressure drop than a pipe bend).

Flareless Fittings

Flareless compression fittings have the advantage that only the pipe ends need to be cut true and square, when the fitting can be assembled in place and tightened up. Components for such a fitting comprise a body, usually of hexagonal shape to accommodate a spanner, a sealing ring or compression sleeve and a union nut or nuts (one for each end). Once the joint has been tightened, the sealing ring is deformed into a 'grip' on the tube, and usually remains on the tube if the joint is subsequently disassembled. Such a joint can usually be broken down and re-made repeatedly, if necessary, without loss of mechanical strength or hydraulic sealing. There are, however, different types of such fittings and couplings. A simple form of flareless compression joint, which is suitable for use on the more ductile tube materials (for example, copper) is shown in Fig 1. Here a shaped collet is tightened up by a nut, producing an annular deformation in the tube wall both to provide grip and lock the tube end within the coupling. Sealing is effected by the fit of the pipe ends in the ferrule. This type of joint is easily made and can be broken down and re-made as required. Similar types of coupling, usually with an internal sleeve, are widely favoured for jointing plastic tubing.

The simple olive type compression joint is shown in Fig 2. This type of joint, properly made, is quite suitable for high-pressure work on small-diameter tubes and does not rely on the tube wall deforming to provide the joint with mechanical strength. It is essential that the tube projects well

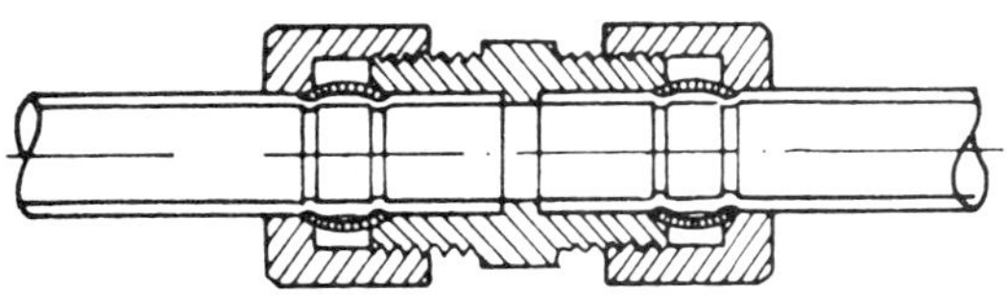

Fig 1 Typical compression type coupling for non-ferrous tubes.

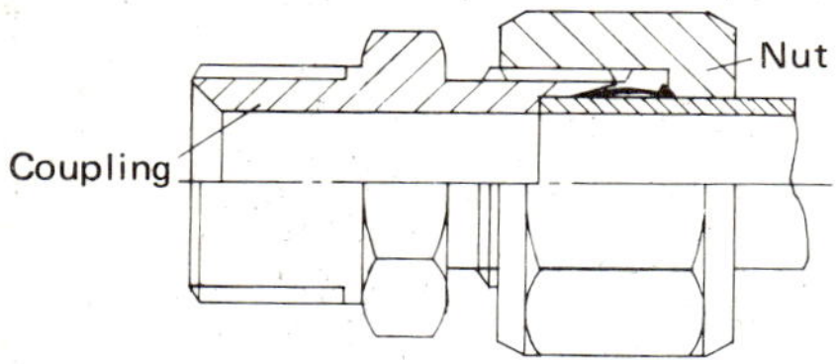

Fig 2 Flareless coupling (compression type) with olive.

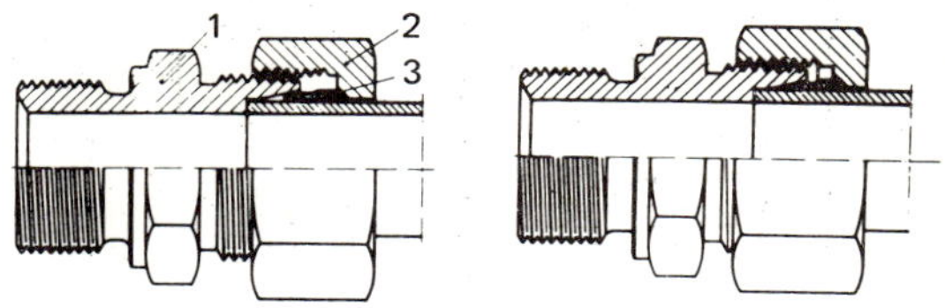

Fig 3 Bite type coupling to DIN 2353 standard. 1—union body. 2—cutting ring. 3—nut in pre-tightening position. (Lucas Fluid Power).

through the olive. Tightening the nut then tightens the joint and provides a compressive 'lock' *via* the olive. In the case of hard tubing considerable torque may have to be applied to the nut effectively to tighten the joint. Again this type of joint can be broken and re-made (preferably using a new olive).

A further type is the 'bite' coupling. In this case the collet or olive is replaced by a hardened steel sleeve, or sealing ring, with a cutting edge — Fig 3. As the nut is tightened, the sealing ring cuts into the pipe wall, throwing up in front a small ridge of displaced metal, which provides a metallic interlock. The depth of cut is controlled by the proportions of the sleeve and the final position is on a surface of freshly cut metal, with the leading edge of the ring firmly embedded in the tube wall. Thus, although the joint is completed without the ring actually compressing on to the tube, positive sealing is provided by the tube end being forced against the central fitting, with positive mechanical locking against axial displacement. The ring also provides a certain 'spring washer' effect between the nut, coupling body and tube, to maintain sealing under conditions of heavy vibration. The joint can be broken, with the ring remaining on the tube, and re-made without affecting its efficiency. DIN 2353 standard covers bite type tube couplings.

Various other proprietary compression-type couplings embody a 'bite' action. The body may be so shaped that sealing is produced by the tube ends butting against shoulders in the body (rather than the tubes themselves butting end to end). In another type, the body entry is tapered so that although the locking sleeve provides a primary seal, the entry of the tube end into a tapered body produces a secondary sealing action. In a further variation, a conical-shaped hardened-steel wedging ring is employed, which is forced against a tapered annular seating on the body when the nut is tightened. Further tightening displaces the tip of the cone downwards until the leading edge cuts into the tube surface, raising a burr which eventually fills the annular space between the cone and body, when the joint is locked. Most of these special forms are intended for use on larger diameter pipes.

Flared Fittings

Flared fittings require the use of a special tool to expand, or flare, the end of the tube. Their application is limited to tube materials which flare readily to the required angle without cracking or excessive hardening. Flare angles may vary from 30° to 90° (inclusive). Flare angles are normally 30° (British practice) or 37° (American practice). Flaring is accomplished with a special tool, both the surface of the tool and the pipe bore being perfectly clean and well oiled. It is also essential that the tube end be cut true and square and deburred if necessary.

Flared-type couplings consist, basically, of a threaded nipple with a centre shoulder, a nut or threaded ferrule for each end, and a sleeve for each end — Fig 4. The end of the tube is cut square and flared to conform to the sleeve, the resulting geometry of the flared end being such that the

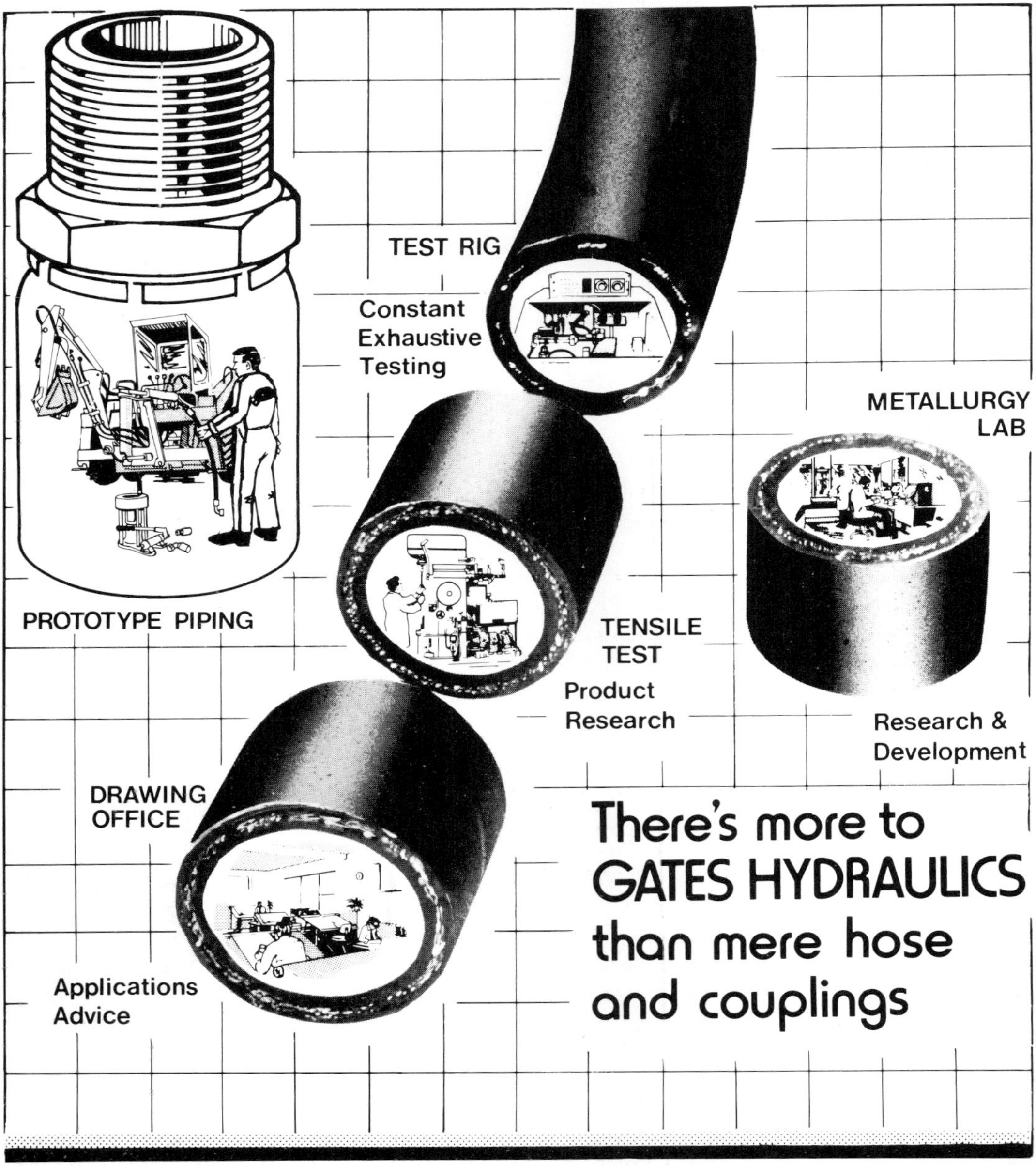
TEST RIG
Constant Exhaustive Testing
METALLURGY LAB
PROTOTYPE PIPING
TENSILE TEST
Product Research
Research & Development
DRAWING OFFICE
Applications Advice
There's more to GATES HYDRAULICS than mere hose and couplings

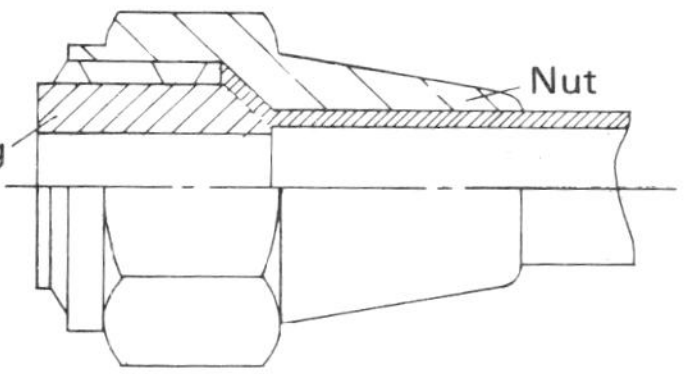

Fig 4 Typical flared coupling.

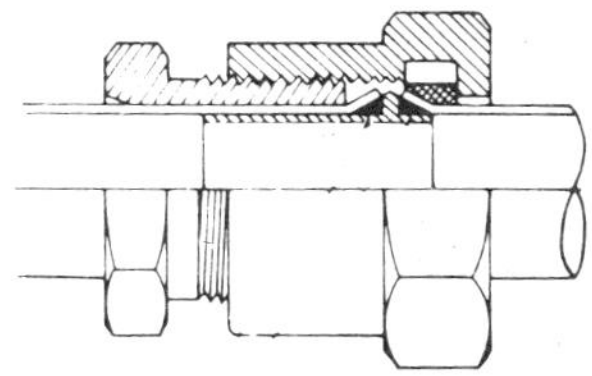

Fig 5 J.R. flared coupling incorporating sealing rings.

tube end projects evenly just beyond the sleeve. When the joint is assembled and the nut tightened, a pressure-tight joint is completed with good mechanical strength. Joints of this type are easily made and produce a good seal.

Variations on this principle are mainly concerned with the shape and form of the sleeve (and also the use of different flare angles). A suitable design of sleeve ensures that the flared end of the tube is not subjected to 'wiping' action as the nut is tightened, and also acts as a lock washer and vibration damper. The sleeve may also take the form of a split collet, which increases its tightening action on being forced into intimate contact with the flared end of the tube.

Sealing rings may also be incorporated in a flared coupling for improved sealing under high pressures. In a typical design an internal sleeve is employed for fitting with a collar – Fig 5. One sealing ring abuts on each side of the collar, so that the annular space is completely filled by the compressed rings when the coupling is tightened. Further advantages offered by this system are that the sealing rings can accommodate imperfections in flaring without affecting the sealing qualities of the joint and also provide a certain degree of self-adjustment against misalignment of the tube.

Suitability for Flaring

Straight carbon steel tubes can normally be flared readily. Other materials, notably alloy steels, may be difficult or impossible to flare to the required degree without cracking. In such cases flareless fittings would normally be specified.

A test on the flaring qualities of any hydraulic tubing is to flare the end to a minimum of 35% increase in diameter without developing any traces of minute surface cracks or notches. Tubing which passes this test can be taken as 'flareable'.

Flaring should always be done with tools specially designed for the purpose and the tools kept clean and in first-class condition. Overworking of the tube end should be avoided during flaring as this will decrease the fatigue strength of the material. The use of heat for flaring (or bending) is generally inadvisable since this may significantly reduce the strength of the material by softening it. Certain tube materials, however, may be specified for annealing before cold working, and subsequently heat-treated to recover their full working properties.

Flared steel tubing with steel end fittings will stand repeated connection and disconnection. Copper tubes will tend to work-harden under such conditions and thus the number of times a flared joint can be broken and re-assembled is strictly limited.

Formed Pipe Fittings

Formed pipe fittings embody the principle of deformation of the tube wall with a compression-type fitting, but call for some pre-shaping of the tube end before the fitting is assembled. They are therefore intermediate between the two types – flareless and flared – embodying the advantages of each, but are normally only used on larger pipe sizes.

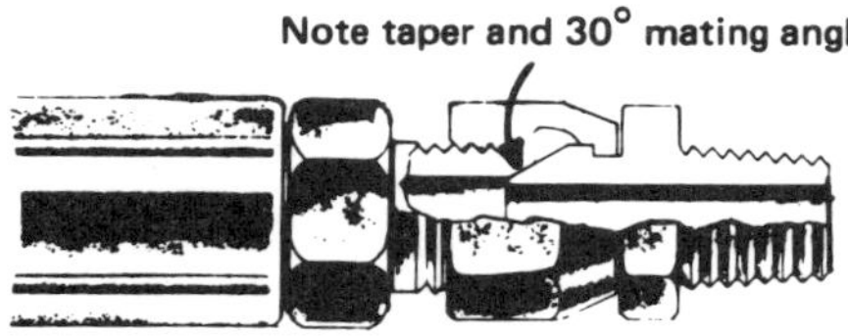

NPTF male coupling with NPSM female swivel adaptor.

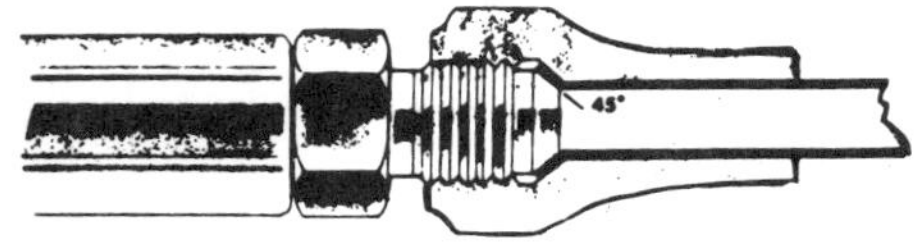

SAE male 45° flare mated with SAE female 45° swivel.

National pipe thread male coupling.

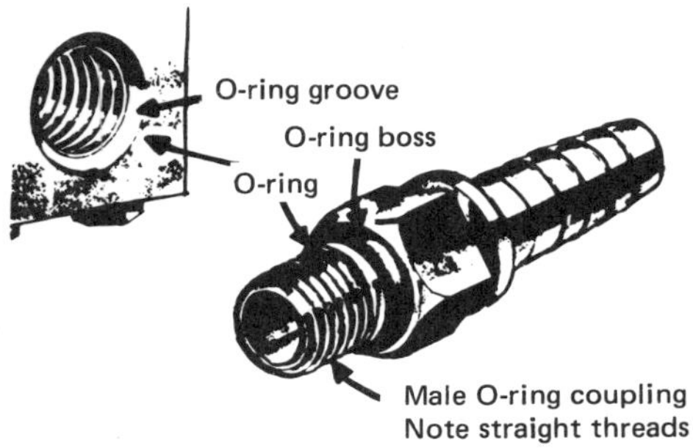

Male O-ring coupling. Note straight threads.

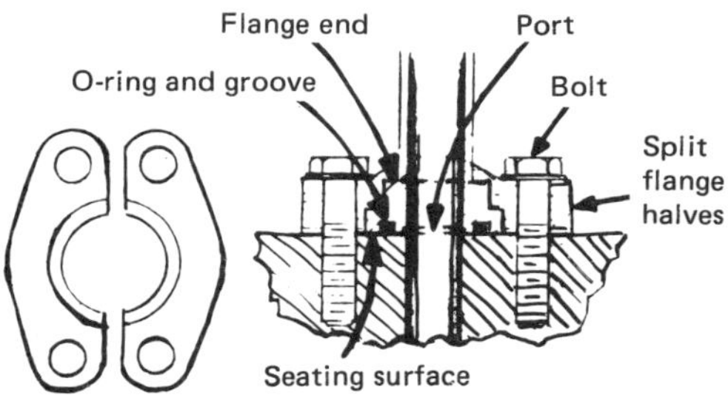

Flanged tube coupling

SAE male 37° flare (JIC) mated with SAE female 37° flare (JIC) swivel.

Examples of Standard American couplings.

In the example shown in Fig 6 the tube is pre-formed with an annular projection rolled outwards. This projection is located well in from the end of the tube, to minimize the risk of splitting or crack deformation, even on less ductile materials. A pressure-tight seal is then produced by tightening the nuts, thus drawing the sleeve into contact with the sealing ring and at the same time compressing the ring, with the conical ends of the sleeves forced into contact with the tapered ends of the nipple. A particular advantage offered by this type of coupling is that it can accept a considerable tolerance on tube diameter since the bead can concertina if necessary. A somewhat simpler form using a sealing ring outside the tube preform is shown in Fig 7.

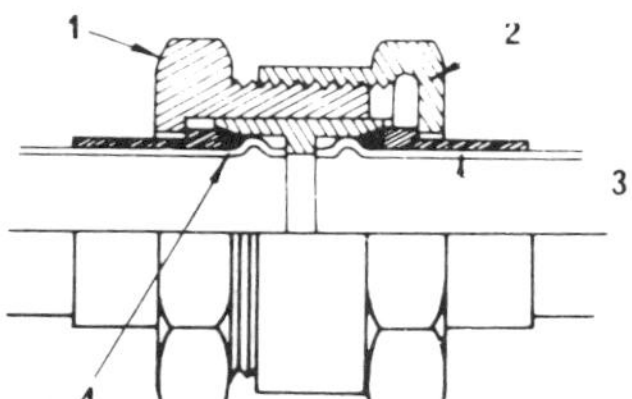

Fig 6 J.R. non-flared type coupling. 1 and 2—nut. 3—seal. 4—sealing ring.

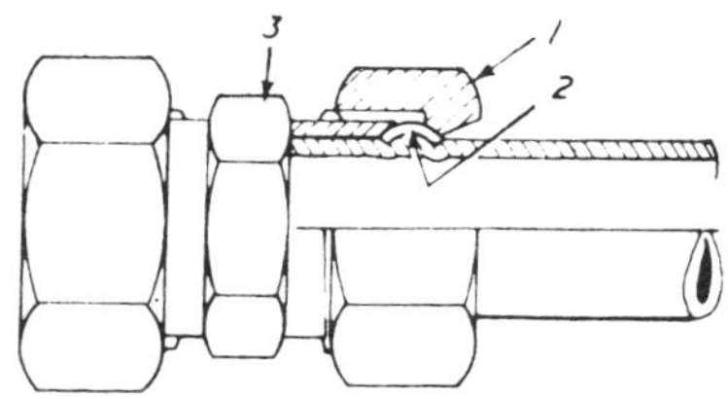

Fig 7 Typical compression type coupling 1—nut. 2—olive. 3—body.

Brazed Joints

Brazed and soldered couplings are normally used with copper pipes and tubes, and are thus strictly limited in hydraulic applications. Brazed connections may, however, be employed with copper-alloy (*eg* tungum) instead of conventional tube fittings.

Good joints of this type depend on capillary attraction to produce penetration of the solder or brazing alloy to both ends of the joint. Couplings of the pre-tinned copper type can be used with soft or half-hard copper tubes, employing solder paint and heating by blowlamp to complete the joint. Another type houses the solder in an annular recess, the amount of solder being just sufficient to complete the joint when melted. Such joints would not normally be used for hydraulic pressures in excess of 14 bar (200 lb/in^2) or, of course, at temperatures which could soften the solder.

Brazed joints in copper alloy tubing for high-duty work can be done with silver brazing, allowing between 0.05 mm and 0.15 mm (0.002 in and 0.006 in) clearance for capillary joints. It is important in the brazing of copper alloy tubes that the parts be raised to a temperature no higher than that which is just sufficient to melt the brazing alloy and the source of heat removed as soon as the joint has been completed. Excessive heating can destroy the properties of high-strength copper alloys. Copper alloys can also be brazed to steel fittings, but the joints in such cases are less reliable.

Welded Joints

Whilst welding can produce the strongest joints both mechanically and hydraulically, the technique is tedious and expensive when applied to small diameter lines. In consequence it is little used. The fact that the joint is permanent and cannot be broken down may also be a disadvantage. For these reasons welded joints are seldom employed in modern hydraulic systems.

The main exception is that of high-pressure, high-temperature systems, employing stainless steel tubing. Stainless steel is difficult to flare and welded joints may be preferred to flareless fittings, particularly when used with synthetic fluids.

Welding also becomes a much more attractive proposition with larger pipe sizes, either for complete (permanent) joints or the fitting of flanges to pipe ends to complete (breakable) flanged couplings.

Screwed Unions

Screwed unions are a 'traditional' form of pipe coupling. They afford a simple, compact form of joint which can readily be made and broken down again, as required. Malleable iron unions are restricted in service to a maximum of about 35 bar (500 lb/in^2) with water (the majority are rated lower) and cast gun-metal unions up to 17.6 bar (250 lb/in^2). Steel unions are employed for higher pressures and may be rated up to 140.6 bar (2 000 lb/in^2) or more: or 210–350 bar (3 000–5 000 lb/in^2) with special designs and included seals. Screwed unions are also used with PTFE wrapped around the threads as a seal.

Tapered threads are best for screwed connections, the most efficient seal being given by having a slightly greater taper on the male than on the female part of the joint (although PTFE tape wrapping will give a similar performance with identical threads). In all threaded joints, the male thread is cut on the outside of the pipes or tubes to mate with a female thread in the union or sleeve fitting. Dies used for cutting threads should always be kept sharp and clean, and those used for metal should not be used subsequently to thread plastic tubes.

The strength of screwed union joints is usually low and so hydraulic applications are normally confined to low-pressure systems (for example, up to 17.5–35 bar (250–500 lb/in^2), although there are exceptions. Strictly limited use may also be made of gas-type couplings (where joint rings, compressed by gland nuts screwing into a sleeve, provide the seal). The pipe does not have to be threaded to take this type of coupling (and both plain and threaded end tubes or pipes can be used); but the pipe ends are not jointed mechanically so the individual pipes must be supported against end loads which could cause them to separate – for example, fluid pressure. Basically, in fact, this form of coupling is a weak compression joint, capable of producing a good seal but low in mechanical strength. Application of such couplings is strictly limited for hydraulic work, although they may be used on small water systems up to 10.5 bar (150 lb/in^2).

Modern thread sealants are free from the tendency of early sealants to degrade and contaminate the system and can be advantageous with threaded couplings. In addition to improved sealing, such sealants increase the strength of the connection and protect it against corrosion or loosening due to shock and vibration.

Such sealants are of the fast-setting anaerobic type and may be 'soft' or 'hard' setting. In the former case the seal is easily broken for disconnection.

For optimum results with anaerobic thread sealants pipe threads should be cut according to DIN 2999 and assembled as shown in Fig 8. On a hand-tight assembly (left) at least 2¾ thread windings should be left unengaged. With wrenched-tight assemblies (centre) threads not fully cut should not be engaged since this would expand the fitting from the pipe and cause leakage. The right-hand diagram shows correct assembly for a wrenched-on fitting.

Two examples of union type couplings are shown in Fig 9. Both incorporate elastomeric seals, which also have the advantage of providing vibration damping as well as sealing. Where the service temperature is too high to use an elastomeric seal, metal wedge seals can be used.

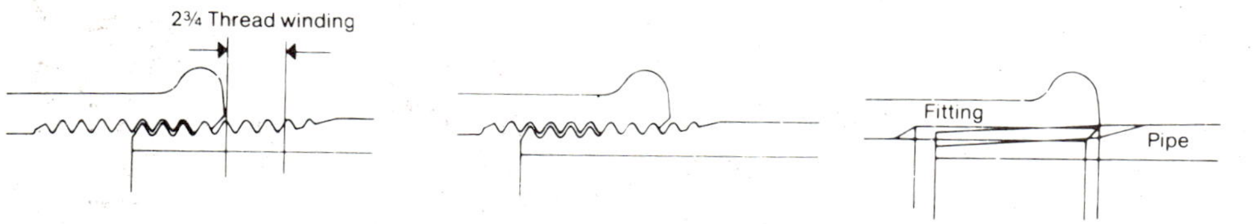

Hand tight Wrenched tight Area of interference on a pipe joint

Fig 8

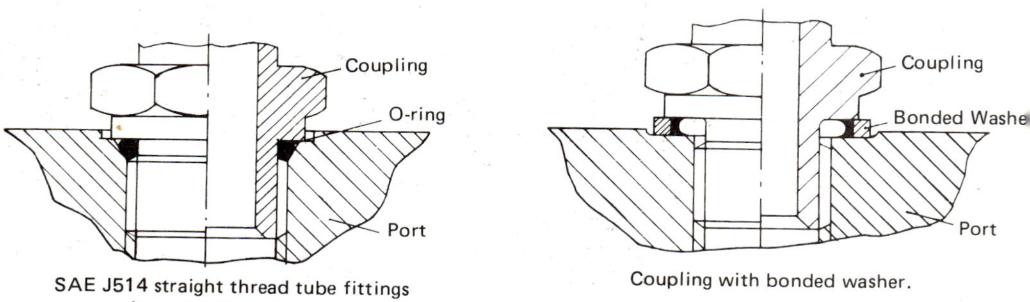

SAE J514 straight thread tube fittings incorporating O-rings.

Coupling with bonded washer.

Fig 9

B. & J. LIGHT ENGINEERING (HOLDINGS) LTD.

629, Spur Road, North Feltham Industrial Estate, Feltham, Middlesex.
Telephone: 01–890 4282/3 Telex: 948227

B&J

FAIR DEAL 1983

HYDRAULIC FITTINGS & ADAPTORS

Do you, as a Buyer get a fair deal?
How often when you only require a limited number of fittings are you told, by your supplier "Sorry, we only supply to special order. Minimum quantity 5000". or "Yes sir, we have these in stock but you have to order box quantity of 200".
Hard luck if you only want say 100 or even 25.
You can of course pay astronomical prices for small quantities.
Fair enough if you only want 1 item but what if you require a couple of dozen different items, then the bill really begins to tot up. How can you justify buying 24 times 200, just to get a good price.
If you are the Managing Director or have plenty of loose cash, (O.K.) but if you are a buyer with the cut costs or, get the chop, what do you do? Tell a little white lie, hoping no one will notice! Wear a bullet proof vest? or get your retirement slippers out early?
WHAT ELSE CAN I DO? HAVE YOU TRIED A FAIR DEAL FROM B. & J. ADAPTORS?
WHAT'S THE GIMMICK?
NO GIMMICK SIR. If you are a Buyer using a substantial amount of fittings per annum, then you can qualify for the B. & J. fair deal. How?
First send for our illustrated catalogue and price list for 1983 or you may have already received one. No catch here. Prices up to date, clearly laid out information. Quantity discounts shown with code numbers.
I THOUGHT YOU SAID THIS WAS A FAIR DEAL. I can get good discounts anywhere if I place bulk orders. I don't want vast quantities of every fitting. I want lots of different types of fittings in small quantities.
That's where the fair deal comes in. B. & J. are able to negotiate maximum discount terms on the majority of all their fittings, provided you are able to agree to a minimum composite order value, then all types and sizes can be lumped together to give the maximum discount on all items ordered from stock.
Why don't you have a confidential chat with our Managing Director, Mr. Charles Collins. Give him a ring today on 01–890 4282/3. You will be surprised at what can be arranged.

FAIR DEAL LIST PRICE

DISCOUNT METHOD 1. (Suited to the smaller irregular buyers).

Total Single Order Value – Discount

£0	–	£199	= Nil
£200	–	£499	= 2½%
£500	–	£999	= 5%
£1000	–	£1999	= 7½%
£2000	–	£5000	= 10%
Over	–	£5000	= 12½%

The Discounts offered above are based on TOTAL SINGLE ORDER VALUE of RANDOM QUANTITIES of VARIOUS TYPES and SIZES OF ADAPTORS.

Discount Method 2. (More suited to the regular buyer).

For those buyers who are prepared to enter into an agreement to purchase a regular MINIMUM Value EACH MONTH, we are prepared to offer the discount listed below, on all invoices.
The discount would depend on the level of purchases agreed, and would be subject to three monthly reviews.

AGREED TOTAL MONTHLY ORDER VALUE	Discount on Each invoice	GROUP Level
£0 – £199	= Nil	A
£200 – £499	= 2½%	B
£500 – £999	= 5%	C
£1000 – £1999	= 7½%	D
£2000 – £5000	= 10%	E
OVER £5000	= 12½%	F

Once a GROUP LEVEL has been agreed these discounts will apply to all orders, irrespective of value and for random quantities of various types and sizes of adaptors.
BULK ORDERS. Large volume of a particular size or type of adaptor, special discounts can be negotiated.

Write or 'phone for full price list
B. & J. Light Engineering (Holdings) Ltd, 629 Spur Road, North Feltham Industrial Estate, Feltham, Middlesex.
Telephone: 01–890 4282/3 Telex: 948227

We seek additional european agents and distributors

Please send me more information on your products and the B. & J. Fair Deal Scheme:
Name . . . .
Position. . . . .
Company name/address . . . .
. . . .
. . . . Telephone . . . .
B. & J. Light Engineering (Holdings) Ltd, 629 Spur Road, North Feltham Industrial Estate, Feltham, Middlesex.
Telephone: 01–890 4282. Telex: 948227

Flanged Couplings

Bolted-up flanged couplings are normally only employed on large diameter pipes. The flanges may be formed integral with the pipe section (for example, with cast pipes), or fabricated separately and welded or similarly mounted on the pipe end. Hubbed flange connections are produced for standard tubes from 75 mm (3 in) bore upwards where the tube end is flared and then compressed onto an olive. The flanges are fitted with dowel pins to ensure accuracy of alignment and a straight pull-up. Flanges are also produced for welding to standard sizes of rigid PVC tubing.

Various methods of flange mating may be employed, *viz*

(i) Ground-face flat flanges. A sealing compound or gasket may or may not be used between the faces.

(ii) Male boss mating, with a recess in a matching female flange. This type normally incorporates a soft copper or aluminium gasket between the mating surfaces, or an O-ring located in a machined groove.

(iii) Recessed flanges, sealing being provided by a gasket or O-ring, or sealing compound.

(iv) One flange-face flat and the other grooved to take an O-ring. This is capable of sealing up to very high pressures without relying on high mechanical compression of the joint.

See also chapters on *Hydraulic Seals* and *Hose Couplings and Fittings*.

Hydraulic Hose

HYDRAULIC HOSE is the general description given to flexible pipes and tubes designed as suitable for the containment of hydraulic fluids under pressure. Specifically, however, hose is taken as referring to reinforced construction of flexible tubes.

Hydraulic hose is the most convenient, if not the only solution, for lengths of hydraulic lines connecting components where movement has to be accommodated. Another primary application is to provide easy coupling or decoupling of lines at particular points in a system. Secondary applications of flexible lines include isolation of the pump from the pipework for noise and vibration damping; use as a shock absorber; and a variety of installation requirements where the use of rigid lines would present unwanted installation, operational or maintenance problems.

Production sizes of hydraulic hose range from 4 mm (3/16 in) up to 100 mm (4 in) inside diameter, except in the case of very high pressure and super high pressure hose which is normally of small diameter only.

Sizes of composite hoses commonly follow SAE standards where sizes braided on the outside cover are designated in sixteenths of an inch by using a 'dash-size' equivalent to the numerator of the fraction. Thus, −12 is 12/16 or 3/4 in size. There are exceptions. For example, aircraft hose size designations are based on the outside diameter of the tubing to which it is connected. Truck hose and hose covered by SAE specifications 100R5 (textile/wire/textile types) also are marked in this manner.

The SAE also specially established hose identification details. They dictate that the entire length of hoses are legibly marked, parallel to the longitudinal axis, using a stripe or stripes that show the respective SAE hose specification number. The i.d. size designation is repeated at least once every 22.8 cm (18 in) and at the manufacturer's option, dash size designation may be included. A coloured yarn incorporated into the cover identifies the manufacturer.

Normal Ratings

Flexible hoses can be generally described by pressure ratings, *viz*

Low pressure hose — working pressure range 17—35 bar (290—500 lb/in^2)

Medium pressure hose — working pressure range 35—210 bar (500—3 000 lb/in^2)

High pressure hose — working pressure range 70—310 bar (1 000—4 450 lb/in^2)

Very high pressure hose — working pressure up to 400 bar (5 700 lb/in^2)

Super high pressure hose — working pressure up to 440 bar (6 250 lb/in^2)

It is a general characteristic in any design pressure range that with simpler construction a smaller diameter will carry a higher pressure, and *vice versa.* Hence pressure ratings for a particular type or construction are given as a range, when actual pressure ratings will be specific to diameter size.

SS50

An entirely new range of thermoplastic wire reinforced hoses from Intec conforming to SAE100 R8/9/10.

MS40

A small bore hose (1.8mm I.D.) with burst pressures of 916 and 1550 bars.

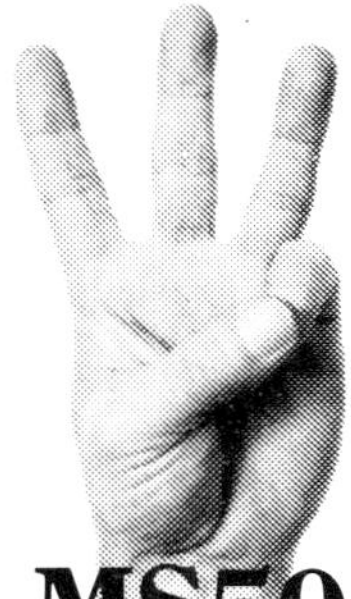

MS50

A 2.2mm bore hose with burst pressures of 986 and 2042 bars.

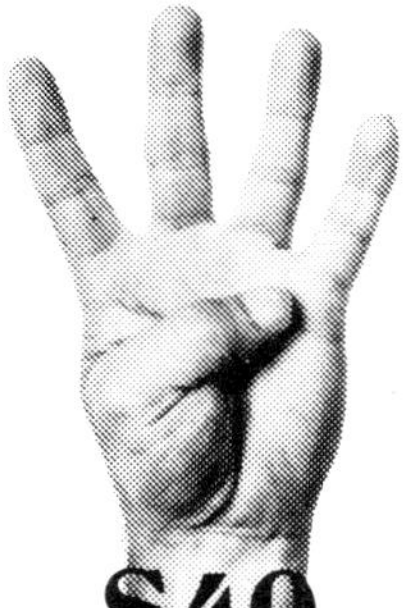

S40

Similar to S40 but with an elastomer core increasing flexibility and giving a tighter bend radius. Also conforms to SAE100 R7.

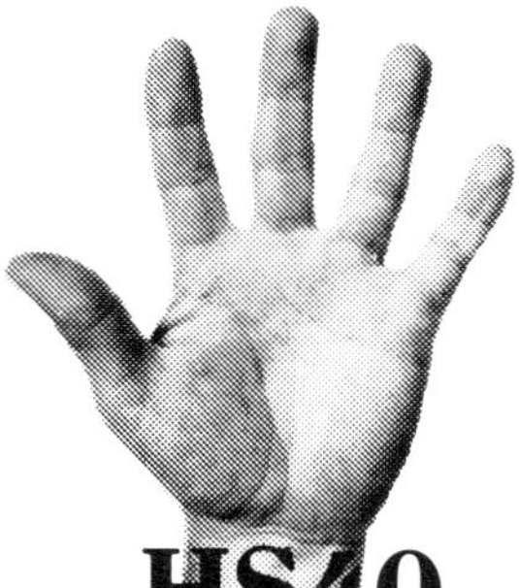

HS40

This thermoplastic hose range conforms to SAE100 R7 (and on performance to SAE100 R1) and is available in bore sizes 3/16" – 3/4".

KS50

A new range of hoses in bore sizes 3/16" – 1/2" and conforming to SAE100 R8 and on performance to SAE100 R2.

If you count on hydraulic hoses you'll be glad of our numbers.

And whatever your hose specifications, we give you our most important number — Intec in Plymouth (0752) 662844.
This is the only number you really need.
Ring us with enquiries, problems, even orders. We're ready to help.

Hydraulic Hoses you can count on from Intec —
A member of the Tecalemit U.K. Group of Companies

A member of the Tecalemit U.K. Group of Companies

intec ltd

INTEC LIMITED, Plymouth, Devon PL6 8LA.
Telephone: 0752 662844. Telex: 45561.

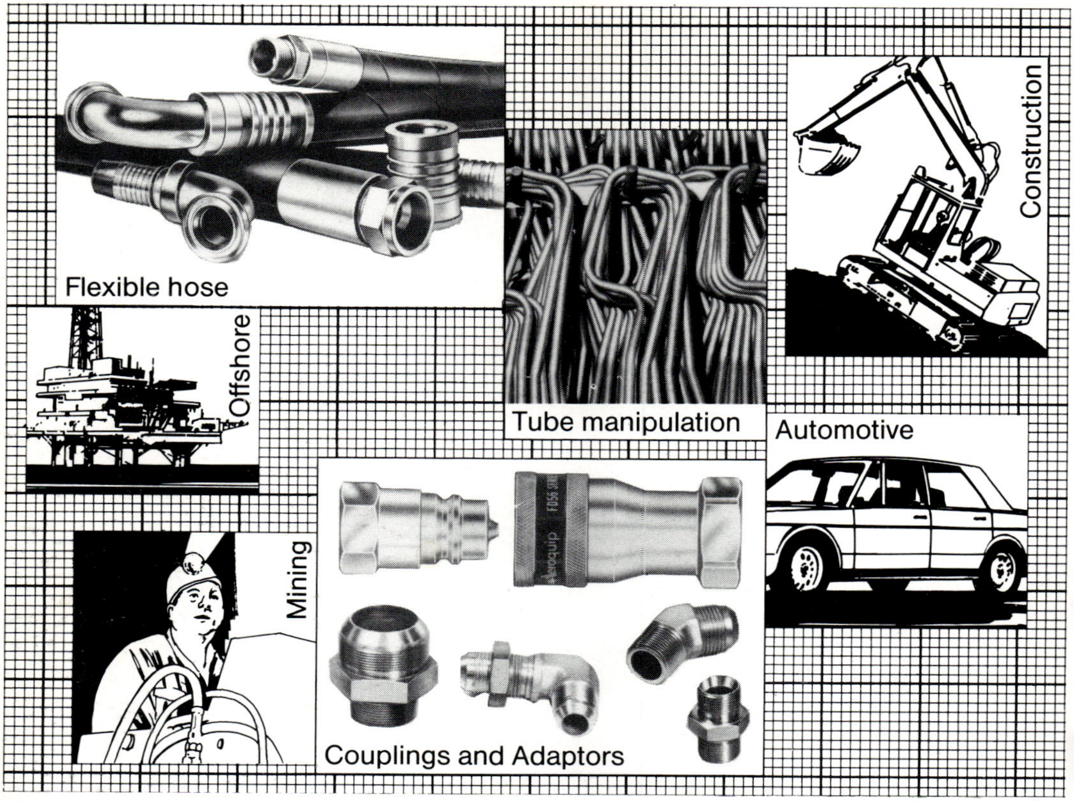
Flexible hose
Construction
Tube manipulation
Offshore
Automotive
Mining
Couplings and Adaptors

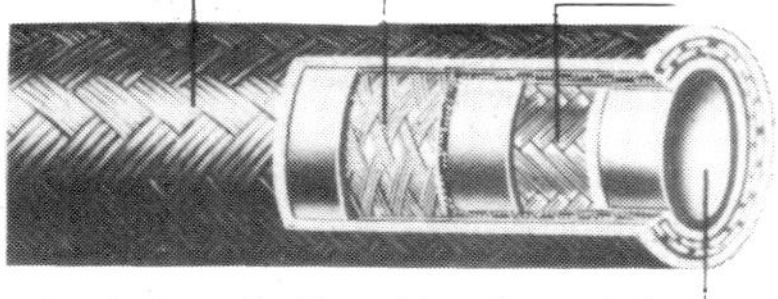

Fig 1 Basic construction of hose. (Aeroquip Corp).

Construction

Basically a flexible hose consists of an inner tube, reinforcement and an outer cover (Fig 1). Each hose component should be tailored to specific performance specifications and operating conditions that provide the longest possible service life:

The Inner Tube. The material must be flexible, withstand high and low temperatures without deterioration, be compatible with the fluid medium and have a smooth surface to ensure free flow. Many types of synthetic rubber tubes are available to accommodate a wide range of fluids.

The Carcass. This strong, flexible, reinforcing, and supporting member prevents the tube from bursting under pressure, or collapsing under suction. Reinforcement materials are knitted, braided or spiral-wound of natural or synthetic yarns and fibres, metal wires, or combinations of these materials.

The Outer Cover. This protects the carcass from abrasion, corrosion, heat, light, moisture, oils and weathering. It also provides identification either by marking, colour or corrugations.

Tube Materials

The inner tube material is normally synthetic rubber (although homogeneous nylon tubing is used in the construction of braided nylon hose — see later). The synthetic rubber components commonly used are:

1. *Nitrile* — a synthetic rubber compound which has optimum resistance to petroleum based oils and will handle temperatures up to 120°C (248°F). However, tubes produced from this basic material are unsuitable for the conduction of ester or vegetable-based oils.
2. *Neoprene* — a compound with an acetylene base and a fairly good resistance to oils, but its main characteristic is its ability to withstand abrasion and sunlight. It is, therefore, particularly useful as an outer cover for hose.
3. *EPD* — a compound mainly used to carry phosphate-ester based fluids. Being prepared from petroleum raw materials, it is incompatible with petroleum-based oils.
4. *PTFE* — a chemically inert semi-rigid material with a wax-like surface which, when extruded into an inner tube, needs to be thin in section. It provides minimum resistance to flow and is compatible with most fluids normally carried. PTFE will operate under a very wide temperature range —20 to +200°C (—40 to +400°F). It is supplied in different grades and the correct grade for a specific requirement needs correct sintering of the material to ensure positive performance and adequate resistance.

5. *AQP* – an elastomer, which combines the advantages of 1, 2, and 3, as it will safely conduct most types of hydraulic fluid, lubricating oil or fuel oil and has a temperature range from –49 to +150°C (–56 to +300°F). Its good high temperature capabilities are, moreover, combined with long life.

Considerable skill is required in the production of inner materials, commencing with the blending and mixing of the basic ingredients; accuracy in curing the final compound is essential.

Reinforcement

Wire braid reinforcement is the type most commonly used for hydraulic applications because it provides the best service life/cost economics. Further, basket weave wire braid provides good dimensional stability and ease of coupling installation.

Each group of reinforcement wires is called a 'plait' and each wire an 'end'. The number of ends in a plait will vary between similar sizes and types of hose depending on the intended use for the hose. During manufacturing, a thin gauge of compounded rubber formulated to provide high adhesion, is applied between each layer or 'ply' of reinforcement. Tension produced in the braiding process pulls the braid into the bonding material, forcing the material through the braid interstices. This provides adhesion between reinforcement layers and helps the finished hose to resemble a multi-ply composite.

Hose is subjected to specific hoop and end forces produced by hydraulic pressures. The angle at which the braid is applied affects the length of pressurized hose. When the hose is pressurized, the

Examples of hose constructions.
(See also Table I).

reinforcement attempts to assume a neutral braid angle to balance the internal forces. When the force vector angle formed is greater than 54° (the neutral angle), the hose elongates and the o.d. decreases. When the force vector angle formed is less than 54°, the hose contracts and the o.d. increases. If the hose has been braided to this neutral braid angle, pressurization produces no change in the net length and radial applications. It may be desirable to design the braid angle to produce a change in length to improve coupling retention.

Cover Variations

The hose cover can be wire or textile braid or rubber. Textile braid covered hoses, known as coverless hoses, are lightweight and permit the escape of gases such as air or oxygen which permeate the inner tube. Coverless hoses are employed as air brake hose, in some Freon containment or transfer systems, and in fuel lines. Stainless steel wire braid covered hose is corrosion-resistant.

Rubber covers that protect against environmental conditions are temperature- and solvent-resistant, as well as weather-resistant (ozone, ultravoilet, humidity). Normally, hydraulic hose systems containing petroleum based fluids utilize neoprene as cover materials.

Hose Standards

Hose construction is covered in various National and International standards including ISO, BSI, SAE, DIN, API and CETOP. Most proprietary hoses conform to one standard or another, the SAE standards being widely followed (see Table I).

Relevant British standards are:

BS 3640: 1963 Type 1

Rubber tube and cover, reinforced with one or two braids of high tensile steel wire.
Size range 3/16 in to 2 in bore.

BS 3832: 1964 Type 2.

Rubber tube and cover, reinforced with one or two braids of high tensile steel wire.
Size range 3/16 in to 2 in bore.
Used on service applications more arduous than those covered by BS 3640.

The world hose industry is represented by ISO/TC45 Working Group 9, which is responsible for recommendations on all aspects of hose types, manufacturing and testing procedure.

Some general parameters based on SAE standard hydraulic hose types are given in Table I.

Pressure Rating

Most hydraulic hoses are constructed to meet established pressure rating standards. When designing hydraulic hoses, three different hose pressure ratings are considered: maximum operating pressure, proof pressure, and minimum burst pressure. The maximum operating pressure is normally one quarter of the minimum burst pressure rating, giving the hose a safety factor of 4:1.

Specification of these data may vary slightly with different manufacturers. Some classes of hose may be rated for a nominal working pressure with a higher than usual safety factor for general applications, and a maximum working pressure rating for approved applications where the safety factor is reduced to 3.

Logically, with controlled manufacture, a safety factor of 3 should be satisfactory for systems working at substantially constant pressure. Where the system is subjected to surge pressures or other adverse conditions generating pressure pulses, a factor of 4 (or higher) is recommended. Required behaviour under impulsive conditions is also specified in various standards, *eg* BS 3832 specifies the minimum number of cycles to failure at peak pressures of 125% of the mean pressure at a rate of 35 pulses per minute — see also Table II.

TABLE I – GENERAL PARAMETERS OF SAE HYDRAULIC HOSES

SAE Standard Hydraulic Hose Types	Hydraulic Fluids	Temperature Range	Diameter Range (inches) INS	OUTS (approx)	Minimum Burst Pressure Range bar (lb/in^2)	Proof Pressure Range bar (lb/in^2)	Maximum Operating Pressure Range bar (lb/in^2)	Minimum Bend Radius Range (inches)
100R1 Steel wire reinforced rubber covered	Petroleum & water based	−40 to + 93°C (−40 to +200°F)	3/16 to 2.0	0.05 to 2.6	840–105 (12000 to 1500)	420–52.5 (6000 to 750)	210–26 (3000 to 375)	3.5 to 25.0
100R2 High pressure steel wire reinforced rubber covered	Petroleum & water based	−40 to + 93°C (−40 to +200°F)	3/16 to 2.0	0.6 to 2.7	1400–315 (20000 to 4500)	700–160 (10000 to 2250)	350–80 (5000 to 1125)	3.5 to 25.0
100R3 Double fibre braid rubber covered	Petroleum & water based	−40 to + 93°C (−40 to +200°F)	3/16 to 1.1/4	0.5 to 1.75	420–105 (6000 to 1500)	210–52.5 (3000 to 750)	105–26 (1500 to 375)	3.0 to 10.0
100R4 Wire inserted hydraulic suction	Petroleum & water based	−40 to + 93°C (−40 to +200°F)	3/4 to 4	1.4 to 4.8	84–10 (1200 to 140)	42–5 (600 to 70)	21–2.5 (300 to 35)	5.0 to 24.0
100R5 Single wire braid textile covered	Petroleum & water based	−40 to + 93°C (−40 to +200°F)	3/16 to 1.13/16	0.5 to 2.2	840–100 (12000 to 1400)	420–49 (6000 to 700)	210–24.5 (3000 to 350)	3.0 to 13.25
100R6 Single fibre braid rubber covered	Petroleum & water based	−40 to + 93°C (−40 to +200°F)	3/16 to 5/8	0.5 to 0.94	140–100 (2000 to 1400)	70–50 (1000 to 700)	35–24.5 (500 to 350)	2.0 to 5.0
100R7 Thermoplastic	Petroleum & water based & synthetic	−40 to + 93°C (−40 to +200°F)	3/16 to 1	0.4 to 1.4	840–280 (12000 to 4000)	420–140 (6000 to 2000)	210–70 (3000 to 1000)	3.5 to 12.0
100R8 High pressure thermoplastic	Petroleum, water based & synthetic	−40 to + 93°C (−40 to +200°F)	3/16 to 1	0.6 to 1.5	1400–560 (20000 to 8000)	700–280 (10000 to 4000)	350–140 (5000 to 2000)	3.5 to 12.0
100R9 High pressure, four-spiral steel wire reinforced rubber covered	Petroleum & water based	−40 to + 93°C (−40 to +200°F)	3/8 to 2	0.85 to 3.0	1260–560 (18000 to 8000)	625–280 (9000 to 4000)	315–140 (4500 to 2000)	5.0 to 26.0
100R10 Heavy duty four-spiral steel wire reinforced rubber covered	Petroleum & water based	−40 to + 93°C (−40 to +200°F)	3/16 to 2	0.8 to 2.85	2800–700 (40000 to (1000)	1400–350 (2000 to 5000)	700–175 (10000 to 2500)	4.0 to 28.0
R100R11 Heavy duty six-spiral steel wire reinforced rubber covered	Petroleum & water based	−40 to + 93°C (−40 to +200°F)	3/16 to 2	0.9 to 3.0	3500–840 (50000 to 1200)	1750–420 (25000 to 6000)	875–210 (12500 to 3000)	4.0 to 28.0

Hose applications that use above rated working pressure will result in shortened service life and premature failure. Failures such as hose rupture or fitting blow-off increase costs by increasing the frequency of hose replacement and also by causing equipment 'downtime'.

The proof pressure, twice the maximum operating pressure, is used during non-destructive inspection testing. Minimum burst pressure rating of a hose assembly is normally four times the maximum operating pressure. Actual burst pressure is the pressure at which rupture of the hose will occur.

TABLE II – PRESSURE IMPULSE RATINGS SPECIFICATION – MINIMUM REQUIREMENTS

Bore Diameter in	BS 3640: 1963	BS 3832 : 1964		SAE 100R1	SAE 100R2	SAE 100R5
		Minimum cycles to failure	Minimum average cycles to failure			
3/16, 1/4, 3/8		100 000				
1/2		75 000	100 000			
5/8	20 000	50 000	75 000	150 000	200 000	150 000
3/4, 7/8, 1		35 000	50 000			
Over 1 inch		To be agreed between purchaser and supplier				

Thermoplastic Hose

Thermoplastic hoses have improved greatly in quality and performance over the past twenty years and are now commonly considered for hydraulic applications, particularly as many may have pressure ratings comparable to that of 1- or 2-wire braid conventional hose. Particular advantages of thermoplastic hoses are lower weight (normally less than half that of conventional wire-braid hose), better flex impulse life and better resistance to abrasion. Properly constructed a thermoplastic hose is also capable of expanding under pressure and will still retain a high burst pressure.

Homogeneous semi-rigid nylon tubing can be rated for maximum working pressure of the order of 90 bar (1 250 lb/in^2) and may, therefore, be considered suitable for feed lines on low to moderate pressure systems. It is, however, more often employed for flexible lines with a braided reinforcement, where it is directly competitive with conventional flexible hoses.

The tensile strength offered by various grades of nylon ranges from about 430 bar (6 000 lb/in^2) up to 850 bar (12 000 lb/in^2). With glass braided fibre reinforcement, a maximum tensile strength of the order of 1 700 bar (24 000 lb/in^2) may be achieved.

A particular advantage offered by nylon is that it is free from fatigue characteristics, and also undergoes relatively little creep or cold flow at elevated temperatures. On the other hand, the strength is markedly temperature-dependent and decreases rapidly with increasing heat. Tensile strength figures are normally quoted for 20°C (68°F) and maximum service temperature for nylon pressure tubing is about 60°C (140°F), at which level the maximum permissible working stress has been degraded by up to 40%.

Coiled nylon tubing hose may also be used for pressure damping, although it is more usual to find it employed in the form of recoil hose. That is, the hose is coiled in the form of a 'spring' length which can accommodate extensive axial movement, merely expanding or contracting as a 'spring' with self-retracting characteristics. This can provide a much neater and less troublesome installation than using a flexible hose to accommodate the same degree of movement.

All-Metal Hose

Metallic flexible hose normally takes the form of a convoluted metallic tube close braided with steel wire – see Fig 2. Depth and pitch of convolution and braid tension are critical factors since these govern the change in length of the tubing under pressure (which should be reduced as far as possible to minimize fatigue characteristics) and also the degree of 'fretting' likely on the convolution crests. As well as measuring strength and protecting the outer surfaces of the convoluted

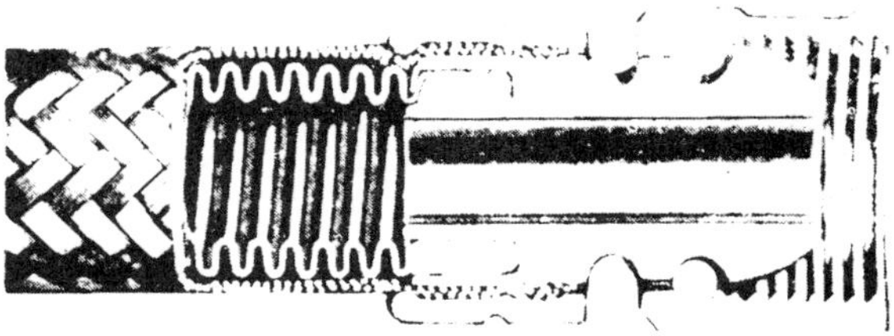

Fig 2 Typical convoluted stainless steel hose with wire braid. (Palmaflex).

tube, the braid layer or layers also serve the useful function of dampening vibration and resonance, whilst containing the end loads imposed by internal pressure.

The principal advantage offered by all-metal tubing of this type is the high working temperature possible — typically in excess of 400°C (752°F) with seamless stainless steel (convoluted) tubes and braid. Such hoses are, of course, flameproof. Pressure rating can be quite high, although this tends to decrease rapidly with increasing bore size.

An alternative form of construction is appreciably heavier and is to form the convolutions from helically wound brass strip, seamed and soldered and finally reinforced with single or double layers of tinned cadmium copper wire braid. Service temperature in this case is reduced to 121°C (250°F) maximum with soldered assemblies, or up to 171°C (340°F) with high temperature solders.

A still more suitable form for high pressure applications forms the convolutions in three-ply stainless steel tube with an outer stainless steel braid. The outer braids are designed to contain extreme pressure end loads and the hose is intended to be used with welded on fittings. Flexible metallic hose of this type retains the high operating temperature characteristics of conventional convoluted stainless steel pipes (for example, up to 500°C (932°F)), with pressure ratings of up to 420 bar (6 000 lb/in^2) possible, depending on bore size.

All-metal flexible hoses can be expected to have a higher frictional resistance than plain bore hoses because of the convoluted inner form. This can be reduced by incorporating a flexible smooth bore lining, although the service temperature of the composite hose will then be reduced to that of the linear material. Also the liner will expand between the convolutions under internal pressure so that the bore diameter will still tend to take the form of a series of ridges. The cost of the hose is increased appreciably in any case, and with a high temperature liner material, such as PTFE, the increase in cost can be considerable.

A more practical form of construction where the use of a smooth bore PTFE liner is required is semi-metallic hose, comprising alternate layers of PTFE tube and steel wire braid. Although this reduces the maximum working temperature possible to the limit for PTFE such a construction does permit high pressure ratings — up to 280 bar (4 000 lb/in^2) with 4.7 mm (3/16 in) bore hose — and the hose is substantially flameproof.

Coiled Tubing

Coiled tubing may be used for high temperature applications in place of metallic flexible hose or swivel fittings, although an important factor to consider in such cases is the fatigue effects resulting from mechanical flexing and pressure cycling. Also a stack of coiled tubes is highly susceptible to sympathetic vibration as a spring mass in the presence of vibration, although this can usually be eliminated or reduced by friction damping. This can take the form of close winding of the coils so that they bear against each other, or interleaving the coils with friction bands.

In practice the form of coiling would appear to affect the fatigue characteristics quite drastically with pressure cycling, depending on the amount of tube flattening produced. The smaller the degree of flattening the better. With mechanical flexure only, flattening during coiling is of less significance.

With a deflection of 7° per coil a fatigue life in excess of 200 000 cycles can be expected for stainless steel tubing working at temperatures up to 100°C (212°F).

Hose Fittings

Hose fittings are of two basic types; permanently attached and reusable, each designed to meet the specific requirements of the installation. Permanently attached fittings are crimped or swaged on the hose ends and are discarded if the hose is, for any reason, replaced. Resuable fittings are screwed or clamped to the hose end and can be removed and used again should the hose have to be renewed. Hose end fittings can be made of steel. Efficient fluid piping brings the several benefits of reducing downtime, minimizing maintenance, reducing stock-holding, cutting costs, speeding up production, or solving some other problem.

Hose Selection

In the case of hydraulic hose, the working pressure is generally the primary parameter, which will normally determine the type (construction) of hose required. Other major factors to consider are the frequency and intensity of surge pressures, fluid temperature, and the amount of flexing which will be required.

Pressure Surges

Resistance of hose to pressure surges can only be determined empirically. General agreement has now been reached that the applied pulsations should be of square wave form as being more capable of reproduction and easier to control within the parameters of the testing equipment currently available. Further, the results from this wave are much more consistent with regard to experimental scatter within a random sample.

Of course other waveforms are still used for certain government, military and defence specifications and also in some national standards where historical data is still very relevant, but at this time these are all exceptions, rather than the rule.

The number of impulses per minute, another very important feature, has also been accepted within ISO, together with the minimum number to failure. All of these items are included in ISO Standard 1436. Recent thinking on impulse testing includes the need for a standardized testing fluid within the testing circuit, and also the possible need to specify the maximum speed of application of the rate of pressure rise. The latter would well apply to hydraulic hose of all spiral-steel wire construction.

Spirally reinforced wire hose is extremely resistant to high impulse pressure surge conditions. This is because there are no crossover points (as in braided construction), and hence no local places of 'rubbing' action — the wires 'slide' over one another. This type of hose is extremely difficult to produce and great care is needed during the various stages of production. Until recently, it has only been possible to produce it in lengths of up to approximately 20 m (65 ft 7½ in), but by applying the techniques of long length flexible mandrels, plus advanced machinery design it is now possible to manufacture it in lengths comparable with those achieved by the established steel braided wire long length processes. This is a major breakthrough in hose design and technology.

Hoses of this type have greater impulse life performance and are capable of operating at higher working pressures than conventional braided constructions.

Hose Sizing

Hose sizing is normally determined by flow rate requirements and acceptable pressure drop. Sizing is normally based on recommended flow rates, *viz*

High pressure feed lines — 2–4.5 metres/sec (7–15 ft/sec)

Intake lines — 0.6–1.2 metres/sec (2–4 ft/sec)

Return lines — 0.6–1.2 metres/sec (2–4 ft/sec)

Theoretical bore sizes can be determined directly as:

$$d = 1.127\sqrt{\frac{Q}{V}}$$

where Q = delivery or flow rate, V = recommended velocity — in consistent units

Theoretical solutions for actual bore size required are, of course, somewhat nominal, since the parameter on which they are based — recommended flow velocity — is also nominal. This provides flexibility to 'adjust' the bore size to match a standard pipe or tube size as necessary. Only in very special circumstances would an exact theoretical size of tube be specified which differed from a standard size, and this would inevitably result in a substantial increase in cost. The general solution, therefore, would be to select the nearest standard size of pipe which should give a satisfactory flow velocity within these two sizes. Alternatively, the theoretical size required could be based on the maximum recommended flow velocity, and final selection then based on the next standard size up. This will result in a nominal 'minimum size' tube.

The adoption of flow velocities within such limits generally ensures that the resultant pressure drop will be moderate and within normally accepted limits. Since such calculations are based on full flow, this does not preclude the possibility of much higher localized velocities being developed which could, of course, seriously modify the total pressure drop realized. This, however, is largely a matter of system design requirements, which could modify pipe sizing based on full flow characteristics.

The pipe size, once established, defines all three parameters — flow rate, flow velocity and pressure drop — for a fluid of specific viscosity. Recommended flow velocities are based on typical fluid viscosities. Flow velocities may have to be reduced in the case of higher viscosity fluids in order to avoid excessive pressure drop, *ie* an increase in line sizes. In the case of high pressure systems, line sizes in excess of 19 mm (¾ in) diameter should not be necessary and will only result in excessive cost, unless the flow rate required is very large.

Where close calculation of pressure drop is required the effect of bore tolerance must be considered. Permitted bore tolerances are more generous in the case of smaller pipes (see Fig 3) and an assessment of likely maximum errors in working to nominal bore sizes is:

Sizes up to and including 6 mm (¼ in) bore — 15%

Above 6 mm (¼ in) and up to 12.5 mm (½ in) bore — 10%

Above 12.5 mm (½ in) and up to 25 mm (1 in) bore — 5%

Above 25 mm (1 in) up to 50 mm (2 in) bore — 2½%

Rubber compressibility and variation in braid angle can generally be neglected as sources of error in pressure drop calculations as these effects tend to cancel out one another (*ie* rubber compressibility will result in a decrease in pressure drop of the same order as the increase in pressure drop caused by braid distortion under the same pressure load).

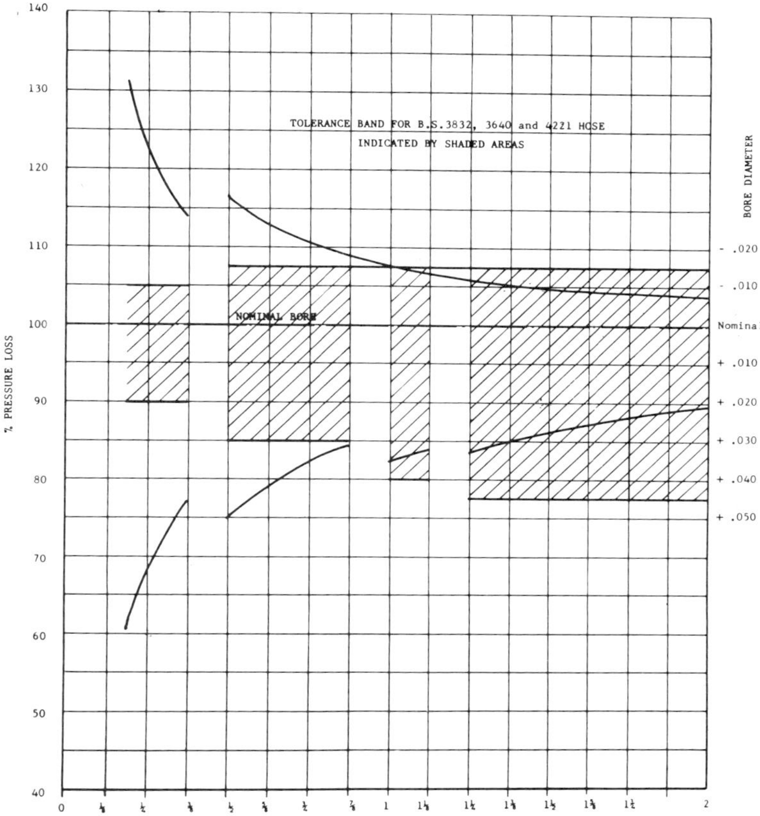

Fig 3 Effect of permitted bore diameter variation on pressure losses. Nominal bore losses indicated as 100%.

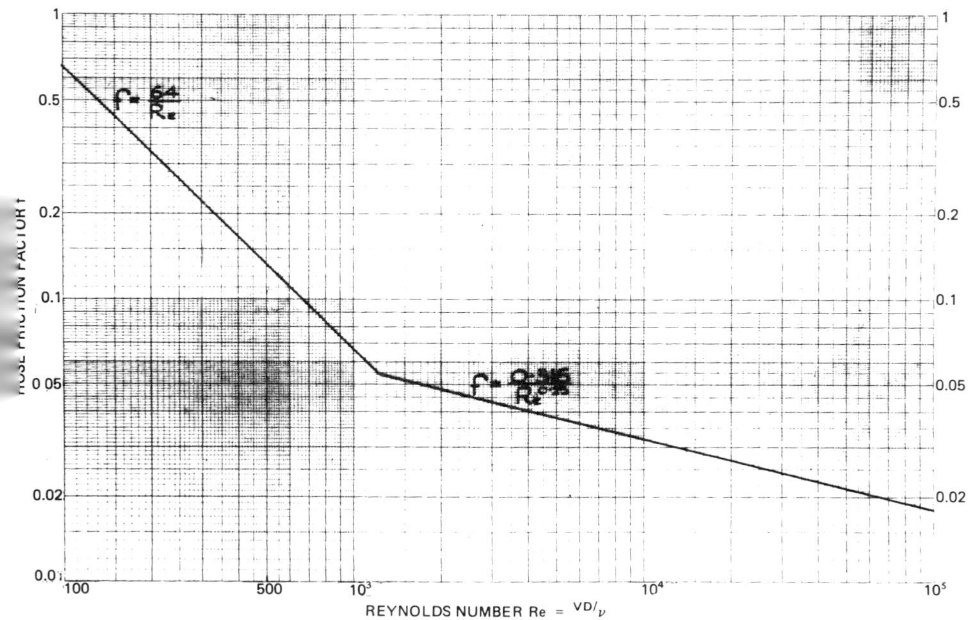

Fig 4

Friction factors for typical high-quality hose are shown in Fig 4, plotted against flow Reynold's number. These can be used for calculating pressure drop from standard flow formulas — see chapter on *Pipework Calculations.*

Compatibility

Compatibility is seldom a problem with metallic tubes or conventional hoses used with oil fluids. Alternative liner materials may, however, be required in the case of hoses used with some fire-resistant fluids or for higher temperature working. In such cases the advice and recommendations of the hose manufacturer should be sought. The majority of elastomeric hoses are suitable for working over a temperature range from –40°C (–40°F) up to 100°C (212°F). Pressure rating will, however, begin to fall off at temperatures much above 20°C (68°F).

See also chapters on *Hose Couplings and Fittings,* and *System Maintenance.*

Hose Couplings and Fittings

HOSE END fittings fall into two broad groups – permanent or re-usable types. Flexible hoses are ommonly supplied as complete assemblies with permanent end fittings, normally of the swaged ype. The end fitting then comprises a body with an external sleeve, suitably shaped with gripping urfaces – Fig 1. The hose end is inserted between the two, and the outer sleeve is then swaged own to grip the hose in place. For higher pressure working the hose itself is split circumferentially, r the outer cover removed, so that the swaged sleeve obtains a 'bite' on the wire braid. On low ressure hose, the hose may simply be clamped onto the body of the fitting with a taper nut – ig 2. Alternatively a socketless hose fitting may be used for working pressures up to 17.5 bar 250 lb/in^2) as in Fig 3. In this case it is only necessary to cut the hose off true and square and crew it onto the fitting until properly located. Simple clamp-type fittings are also used on low ressure hoses, incorporating a metal band or wire to clamp the hose onto a nipple.

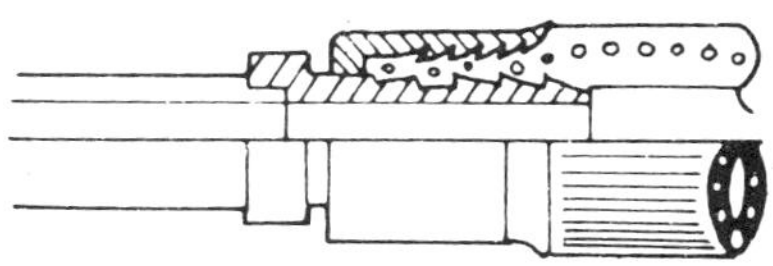

Fig 1

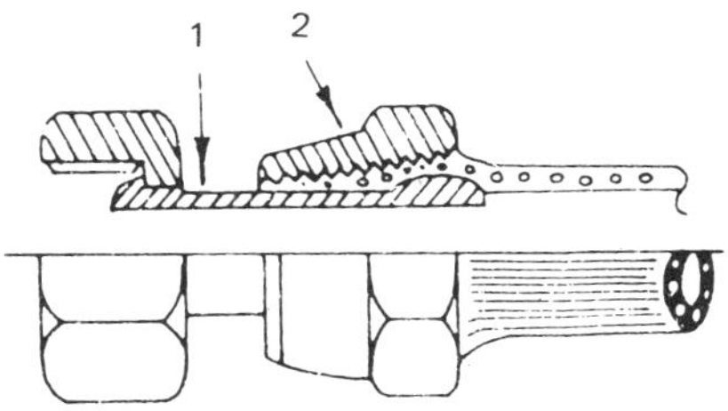

Fig 2 Swaged end fitting for low pressure hose.
1–sleeve. 2–taper nut.

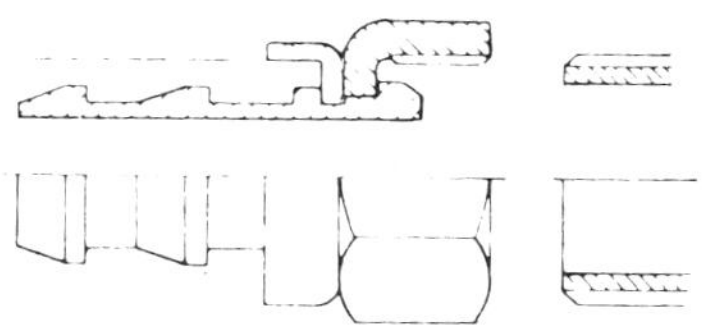

Fig 3 Socketless hose end fitting.

For high pressure hoses, lip-seal fittings are generally preferred. In the example shown in Fig 4, e nipple is machined to a sharp edge which is forced into the hose and the outer braiding is ndwiched between the nipple and socket, or outer ferrule. A lip on the nipple provides a positive rip against tensional loads, when the fitting is tightened up or swaged in place. Alternatively, the ipple may be forced on the hose and the outer braiding clamped in place under compression by eans of a ductile sleeve.

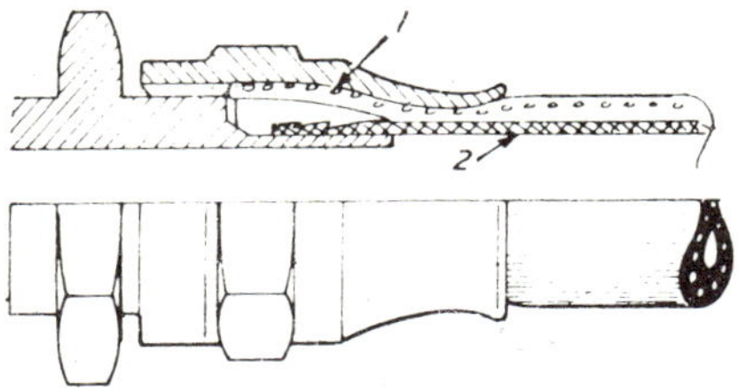

Fig 4 One form of lip-seal end fittings. 1—braid. 2—hose inner lining.

Simple *re-usable fittings* are assembled by removing the outer cover of the hose and screwing the sleeve onto the exposed braid. The nipple is then screwed through an opposite hand thread in the top of the sleeve, providing a firm compression and grip for the braid.

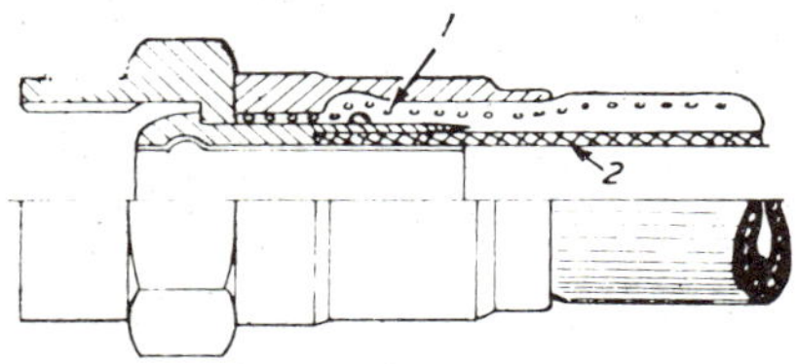

Fig 5 Typical form of detachable end fitting for high pressure hose. 1—braid. 2—hose inner lining.

Fig 5 shows a fitting where both the inner hose lining and braid are gripped compressively. Hose fittings of this type have the advantage that they require no special tools and can be assembled by hand, as well as being re-usable. They enable a hose to be made up to length on the job, instead of having to work to stock lengths of hose with integral end fittings. Some designs of re-usable hose fittings, however, demand the use of an arbor press to complete assembly.

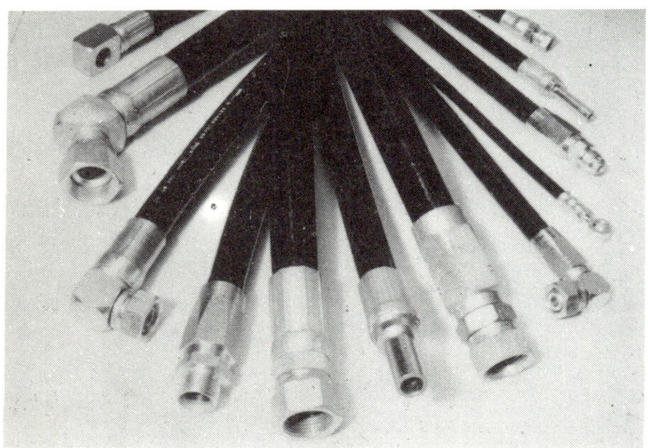

Selection of Intec hose end fittings.

Self-Sealing Couplings

Self-sealing couplings are particularly useful where some part of the system connections may require periodic removal or change of position. They are based on providing a two-part seal — the first provides a seal between flow and atmosphere and the second a complete seal for through-flow when the coupling is fully assembled. The action is reversed when the coupling is disassembled.

Such couplings incorporate a self-sealing valve in either the plug or socket connections of the coupling, or in both. In the first two cases, a poppet valve is normally used; in the last a poppe

HANDBOOK OF VALVES, PIPING & PIPELINES 1ST EDITION

OVER 500 PAGES, HAND BOUND, HARD CASE COVER, GOLD BLOCKED, 2,000 DIAGRAMS, TABLES, CHARTS, ILLUSTRATIONS.

Contents include: Sect 1 — Introductory:- Piping Services, General Summary and Classification of Valves, Valve Nomenclature. Sect 2 — Types of Valves. Sect 3 — Valves — Design and Construction. Sect 4 — Valve Actuators. Sect 5 — Performance, Fluid Transport, Frictional Losses, Pressure Drop. Sect 6 – Pipelines. Sect 7 — Services (Valves). Sect 8 — Services (Pipelines). Sect 9 — Materials. Sect 10 — Guides and Standards. Sect 11 — Data. Sect 12 — Buyers Guide.

TRADE AND TECHNICAL PRESS LTD., CROWN HOUSE, MORDEN, SURREY, SM4 5EW, ENGLAND.

Power to all cylinders

We have the power! The power to keep the pressure on through constant research, development and quality control. The Europower range of hydraulic fittings is one of the world's largest and most comprehensive and our distribution system, impressive. So put the pressure on, call us, we'll send you our latest catalogue with your nearest source of supply.

EH EUROPOWER HYDRAULICS LTD.

High Street, Market Weighton, York YO4 3AD. Telephone (0696) 72361. Telex 52582.

INTERNATIONAL STANDARDS FOR HOSE COUPLINGS

METRIC

METRIC FEMALE DIN 7608 DKM	METRIC MALE DIN 7611 AGM	GERMAN METRIC MALE LIGHT CEL	GERMAN METRIC MALE HEAVY CES	GERMAN METRIC FEMALE LIGHT DKL	GERMAN METRIC FEMALE HEAVY DKS	GERMAN METRIC STAND-PIPE LIGHT RSL	GERMAN METRIC STAND-PIPE HEAVY RSS	90° METRIC FEMALE DIN 7608 TUBULAR ELBOW	45° METRIC FEMALE DIN 7608 TUBULAR ELBOW	FRENCH METRIC FEMALE DKF	FRENCH METRIC MALE	FRENCH METRIC STAND-PIPE

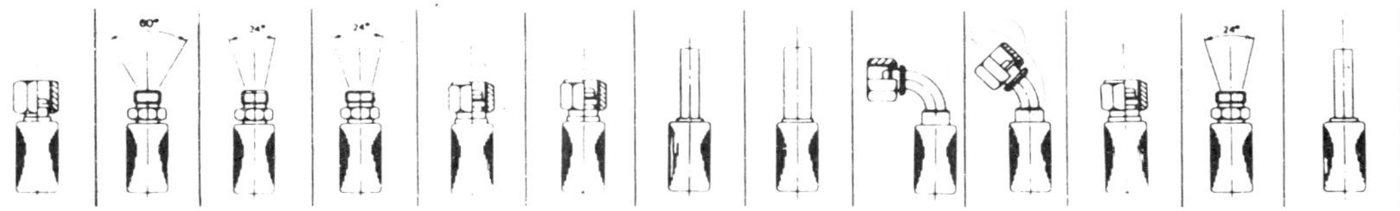

ENGLISH

BSP FEMALE	BSP MALE	BSP(T) MALE	STAND-PIPE o.d.	STAND-PIPE n.b.	90 BSP FEMALE TUBULAR ELBOW	45 BSP FEMALE TUBULAR ELBOW	90 BSP FEMALE FORGED ELBOW	45 BSP FEMALE FORGED ELBOW

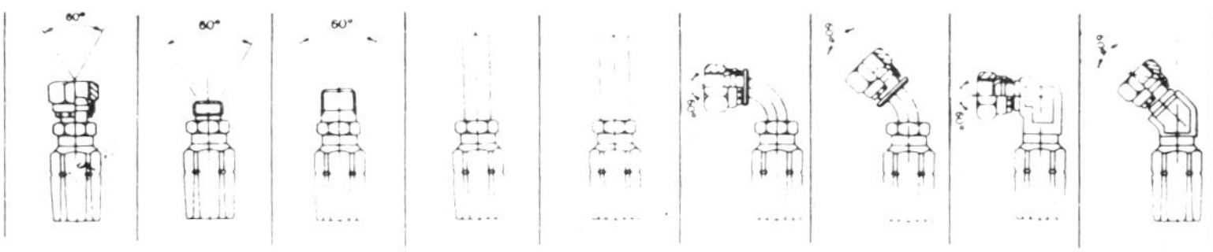

AMERICAN

JIC FEMALE	SAE FEMALE	NPSM FEMALE	JIC MALE	SAE MALE	O-RING BOSS MALE	NPTF MALE	SAE O-RING FLANGE

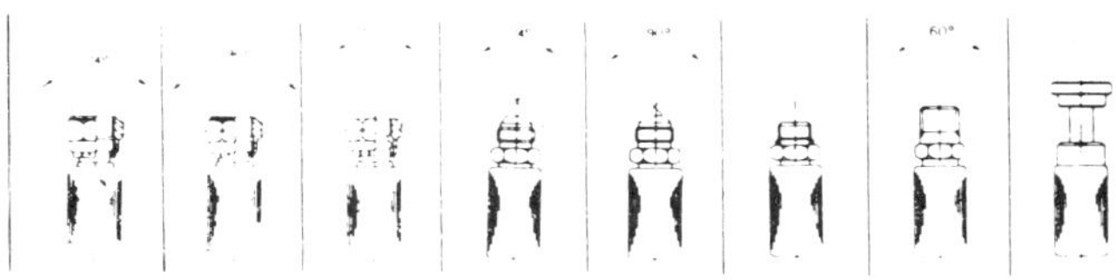

90 JIC FEMALE TUBULAR ELBOW SHORT LEG	90 JIC FEMALE TUBULAR ELBOW LONG LEG	90 SAE FEMALE TUBULAR ELBOW	90 SAE O-RING FLANGE TUBULAR ELBOW	45 JIC FEMALE TUBULAR ELBOW	45 SAE FEMALE TUBULAR ELBOW	45 SAE O-RING FLANGE TUBULAR ELBOW	SAE SPLIT FLANGE PAIR

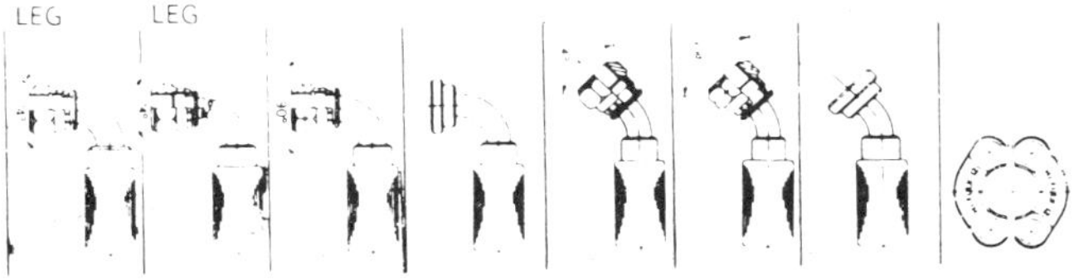

valve in one half and a sliding sleeve in the other, or a poppet valve in each half – this usually being the cheaper solution. Connection of the two components may be by screwing, bayonet fitting, or push-pull quick release action. The main requirement is that they should be easy to part and re-assemble. Flow must be immediately shut off when the line is broken, and immediately re-established when the coupling is assembled. With a fully satisfactory design, no fluid is lost and no air is introduced into the system. This is not usually possible with one-valve or double poppet valve designs, and so the poppet valve associated with a sliding sleeve is generally preferred.

The bore of a self-sealing coupling is usually made larger than the corresponding line diameter, to leave a free area around the valve, equal to the full bore area of the line. This results in minimum pressure drop through the coupling. With high pressure systems, it is usually necessary to lower the line pressure before uncoupling or recoupling, to reduce hydraulic load on the seals and cut down the effort needed to complete the coupling (which would otherwise be considerable in the case of a large coupling). This limitation can be overcome by hydraulically balancing the coupling design so that the sealing valves in each coupling half are subject to only small hydraulic thrusts, regardless of the actual line pressure. Such couplings can be broken and re-made at full system pressure with a minimum of manual effort.

TABLE I – EQUIVALENT HOSE AND THREAD SIZES*

Nominal Size	METRIC	HOSE i.d.		BSP	NPTF & NPSM	JIC 74° CONE	SAE 90° CONE	DASH SIZE	FRENCH METRIC	FRENCH METRIC S'PIPE NOM o.d. mm	GERMAN METRIC HEAVY	GERMAN M.S.PIPE (LIGHT) NOM o.d.mm	GERMAN METRIC LIGHT	GERMAN METRIC S'PIPE (HEAVY) NOM. o.d. mm
		ins	mm											
1/8				1/8–28	1/8–27	5/16–24 UNF	5/16–24 UNF	–2	M16 x 1.5	10				
3/16	M12 x 1.5	3/16	5			3/8–24 UNF	3/8–24 UNF	–3			M12 x 1.5	6	M16 x 1.5	8
¼	M14 x 1.5	¼	6.3	¼–19	¼–18	7/16–20 UNF	7/16–20 UNF	–4	M20 x 1.5	13 G	M14 x 1.5	8	M18 x 1.5	10
5/16	M16 x 1.5	5/16	8			½–20 UNF	½–20 UNF	–5			M16 x 1.5	10	M20 x 1.5	12
3/8	M18 x 1.5	3/8	10	3/8–19	3/8–18	9/16–18 UNF	5/8–18 UNF	–6	M24 x 1.5	17 G	M18 x 1.5	12	M22 x 1.5	14
½	M22 x 1.5	½	12.5	½–14	½–14	¾–16 UNF	¾–16 UNF	–8	M30 x 1.5	21 G	M22 x 1.5	15	M24 x 1.5	16
5/8	M26 x 1.5	5/8	16	5/8–14		7/8–14 UNF	7/8–14 UNF	–10			M26 x 1.5	18	M30 x 2	20
¾	M30 x 1.5	¾	19	¾–14	¾–14	1.1/16–12 UN	1.1/16–14 UNS	–12	M36 x 1.5	27 G	M30 x 2	22	M36 x 2	25
7/8		7/8	22	7/8–14		1.3/16–12 UN	1¼–12 UNF	–14						
1	M38 x 1.5	1	25	1–11	1–11½	1.5/16–12 UN	1.3/8–12 UNF	–16	M45 x 1.5	33 G	M36 x 2	28	M42 x 2	30
1¼	M45 x 1.5	1¼	31.5	1¼–11	1¼–11½	1.5/8–12 UN		–20	M52 x 1.5	42 G	M45 x 2	35	M52 x 2	38
1½	M52 x 1.5	1½	38	1½–11	1½–11½	1.7/8–12 UN		–24	M58 x 2	49 G	M52 x 2	42		
2	M65 x 2	2	51	2–11	2–11½	2½–12 UN		–32						

* Sizes and specifications are largely the same as the American standards SAE 100R1 and SAE 100R2

Choice of Type

The choice between permanent and re-usable fittings is often one of cost. On average, re-usable couplings cost more than the permanent type, but the attachment procedure usually takes longer. When large numbers of hydraulic hoses are replaced, maintenance generally opts for the permanent style fitting.

Another style of re-usable fitting, the push-on type, is perhaps the easiest to attach and requires no special tools. This fitting is generally used for low pressures (up to 250 lb/in^2) and does not have an external clamp. When removing the fitting for use on a replacement hose, care should be taken not to nick the barbs on the insert.

SECTION 2E

Hydraulic Fluids

IN HYDROSTATIC systems the prime requirement of the fluid is to transmit pressure, hence low viscosity and low compressibility are important features. Other factors which must be considered are oxidation stability, load-carrying ability, corrosion protection, cleanliness, seal compatibility, air entrainment and foaming, viscosity index, pour point and filterability.

The viscosity level of an hydraulic fluid is normally selected on the basis of the speed and discharge pressure of the pump which must be satisfactorily lubricated for optimum performance. Too low a viscosity will cause pump slip, leakage and inadequate lubrication, whilst too high a viscosity will cause overheating and cavitation. Probably one of the most common problems associated with hydraulic equipment is leakage, which is the prime cause of high fluid consumption. This is due to high pressures within the system and seal failure due to both quality and incompatibility. A further problem area is cross-contamination with other fluids such as soluble cutting oils and compatibility between the two fluids would be an added advantage.

Types of Fluids

Hydraulic fluids can be grouped into four distinct types, *viz*

(i) *Mineral oils* — either uninhibited or treated with additives;

(ii) *Emulsions* — water-in-oil or oil-in-water;

(iii) *Water-based glycols;*

(iv) *Synthetic fluids* — (a) phosphate esters, (b) carboxylate esters, (c) chlorinated hydraulics, (d) poly alpha olefines, (e) poly glycols, (f) silicone fluids.

Of these groups (ii) and (iii), and (a), (b) and (c) from group (iv) are also classified as fire-resistant fluids.

Table I compares the general characteristics of the chief types of fluids. Table II indicates their normal applications.

Mineral Oils

Mineral oils are the normal choice for industrial hydraulic systems, with the advantage of offering nearly all the requirements of an 'ideal' hydraulic fluid except for fire-resistance. Straight mineral-oil lubricants can be regarded as suitable for hydraulic systems working under ideal conditions, particularly with low fluid temperatures and in perfectly clean systems. As a general rule, however, such oils are compounded with special additives to produce hydraulic oils specifically intended for use in all practical hydraulic systems. The cost of hydraulic oil is more than justified and it should be regarded as a normal choice. There are exceptions, but these are mainly concerned with con-

TABLE I – GENERAL CHARACTERISTICS OF HYDRAULIC FLUIDS

	Mineral Oil	Phosphate Ester	Water-Glycol	Water-in-oil Emulsion	Oil-in-water Emulsion	Chlorinated Aromatics
Specific gravity (typical)	0.864	1.275	1.060	0.916–0.94	Less than 1.0	1.43
Maximum service temperature °C	110	150	65	65	65	150
Water content (%)	None	None	45	40	95	None
Fire-resistance	Poor	Good	Excellent	Good	Good	Good
Viscosity index	High	Low	Very high	High	Low	Low
Lubricating properties	Excellent	Very good	Good	Fair	Fair	Fair-good
Special seals	No	Yes	No	No	No	Yes
Special paints	No	Yes	Yes	No	No	Yes
Rust prevention	Very good	Fair	Fair	Good	Fair	Fair
Toxicity	None	Slight	None	None	None	Slight

TABLE II – TYPES AND APPLICATIONS OF HYDRAULIC FLUIDS
(See also Table III)

Fluid Type	Application
Non-inhibited mineral oil	Hydraulic couplings
Mineral oil with anti-rust and anti-oxidation properties	Circulating systems and piston pumps
Mineral oils with anti-rust, anti-oxidation and anti-wear properties	Systems including highly loaded components, gear and vane pumps
High VI oils with anti-rust and anti-oxidation properties	Circulating systems and piston pumps operating over wider temperature ranges
High VI oils with anti-rust, anti-oxidation and anti-wear properties	Systems hincluding highly loaded components, gear and vane pumps, operating over wider temperature ranges
Synthetic fluids with no specific fire-resistance properties	Hydraulic systems where special properties are required
High VI oils with anti-rust, anti-oxidation, anti-wear and anti-stick/slip properties	Hydraulic slideway systems where intermittent sliding at slow speed has to be minimized
Oil-in-water emulsions Chemical solutions in water Water-in-oil emulsions Water polymer solutions* Phosphate esters* Chlorinated hydrocarbons Mixtures of phosphate esters and chlorinated hydro-carbons Other synthetic fluids	Hydraulic systems requiring a fire resistant fluid, *eg* high temperature plants or where leakage on to hot surfaces can occur. * Particularly suitable for aircraft hydraulic systems.

venience. Thus the hydraulic system associated with an internal combustion engine might employ the same fluid as specified for the engine crankcase. Also on certain types of hydraulic equipment, particularly those associated with agricultural machines, the designer has to allow for the possibility that the hydraulic side may be filled by the user with almost any lubricating oil which is readily to hand. Specialized oil-hydraulic equipment is, however, invariably designed around an oil of a specific viscosity (primarily to suit the requirements of the pump or motor), and a hydraulic oil is always implied in such cases.

Additives for Mineral Oils

Most modern hydraulic oils are compounded with additives, notably oxidation inhibitors, corrosion inhibitors and anti-foam agents. Some oils may have less and others more (*eg* film strength improvers or anti-wear additives can be advantageous where high bearing loads are involved, and pour point depressants for fluids used in systems operating at very low working temperatures, or starting up from cold in very low ambient temperatures). A separate additive is also commonly employed to improve the viscosity index of the oil (see later).

The main cause of deterioration with a straight mineral oil is oxidation. The rate of oxidation is enhanced by heating (*eg* high working temperatures for the oil), agitation (which is present in most hydraulic systems to some extent or other) and the presence of contaminants which can act as catalysts (notably metal particles).

Apart from the loss of lubricating properties, the onset of oxidation is accompanied by the formation of soluble and insoluble degradation products, the latter being deposited in the system in the form of sludge. The oil also loses its ability to separate from water and air, both of which contaminants are invariably present; and will tend to become increasingly acid, which can lead to corrosion.

Oxidation inhibitors work by showing a preferential absorption for oxygen and thus remain effective as long as there is active additive remaining. Some, such as the phosphorous and sulphur compounds, also possess marked anti-wear and anti-corrosive properties and are thus multi-purpose additives.

Oxidation additives are usually added in concentrations of up to 5%, this being the maximum figure for which such additives are fully effective. Higher proportions will not normally give any increase in oil life and may even have undesirable effects.

Very marked improvements in oil life have been achieved with specially treated base oils and oxidation inhibitors, compared with figures realized less than a decade ago. It must be emphasized,

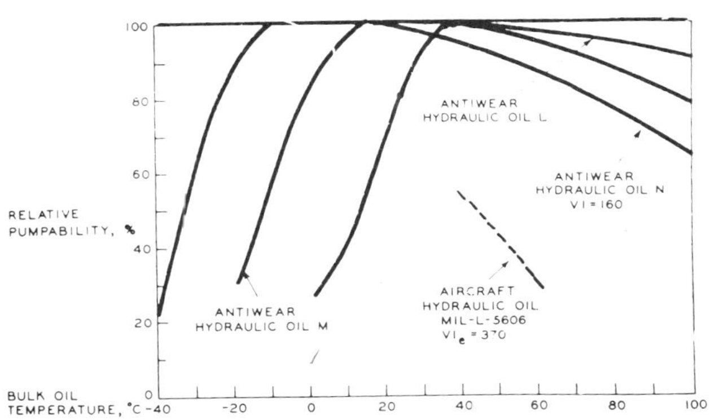

Flow characteristics vs temperature for three low temperature hydraulic oils (vane pump — 70 bar [1000 lb/in^2]).

however, that ultimate life in a particular system will still depend largely on the operating conditions, particularly the oil temperature and the cleanliness of the system.

Corrosion inhibitors are essentially rust inhibitors capable of adhering strongly to metallic surfaces and 'passivating' the surface, or isolating it from contact with air and moisture. The selection of a suitable additive is quite critical, however, both to meet the service conditions concerned and avoid interaction with other additives. In particular, certain types of rust inhibitors have a degrading effect on oxidation inhibitors, whilst others may have a secondary effect of working as an emulsifying agent tending to emulsify any free water present in the oil.

Anti-foam agents are added to ensure effective release of entrained air from the oil surface in the tank without excessive foaming developing at the surface. Basically they are 'foam breakers', causing an early disruption of the air bubbles as they appear. By this means, air normally dissolved in mineral oil and released at lower pressures, or any entrained air, is released with no adverse effects on the working of the system.

Anti-wear additives are basically film-strength improvers, greatly assisting in maintaining the full lubrication properties of the oil. A typical anti-wear additive is zinc dithiophosphate in proportions up to about 1%.

Viscosity Index Improvers

The viscosity index of an oil can be raised with additives. The additives used are normally polymerized methylacrylate esters, or butane or styrene olefines, in proportions of from 4% to 8%. All such additives are susceptible to shear break-down and so the initial VI index achieved is seldom maintained in practice, the extent of the break-down being dependent on the rate of shear experienced by the oil. In general, an initial loss may be expected during the first few hours of working in the system, after which the viscosity index should remain appreciably constant through the useful life of the oil, unless continually subjected to shear stresses in a particular part of the system.

Modern polymeric viscosity index improvers show very much better viscosity retention at even high shear rates than their earlier counterparts but still have limitations at extremely high pressure (*eg* 350 bar or 5 000 lb/in^2). For very high pressure applications where no shear loss can be tolerated, synthesized hydraulic oils are to be preferred as they do not use polymeric thickeners and show no shear loss — *eg* see Fig 1.

Emulsions

Water-in-oil and oil-in-water emulsions are similar in general behaviour, the particular difference being that in the former water is distributed in droplet form through an oil medium, and in the

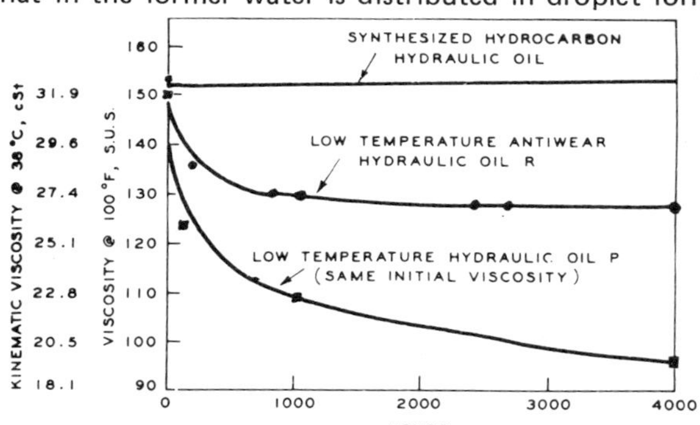

Fig 1 Vane pump shear stability test results 70 bar (1 000 lb/in^2).

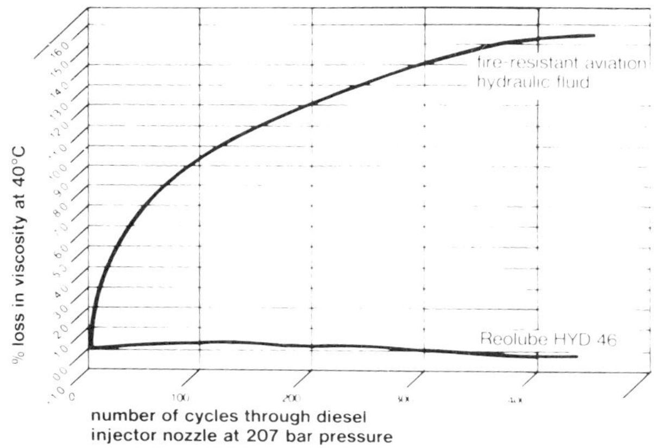

Shear stability of Ciba-Geigy Reolube HYD 46.

latter oil droplets are distributed through a water medium. As a consequence, oil-in-water emulsions (with a water medium) are rather more fire-resistant than water-in-oil emulsions. On the other hand, the latter (with an oil medium) generally have better lubricating properties.

Water-in-oil emulsions are more widely used and may have a water content of up to 60%, fire-resistance being directly related to water content.

Similar additives may be included as for mineral oils, notably oxidation inhibitors, anti-wear and anti-corrosion additives, and also emulsifying agents to maintain the emulsion in stable form. Viscosity index improvers are not used since water-in-oil emulsions are non-Newtonian fluids with no viscosity index as such. Their actual viscosity is dependent on the rate of shear, and at very high shear rates reverts to that of the oil content itself. This generally limits their application to systems or components which do not produce high localized rates of shear, *eg* such emulsions would generally be unsuitable for use with high-speed vane pumps or with rolling bearings.

Properly formulated, water-in-oil emulsions can be quite stable, although some separation may occur if the fluid is allowed to stagnate. This will tend to result in an oil-rich layer or emulsion, or even a pure oil layer forming at the top of the tank level. This need not be significant, for there is usually sufficient agitation on re-starting to re-form a consistent emulsion as the fluid is circulated through the system. A more likely cause of trouble is where separation is caused by the presence of contaminants, as such separation may be more permanent (largely because the emulsifying agent has probably been exhaused by the presence of contaminants). Basically, therefore, water-in-oil emulsions are most reliable in clean systems; they also have strictly limited working temperatures, in common with other water-based fluids, and the possibility of water loss through evaporation and subsequent modification of the fluid make-up.

In general, if there are no leaks in the system, any loss of fluid volume in a water-in-oil emulsion can be replaced by topping up with water. However, if there is fluid leakage, topping up must be done with water/oil mixture, otherwise the balance will be upset and the lubricity of the fluid may be adversely affected, (or the fire-resistance adversely affected if topped up with oil). Manufacturers of this type of fluid generally supply a topping-up concentrate (which will usually also contain an anti-corrosion additive) and specific instructions for its use.

Oil-in-water emulsions are essentially water, containing approximately 2–5% of an emulsifier system to provide limited lubrication and anti-corrosion properties. Their use is limited to massive systems, *ie* which discharge to waste; and in pressure transmitters such as pit-props where lubrication is not very demanding. Their main limitation has been their inability satisfactorily to lubricate pump bearings.

Bacterial Contamination

Surprisingly, mineral oils and oil-water emulsions are prone to contamination by micro-organisms. The necessary environment to introduce bacteria is free water plus a nutrient. Water is always present in oils and the necessary nutrient may be provided by the additives used in hydraulic oils. Basically, as the number of organisms multiplies, additives are consumed, decreasing their percentage and effectiveness. Replacement of oil losses with fresh oil again provides nutrient promoting further multiplication of the micro-organisms.

In a hydraulic system contaminated with bacteria the oil becomes turbid and a slimy sediment is also formed, further increased by depletion of anti-rust and anti-wear additives. Under static conditions the organisms will become concentrated in the interface between the fluid and a water layer. When the system is working, with the fluid in circulation, the bacteria have the effect of acting as an emulsifying agent providing a water-in-oil dispersion. The bacteria themselves can also accumulate in certain parts of the system to cause blockage in valves and filters.

The most effective way of removing bacteria from a system is to drain it completely, flush through with an oil soluble biocide followed by conventional flushing oil and refill with fresh sterile fluid. Filtering or batch filtering is no cure, for whilst this will remove solid particles it will not remove bacterial infection and water.

Sterile oil is produced by re-refining the original oil to include complete sterilization, removal of products of oxidation and fine filtering. Provided this is done comprehensively it can be fully effective and the cost can be as little as one half that of new oil.

Water-based Glycols

Water-glycol fluids originated as straight water-glycerine mixtures, with the glycerine content adjusted to give the required degree of protection against freezing in water-hydraulic systems. The glycerine content employed can range up to 50%. A secondary advantage offered by such mixtures is a raising of the viscosity of the fluid and an improvement in viscosity index. Cost, however, is relatively high, nullifying one of the basic advantages of using water as hydraulic fluid. Thus oil-hydraulic systems are preferable to water systems for low-temperature applications for all general uses, even where a large bulk of fluid is involved.

Water-polyglycol mixtures have, however, been further developed as industrial fire-resistant fluids, mainly around Hydrolube H-2, offering superior protection to water-in-oil emulsions, and lower cost and minimal compatibility problems compared with phosphate esters. Nitrile rubber seals, in fact, are equally suitable for mineral-oil fluids and all water-based fluids. The lubricating properties of these mixtures are greatly improved by the incorporation of anti-wear and load-carrying additives to provide satisfactory lubrication under boundary film conditions; and their viscosity is increased by the addition of polymer type thickeners, which also provide a high viscosity index. They are reasonably stable, although they do need close control and regular checking of water and alkaline content.

The water content controls the fire-resistance of the fluid (increasing with increasing water content). Evaporation and loss of water are likely during service, more especially at higher system temperatures, and so the system design should aim to minimize such losses. Topping-up is normally done with a pre-mixed water/glycol solution, following specific recommendations given by the fluid manufacturer. Water alone, or glycol alone, should not be used to top up a water/glycol mixture to compensate for volumetric loss.

The use of a water-glycol fluid inevitably calls for bearing loads to be de-rated, a typical figure being about one-third of the rating for an oil-lubricated bearing. The use of such fluids is not generally recommended with rolling bearings, or for gear pumps operating at pressures above 35 bar (500 lb/in^2), or components with close clearances relying on boundary film lubrication. Also the maximum service temperature of water-glycol fluids is generally low in order to avoid evaporation and loss of water content, with the frequent need for checking and topping up.

High Water Base Fluids

The category of high-water-base hydraulic fluids falls into two groups under the ISO classification (see also Tables III and IV).

HFAE emulsions of oil in more than 80% water,

HFAS chemical solutions in more than 80% water.

HFAS class fluids, also known as HWBF or 5/95 fluids, normally contain a nominal 95% of water, usually mixed at the point of use with 5% of concentrate (commonly polymers). The water used must be within the hardness limits given by the manufacturer of the concentrate.

Because of the low viscosity of HFAS fluids — 1 centistoke, or the same as water — hydrodynamic bearing films in pumps and motors are much thinner than with mineral oils. Metal-to-metal contact can thus be more frequent, aggravating wear and also generating contaminants. Modification of the detail design of the pump may be necessary to accommodate the former; and filtration down to 10 μm in pressure lines and 25 μm in return lines is recommended to remove wear products. Filters and elements must be of suitable type. Corrosion and erosion can also be a problem and in general all HFAS fluids are not compatible with zinc and cadmium plating, aluminium (unless anodized), cork gaskets, and normal paints and sealants. Stainless steel or GRP is recommended for reservoir construction. For seals, nitrile, neoprene and Viton are suitable elastomers.

Piston pumps in particular seem prone to increased wear and leakage rates operating with HFAS fluids; also showing lower volumetric efficiency (partially offset by slightly higher mechanical efficiency). Many standard piston pumps will operate satisfactorily with HFAS fluids however, without modification at moderate speeds and pressures provided pump inlet pressure is maintained at or above atmospheric pressure to eliminate cavitation, *eg* using a flooded inlet with no filter or strainer on the inlet side.

Standard vane pumps, in general, are unsuitable for use with HFAS fluids because of low volumetric efficiencies and rapid wear due to adverse vane tip loading. These limitations can be overcome with special designs, *eg* with hydraulically balanced vanes with special tip shapes and modified timing and porting. Such pump designs, however, are not suitable for working as motors on HFAS fluids.

Standard gear pumps are not suitable for use with HFAS fluids, nor are they readily modified for such duty.

HFAS fluids can also produce problems with valves, notably leakage due to the low fluid viscosity. In this respect, poppet valves are better than sliding spool valves; but all types of standard valves can give acceptable service (except those with wetted components in aluminium).

Theoretically, at least, an increase in leakage should also result in an increase in heat generation, although this is more than offset by the higher specific heat and thermal conductivity of the fluid. Thus HWBF systems tend to run cooler than mineral oil systems in actual practice.

TABLE III – ISO CLASSIFICATION OF HYDRAULIC FLUIDS

Particular Application	Specific Application	Composition and Special Properties	Symbol 150 – L	Typical Application	Remarks
Hydrostatic	General	Non-inhibited refined mineral oils	HH		
		With improved anti-rust anti-oxidation properties	HL		
		HL with improved anti-wear properties	HM	Highly loaded components	
		HL with improved viscosity temperature properties	HR		
		HM with improved viscosity temperature properties	HV	Construction and marine equipment	
		Synthetic fluids with no specific fire-resistant properties	HS		Special properties
	Hydraulic slideway systems	HM with anti-stick slip properties	HG	Machines with combined hydraulic and oil-way lube systems	
	Applications where fire-resistant properties required	Oil-in-water emulsions	HFAE		Typically 80% + water
		Chemical solutions in water	HFAS		Typically 80% + water
		Water-in-oil emulsions	HFB		
		Water polymer solutions	HFC		Typically 80% – water
		Phosphate esters containing no water	HFDR		
		Chlorinated hydrocarbons no water	HFDS		Possible environment or health hazards
		Mixtures of HFDR and HFDS with no water	HFDT		
		Other synthetics with no water	HFDW		
Hydro-kinetic	Automatic trans-mission		HA		
	Couplers and converters		HN		

TABLE IV – MINERAL OILS AND WATER-BASED FLUIDS COMPARED

Physical properties of mixed fluid	**Petroleum oils**	**HFC Water-glycol fluid**	**HFB Water-oil emulsions**	**HFA 95/5 Fluids**
Density	0.85–0.9	1.05	0.95	1.00
Typical viscosity cSt at 40°C	Very low to very high	(40% H_2O) 43 cSt 40°C	(40% H_2O) 65 cSt 40°C	(95% H_2O) 1 cSt 40°C
Vapour pressure	Low	High	High	High
Corrosion resistance				
Liquid phase	Good	Good	Good	Good
Vapour phase	Fair	Fair to poor	Fair to poor	Fair to poor
Storage stability				
Room temperature	Excellent	Good	Requires care	Requires care
Low temperature	Excellent	Good	Poor	Poor
Lubricity	Excellent	Good	Good to limited	Limited
Bulk fluid temperature limits °C	–30 to +80	–32 to +65	–10 to +65	+5 to +50
Monitoring requirements	Viscosity, neutralization number	Viscosity, water content, pH	Viscosity, water content	Water content, pH, bacteria check
Fire resistance	Non-resistant	Good	Good	Excellent

Stability – HFA fluids are stable from +5°C to +50°C. Below 0°C freezing occurs which can separate the fluid. Above 50°C evaporation is accelerated.

Lubricity – HFA have limited lubricating properties as indicated by the low viscosity. Additives can improve this so lubricity therefore varies with the fluid manufacturer.

Acidity – To retard growth of bacteria HFA fluids should be alkaline at about 8 to 9.5 pH.

Synthetic Fluids

Potential advantages of synthetic fluids are superior oxygen stability, high viscosity index, lower viscosity, lower pour point and good lubricating properties (considerably better than water-based fluids, but not necessarily better than mineral oils).

Phosphate esters are the best known – and the most widely used type. The performance of modern phosphate ester fluids is more or less directly comparable with that of mineral oils, particularly as they can be rendèred in a wide range of viscosities. Their viscosity index is lower than that of mineral oils, but can be enhanced by viscosity index improvers. Bulk modulus is higher, however, which means that phosphate ester fluids are superior to mineral oils as regards compressibility effects at higher pressures. (See also Fig 2).

The chief disadvantage of phosphate ester fluids is their very high cost, followed closely by their complete lack of compatibility with conventional elastomers and paint finishes. Until com-

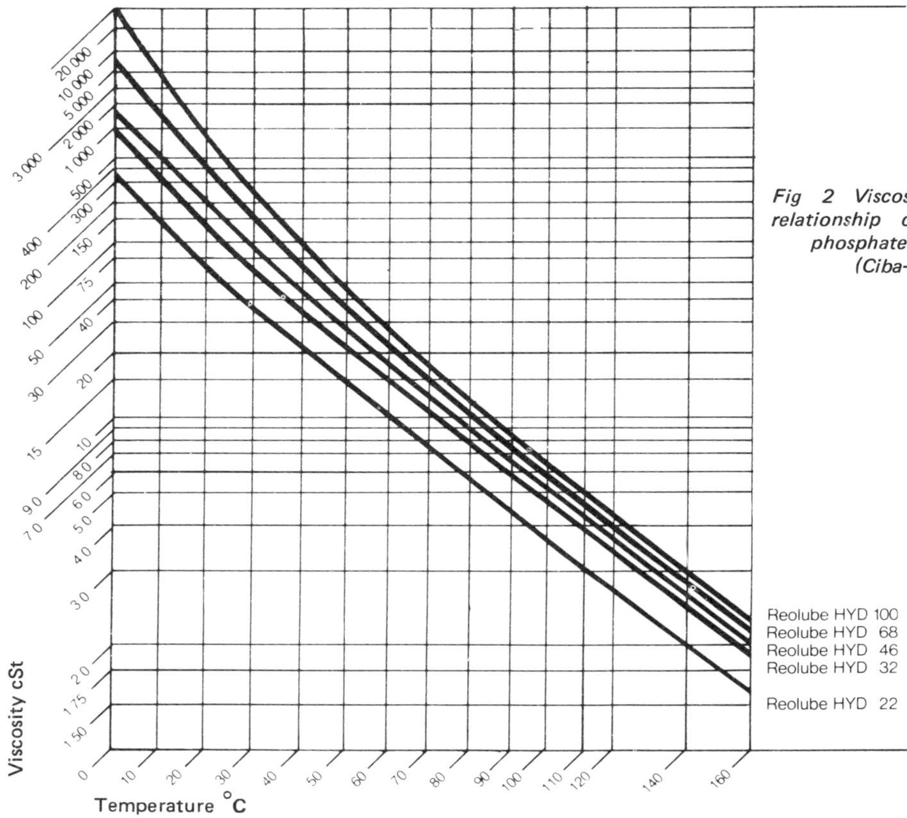

Fig 2 Viscosity/temperature relationship of a range of phosphate ester fluids. (Ciba-Geigy).

paratively recently butyl was the first choice material for elastomeric seals and packings, with possible alternatives in the more expensive Viton and silicone rubbers. Ethylenepropylene rubbers have replaced butyl and are the standard choice elastomer for use with phosphate ester fluids.

About the only paints which are compatible with phosphate ester fluids are epoxy based or polyurethanes. The latter have somewhat limited compatibility and so the former are preferred for painting reservoirs, *etc,* in which the fluid is used. It should be noted that the conventional paints used for external finishes are readily stripped by spilt or leaking phosphate ester fluids.

A rather more minor disadvantage of phosphate ester fluids is their higher specific gravity compared with mineral oils. Maximum service temperature is generally higher and phosphate ester fluids can be worked at temperatures up to about 150°C (300°F) without degradation of the fluid.

If a change is made from a system using a mineral oil fluid to a phosphate ester fluid, a complete change of seals is necessary, as well as a change of paints used on the reservoir. Fluid manufacturers can specify the change-over procedure necessary.

Chlorinated Hydro-carbons

Chlorinated hydro-carbons can be classed as mild extreme-pressure lubricants, with gravity at least comparable to, and probably better than, phosphate esters. Not all chlorinated compounds are good lubricants, however; polychlorinated diphenyls have a relatively poor performance in this respect. Specific gravity is again high (1.43) but such fluids can be produced with a wide range of viscosities. The cost is very high.

Chlorinated aromatics are not used on their own as commercial fire-resistant fluids, but normally mixed with phosphate ester fluids as phosphate ester 'blends', together with such additives as may be thought necessary (notably viscosity improvers).

Silicone Fluids

Silicones are another class of high-cost fluids – prohibitively so for all but highly specialized applications. Their chief attraction is their suitability for working at high service temperatures up to the order of 360–370°C (600–700°F), with the added virtue of an extremely high viscosity index so that reasonable viscosity values are maintained up to very high temperatures.

The performance of silicones as lubricants has been considerably enhanced by the introduction of improved silicone fluids, although these may show some slight loss of high temperature properties. They remain the sole commercial fluids available for working at temperatures in excess of 150°C (300°F).

All silicone fluids are, of course, fire-resistant, but would not normally be selected on this basis alone because of their high cost, compared even with phosphate esters.

Fire-Resistant Fluids

Fire-resistant fluids fall into two distinct categories:

(i) Water-based fluids where fire-resistance is derived from their water content.

(ii) Synthetic fluids where fire-resistance is derived from the chemical composition.

The distinction between a 'non-flammable' and 'fire-resistant' fluid is not exact, since only water is a true non-flammable fluid. Most water-based fluids are fire-resistant rather than non-flammable, with a strong 'snuffing' action. Synthetic fluids range in characteristics from being substantially non-flammable to having limited flammability. All such special fluids are thus most accurately classified as being fire-resistant, rather than non-flammable.

Whilst the advantages of using a fire-resistant fluid in such cases are obvious from the safety point of view, there can be disadvantages in their adoption, notably higher cost, some loss of lubricating properties and problems of seal compatibility, compared with mineral-oil fluids. Their application is, therefore, still far from universal even in systems where a fire risk is present, although they may be mandatory in some situations.

The main types of fire-resistant fluids are:

(i)	Aqueous-based:	oil-in-water emulsions, water-in-oil emulsions, water-glycol mixtures.
(ii)	Non-aqueous based:	phosphate esters, phosphate ester mixtures, halogenated aromatics, silicones.

Specific advantages of aqueous-based fluids are ready availability and relatively low cost. Water-glycol mixtures are appreciably more costly than oil/water emulsions, but have the further advantage of higher viscosity and excellent viscosity-temperature characteristics. Disadvantages are generally poor lubricating properties, working-temperature limitations imposed by the water content and the possibility of phase separation.

Synthetic fluids offer a nearer approach to oil-like qualities and availability in a wide range of viscosities, but may have other disadvantages apart from high cost. Thus halogenated aromatic compounds and silicones are poor lubricants (without additives) and chlorinated aromatics are toxic.

The specific applications for fire-resistant fluids are in high-pressure industrial hydraulic systems operating in close proximity to naked flames or high-temperature sources, or specifically in hazardous surroundings where fire and/or explosion risk must be reduced to a minimum. The other important field of application is for aircraft hydraulic systems, where system pressures are normally higher than in industrial practice, and the consequences of fire even more drastic. The primary risk, in all cases, is that failure of, or damage to, a pressurized system can lead to the release of a fine spray of fluid which can be ejected a considerable distance; and in the case of an oil fluid this spray is highly combustible.

Compatibility

The main problem as regards compatibility is the choice of suitable elastomeric materials for seals, *etc.* Certain fluids may, however, be incompatible with certain metals; and similarly certain additives may attack certain metals.

Specifically, metals to be avoided in contact with different types of fluids are:

mineral oils	—	none
water-in-oil emulsions	—	cadmium and zinc
water-glycol	—	cadmium and zinc
phosphate esters	—	aluminium bearings
chlorinated hydro-carbons	—	aluminium bearings

Lack of compatibility in the case of additives is usually discovered as corrosion of bronze parts — *eg* in vane pumps and bronze piston slippers in piston pumps.

Filters and Fluid Protection

MOST MODERN hydraulic systems operate at fairly high pressures *eg* commonly 140–210 bar (2 000–3 000 lb/in^2); with pumps having close clearances and control valves and servos even smaller clearances. A combination of high pressure and reduced clearance calls for the elimination of particles from the system which could cause clogging, requiring protection by fine filters. The general use of smaller reservoirs also gives more rapid circulation of fluid and thus less opportunity for particles to settle out. High fluid-operating temperatures can mean reduced oil viscosities and less protection against wear, contributing to increased contamination generated within the system. Erosion can be a very real problem in high-pressure, small-bore systems as localized fluid velocities may be as high as 175 m/s (600 ft/sec) or more. This can lead to rapid wear on hardened and polished surfaces adjacent to high velocity fluid streams when particles of only 3 to 5 μm are present in the fluid.

In fact, something over 70% of hydraulic system failures are due to contamination or poor fluid condition. Essentially, therefore, filters are necessary in modern hydraulic systems to provide a particular or specified level of contaminant removal. This can vary with the type of system, types of components involved, application and duty cycle. In the case of mobile hydraulics, for example, failure rates due to fluid deterioration can be as high as 90%.

Contaminants

Contaminants likely to be present, or generated, in hydraulic systems can be generally classified as:

(i) Soluble or non-soluble,

(ii) Abrasive or non-abrasive.

Solid non-soluble contaminants smaller than the clearance spaces can silt them up, resulting in 'sticking'. Larger non-soluble contaminants can produce 'blocking'. Abrasive contaminants of the same sizes as the clearances can lead to high rates of wear. Very much smaller solid particles can also lead to erosion in parts of the system having high fluid velocities. It is therefore essential to incorporate filtration in any closed hydraulic system in order to remove solid contaminants of all types above a specified particle size, governed by the cut-off rating of the filters employed. This is further emphasized by the fact that the presence of solid contaminants in the system will inevitably generate more contaminants, often at an accelerating rate.

Contaminants within a typical hydraulic system may be derived from both *external* and *internal* sources. Contaminants from external sources include those introduced during manufacture of components and assembly of the system. These may include casting sand, drawing compounds, weld splatter, machining chips and loose burrs, elastomeric particles from seals and gaskets, assembly compounds, adhesives and even paint particles. Such particles may range in size from 1–500 μm.

Additionally airborne dust and dirt, rust and other forms of corrosion may be further contaminants with a similar size range.

The environment is also a source of external contaminants during the working of a system, typical entry points for such contaminants being:

(i) Air breathers,

(ii) Rod seals on hydraulic cylinders (particularly in dirty atmospheres and/or with increasing seal wear),

(iii) Access plates and other detachable items.

Silt and Chip Control

The modern way of classifying contaminants is as *silt* or *chips. Silt* refers specifically to fine particles ranging in size from 5 μm down to sub-micronic size. They are responsible for clogging if allowed to build up; or more usually erosive wear and general degradation of the system if the particles are abrasive. *Chips* are larger particles which can cause catastrophic failure, *eg* by forcing open a check or relief valve or causing binding in a pump or motor. The treatment of these contaminants is referred to as *silt control* and *chip control,* respectively.

Clearances and Protection Levels

Theoretically, at least, no particles smaller than the clearances in system components should cause clogging or catastrophic failure through jamming. Typical clearances are given in Table I, which is a nominal measure of the degree of protection required from a system filter, *eg*

	Filter Rating μm
Low pressure systems with generous clearances	25–40
Low pressure heavy-duty systems	15–25
Typical medium pressure industrial systems	12–15
Mobile hydraulic systems	12–15
General machine tool and other high quality systems	12–15
High performance machine tool and other high pressure systems where reliability is critical	3–5
Critical high pressure systems and controls using miniature components	1–2

However such recommendations do not necessarily provide adequate silt control, particularly as the traditional types of filters have a nominal rather than an absolute or actual rating.

Filter Ratings

A *nominal filter rating* is an arbitrary value determined by the filter manufacturer and expressed in terms of percentage retention by weight of a specified contaminant (usually glass beads) of given size. It also represents a *nominal* efficiency figure, or more correctly a degree of filtration. Figures normally used are 90%, 95% or 98% retention of a specified contaminant size. The only standards relating are MIL-E5504A and MIL-E5504B, where Version A defines a 10 micrometre filter as being capable of removing 98% by weight of test dust larger than 10 μm at a certain high concentration; and Version B defines a 10 micrometre filter as being able to remove 95% of 10–20 μm glass beads at a high concentration.

TABLE I – TYPICAL CLEARANCES ON SYSTEM COMPONENTS

Component		Critical Clearance μm
Gear pumps:	Gear to side plate	0.5–5
	Gear tip to case	0.5–5
Vane pumps:	Tip of vane	0.5–1
	Sides of vane	5–13
Piston pumps:	Piston to base	5–40
	Valve plate twin cylinder	9.5–5
Actuators:		5–250
Servo-Valves:	Orifice	130–450
	Flapper wall	18–63
	Spool-sleeve	1–4
Control valves	Orifice	130 upwards
	Spool-sleeve	1–73
	Disc type	0.5–1
	Poppet types	13–40
Bearings:	Sleeve type (plain)	0.5
	Anti-friction (ball & roller)	0.5

Many filter manufacturers use similar tests, but due to lack of uniformity and reproducibility of the basic method the use of nominal filter ratings has fallen into disfavour.

A *mean filter rating* is a measurement of the mean pore size of a filter element. It is far more meaningful than a nominal rating, and in the case of filter elements with varying pore size, more realistic than an absolute rating. It establishes the particle size above which the filter starts to be effective. This is relatively easy to establish by the bubble test.

The *Beta ratio* is a comparatively new rating system introduced with the object of giving both filter manufacturer and user an accurate and representative comparison amongst filter media. It is determined by a Multi-Pass Test which establishes the ratio of the number of upstream particles larger than a specific size to the number of downstream particles larger than the specified size, *ie*

$$\beta x = \frac{Nu}{Nd}$$

where βx = Beta rating (or beta ratio) for contaminants larger than Xμm

Nu = Number of particles larger than Xμm upstream

Nd = Number of particles larger than the Xμm per unit volume downstream

It follows that the higher the β ratio the more particles are retained by the filter, and hence the greater the efficiency of the filter. Efficiency for a given particle size (Ex) can be derived directly from the β ratio by the following equation:

$$Ex = \frac{\beta x - 1}{\beta x} \times 100$$

Types of Hydraulic Filters

Examples of different types of filter elements used in hydraulic filters are given in Table II. Paper (cellulosic media) and wire mesh are widely favoured but can only provide partial silt control.

TABLE II – TYPES OF ELEMENT CONSTRUCTION

Element	Approximate Filtration Range	Remarks
Felt	25–50 µm	Subject to element migration unless resin-impregnated.
Paper	Down to 10 µm or better	Low permeability, low element strength. Subject to element migration.
Fabric	Down to 20 µm	Higher permeability than papers. Higher strength with rigid back-up mesh, etc.
Wire gauze	Down to 35 µm	Suitable for suction strainers, tank strainers, etc.
Wire wound	Down to 25 µm	Good mechanical strength.
Wire cloth	Down to 10 µm	Expensive, but well suited to high pressure systems with high strength and freedom from migration.
Edge type (ribbon element)	40–70 µm	Low resistance to flow with reasonable strength (self-supporting).
Edge type (paper disc)	10 down to 1 µm or better	Degrees of filtration variable with compression. High resistance to flow. Clogs readily.
Edge type (metal)	Down to 25 µm	Very strong self-supporting element; suitable for high temperatures, ie full system pressures.
Sintered woven wire cloth	10–20 µm usual	High strength, suitable for high temperatures. Complete freedom from element migration. High cost. Low dirt capacity.
Glass microfibre	Down to 3 µm or better	Modern preferences for silt control filters – superior performance to cellulosic fibres and wire meshes.
Asbestos fibre	Down to 3 µm or better	Effective as silt control filters, but decreasing application because of possible carcinogenic rating.
Sintered porous metal	Down to 2½ µm	Good mechanical strength, self-supporting and suitable for high pressures and temperatures. Low dirt capacity. Element migration not entirely eliminated under severe conditions.
Sintered porous metal with woven wire reinforcement	Down to 2½ µm	Very high strength. Suitable for full line pressures.
Sintered PTFE	5 to 25 µm	High cost, subject to element migration. Strength improved by reinforcement.
Sintered polythene	30 µm	Low resistance to flow and freedom from element migration. Not suitable for temperatures above 60°C (140°F).
Sintered metal felts	Down to 5 µm or better	High cost, but freedom from element migration. Difficult to clean (elements usually replaced).
Magnetic filters	Ferrous particles only	Little or no resistance to flow.

Sintered porous metal, asbestos fibre and glass micro-fibre media are capable of providing full silt control. The latter type (glass micro-fibre) is a recent development, now used as a medium for both chip control and silt control filters. It offers good pore size distribution, greater open area than cellulose or wire mesh media because of the smaller fibre diameter, and better dirt holding capacity. Its chief disadvantage is that it is low in strength, although this can be enhanced with resin treatment. It is normal practice to support glass micro-fibre elements on both the upstream

TABLE III – FILTER ELEMENT TYPES AND CHARACTERISTICS

	Minimum Particle Size Retained (μm) Nom	Minimum Particle Size Retained (μm) Absolute	Pore Size Control	Flow Capacity	Dirt Holding Capacity	Resistance to Migration	Mechanical Strength	Consistency	Cost
Wire gauze		60–100	Fair	High	Poor	Very good	Very good	Very good	Low
Shaped wire		60–80	Fair	High	Fair	Very good	Very good	Very good	Moderate
Metal disc (stack)		60–100	Fair	High	Fair	Very good	Very good	Very good	Moderate
Felt (pad)	30–40		Poor	Moderate	Good	Poor	Poor	Fair	Low
Felt (pleated)	25–35		Poor	Moderate	Good	Poor	Low	Fair	Low
Paper (pleated)	10–20	25–35	Fair	Very low	Fair	Poor	Poor	Fair	Low
Impregnated (papers)	10–15	10–35	Fair	Very low	Fair	Poor	Poor	Fair	Low
Paper ribbon	50–100		Fair	Moderate	Good	Poor	Fair	Good	Moderate to low
Paper disc	5	20–25	Fair	Low	Good	Fair	Low	Fair	Moderate to low
Woven wire mesh	5	10	Very good	Moderate	Poor	Very good	High	Good	High
Woven wire cloth	5	10	Very good	Fair	Poor	Very good	High	Very good	Very high
Wire cloth and paper		10–25	Good	Low	Poor	Good	Fair	Good	High
Glass microfibre	1,2,3	1,2,3	Moderate	High	High	Poor	Low	Very good	Low
Sintered metal	2,5 or 10	2,5 or 10	Very good	Moderate	Fair	Very good	Very good	Very good	Moderate to low

Miniature filters and strainers. (Microfiltrex Ltd).

Precision miniature filters and strainers. (Microfiltrex Ltd).

and downstream sides. Asbestos fibres can have an even smaller diameter than glass micro-fibres, making them even more effective for silt control filtering. The material is, however, regarded as hazardous to health and also subject to migration.

Particular characteristics of different filter types are summarized in Table III. Plain wire mesh is generally suitable for chip control down to around 75 μm (200 mesh), and other elements with a 10 μm nominal rating for finer chip control. For silt control, filters with an absolute rating of 6 μm or 5 μm are required.

Pall 8300 Series low pressure filters provide flows to 2200 lit/min; also available in duplex designs for non-stop system protection.

Pall 9800 Series high pressure filters for working pressures to 420 bar. The housings are non-welded.

Pall Spin-on filters handle fluid flows to 400 lit/min; also suitable as reservoir air breathers.

Pall Ultipor filter elements. The filter medium consists of inorganic fibres impregnated with epoxy resins for maximum temperature and fluid compatibility.

Cleanable *vs* Disposable Filter Elements

In general low cost filter media are regarded as disposable (replaced with a new element) as the cost of re-cleaning may be as high or higher than the replacement element. Certain media may be difficult or almost impossible to clean anyway, or be susceptible to damage if re-cleaned. The more robust media — *eg* wire mesh and sintered porous metal elements — are more economically re-cleaned, when they may be used a number of times. Three to seven cycles of cleaning and re-use are typical, but this depends on the service conditions and cleaning facilities available. The most suitable cleaning system for such types is an ultra-sonic bath.

Filter Location

Filter *location* in a system is equally as important as silt control and chip control. General practice is to locate filters on the pressure side of the pump. A filter on the suction side of the pump can be less robust and less costly, but would have to be of large size and with a relatively high cut-off to avoid the possibility of restricted flow starving the pump and causing cavitation.

Possible positions for filter location are:

(i) In the delivery line immediately downstream of the pump.

(ii) At the inlet point to each critical component, if necessary.

(iii) In the return line to the reservoir.

(iv) In the pump-case drain line.

(v) External (batch filtering).

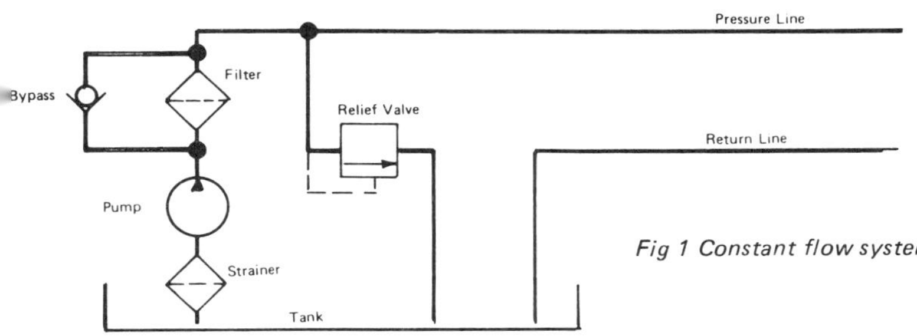

Fig 1 Constant flow system.

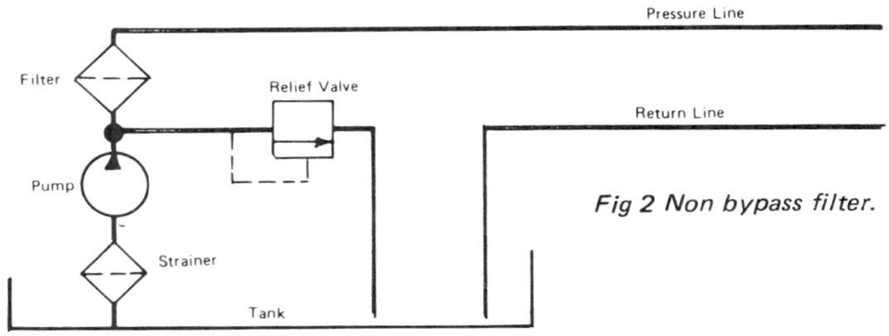

Fig 2 Non bypass filter.

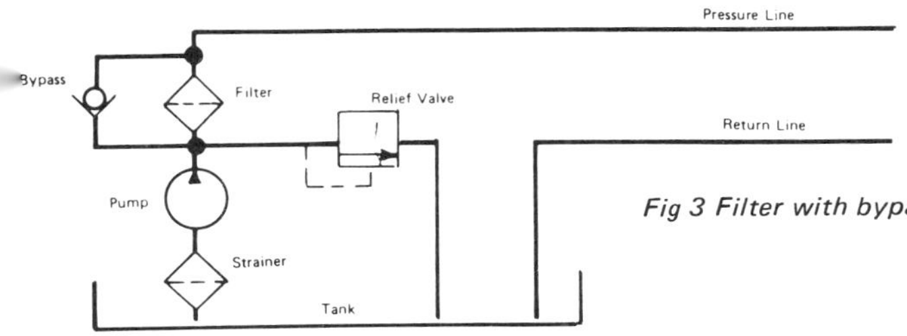

Fig 3 Filter with bypass.

Pressure-Line Filtering

A pressure-line filter is located on the delivery side of the pump and is thus exposed to full system pressure. It will protect the following system from pump-generated or pump-passed contaminants, but not from any contaminants generated downstream of the filter.

Three possible filter configurations are shown in Figs 1–3. Locating the filter before the relief valve gives constant flow through the filter (Fig 1). Located downstream of the relief valve the flow through the filter will depend on system demand; and in off-load periods will have leakage low or full flow depending on whether the control valve is of blocked-centre or open-centre type, respectively (Fig 2). Such positioning thus makes it more difficult to estimate the varying flow rates to which the filter may be subjected and so the former system is normally preferred. In this case a bypass across the filter is essential to eliminate excessive pressure build-up against the pump should the filter become clogged (Fig 3).

Additional protection for the system can then be provided by further filters preceding critical components, or point-of-use filters (see Fig 4). Filter requirements can be selected in a number of different ways, depending on how critical protection is for each component. If the first filter following the pump) provides the necessary fine filtering, the first component in the system is

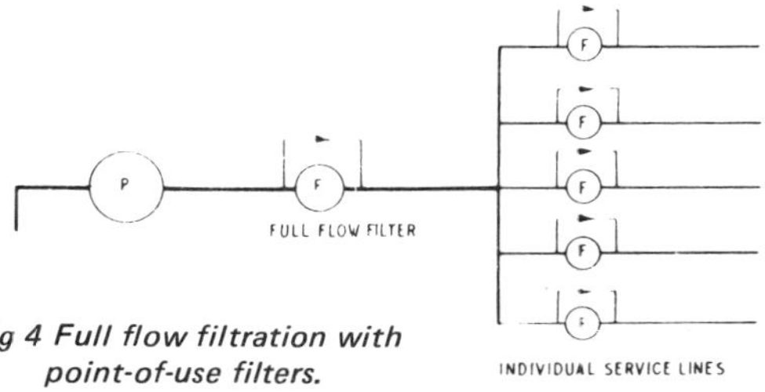

Fig 4 Full flow filtration with point-of-use filters.

protected. Subsequent components *needing* protection from contaminants which may be generated by the first component can be preceded by an additional fine filter. A component with more generous clearances not needing such protection need not have a point-of-use filter preceding it.

Equally a point-of-use filter (chosen with a suitable rating) could precede *each* component needing protection, when the first filter following the pump could be of a coarser type for lower resistance to full pump flow.

Return-Line Filtering

A return-line filter is located downstream of the last working component in the system, but upstream of the reservoir — Fig 5. It thus removes all contaminants (down to its rating level) ingested or generated by the pump and system components before the fluid is returned to the reservoir. It has the advantage that it is not likely to be subjected to large pressure surges as can occur in pressure lines, but it can be subject to unsteady flow conditions. It thus needs to be robust enough to accommodate flow surges.

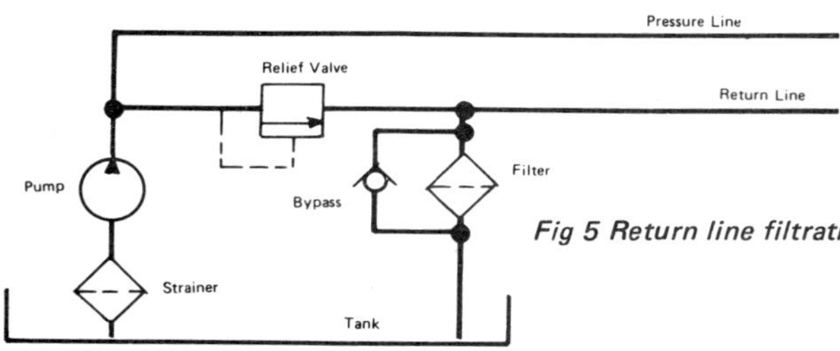

Fig 5 Return line filtration.

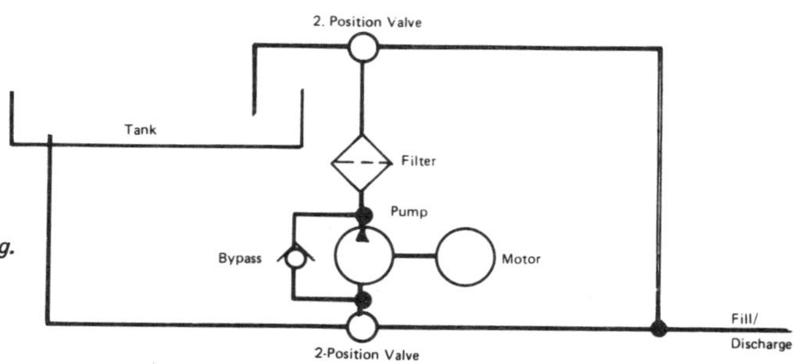

Fig 6 Off-line filtering.

Bypass or Off-Line Filter

A bypass filter is located in a separate loop between the pump and reservoir, this loop operating independently of the main system. Its purpose is to provide a means of cleaning the fluid contained in the reservoir only (Fig 6).

Its particular use is for overall fluid cleaning at suitable maintenance intervals. It can, if necessary, be operated when the main system is in use. It does not, of course, dispense with the need for filter(s) in the main system since it only cleans the amount of fluid present in the reservoir.

Suction Filters

The only provision usually made for filtering on the suction side of the pump is a simple strainer fitted to the suction line in the tank or reservoir. The suction line connecting to the tank should be located as far as possible from the turbulent flow induced by the return line, and the two should be separated by baffles providing a sharp-edged weir barrier, if possible, so that sediment cannot be carried across from the 'return' side to the 'suction' side of the tank. Sometimes a strainer may also be fitted on the return line to prevent coarse particles being released into the tank, but this should not be necessary in the case of static installations as effective 'separation' can be achieved by good tank design.

Hydraulic multi-pass filter performance test rig. (Pall Industrial Hydraulics Limited).

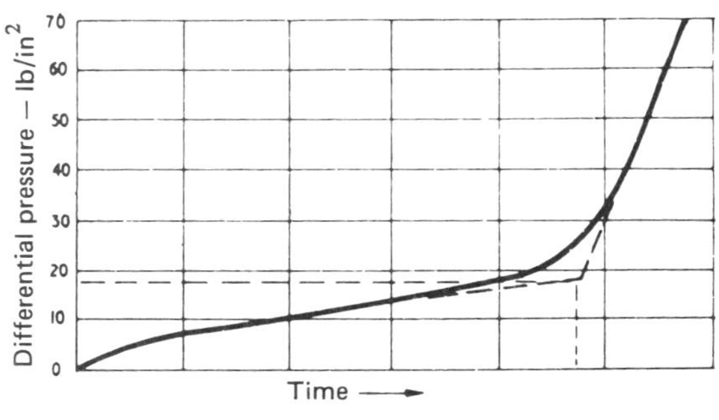

Fig 7 Typical time pressure drop characteristic curve.

Flow Capacity and Pressure Drop

The pressure drop across a filter is caused partly by a velocity head loss which is dependent only on the size of the filter and remains constant for a constant flow rate; and partly by a viscous loss which depends on fluid viscosity and the permeability of the filter medium. This latter loss increases as the element becomes more clogged, leading to a gradual rise in pressure drop as contaminants collect on the element. Eventually the pressure drop will rise sharply after the fashion shown in Fig 7. The apex of the 'knee' so formed by such a characteristic curve determines the useful life of the element, in terms of an 'acceptable' level of differential pressure increase, and the period over which this increase is substantially low and linear in characteristics. The differential pressure can be expected to rise very sharply with further time in circuit.

The filter bypass would normally be set to this 'knee' pressure, or to the safe limiting pressure for the strength of the element, whichever is the more significant. A 'safe' pressure would not be more than about 50% of the 'bursting' or disrupting pressure of the element, although the bypass would normally start to open below this figure to give some differential pressure release.

The differential pressure developed across the filter element, is of course, independent of the system pressure, which explains why relatively weak filter elements, with a disrupting pressure of perhaps 7 bar (100 lb/in^2) or less can perform satisfactorily in systems operating at a pressure of 210 bar (3 000 lb/in^2) or more. The element is never subjected to more pressure than the actual differential developed across it, unless it becomes clogged and is not pressure-relieved by a bypass valve. The inertia of such a valve may, however, make it susceptible to damage under rapid surge pressure conditions mainly because of the instantaneously modified flow rates and changes in velocity head.

The limiting differential pressure depends on the filter medium construction and may vary widely with different types. As a general rule, however, the finer the filtering the lower the permissible differential pressure, which in turn means the larger the size of filter likely to be required.

The relationship between pressure drop and flow rate is largely determined by the design and size of complete filter, not the element, and by the fluid viscosity. Characteristic curves in this case can be calculated for a specific fluid and specific fluid temperature (*ie* specific fluid viscosity). Performance curves of this type are normally available from manufacturers for each filter type and size.

The general relationship between pressure drop, flow rate and fluid viscosity can be expressed in the form:

$$\text{pressure drop} = K\,A\,Q\,\nu$$

where A is a factor dependent on element type and area
Q is the flow
and ν is the fluid viscosity at the temperature considered

K is an empirical factor which can only be determined individually for a particular design of filter, and is virtually a measure of its overall permeability or porosity. Typically, filters are designed so that the 'K' value for a new element is consistent with a pressure drop of not more than 1 bar (14.3 lb/in^2) at maximum flow rate with a specific fluid viscosity (normally a typical mineral oil at a typical working temperature). The 'K' value can be calculated directly from performance data for an individual filter. From this it is possible to calculate the pressure drop at any other fluid viscosity from the formula.

For a given design and size of filter:

$$\Delta P = K_1\,Q\,\nu$$

A more general relationship is:

$$\Delta P = K_2\,A\,D\,Q\,\nu$$

where D is an additional parameter dependent on the filter design.

Puralator high pressure filters.

Filter Sizing

A suitable size of filter is selected from the flow rate/pressure drop characteristics, bearing in mind that these must be corrected for fluid viscosity if it is different from the empirical figures or quoted viscosity. Where size and weight are not important it is generally best to use an oversize rather than a marginal size filter, unless the cost factor is critical. Alternatively, two or more filters can be used in parallel, although this will increase both the number of connections required and the pressure drop through the fittings. An oversize unit will give a lower pressure drop for the same flow rate and a longer period between servicing.

Compatibility

Compatibility with the system fluid must relate to the system temperatures involved to ensure that no degradation of the element or its seals occurs during its normal service life. Degradation can occur through:

(i) Absorption of fluid into the filter medium or binder causing swelling (increased pressure drop and choking); or migration of particles downstream.

(ii) Hardening or embrittlement of the filter element which can cause cracking and breakdown of the material.

(iii) Disintegration of the element.

In general asbestos, glass micro-fibres and wire mesh are fully compatible with all hydraulic fluids (provided the complete filter does not include parts in aluminium, cadmium, magnesium or zinc which are attacked by water-in-oil fluids). Cellulosic media tend to swell in water and are not generally suitable for water-in-oil and water-glycol fluids. Filters with active media cannot be employed as these are capable of removing additives commonly used in hydraulic oils.

Compatibility with other system components needs little comment, other than that the filter should be readily fitted and coupled to existing units and the fact that filters of the required size are available to fit standard line sizes, *etc.* It is also desirable that the form of the filter is such that it is readily accessible for removal of the bowl or body and element for cleaning or element replacement.

Magnetic Filters

Magnetic filters range from simple magnetic plugs normally intended for fitting in reservoirs, to versions of more or less conventional filters incorporating a permanent magnet element. An example of a magnetic filter is shown in Fig 8.

A magnetic filter will attract and collect only ferrous metal particles, such as wear particles. A proportion of non-magnetic particles may also be retained, however, by a mechanical entrainment process, although performance in this respect is unpredictable. Also there is a possibility of particles held by the magnet being swept off and back into the flow during cold starting when viscous forces are at a maximum or should the magnetic element become over-loaded.

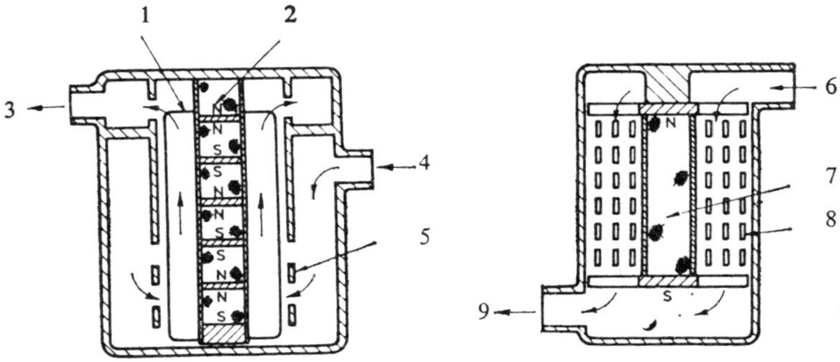

1—fins.
2—magnets in tube.
3—outlet.
4—inlet.
5—perforated tube.
6—inlet.
7—magnets in tube.
8—magnetized iron cage.
9—outlet.

Fig 8 Alternative arrangement of magnetic filter assemblies.

Magnetic filters, therefore, can only be regarded as secondary filters for the specific purpose of removing ferrous metal particles or initial wear products. They can do a useful job when fitted on the delivery side of the pump in a new system during running-in, for example. They in no way replace other types of filter for hydraulic circuits, although some designs may be a combination of both types in a single unit.

Flexible Separators

Conventional breathers fitted to the tank inlet provide only limited protection against atmosphere-borne contaminants entering the tank since their cut-off point is normally no better than 15–20 μm. Certainly particles smaller than this entering the fluid may be removed by the system filter(s), but in systems requiring ultra-clear fluid the conventional tank breather may be replaced with a flexible separator to ensure that no contaminant enters the system at this point.

The principle of the flexible separator is to provide a permanent flexible non-porous barrier between the atmosphere and the system fluid, without affecting the operational functions of the system components. The flexible separator takes the form of a synthetic rubber bag, totally enclosed, save for a metal stem giving access to the bag interior. The separator is introduced into the fluid reservoir with the stem protruding through the tank lid (see Fig 9), and providing the tank, including the breather, is effectively sealed, as the fluid level rises and falls air flows in and out of the separator only. Thus all air and its associated contaminant is prevented from contacting the fluid.

Alternatively, where internal installation is impracticable, the separator may be connected externally to the reservoir (see Fig 10). In this case, air trapped in the reservoir flows in and out of the separator as the fluid level rises and falls. Thus all air and its associated contaminant is eliminated.

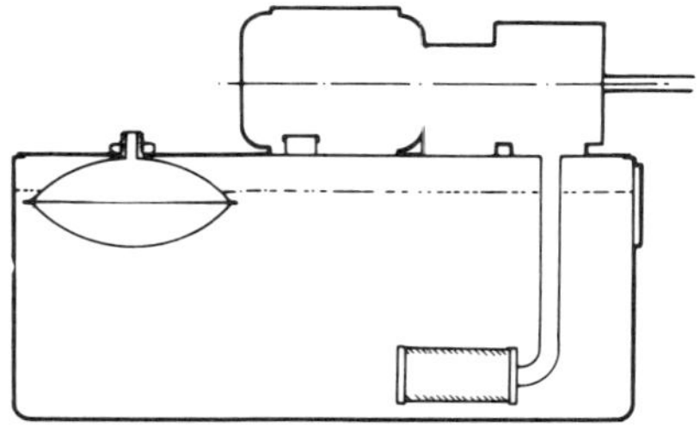

Fig 9 Pronal breather flexible separator installed inside fluid power reservoir.

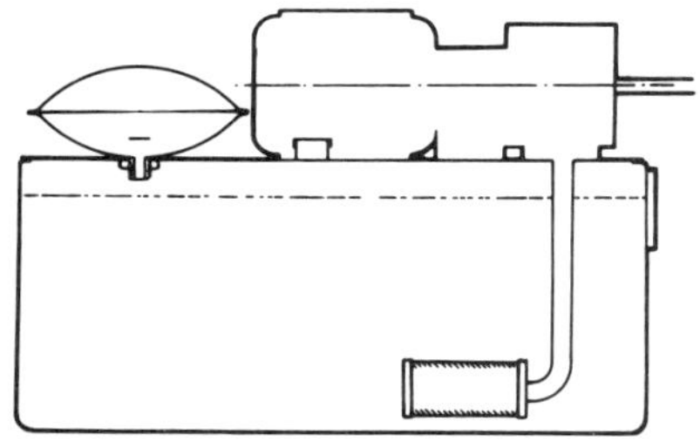

Fig 10 Pronal breather flexible separator installed externally.

In both the described methods, the fluid in the reservoir is always at atmospheric pressure.

A further possibility is to seal the reservoir, which is then slightly pressurized, but in this case, the separator, installed in the reservoir, is completely enclosed and contains a slight positive pre-charge pressure. The reservoir then functions in a similar way to a low-pressure accumulator. This system has the added advantage of giving the pump a positive head, but of course must be stressed with a positive pressure in mind.

For non-pressurized systems, the separator is sized on the normal fluid displacement from the reservoir plus an additional capacity to compensate for system losses; this additional leakage compensation should normally be not less than 25% of the displaced volume.

In practice, the separator volume is usually quite small in relation to the tank capacity. In hydro-static applications, for example, the volume displaced by the cylinder rod needs to be considered. If accumulators are being used, then their normal storage capacity should, of course, be known.

Having over-sized the separator capacity by 15%, there is enough additional capacity available to cater for normal system leakage. However, in the case of serious maintenance neglect, where a system is not topped up over a long period, or in the event of catastrophic failure when the fluid level drops dramatically, the maximum separator volume could be exceeded. In this event, as no air can enter the reservoir and the separator is fully expanded, a depression in pressure within the reservoir will result, which could cause the pump to cavitate. There are two simple and practical solutions to this problem. The first is to employ a low level alarm system; this may consist of an audible or visual warning, which if not acted upon is followed, after a further drop in fluid level, by automatic pump switch-off. The second is the use of a normally sealed filter cap which has a built-in suction-relief valve opening at 0.5 lb/in^2 depression. Thus, when the fluid level drops catastrophically, air is allowed into the reservoir.

System Cooling

IN ANY hydraulic system it is desirable to maintain the fluid temperature at, or preferably below, the recommended or specified maximum working temperature for continuous duty. This will help prolong fluid life, maintain a satisfactory fluid viscosity, and also prolong seal life. Also in the case of water-based fluids, low operating temperatures reduce fluid loss through evaporation.

Sources of Heating

The principal source of heat generated in a hydraulic system is power losses at the pump. Heat is also generated by friction losses in the pipework and components so that self-heating is proportional both to flow rate and pressure drop. The whole system may also be subject to additional heating (or cooling) from external sources.

In practice only heat generated by the pump, pressure drop through valves and hydraulic motor power losses are likely to be significant. Heat losses generated by line friction and cylinders are usually dissipated from the surface of those components. In fact, in most small circuits heat losses are balanced at an acceptable temperature level by natural heat transfer from the fluid to the pipes, actuator and reservoir, and subsequent dissipation to the air. To ensure adequate dissipation in this manner the reservoir should be free standing and with the sides and bottom fully exposed and preferably in a position where it receives ventilation by air draught. Actuators should be similarly placed, to avoid localized overheating of the fluid at these points.

In some cases the source of heating can be a localized one, which is easy to overlook. An overall guideline is: the difference between the power input to a hydraulic circuit and the mechanical power output developed by the actuators represents the energy transformed into heat. Where this difference is less than about 2 hp overheating troubles are not likely to occur, provided the whole system is reasonably well ventilated or exposed to open air. For larger differences, or where it is necessary to hold working temperatures to specific levels to ensure consistent operation (*eg* in the case of machine tool actuators), the cooling requirements may need analyzing in detail.

Effects of Viscosity

Control of fluid temperature is highly desirable to avoid large viscosity changes when starting from cold and running up to working temperature, and for reasonably accurate predetermination of the working viscosity of the fluid so that the pump receives adequate lubrication and operates at high volumetric efficiency. Changes in oil viscosity can also seriously affect machine efficiency and performance. The higher the system pressure the more heating the fluid is likely to receive when the system is working and the greater will be the change in viscosity. The volume of fluid in the system can also be a critical factor, the lower the volume the more heat received per unit volume for a given amount of work (a proportion of which is inevitably converted into heat). Marked

THE BEST EXCHANGE RATES IN THE BUSINESS...

. . . that's what we are offering to manufacturers who have a heat transfer problem to solve.

And you get value for your money in two ways — firstly, SHE is acknowledged to have the most efficient range of air cooled heat exchangers currently available, and secondly they have an established record for long, trouble-free service.

So your problem is solved — and will stay that way.

Selection of hydraulic oil coolers with electric motor driven fans.

SHE Ltd, Lincoln, England

SHE

SPECIALIST
HEAT EXCHANGERS
LIMITED

Freeman Road, North Hykeham, Lincoln LN6 9AP
Telephone: Lincoln (0522) 683123 and 694545
Telex: 56133

changes in ambient temperature can also affect the actual working temperature of the fluid, but this is a problem more usually confined to aircraft hydraulics than industrial hydraulics.

Cooling

In most systems some degree of cooling is automatically provided by the reservoir, and also by heat radiated from pipework (although this is generally negligible). Where additional cooling is required, a heat exchanger (oil cooler) can be incorporated in the system. This can be water-cooled or air-cooled and is generally placed in the return line to the reservoir on the low pressure side of the relief valve.

AM13 Guided flow hydraulic oil cooler. (Serck).

Reservoir Cooling

The cooling provided by a reservoir is primarily dependent on its surface area (A), its location and the difference between fluid temperature and ambient temperature. The effective coefficient of heat transfer is not just that of the reservoir walls. It is modified by the presence of a boundary layer of fluid and the degree of turbulence present inside the reservoir. It will be further modified by any airflow past the exterior walls of the reservoir.

For the average design of reservoir, positioned in substantially static air, an empirical formula for estimating a suitable size of reservoir is:

$$A = \frac{kQP}{\Delta T}$$

where

P = system pressure drop
Q = flow rate
ΔT = temperature difference (fluid and ambient)
k = a coefficient

For A in m^2, P in bar and T in °C, k = 0.00785.
For A in ft^2, P in lb/in^2 and T in °F, k = 0.015

TABLE I – TYPICAL EMISSIVITY OF RESERVOIR SURFACES

Surface or Surface Finish	Emissivity
Aluminium sheet, untreated	0.04–0.05
Steel sheet untreated	0.65–0.82
Aluminium paint	0.27–0.67
Oil paints	0.92–0.96

An effective way of increasing the surface area of a reservoir without increasing its internal volume is to add fins. The cooling performance of a reservoir will also be improved by good ventilation and suitable choice of reservoir wall material and the use of a paint finish with high emissivity to increase radiation of heat — see Table I.

Water-cooled Heat Exchangers

Cooling by water generally results in a smaller size unit than air-cooling for the same thermal performance because of the lower initial temperature available when raw mains water is available for cooling. The possible disadvantages of water-cooling are the increase in water costs and the fact that it may be inconvenient to pipe water to the installation.

The most compact and generally the most effective type of water-cooler is a cold water counter-flow heat exchanger, using a nest of copper or copper alloy tubes. In simple installations a return circuit is used for feeding the cooler, and it is often convenient to connect the cooler to the relief valve exhaust. In big plants it is generally best to provide an auxiliary circuit for feeding the cooler (and possibly the filters). The auxiliary circuit can then be connected or disconnected freely from the main circuit.

Where reasonably exact calculations are required it is necessary to determine the value of the temperature difference involved (Δt) fairly accurately. This may be complicated by the fact that the temperature difference is continually changing — *eg* with a counterflow heat exchanger Δt changes as the hydraulic fluid is cooled and the cooling water gets warmer. In such cases it is necessary to work with the logarithmic mean temperature difference, defined by:

$$LM\Delta t = \frac{(T_1 - T_4) - (T_2 - T_3)}{\left(2.3 \log \dfrac{(T_1 - T_4)}{(T_2 - T_3)}\right)}$$

where T_1 = temperature of incoming hydraulic fluid

T_2 = temperature of hydraulic fluid leaving the heat exchanger

T_3 = temperature of incoming cooling water

T_4 = temperature of water leaving the heat exchanger

A simplified formula can be used for convenience and will generally introduce little error:

$$LM\Delta t = ½\left(T_1 + T_2) - (T_3 + T_4\right)$$

Air Blast Coolers

Air blast cooling is becoming increasingly popular to avoid the cost of water and possibly secondary coolers. It offers a simpler (although more bulky) system as an air-cooled heat exchanger normally requires only two pipe connections to fit into the circuit concerned. It can also be designed to provide efficient cooling to within 5 degC of the surrounding atmosphere or even less if required. Air-cooled heat exchangers do, however, require electrical, mechanical or hydraulic power to drive the fan, but can very often be made self-contained on the hydraulic power pack.

The basic requirements for efficient air to oil cooler design include:

(a) Matrix construction with good heat transfer characteristics which are obtainable by using extended heat transfer surfaces both in the bore of the cooler tubes and on the outside of the tube walls.

(b) The matrix should be soundly constructed to withstand vibration and pressure requirements, and should include brazed or welded tube/header joints.

(c) The cooling air should be provided by an efficient fan running at a reasonably low speed to avoid objectionable noise levels and also to reduce motor power to a minimum.

The power source for the fan is usually provided by an electric motor for stationary installations, but hydraulic drives are becoming more popular for mobile applications and also for underground use or where fire and explosion hazards exist.

Air to oil cooler with hydraulic motor driven fan. (Specialist Heat Exchangers Ltd).

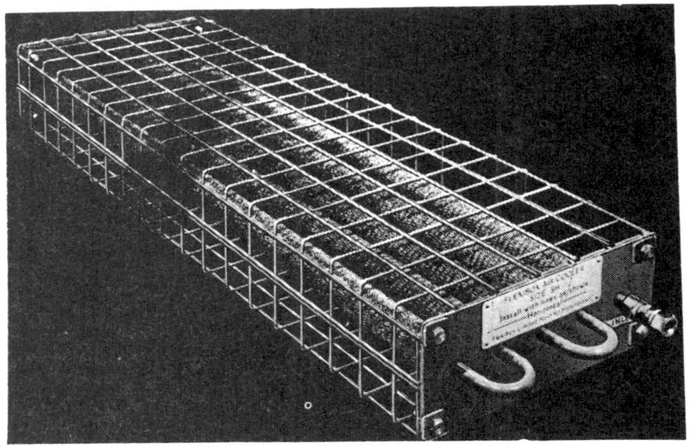

Flexibox air cooler.

Service Comparisons

Air blast cooling has the advantage that the only possible malfunction – the fan not running – is immediately apparent and calls for remedial attention. It also allows quick visual checking for correct functioning and the state of any fouling. By contrast, particularly in hard water areas, it may be necessary to strip and de-scale the tubes of a water-cooler every three months or so.

Considerations of maintaining oil flow are the same for both forms of cooling, although many larger systems and power packs are now fitting fan blown coolers incorporating a circulating pump, thus making the cooling function 'off-line', and independent of the main system flow fluctuations. This arrangement has the definite advantage of removing the cooling elements from the rapid pressure peaks and 'hammering', which is often found in the return line of large complex systems or servo systems.

If the oil passage walls suffer from fatigue or rupture, the air blast cooler has maintenance advantage. The leak can be seen and causes no lasting damage, since oil escapes to atmosphere. The consequences of an oil/water rupture can be much more serious, especially if water reaches the pump intake. The resulting damage and rectification costs may even involve a complete system strip and rebuild.

Hydraulic Seals

SINCE HYDRAULIC systems may operate at pressures from 100 bar (1 500 lb/in^2) to 210 bar (3 000 lb/in^2), all seals used need to be of high-pressure type. Even higher pressures up to 700 bar (10 000 lb/in^2) may be used in heavy-duty hydraulic machines, such as presses, representing even greater demand on the types of seals used, relatively often aggravated by unfavourable working conditions.

Static Seals

The majority of static-seal requirements on hydraulic components can be met by O-rings. These can be assembled in standard grooves, dimensioned to match the ring section diameter. The pressure rating of the ring increases with reducing clearance gap. Thus, with a normal clearance gap, effective sealing is obtained up to about 100 bar (1 500 lb/in^2), with the possibility of extrusion at higher pressures. Provided the mating surfaces are reasonably smooth and the squeeze is applied uniformly around the circumference of the ring, further reduction in clearance gap (*ie* tightening up of the joint) up to the limit of nominal metal-to-metal contact can effectively seal pressures in excess of 7 000 bar (100 000 lb/in^2) with no extrusion of the ring.

Compression-type couplings, widely used in hydraulic services, commonly rely on metal-to-metal contact to provide a seal. A number of manufacturers, however, also produce couplings with either an O-ring or profiled elastomeric ring seal — *eg* see Fig 1. Most modern aircraft hydraulic systems also employ couplings with trapped O-rings. Metal wedge seals are an alternative to trapped O-rings and preferred where higher fluid temperatures are involved. Equally, of course, they can be used for static sealing well beyond the usable temperature range of elastomeric O-rings. Alternative types of all-metal seals are also used for higher working temperatures (up to 300°C) and are effective under both standing and pulsating pressure up to 345 bar (5 000 lb/in^2).

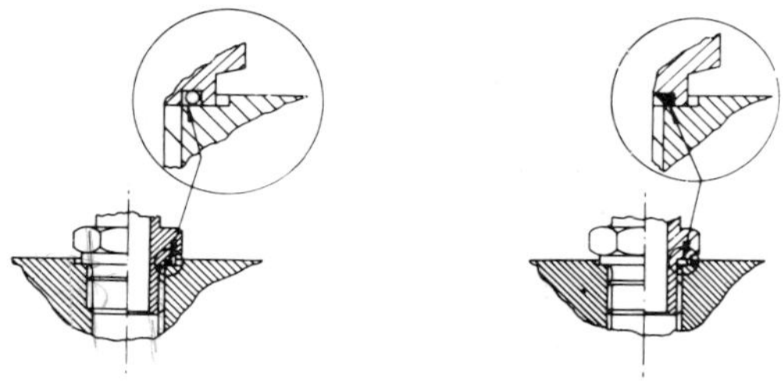

A coupling with an O-ring face seal. An elastomeric profiled ring.

Fig 1

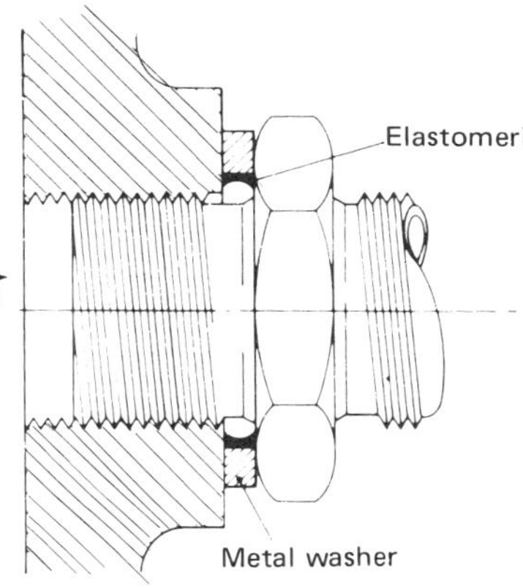

Fig 2 Bonded seal.

The bonded washer (Fig 2) is another example of a static seal commonly used on systems operating at pressures of the order of 200–300 bar (3 000–4 500 lb/in^2). Bonded washers for use in oil-hydraulic systems have the sealing element in nitrile rubber of *circa* 90 degrees hardness, with cadmium plated steel outer ring. Those for phosphate ester fluids are composite EPDM/metal washers, in this case not bonded for phosphate ester would destroy the bond.

Gaskets

More limited use is made of conventional gaskets for static hydraulic seals. There are many developments in modern gasket materials and designs, however, (such as PTFE envelope gaskets) which can make them worth considering for particular duties. Again, too, there are many types of all-metal gaskets suitable for high-temperature services.

Printed Gaskets

Printed gaskets are another modern type of flat seal. Using this technique a gasket can be produced with the sealing medium exactly where required incorporating the correct thickness, resilience and environmental resistance to suit the joint under consideration. Further, it is possible to extend the role of such a gasket to encompass other functions. Complex configurations, impossible with O-rings, are also easy to produce – *eg* a single printed gasket could be used to replace a multiplicity of O-rings and associated grooves.

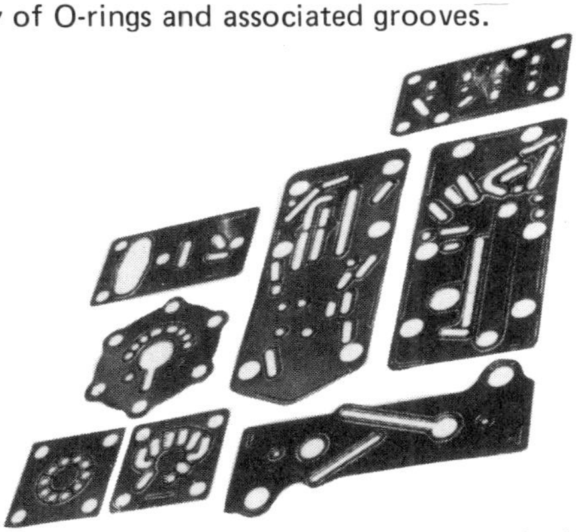

Fig 3 'Dowprint' gaskets for high pressure hydraulic circuit.

Printed gaskets for hydraulic services are normally made with selective deposits of polyurethane or similar polymers on a metal, phenolic or composite substrate and have a pressure rating up to 405 bar (6 000 lb/in^2), with a maximum service temperature of 90°C. A metal substrate would be the normal choice for high-pressure hydraulics, when slots and holes for valve porting, *etc* can be sealed with sealant beads on both sides of the plate. An example of such an application to high-pressure hydraulics is given in Fig 3.

Dynamic Seals

The choice of a high pressure hydraulic seal for dynamic duties depends both on the application and size of component involved. Relatively simple seal rings may provide satisfactory performance on precision-made components at pressures up to 350 bar (5 000 lb/in^2), but prove unsuitable on larger industrial components operating at much lower pressures where the working conditions are generally unfavourable. For heavy-duty applications in fact, and particularly with larger machines or components where fairly generous tolerances are involved, seal sets are normally preferred to simple seal rings.

For reciprocating (dynamic) seals the choice lies between pressure-energized or lip seals, and squeeze seal rings (the modern counterpart of compression packings). Lip seals are suitable for both piston and rod seals, but although lip seals are generally favoured for piston seals there has been an increasing trend towards the use of squeeze seals for rod applications, except in heavy-duty applications or cylinders which may be subject to side loads and thus need a more substantial rod seal contact area. Equally, there is an increasing application of combined or composite seals of individual proprietary design, based on providing both pressure-energized and 'squeeze' sealing action in an easily-housed seal ring.

O-rings are also widely used both as rod and piston seals, although for this application they are mainly suitable for small-diameter short-stroke cylinders and moderate system pressures. They are, however, often used for higher pressures by certain industries, notably the American aircraft industry, in which they have established a particularly good reputation.

There are, however, definite limitations with any form of simple elastomeric seal ring used in high-pressure hydraulic cylinders operated under unfavourable or severe conditions — *eg* a susceptibility to extrusion and localized wear. Extrusion may be controlled by back-up rings, but seal wear can be pronounced. Simple ring seals specifically designed for medium- and high-pressure hydraulic service (particularly cylinder seals) are thus normally made of reinforced materials, or have reinforcements which prevent extrusion. Some also incorporate low friction (PTFE) rubbing surfaces to minimize wear.

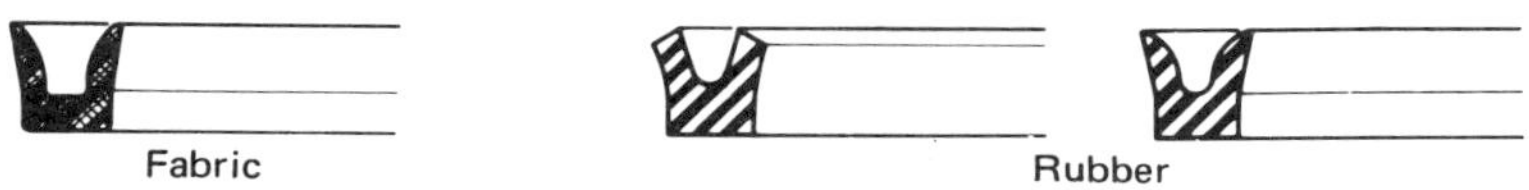

Fig 4 Typical modern U-ring sections.

U-Rings (Fig 4)

The U-ring is the basic form of lip seal produced in numerous varieties of lip shape and section solidity. It has the advantage of taking up very little space and requires only a simple machined housing. In general U-rings are inexpensive, efficient seals for medium-duty hydraulic cylinders at pressures up to about 350 bar (5 000 lb/in^2).

U-rings are used singly (or back-to-back for double-acting seals); or in sets. Simple U-rings are usually supported and located by lantern or positioning rings, especially where large sections are involved. U-ring sets are usually assembled with header rings.

For cylinders subject to rough usage, shock loads or high pressures, the type of U-ring fitted up with (or incorporating) an anti-extrusion ring is recommended.

V-Rings (Fig 5)

The V-ring or chevron seal is invariably used in sets, together with headers and with a light preload applied on assembly. A minimum of three rings is usual, suitable for pressures up to 35 bar (500 lb/in^2). Very much higher pressures can be accommodated by increasing the number of rings, and also by the use of rubber-impregnated fabric for their construction.

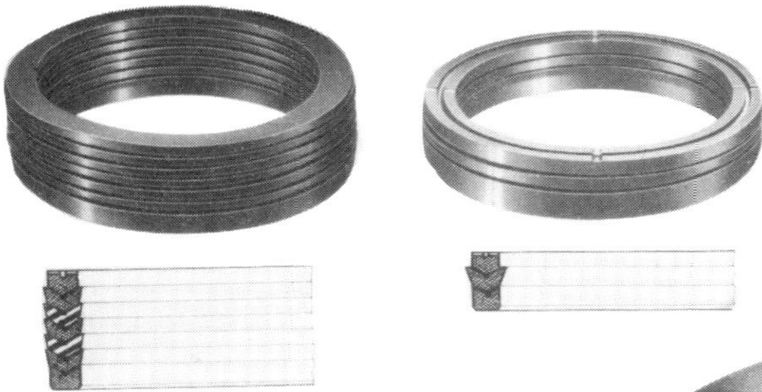

Fig 5 Examples of modern V-ring seal sets for hydraulic cylinders. Rod gland seal (left) and piston seal (right).

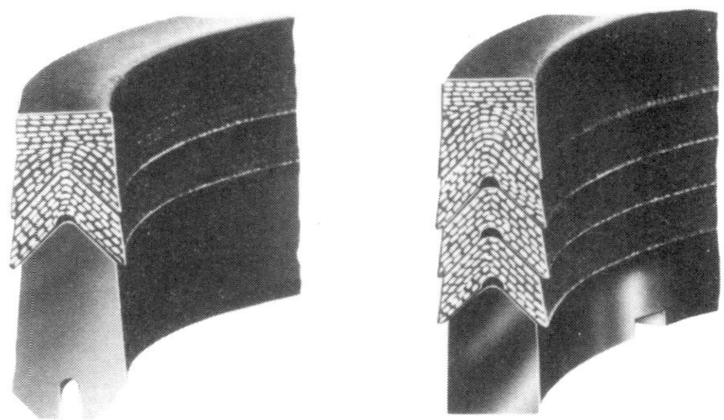

Conventional V-pack, or chevron seal (right) with Walker 'Lofilm' chevron (left).

Increasing the number of rings will tend to increase wear. Careful adjustment is also necessary if high friction and high wear rates are to be avoided.

Performance can only be assessed accurately by taking into consideration the operating conditions, when the pressure range is almost unlimited – *eg* it can be as high as 7 000 bar (1 000 000 lb/in^2).

To increase low pressure efficiency a rubber V-ring can replace one (or more) rubberized-fabric rings in the set. For heavier duties hard end rings of special fabric are normally incorporated. An anti-extrusion ring or anti-wear ring may be built into the female end ring.

V-ring sets are widely used, both for piston and rod seals, on larger sizes of components designed for medium to heavy duties. Their chief limitation is that they require a fairly substantial gland length to accommodate them.

Cups (Fig 6)

The cup seal with a flared lip is essentially a single-acting piston seal, but when used in hydraulic services is suitable only for low pressures. Cups have low friction, especially on starting, and will generally tolerate a wide range of temperatures for long periods without wear.

Fig 6 Typical cup sections in fabric or rubber.

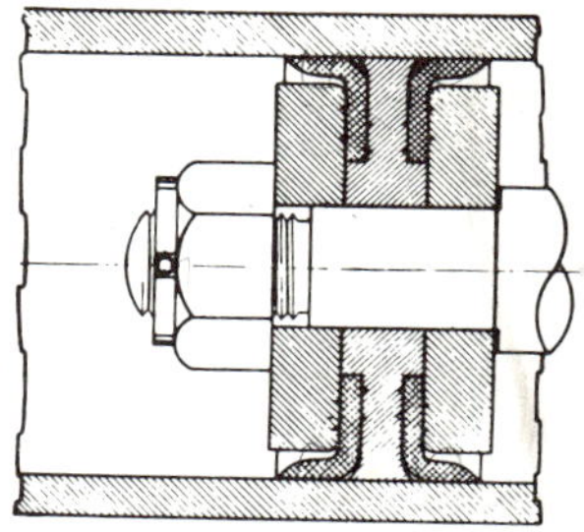

Fig 7 Double-acting piston assembly with cup seals.

Cup seals are made in rubber or fabric and normally assembled with a matching support plat and junk plate to stabilize the section. They can be used satisfactorily in cylinder bores which would cause early failure with other types of seal because of excessive clearance between pistor and cylinder, or generous tolerances on internal diameter. To perform as double-acting seals cup must, of course, be mounted back-to-back in a single assembly. The actual method of mountin depends on the pressure to be accommodated — *eg* see Fig 7.

Collars (Fig 8)

Collars, also known as flange seals or hat packings, feature a deep flexible lip. This type of seal ha application as a gland seal or rod wiper in hydraulic cylinders, sealing effectively at both low anc high pressures. It is more compact than a U-ring, but less efficient as a seal.

Fig 8 Modern collar sections.

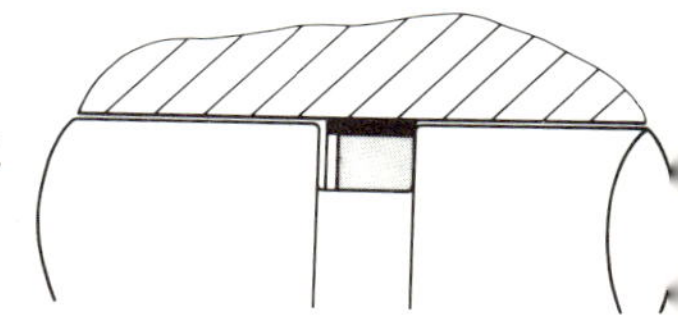

Fig 9 Composite of co-axial seal.

Combination Ring Seals (Composite Rings)

This description embraces simple seal rings combining an elastomeric ring with a rigid back-up ring in an integral construction. A typical form is a square or rectangular elastomeric ring in combina tion with a thinner ring of PTFE to produce a composite seal. The PTFE ring is on the inne diameter in the case of an internal seal; and on the outer diameter in the case of an external seal

Such a seal can be located in a simple groove in a similar manner to O-ring fitting, but with groove dimensions specific to the size of composite ring. This governs the degree of squeeze anc thus the radial pressure on the PTFE ring, and also locates this latter ring axially. Besides providin a bearing surface this PTFE ring also confines the elastomeric section within its groove, eliminatin any possibility of extrusion so that no back-up rings are required, even at pressures of the order o 350 bar (5 000 lb/in^2), (Fig 9).

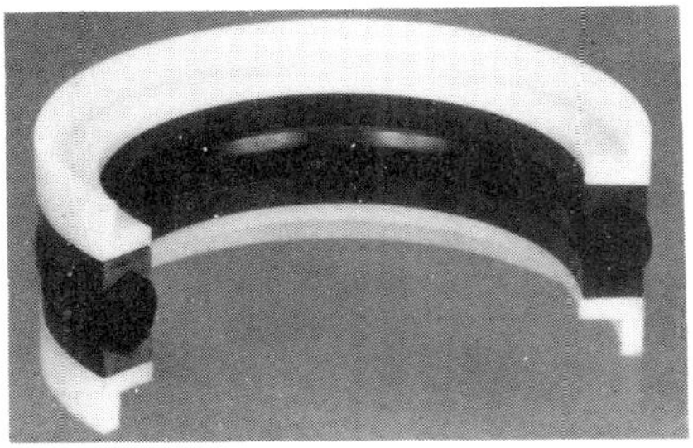

Walker 'Spoolpak' five-component composite seal incorporating anti-extrusion rings and two bearing rings.

Numerous integral combinations have been developed as proprietary seal rings, the majority of which are designed to be accommodated in simple grooves. Groove dimensions are specific to the geometry and construction of the particular section in order to provide the required amount of squeeze on the elastomeric section for satisfactory sealing at low pressures. Sealing at higher pressures then depends on further deformation of the elastomeric section maintaining a seal contact pressure greater than the applied pressure.

Proprietary seals of this type range from quite simple compact sections to large heavy-duty sections, and cover the full range of working pressures normally required in hydraulic cylinders and other components. All offer superior performance to O-rings with similar simplicity of fitting, although at increased cost. They are produced both as single-acting and double-acting seals. In the latter case the elastomeric section forms the middle part of the ring, with back-up rings bonded to it on either side. These also act as bearing rings although some designs incorporate additional bearing rings.

A particular advantage of composite ring seals is their compactness, reducing the size of housing required. Piston seals incorporating bearing rings can also enable the length of the piston to be reduced. Many seals of this type are also designed for simple assembly on a one-piece piston head which is both stronger and more cost-effective than the traditional method of assembling piston seals on a built-up piston head.

Another type of modern seal is the multi-lip seal with separate fabric headers and anti-extrusion wear rings which can directly replace V-packs. There have, too, been changes in the fabrics used for reinforcement in rubber-impregnated seals, and for header rings. One of the latest materials is polyester/elastomer comprising a dense mat of randomly oriented polyester fibres bound together in matrix with synthetic rubber. The high content of uniformly distributed fibres gives this material high strength and at the same time virtually eliminates the possibility of delamination or fraying.

Table I presents a survey of typical modern combination ring seals, or unit seals as they are sometimes called.

Friction and Wear

The general effect of pressure on all medium- to high-pressure dynamic seals is increased contact pressure together with overall deformation of the 'elastic' section and a tendency to provide extrusion into any clearance space. Increased contact pressure can lead to increased friction and wear, although this effect can be minimized by adequate lubricant being present. Deformation of the section is a necessary feature of the working of the seal and is allowed for in its design. Extrusion can be avoided by proper fitting and support for the section (either in the actual construction, or with separate anti-extrusion rings).

TABLE I – EXAMPLES OF PROPRIETARY COMPOSITE SEALS

Name and Section	Type	Remarks
Dowty DP1 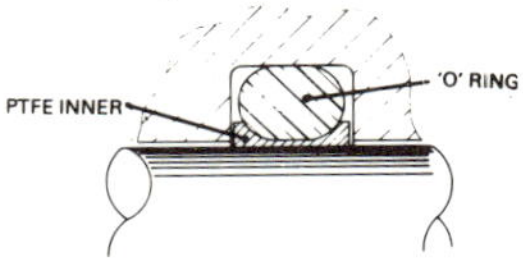	Elastomeric O-ring with PTFE element.	Direct alternative to O-rings, fitting standard O-rin grooves. Internal and external types.
Dowty DP2 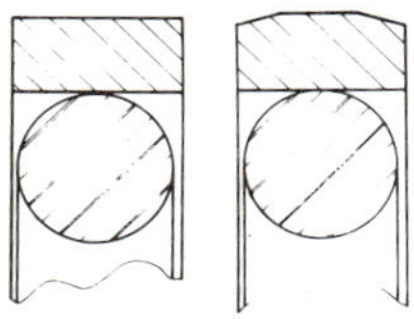	Elastomeric O-ring with filled PTFE outer ring.	Low friction co-axial seal. Ring section rectangula or chamfered.
Dowty DP3 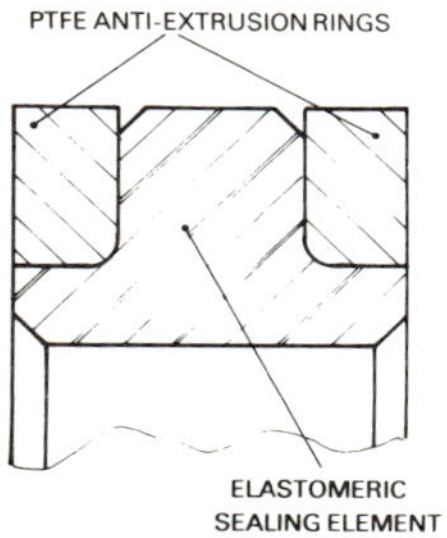	Elastomeric sealing element with PTFE anti-extrusion rings.	Aerospace series seal for double-acting piston rod and piston head sealing. Assembles in MIL-G-5514 housings.
Dowty DP4 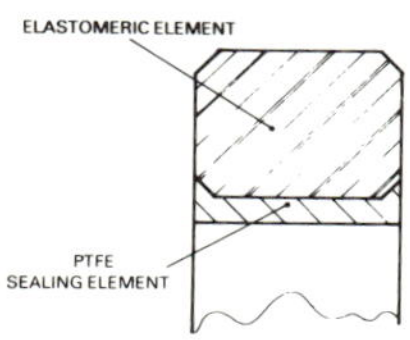	Rectangular section elastomeric ring with PTFE element.	Double-acting seal in aerospace series for piston rc and piston heads. Assembles in MIL-G-5514F housings.
Dowty PTFE 'V5' 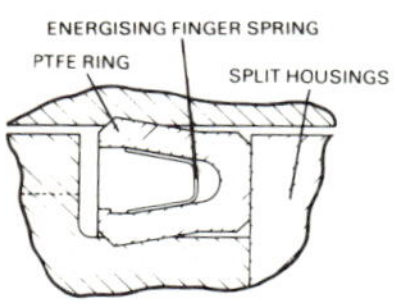	Modified V-ring form in PTFE with energizing finger spring.	Available in internal (rod) and external (piston) seal forms.
Walker Argo Seal 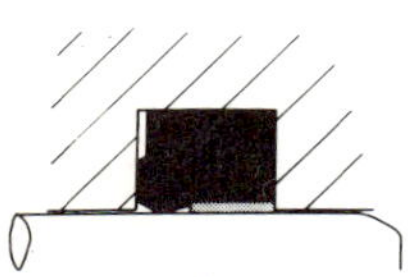	Elastomeric lip seal with PTFE sleeve seal.	Developed as a shallow, low friction cylinder rod seal.

*Dowty Polypac

cor

TABLE I – EXAMPLES OF PROPRIETARY COMPOSITE SEALS (contd). . .

Name and Section	Type	Remarks
Walker Composite 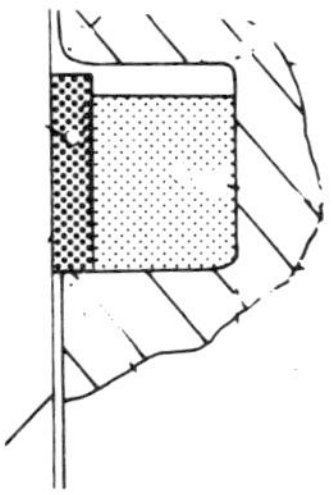	Co-axial seal with endless elastomeric ring and endless PTFE ring.	Hydraulic cylinder seal designed to suit cylinders conforming to ISO tolerance Class H9.
Walker Solosele 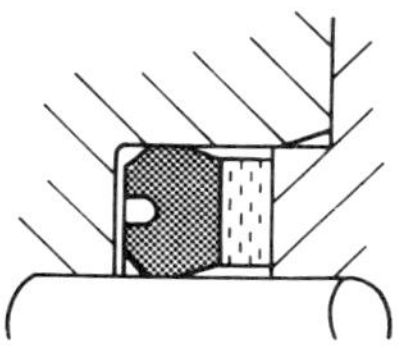	Profiled rubber seal with fabric anti-extrusion element.	Primarily a rod or single-acting piston seal (Type G). Also available as double-acting seal (Type S).
Walker Hypak 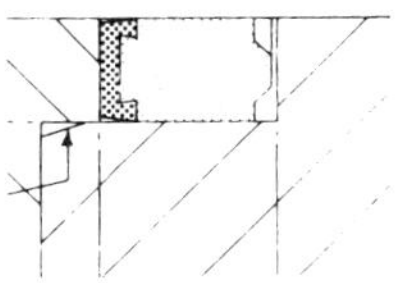	Synthetic rubber sealing element with nylon anti-extrusion base ring.	Primarily a rod or single-acting piston seal (Type G). Also available as a double-acting seal (Type S).
Walker Spoolpak 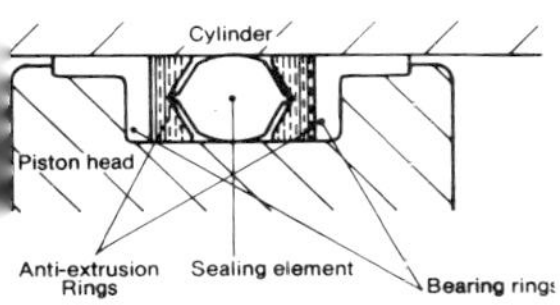	Elastomeric sealing element with anti-extrusion rings and bearing rings.	Five component double-acting seal designed for simple fitting to single piece spool type piston heads.
Balsele* 	Elastomeric sealing section with acetal resin anti-extrusion or wear rings.	Three basic versions – single-acting, double-acting, and grooved. Specially designed for hydraulic cylinders.

Dowty Polypac

cont...

TABLE I – EXAMPLES OF PROPRIETARY COMPOSITE SEALS (contd). . .

Name and Section	Type	Remarks
Selemaster*  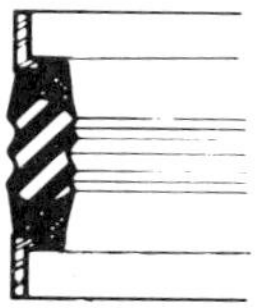	Rubber sealing element with V-type header ring(s) in acetal resin.	Single- and double-acting versions, designed for high pressure cylinders.
Balmaster*  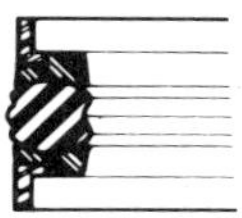	Multiple-lip rubber section supported by acetal resin anti-extrusion ring(s).	Designed for heavy duty hydraulic cylinders. Single- and double-acting versions.
Orpac* 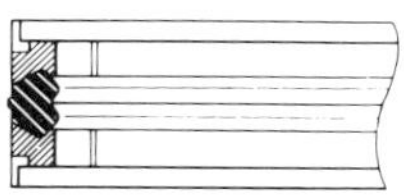	Elastomeric seal section with either anti-extrusion or anti-extrusion/bearing rings.	Compact double-acting cylinder seal which can be fitted to one-piece pistons.
OP70* 	Nitrile rubber seal section with two endless anti-extrusion rings in acetal resin.	Developed for use on one-piece pistons in medium duty hydraulic cylinders.
Hallite Piston Seals		
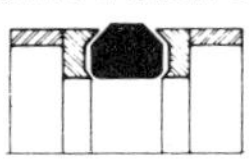	NBR, Polyamide, Polyacetal	
	NBR, PTFE.	
	NBR, Rubberized fabric.	Available in three ranges: Light duty – pressures up to 160 bar; Medium duty – pressures up to 250 bar; Heavy duty – pressures up to 400 bar
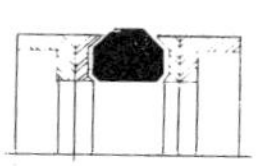	NBR, Glass reinforced. Polyamide	
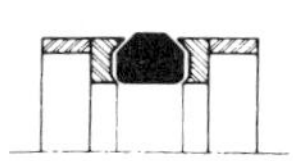	NBR, Polyamide, Polyacetal.	

*Dowty Polypac

cont

Name and Section	Type	Remarks
Hallite Rod Seals 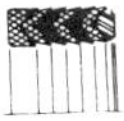	NBR, Rubberized fabric.	
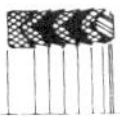	NBR Rubberized fabric. Polyacetal.	
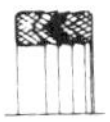	NBR, Rubberized fabric. Polyacetal.	
	NBR, Rubberized fabric.	Available in three ranges: Light duty – pressures up to 160 bar Medium duty – pressures up to 250 bar Heavy duty – pressures up to 400 bar
	NBR, PTFE.	
	NBR, Rubberized fabric	
	NBR, Rubberized. fabric. Glass reinforced polyamide.	

*Dowty Polypac

Friction can vary in a complex manner. As a generalization, friction is relatively high on start-ıp, then progressively decreases to a minimum value with increasing rubbing speed, after which it ncreases again – Fig 10. The actual value of friction (or friction coefficient) depends on the hard-ness and nature of the elastomer, its rubbing area, and the finish of the surface against which it ubs.

Friction also varies with temperature. Here the general effect is for friction to increase with ncreasing temperature with a shift of the minimum friction point to a higher rubbing speed (also hown in Fig 10).

Some actual friction values derived for a representative range of modern hydraulic seals as arying with pressure are given in Fig 11. Losses due to friction are expressed as a percentage of he total theoretical power of the cylinder in which they were tested.

Fig 12 extends these data to actual frictional resistance at different working pressures and peeds, where optimum rubbing speeds (*ie* for minimum friction) are quite clearly defined. This hows that rubbing speed, as much as pressure, can have a marked effect on the choice of seal for ptimum performance (minimum friction loss).

urface Finish

t is generally accepted that the better the surface finish against which the seal rubs the lower the riction and the longer the seal life. A certain degree of 'residual' surface roughness is not necessary or retention of lubricant. However, the finish of the rubbing surfaces, if 'relatively rough' to start vith, tends to improve during the running-in period (typically around 3 000 cycles).

Recommended values of surface finish for modern elastomeric seals are R_a = 0.4 μm for ylinders and R_a = 0.2 μm for rods, although up to three or four times these values may be

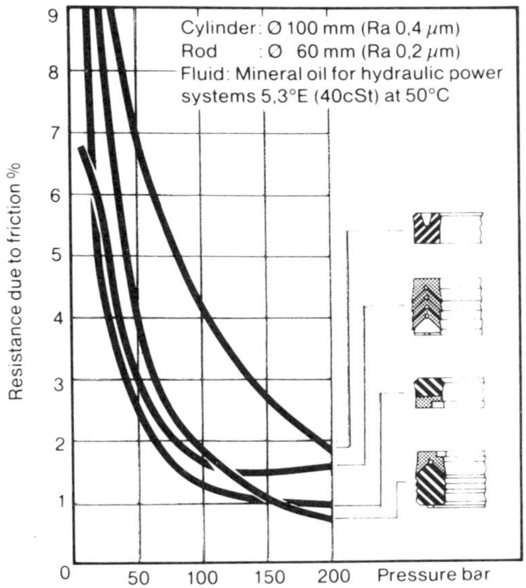

Losses due to friction as a % of the total theoretical power of the cylinder.

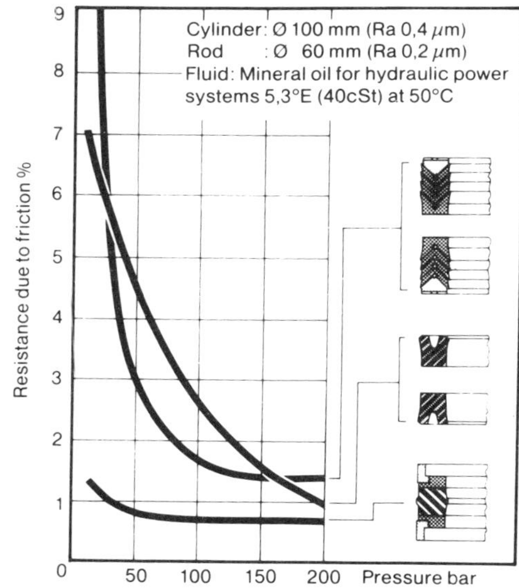

Losses due to friction as a % of the total theoretical power of the cylinder.

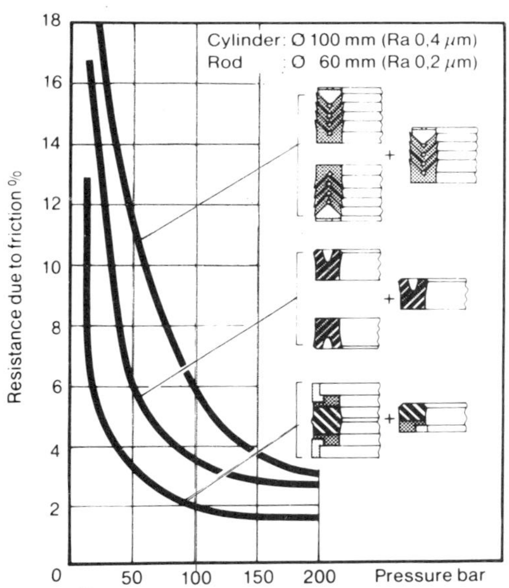

Losses due to friction as a % of the total theoretical power of the cylinder.

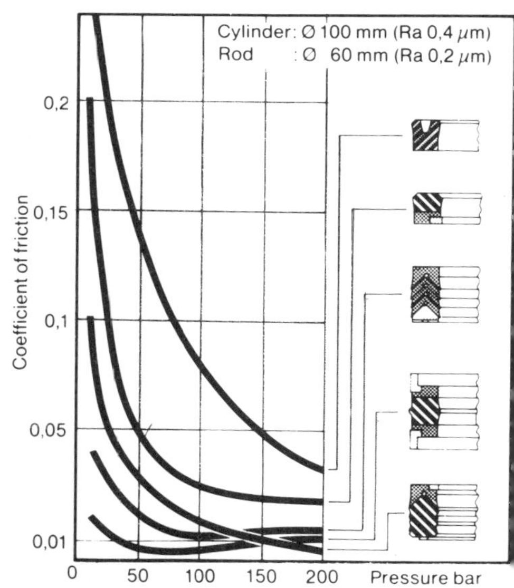

Friction coefficient for seals made with different profiles and materials.

Fig 10
(Dowty Seals Ltd).

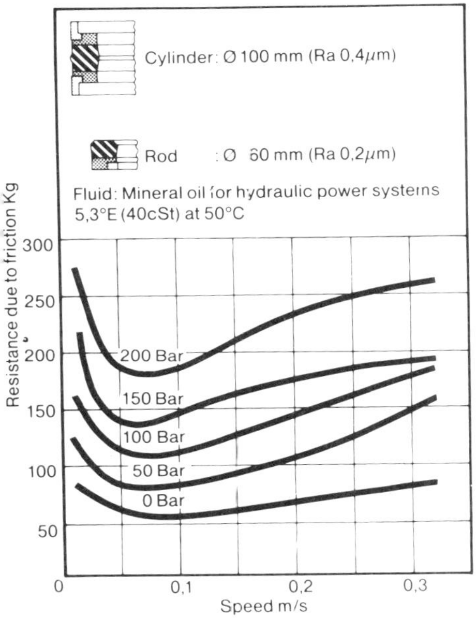

Fig 11 Losses due to friction of Dowty Polypac Balsele type D11W seals on piston and type B on rod, at various working pressures. (Dowty Seals Ltd).

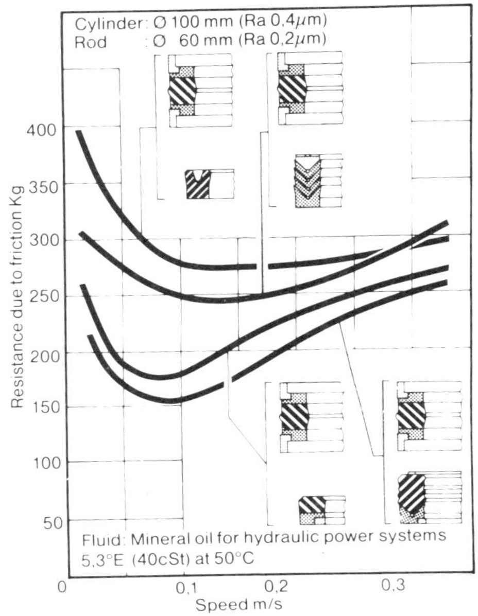

Fig 12 Losses due to friction of cylinders fitted with Balsele D11W on the piston and type B 'Selemaster' 'Veepack' or polyurethane 'U' ring rod seals. (Dowty Seals Ltd).

tolerable for certain types of seal with no undue reduction of life (but not for O-rings). However, actual frictional values realized, and seal life, have been shown by tests to be closely related to surface roughness values for surfaces obtained by cold drawing or honing, but less so in the case of rolled surfaces.

Stick-slip characteristics (*ie* breakout friction) are only partly dependent on surface finish. They are mainly related to the friction characteristics of the seal material. Here polyurethane rubbers tend to show greater breakout friction than other elastomers.

Double-acting 'Polyseal' fabric reinforced seals with polyacetal anti-extrusion/wear rings.

TABLE II – SUMMARY OF SEAL RUBBER PROPERTIES

Elastomer	Main characteristics	Recommended temperature range in air °C	Hardness range
Natural rubber	Excellent strength, wear and resilience. Poor resistance to mineral oils.	–25 +80	30–90°BS
Polychloroprene	Excellent air ageing and strength. Fairly good resistance to mineral oils.	–20 +100	35–90°BS
*Nitrile (low)	Good mineral-oil resistance.	–50 +80	40–90°BS
*Nitrile (medium)	Good mineral-oil resistance.	–30 +90	40–90°BS
*Nitrile (high)	Excellent mineral-oil resistance.	–10 +100	40–90°BS
Polyurethane	Excellent strength and abrasion resistance. Poor heat and hot water resistance.	–20 +70	65–90°BS
Butyl	Excellent air ageing and chemical resistance. Good resistance to phosphate ester oils. Poor resistance to mineral oils.	–40 +120	30–85°BS
Ethylene propylene	Excellent resistance to phosphate ester oils. Poor resistance to mineral oils.	–40 +120	35–95°BS
Silicone	Excellent high- and low-temperature resistance. Low strength and fair oil resistance.	–100 +250	35–85°BS
Fluoroelastomer (*eg* Viton)	Outstanding oil and fuel resistance. Excellent high-temperature resistance.	–20 +200	60–90°BS

*Nitrile rubbers are classified according to the percentage of the acrylonitrile content of the base co-polymer.

Seal Materials

Mineral-oil fluids are compatible with a wide range of synthetic rubbers, but cause excessive swelling, deterioration and eventually cracking of natural rubbers. Nitrile rubbers are the normal choice of elastomer for seals, unless high-temperature working demands the use of more exotic synthetics. The maximum working temperature of nitrile rubber is, however, higher than the maximum working temperature normally acceptable for oil fluids. (See also Table II).

Nitrile rubbers are designated low, medium and high nitrile, depending on ascending acrylonitrile content. Medium nitriles are most widely used since they combine excellent compatibility with a useful working temperature range, typically from –30°C (–22°F) to +120°C (+248°F). The lower the acrylonitrile content the better the low temperature performance, down to approximately –40°C (–40°F), but the poorer the hydrocarbon fluid resistance.

For proofed-fabric materials both nitrile and neoprene compounds are used as standard.

Polyurethane compounds feature outstanding tensile strength and abrasion resistance and are suitable for a wide range of mineral oils and fuels. The physical performance of polyurethane elastomers continues to be improved. In addition, further polymers are still being developed and may well influence choice for further seal materials.

High-Temperature Seals

The maximum working temperature of elastomers is limited to about 80–120°C (175–250°F), depending on type. Seals in nitrile rubber are generally usable at temperatures up to 100°C (212°F). Fluorocarbon is the most satisfactory material for use in the temperature range 100–200°C (210–400°F) and has excellent compatibility with hydraulic fluids, but lacks 'elasticity'. Silicone rubbers offer an even wider range of working temperatures with good elastomeric properties, but are weaker mechanically and have poorer compatibility with some hydraulic fluids. Either material can, however, prove quite satisfactory for high-temperature static seals.

In the case of dynamic seals, automotive-type metallic piston rings are an obvious choice for high-temperature piston seals. If necessary these can be energized by wave springs. Alternative designs to consider are:

(i) Wedge seals based on a split PTFE ring bearing on a continuous PTFE ring, energized by a metallic spring ring.

(ii) Metallic cup seals energized by a metal garter spring or finger springs.

See also chapter on *Leakage.*

SECTION 3

Basic Hydraulic Circuits

THE OBVIOUS aim of hydraulic-circuit design is to provide the functions required from the system in an unambiguous way (and particularly avoiding any 'locking' conditions which could develop). It is equally important that the efficiency of the circuit should be as high as possible — e any potential sources of excessive fluid friction and pressure drop or other energy losses must be excluded. At the same time components must be 'protected' where necessary.

Starting with the pump, a pressure relief valve in parallel with the pump will ensure that pressure generated in the system is limited to a safe level, and the pump itself (if continuously driven) can be unloaded when there is no system demand — Fig 1. Both actual delivered pressure and flow can also be controlled, or set, by suitable valves.

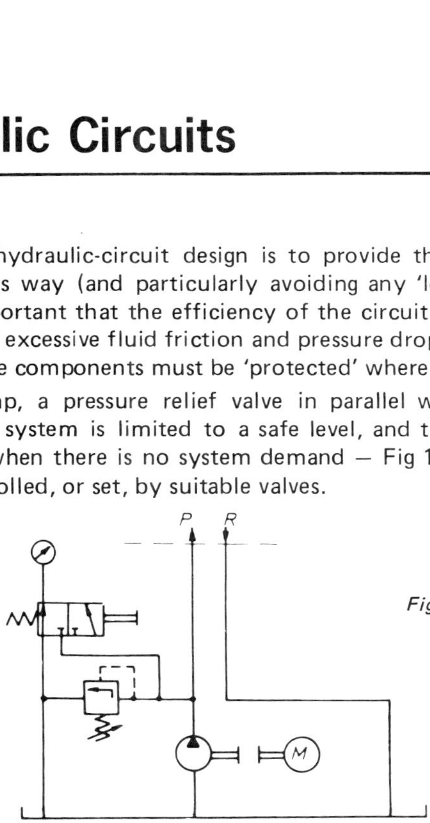

Fig 1

Supply is normally selected by a directional-control valve (selector) of suitable configuration, nd of suitable size to match the rated delivery of the pump. Two approaches are possible — a losed-centre system or an open-centre system. With closed-centre the valve cuts off the delivery vhen the valve is in the normal 'off' position. Under these conditions all the pump flow is directed hrough the relief valve set at maximum operating pressure — Fig 2a. Such a system can be paricularly wasteful of input power as well as generating considerable fluid heating. Thus it would ormally only be used with a pressure-compensated pump where the output is automatically educed to zero when system pressure increases to a pre-set level. The energy loss of the pump perating under these conditions can be quite low. Another solution is an electrically operated ilot-controlled pressure-relief valve when only the control fluid passes through the directionalontrol valve.

The alternative is an open-centre system (Fig 2b) where flow is bypassed to the tank, resistance o flow being minimal and thus the pressure generated by the pump quite low (equal to that of the ressure drop through the valve and associated lines). One possible limitation of this system is

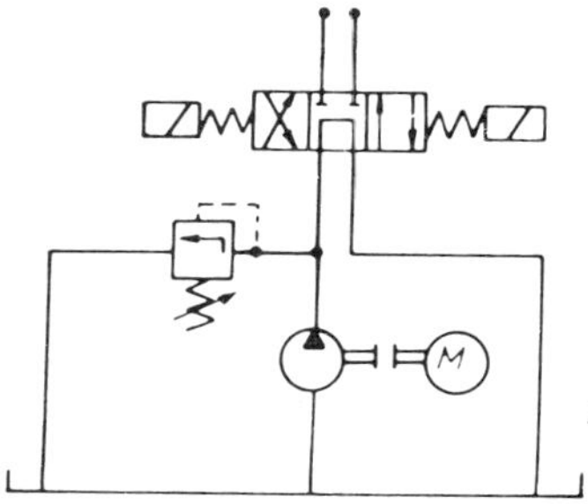

Fig 2a

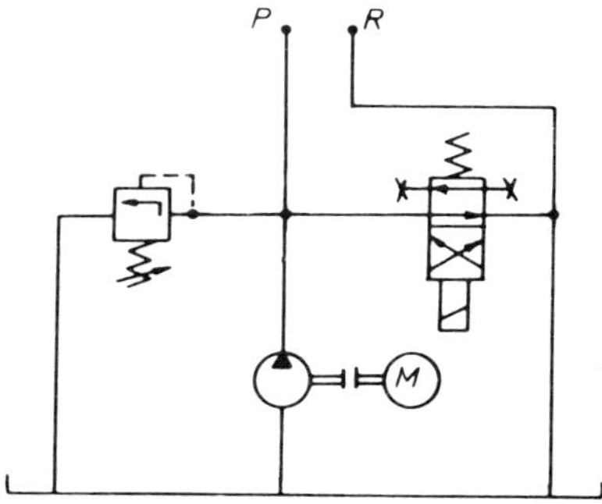

Fig 2b

that the pump delivery remains the same, so energy loss is small only as long as the pressure drop through the bypass circuit remains low. If a number of different valves are enclosed in the bypass circuit the pressure drop can be relatively high.

Neutral-position flow losses can also be higher than anticipated with this system if the pump is not driven at constant speed. If the driver is likely to vary in speed, the pump delivery must be sufficient to meet the demand at the lowest running speed anticipated. At higher speeds there will be higher losses when flow is bypassed. In such a case considerable saving in energy could result from fitting a flow-control valve in the bypass, limiting bypass flow to an acceptable level. Any excess flow is then unloaded by the relief valve directly to the tank with a comparatively low loss.

There are, however, other working conditions which can affect the choice of a closed-centre or open-centre system. Open-centre valves are normally designed with over-lapping characteristics – *ie* with the bypass gradually closing whilst the cylinder ports are already opening. This makes the system sensitive to load, so that metering becomes more and more difficult as load increases (metering refers to the control of flow by gradual opening of the valve). This will result in a dead band and can call for the inclusion of a check valve in the circuit (*eg* when the cylinder must hold against a load).

The characteristics of open-centre and closed-centre systems are compared in Tables IA, IB and IC, from which it is obvious that both system performance and energy loss considerations (efficiency) are involved in deciding on the more suitable type. Many other factors may also have to be considered, in particular the actual period over which the pump will be operating at full pressure, since this will affect the life of the pump. Energy conservation *via* power saving has considerable significance in present day times — and hydraulic systems as a whole have been subject to considerable criticism in that the pump is usually run continuously, with inevitable power wastage. Realistically, a stop-start system would probably result in even more wastage, but much can be done in circuit design to reduce power losses to a minimum.

Flow Control

In a constant-volume system flow from the pump to the actuators is controlled by means of valves. A variable-volume system is controlled by varying the pump delivery, and the pump can deliver fluid directly to the actuators. Time, fluid and system elasticity, and leakage are, however, further factors which may have to be considered.

As a generalization, too, there is a minimum flow rate which can be satisfactorily controlled. For oil fluids this is of the order of 80 cm^3/min (5 in^3/min). Thus for any size of cylinder there is a minimum stroking speed required in order to maintain satisfactory control of flow — see Table II

TABLE IA – CHARACTERISTICS OF SINGLE VALVE SYSTEMS

Parameter	Open-Centre	Closed-Centre
Pump	Any	Pressure-compensated type needed.
Pressure	Minimum	Rating more critical but full pressure always available.
Flow	Maximum	
Power loss (neutral position)	Small but not negligible	Pump runs at maximum pressure.
Deadband	Unavoidable	Can be made quite small.
Hold	Needs check valve.	In neutral position.
Cylinder at end of stroke	Rapid rise in oil temperature	Lower oil temperature rise.
Pressure loss	Less than closed-centre at high flow rate and low pressure	
Bypass (directional valve)	Required	Not required.
Relief valve	Required	Not essential but desirable.
Metering	Sensitive to load	Smaller power loss than open-centre at large loads, less load sensitive.
Metering characteristics	Meters at maximum flow	Meters at maximum pressure and minimum flow.

TABLE IB – CHARACTERISTICS OF VALVES IN SERIES CIRCUITS

Parameter	Open-Centre	Closed-Centre
Pressure	Split amongst loads	
Flow	Same to all cylinders (actuators)	
Metering	Simultaneous	Not applicable
Inter-lock	One valve in open position prevents operation of other valve circuits	

TABLE IC – CHARACTERISTICS OF MULTIPLE VALVES

Parameter	Open-Centre Series – Parallel	Closed-Centre Parallel
Pressure	Maximum can be applied to all loads	Maximum can be applied to all loads.
Flow	Cylinder (actuator) with smallest load moves first	Cylinder (actuator) with smallest load moves first
Metering	Simultaneous	Simultaneous.

The flow rate in suction lines can affect control, although this will not be significant provided he normal recommended maximum flow rate of 1.2–1.5 m/s (4–5 ft/s) is not exceeded. Suction ne sizing can be very important in cases where a relatively large piston area is moving at a high roking rate, since, if the resultant flow through any part of the suction side reaches a value of m/s (20 ft/s) or more, it is almost certain that suction oil will not fully follow up the movement nd control will be affected – see also Table III.

TABLE II – MINIMUM PISTON AREA FOR FLOW CONTROL

Speed cm/sec	Min. piston area cm^2
1	1.36
2	2.72
3	4.08
4	5.44
5	7.80
6	8.16
7	9.52
8	10.88
9	12.24
10	13.6

Speed in/sec	Min. piston area in^2
1	0.083
2	0.042
3	0.028
4	0.021
5	0.017
6	0.014
7	0.012
8	0.0104
9	0.0093
10	0.0083
11	0.0075
12	0.0070

Speed	Min. piston area in^2
1	5.00
2	2.50
3	1.67
4	1.25
5	1.00
6	0.833
7	0.714
8	0.625
9	0.556
10	0.500
11	0.455
12	0.417

TABLE III – MAXIMUM BORE AREA FOR SUCTION FLOW
Bore area to maintain flow velocity below 6.1 m/sec (20 ft/sec)

Flow cm^3/sec	Bore Area cm^2
1	0.0016
2	0.0033
3	0.0049
4	0.0066
5	0.0082
6	0.0098
7	0.0114
8	0.0131
9	0.0148
10	0.0164
20	0.0328
30	0.0492
40	0.0656
50	0.0820
60	0.0983
70	0.1144
80	0.1311
90	0.1475
100	0.1639

Flow in^3/sec	Bore Area in^2
1	0.00417
2	0.00834
3	0.01251
4	0.01668
5	0.02085
6	0.02502
7	0.02919
8	0.0336
9	0.03753
10	0.0417
20	0.0834
30	0.1251
40	0.1668
50	0.2085
60	0.2502
70	0.2919
80	0.3336
90	0.3753
100	0.4170

Flow lit/min	Bore Area cm^2
1	0.0273
2	0.0546
3	0.0819
4	0.1093
5	0.1365
6	0.1638
7	0.1911
8	0.2186
9	0.2459
10	0.2732
11	0.3005
12	0.3275
13	0.3552
14	0.3822
15	0.4095
16	0.4372
17	0.4644
118	0.4918
20	0.5464

Flow gal/min	Bore Area in^2
0.1	0.00193
0.2	0.00386
0.3	0.00579
0.4	0.00772
0.5	0.00968
0.6	0.01158
0.7	0.01351
0.8	0.01544
0.9	0.01737
1	0.0193
2	0.0386
3	0.0579
4	0.0772
5	0.0965
6	0.1158
7	0.1351
8	0.1544
9	0.1737
10	0.1930

Single-Acting Cylinders

Basic control circuits for a single-acting cylinder are shown in Fig 3, using 2- or 3-way, 2-position selectors. If the cylinder needs to be held in an intermediate position, then a 3-position, 3-way selector must be used (Fig 4).

Double-Acting Cylinders

Double-acting cylinders can be controlled by a 4-way, 2-position selector where no intermediate hold position is required – Fig 5; or a 4-way, 3-posit on selector for intermediate hold – (Fig 6).

Further refinement of these circuits may be necessary if the piston has to be held in a particular position for long periods, *eg* check valves could be included at a hold position to eliminate movement due to internal leakage, or for speed control.

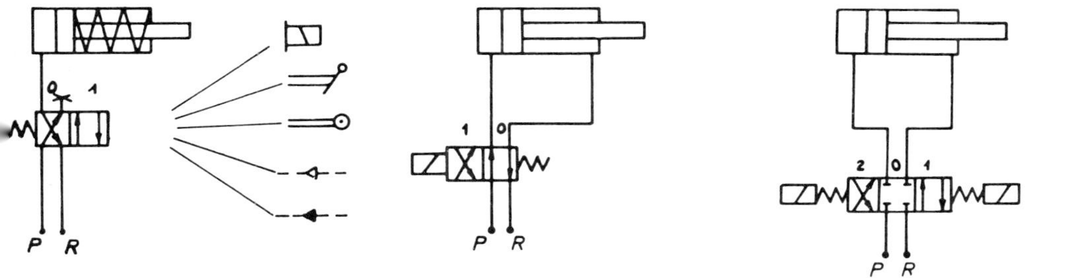

(a)
Forward: Position 1.
Return: Position 0.
Can only be stopped at the end positions.

(b)
Forward: Position 0.
Return: Position 1.
Can only be stopped at the end positions.

(c)
Forward: Position 1.
Return: Position 2.
Can be stopped in any desired position: Position 0.

Fig 3

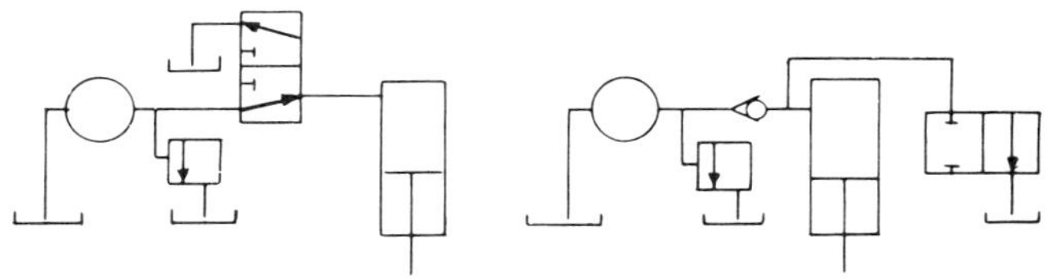

Fig 4

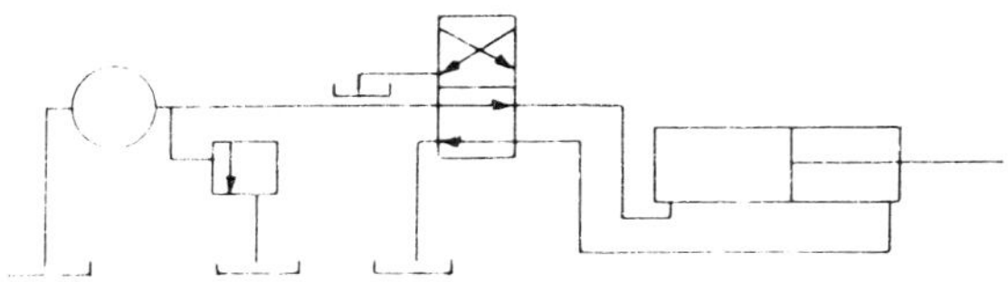

Fig 5

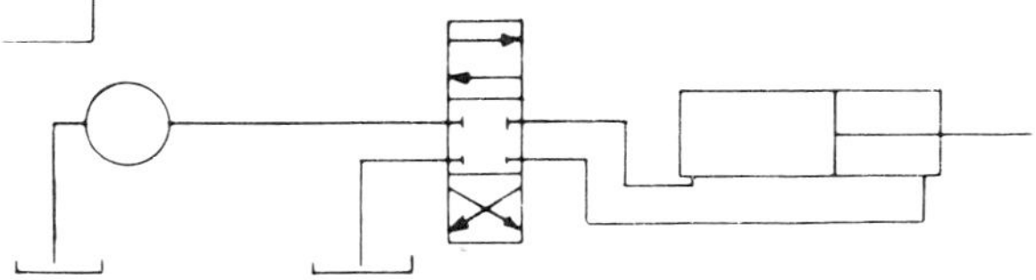

Fig 6

Cylinder Speed Controls

Simple speed control of cylinders can be realized with an adjustable restrictor or flow-control valve in either the supply or exhaust line – Fig 7. Speed control is effective with both directions of motion but forward and return velocities will vary in relation to the different effective piston areas in the cylinder.

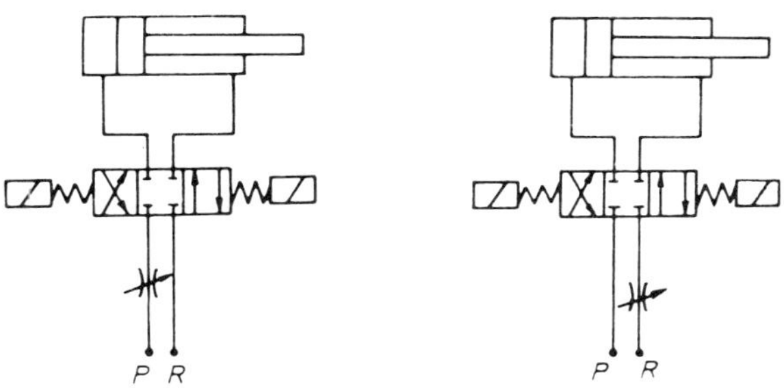

Fig 7

If the restrictor is placed in the return line after the directional-control valve the piston is 'braked' between supply and lock pressure, resulting in an even thrust and preventing the piston running away.

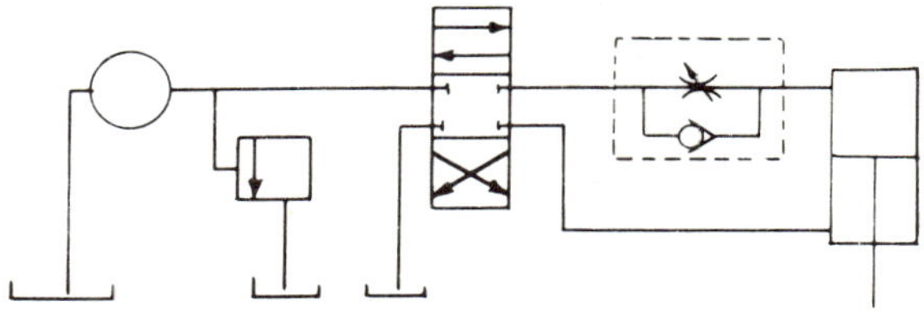

Fig 8

Meter-in speed control (Fig 8) can be used where the cylinder is always under load on the controlled stroke. If the load is highly variable or has a tendency to run away, the meter-out speed control (Fig 9) can be used. Speed control can be applied to either the inward or outward stroke, the metering valve being on the appropriate 'on' or 'out' line respectively.

Both of these straightforward control systems suffer from relatively high energy losses (and high fluid heating) since the pump is operating at maximum pressure all the time and excess flow is discharged through the relief valve. Such losses can be minimized by adopting bleed-off speed control.

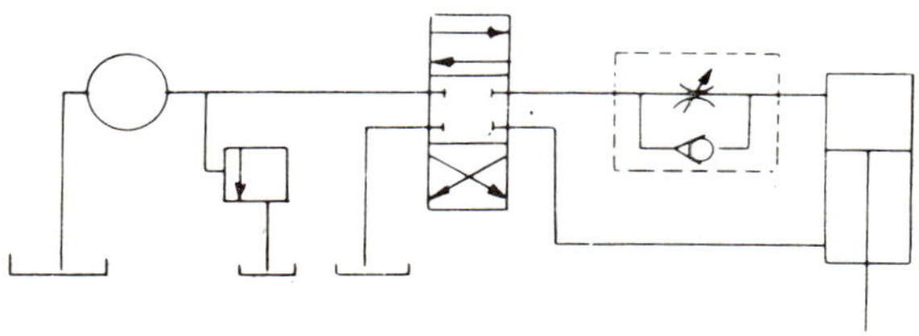

Fig 9

In the bleed-off circuit (Fig 10) the throttle valve is located in the line to the cylinder, bypassing to the reservoir. This valve is operative only when that particular line is pressurized, and in this case operating speed is increased in proportion to bleed flow. All throttling losses are linked to this bleed flow and pump pressure automatically adjusts to load. The main disadvantage of such a system is that flow control (speed control) is indirect and will vary if the actual delivery of the pump varies with load.

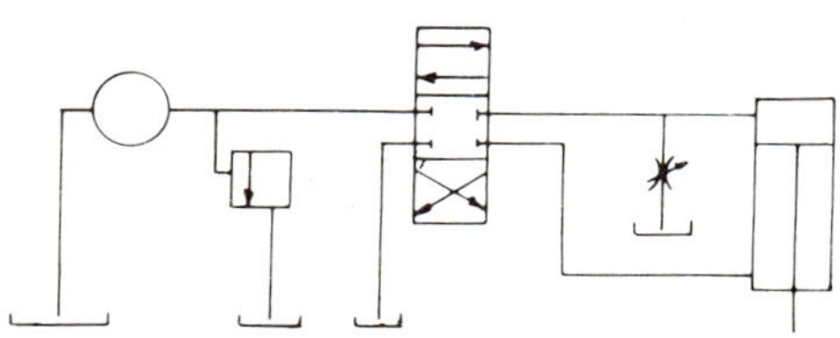

Fig 10

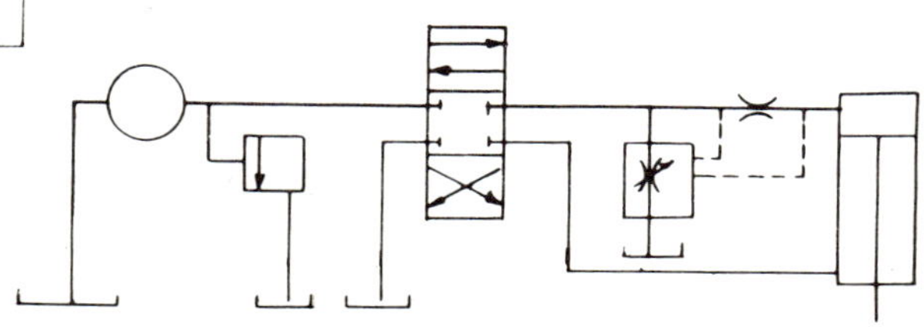

Fig 11

For this reason the bypass speed control (Fig 11) is commonly preferred where losses are to be minimized. A restrictor is used in one line to sense the flow rate and set the bypass flow accordingly. This gives the same positive speed control as meter-in or meter-out. Both the bleed-off and bypass circuits are basically meter-in systems. They would have no effect on cylinder speed if connected to the exhaust line from the cylinder.

There are various other ways in which speed control can be realized, *eg* to give different or selectable piston velocities.

Rapid-Motion Valves

Rapid-motion valves are designed to provide a change-over from slow to rapid forward motion by automatic change-over at increasing piston load. A typical circuit is shown in Fig 12, where both sides of the cylinder are connected in the valve centre position. Pump pressure then acts on the differential piston area providing slow motion. After a pre-determined pressure has been reached the control piston of the rapid-motion valve switches to the through-flow position and full pressure is applied to give the piston rapid motion. There are various possible modes of working depending on the rapid-motion port inter-connections.

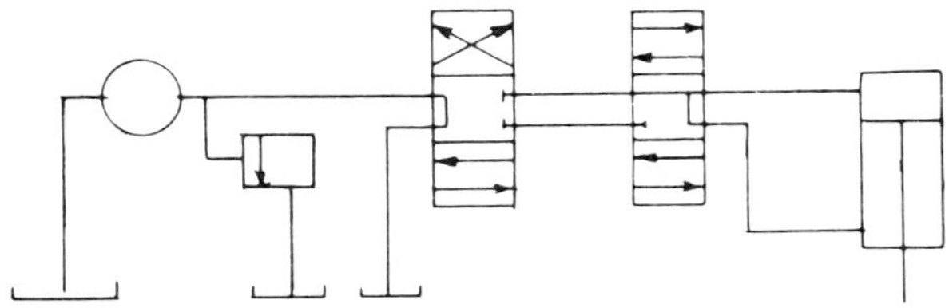

Fig 12

Pressure-Control

Some examples of control by pressure-control valves are given in Figs 13–16. In the circuit of Fig 13 a pressure-reducing valve is used to give a lower working pressure for one cylinder. In the circuit of Fig 14 an unloading valve is incorporated to 'constrict' the cylinder motion – *ie* the pump is unloaded when a pre-determined pressure is realized at a particular point of the stroke.

Figs 15 and 16 are sequential circuits providing operation of a second cylinder when a pre-determined pressure has been built up in the first cylinder.

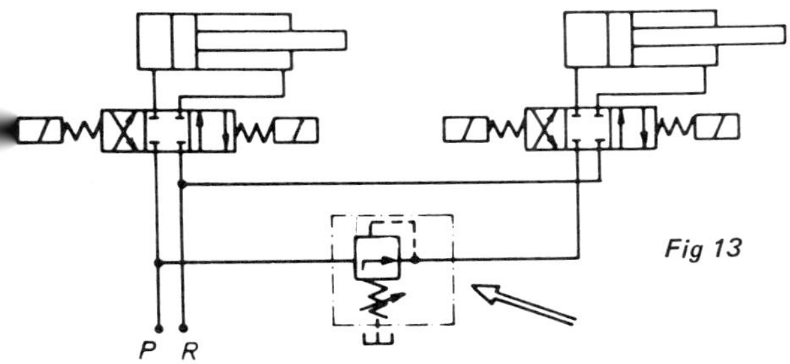

Fig 13

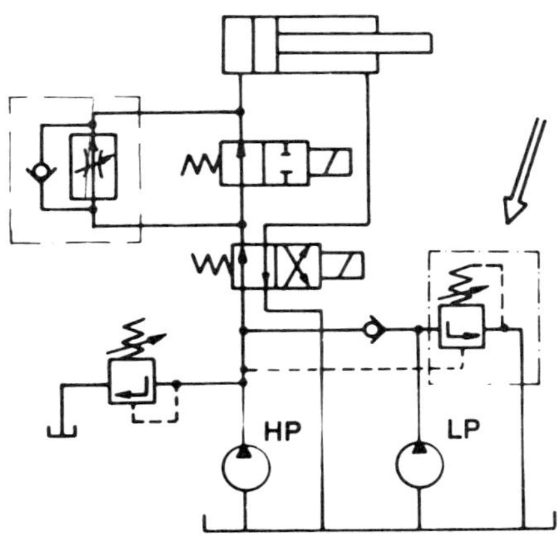

Fig 14

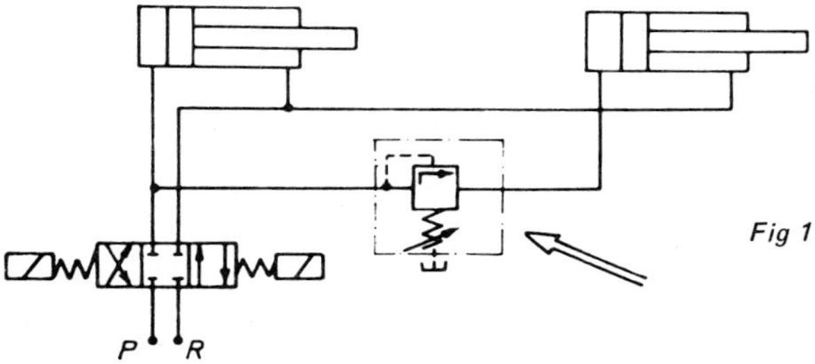

Fig 15

Fig 16

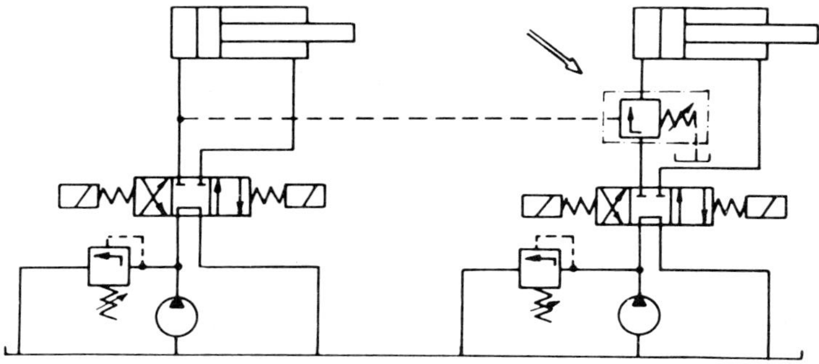

Synchronization

When two or more cylinders are connected in parallel to the same selector, the cylinder with the heavier load will move first and complete its stroke; the next cylinder will then start to move, and so on. Even if the loads are meant to be equal they are never likely to be *exactly* so, and this form of sequential motion will always occur.

Synchronization of cylinders can be achieved mechanically, electrically (*eg* by feedback to electro-hydraulic servo-valves), or hydraulically (*eg* by suitable circuit design). Alternative systems need to be assessed on merit, simplicity, cost, *etc.* Only hydraulic synchronization will be considered here.

Where the cylinder loads do not differ appreciably, satisfactory synchronization may be achieved simply by incorporating restrictors in each of the cylinder lines, adjusted to provide synchronous speeds of operation — Fig 17. The accuracy of synchronization obtained depends on

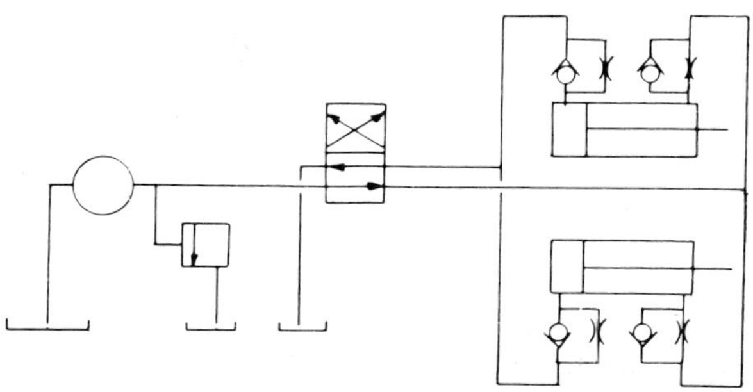

Fig 17

he loads remaining constant and even then it is usually only poor. Another limitation is the elatively large energy wastage and fluid leakage. Also, the available cylinder forces are reduced.

Series connection of cylinders is another simple method of synchronization (Fig 18), but again he degree of synchronization likely to be offered is poor. It is best if the cylinders are both of the hrough-rod type, otherwise synchronization will only be achieved if the differential (annular) area of the first piston is equal to the full piston area of the second cylinder. Small differences in this espect can be taken up with compensating valves, but again only low accuracy of synchronization s usually achieved. A basic limitation with this arrangement is that each cylinder in the system nust be proportionately smaller in size.

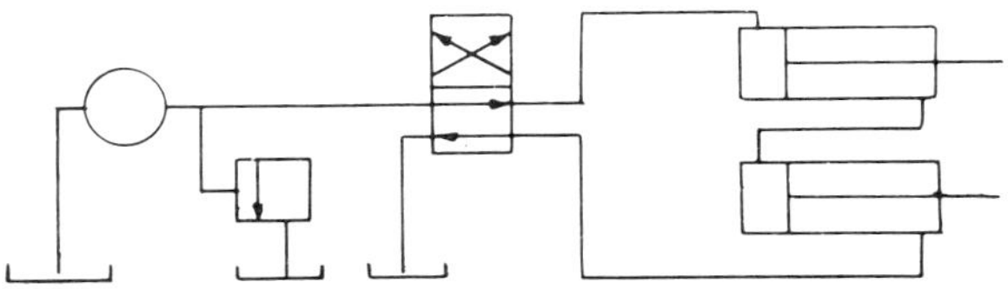

Fig 18

Better synchronization of two equal-size single-rod cylinders is possible if one cylinder is etracting when the other is extending. The system shown in Fig 19 can then be used. Again, the ccuracy of synchronization can be improved with compensating valves. Flow regulators can also e used to obtain synchronization of cylinder movements. There are numerous possiblities in this irection, one circuit example being shown in Fig 20. In general, if loads are liable to fluctuation etter synchronization is usually achieved by locating the flow regulators in the return lines rather han in the inlet lines.

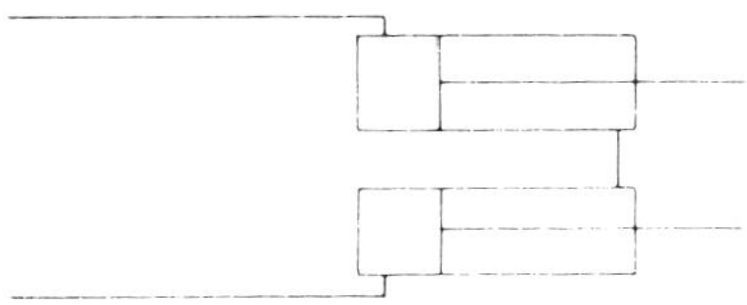

Fig 19

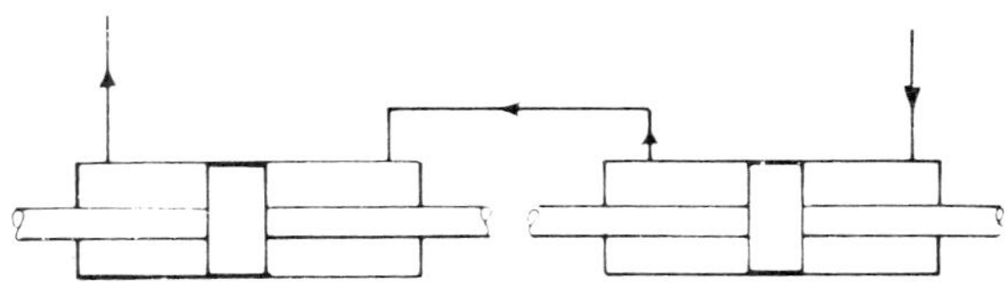

Fig 20

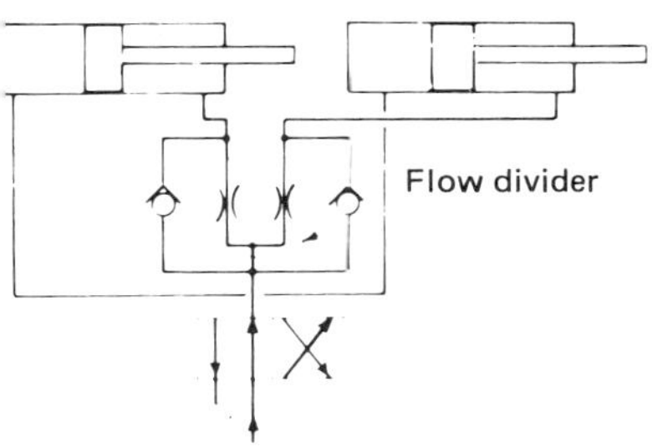

Fig 21

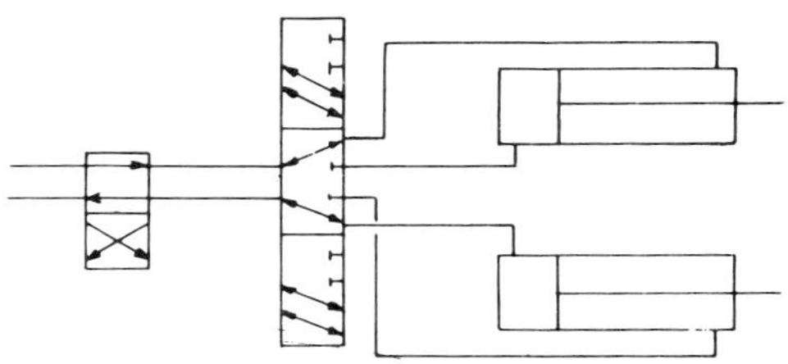

Fig 22

Synchronization can also be achieved with flow dividers — two typical base circuits being hown in Fig 21. In all such circuits, even if they are initially synchronized quite accurately, rogressive loss of synchronization will develop due to internal leakage. It is thus necessary to include some method of re-synchronization in the circuit. This can be manual, *via* two additional valve positions enabling each cylinder to be bottomed-out in turn, or automatic. An automatic re-synchronization circuit is shown in Fig 22. Here the cylinders are synchronized at the end of each stroke.

Where re-synchronization is critical, better accuracy is achieved by using two hydraulic motors with stators coupled together so that they run at identical speed as in the circuit shown in Fig 23. These motors are 'driven' by the cylinder flow and, being mechanically coupled, provide equal flow in each cylinder line. Restrictors across each motor permit flow to pass once one cylinder has bottomed-out, thus allowing the cylinders to re-synchronize at the end of each stroke.

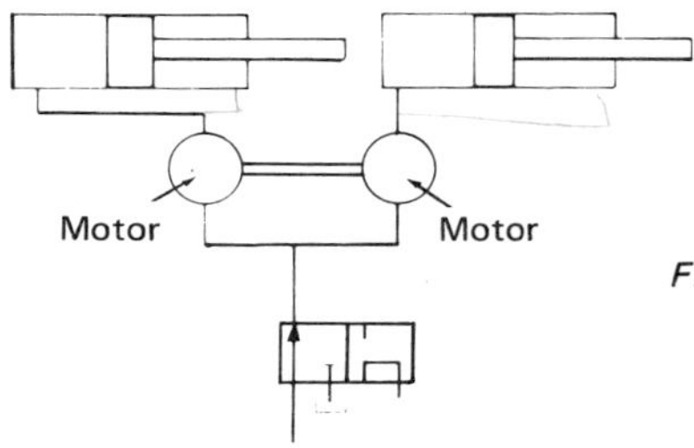

Fig 23

Sequencing

Commonly, when selection is made, two or more jacks are called upon to operate in a pre-determined sequence. There are various ways in which this can be done, the simplest, if the sequences are relatively straightforward, usually being to incorporate a separate sequence valve or valves in the circuit.

A single flow-sequencing circuit is shown in Fig 24; this configuration provides for both jacks to extend simultaneously but delays retraction of the upper jack until the lower jack has completed its retraction stroke and operated the sequencing valve *via* the mechanical trip. The sequencing valve could be placed in the return line, instead of the forward line to the upper jack, with modified system characteristics (that is, the upper jack would then be locked in the extended position until released by retraction of the lower jack).

A simple system of double-sequencing is shown in Fig 25, again with the sequencing valves in the forward flow lines. This would allow, for example, a sequence operation of under-carriage doors opening and closing, with under-carriage retracting and extending. Again the sequencing valves could control return flow rather than forward flow for positive (hydraulic) locking of one jack.

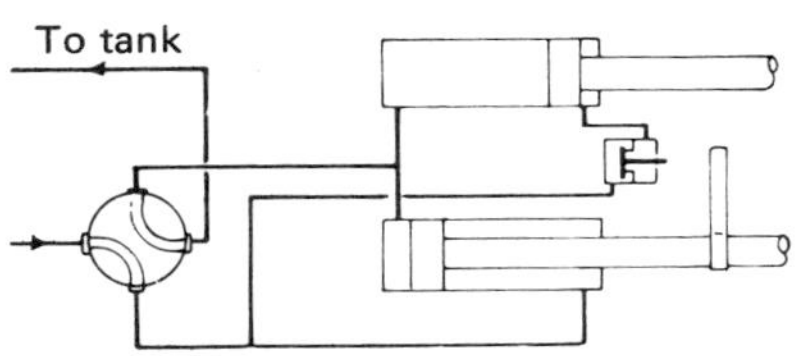

Fig 24

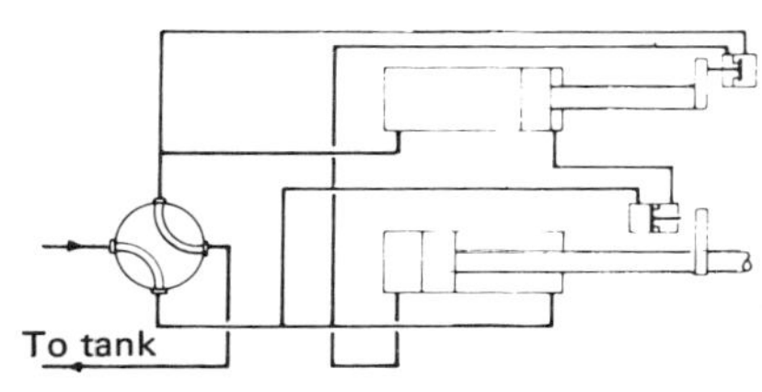

Fig 25

Multiple sequencing is more conveniently controlled by limit switches operating solenoid valve — *ie* electro-hydraulic operation rather than pure hydraulic. Electro-hydraulic sequencing may als be preferred for simple sequencing. Alternatively, mechanical sequencing systems may be pre ferred, where mechanical movements are suitably linked to move appropriate selectors in the required sequence. This has a certain advantage in that mechanical locks can readily be incorporated as part of the system rather than as separate items.

Energy-Saving Accumulator Circuits

A basic accumulator circuit is shown in Fig 26, where the accumulator is first charged by the pump and then the pump shuts down. Circuit pressure is supplied from the accumulator until the pressure drops to a pre-determined minimum, when the pump is cut in again. The control element used is a pressure switch which senses the accumulator pressure and controls the electric motor switching. This stop-start circuit can provide substantial savings in energy where the hydraulic circuit is used only intermittently.

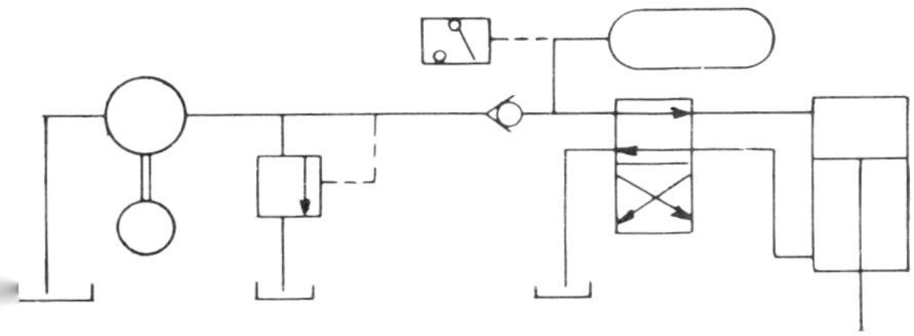

Fig 26

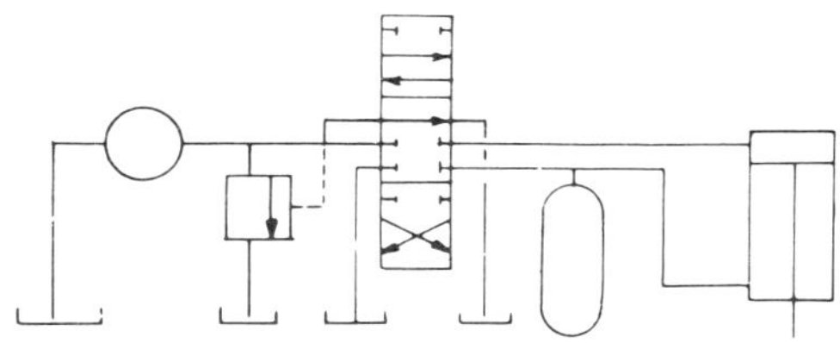

Fig 27

Another energy-saving circuit is shown in Fig 27 where a 6-way, 3-position selector allows the accumulator to pressurize the circuit with the selector in the neutral position. This can provide 'hold' conditions for long periods, during which the pump can be switched off, and also be worked like the previous circuit for intermittent duties (*ie* system pressure supplied by the accumulator with the pump re-charging the accumulator during idle periods and then being switched off by a pressure switch).

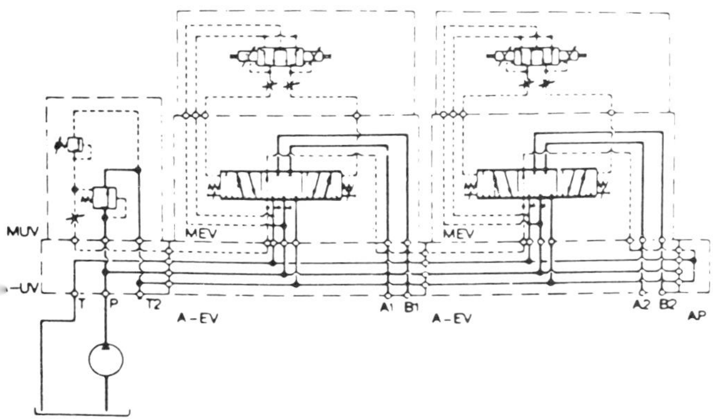

Fig 28 Proportional control circuit with constant delivery pump and two double-solenoid pilot operated directional control valves, mounted on sub-plates, together with a hydrostat, which gives pressure relief, pump unloading and 3-way pressure compensated flow control

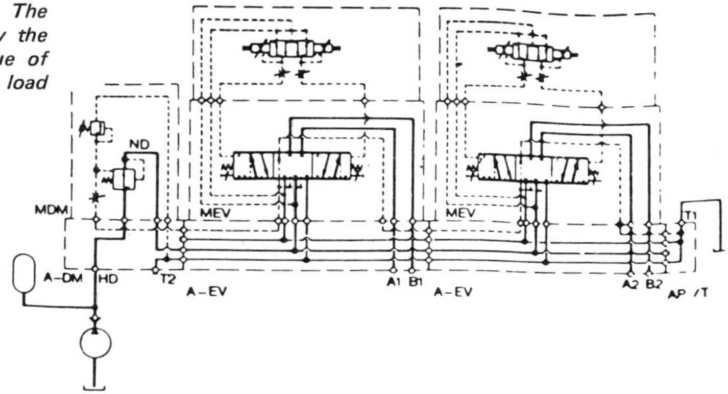

Fig 29 A similar circuit with constant-delivery pump and accumulator. A MDM 2-way flow-control block without the unloading feature is used. The accumulator pressure is reduced by the pressure balance piston to a value of 100 lb/in^2 above the existing load pressure.

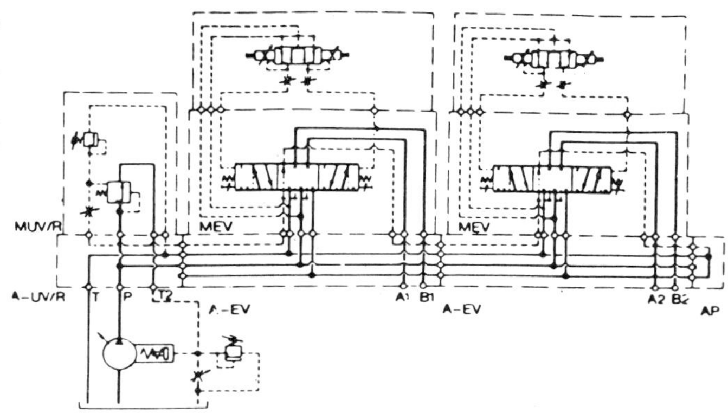

Fig 30 Circuit with special entry block MUV/R for the control of a variable-delivery pump. Unused oil is led to port T2 and hence to a simple throttle causing a back-pressure which is applied to the control cylinder of the pump. Pump pressure thus corresponds to load pressure and delivery to the load speed giving an effective power match.

Examples of circuits providing proportional controls are shown in Figs 28, 29 and 30.

Servo-Systems

THE CORRECT definition of a *servo-system,* as applied to hydraulics, is a system which provides both power amplification and automatic correction for deviation, *ie* a *closed-loop* system with feedback — Fig 1. The feedback is essentially an error signal, re-positioning the control valve to adjust the output to remove the error signal. The controlling valve is then specifically referred to as a *servo-valve* and is capable of accepting both an input signal and a feedback signal (error signal). The two are unbalanced as long as there is a positive or negative error signal. Thus the servo-valve continues to respond to the combined signal, until the error signal falls to zero — *ie* the output quantity corresponds to that of the input signal quantity.

This is a true servo-system, the main difference between it and an ordinary actuator circuit being on the control valve employed, *ie* it is a *servo-valve* capable of providing continuously variable flow with changing input signal. The latter is normally an electrical signal, but feedback signals may also be derived hydraulically or mechanically.

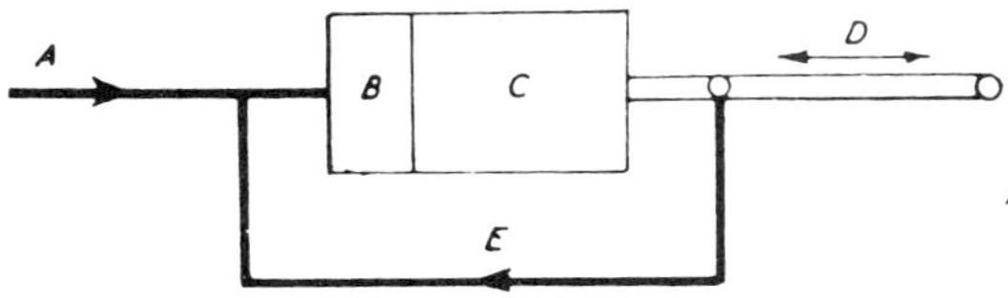

Fig 1 Elementary closed loop with feedback.
A—input. B—servo-valve. C—actuator. D—output. E—feedback.

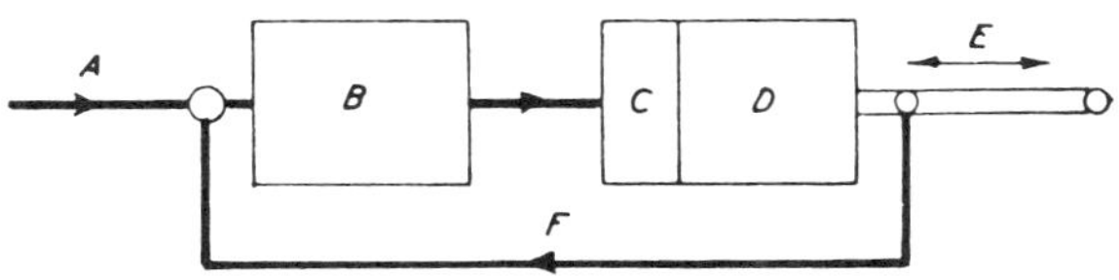

Fig 2 Basic circuit of elementary servo mechanism.
A—input. B—amplifier. C—servo-valve. D—actuator. E—output. F—feedback.

As with simpler circuits, additional amplification may be introduced to accommodate a relatively weak input signal — Fig 2. This amplifier is incorporated within the closed loop, to accept both the command and error signals. This system now provides both two-stage amplification and computation of position or linear response, with self-correction *via* feedback. Thus it will give true proportional response to the input signal, regardless of the effects exerted by external load. It also offers the advantage over the simpler system of being able to control high power levels at low input (signal) levels, both for the command and feedback signals. The combination of a servo-amplifier, servo-valve and actuator with a closed-loop feedback is generally referred to as servo-mechanism.

Electrical Feedback

Where the control valves are of the electro-hydraulic type the feedback signal can also be electric. The simplest way to do this is to connect the output movememt to a potentiometer to generate a feedback signal directly proportional to the movement, which basically provides an analogue-type control. More sophisticated feedback signalling can be devised if required *via* a suitable transducer to eliminate the basic limitation of a simple analogue response (*ie* a tendency to hunt); or to provide an alternative function, *eg* compensation based on rate of change of either output to prevent over-shooting (derivative control), or of some other characteristic in the system (integral control).

This more sophisticated treatment is needed in modern hydraulic machine controls where forces or pressure must be infinitely adjustable, with regulated flows and speeds, controlled speed changes and exact cyclic repeatability. These features cannot be achieved in traditional hydraulics because manual and semi-automatic valves are severely limited in their flexibility by the relatively small number of control levels and by the presence of harmful transients normally associated with level changes. This is concerned with electro-hydraulic controls as a whole working on a closed-loop basis (Fig 3), rather than servo-operation as such.

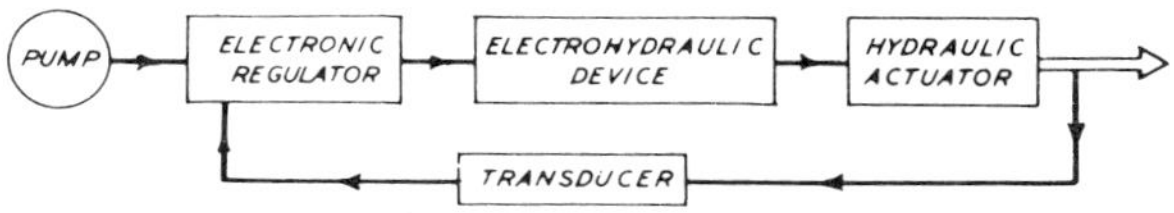

Fig 3

Electronic Feedback

The great advantage of electronic controls is that signalling rates are very fast. Also such controls are appreciably simpler to design and far more compact than mechanical or hydraulic feedback systems, especially when an increasing degree of sophistication is required. Proportional signals (voltage error signals) are readily derived from potentiometers, and simple resistor-capacitor networks can provide any necessary time delays or integral-derived signals.

Alternative control methods available for more sophisticated systems include:

(i) Pulse-length modulation – digital modulation of bistable switching valves.

(ii) Differential pulse-length modulation – which avoids the high power loss inherent in pulse-length modulation systems and provides a closed-centre system adaptable to fluid valving techniques.

(iii) Numerical control systems – involving the use of logic elements.

Whichever method is used, signals are usually amplified to yield a final output of the order of 0–15 milliamps d.c. for application direct to a solenoid-operated control valve or pilot valve.

Mechanical Feedback

Purely mechanical feedback systems are normally based on a differential lever mechanism, with manual input. Both the input and the actuator output movements are connected to the lever which is free to float under the action of feedback movement, automatically adjusting the servo-valve position to compensate.

Actual synchronization of movement may be achieved more or less instantaneously, depending on the characteristics of the external load. Correction will also be initiated after movement has stopped, should the output member again be displaced by external load, with the input member held in its original (signal) position.

Mechanical feedback systems are relatively easy to design, but have distinct performance limitations where precise or complex response is required, and also tend to be cumbersome as regards the length and complexity of the mechanical linkages involved. It is possible to simplify the linkage — *eg* by mounting the valve on the actuator with the differential lever moving with the actuator or mounting the valve on the moving part of the actuator and so dispensing with the differential lever entirely. In the latter case the input member is connected directly to the valve unit which is designed so that axial displacement of the input member produces an axial displacement of the output member in the same direction, thus automatically correcting for 'feedback' movement. An example of this compact form of mechanical system is shown in Fig 4.

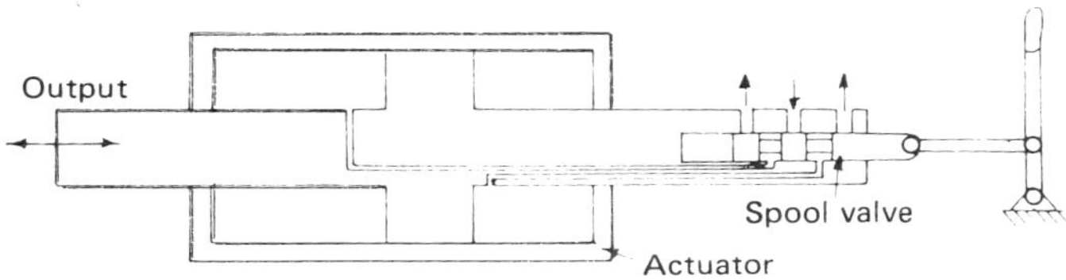

Fig 4

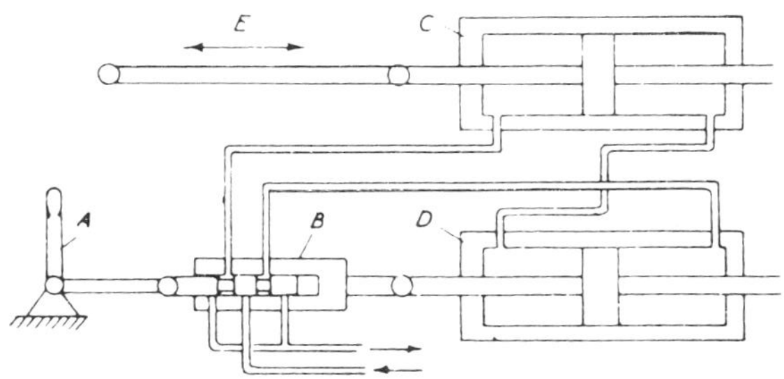

*Fig 5 A simple servo mechanism with hydraulic feedback.
A—input. B—servo-valve. C—actuator. D—auxiliary cylinder. E—output.*

Hydraulic Feedback

Hydraulic feedback systems tend to be even bulkier since two cylinders of equal size are required, one being the actuator and the other an auxiliary cylinder which provides feedback — Fig 5. The body of the control valve is connected rigidly to the piston of the auxiliary cylinder. Thus any movement of this piston displaces the valve body in the same direction as the initial displacement of the valve plunger which is controlled by the input member.

Initial movement of the input member displaces the valve plunger to direct fluid into one end of the main actuator, and an equal amount of fluid is displaced from this actuator into the auxiliary cylinder. This results in movements of the auxiliary cylinder piston, maintaining flow to the main actuator until a 'synchronized' position is reached, when the servo-valve is closed.

The basic advantage offered by hydraulic feedback is that it can be adapted to any size of actuator, is completely free from mechanical backlash and wear, and sensitivity and response can be very high. Lag can be reduced to as low as 1%, although a somewhat higher figure may have to be accepted in the case of fast movements, valve design being the critical factor.

Electro-Hydraulic Servo-Control Systems

These are two main types of electro-hydraulic servo-control systems in use, based on the employment of servo-valves. One is where *positioning* control is required. The other is based on *load monitoring.*

A *positioning* system works on the closed-loop system described previously, with feedback. The performance of such a system can be improved by the addition of a velocity-control network to give positioning accuracy of the order of 0.025 mm (0.001 in) or better, depending on the resolution of the feedback components. It is also a relatively simple matter to extend such a system to multi-axis working with each axis being driven by its own servo-actuator controlled from separate channels.

A *load-monitoring* system is one based on monitoring the pressure in the actuator to maintain a pre-determined load (rather than position), regardless of actuator position. The sensor in this case is a transducer, sited on the load or, preferably, in the actuator. The transducer then feeds the servo-amplifier with a voltage signal, the value of which corresponds to the load. This is compared with a smaller reference signal. Any difference represents an error signal, with adjustment of movement until this is reduced to zero.

It is also possible to accommodate two variable functions in such a system — load variable (as described) and frequency variable. The latter is achieved by super-imposing frequency oscillation to give varying frequencies which may vary from one per minute to thousands of cycles per second. Under such conditions, although the load is held steady, the cyclic amplitude of the actuator decreases with the increase in cyclic frequency for a given power potential.

Load-monitoring control systems have a particular application for fatigue testing materials and components.

Servo-Valves

A basic difference between an ordinary solenoid-operated control valve and a servo-valve is that whereas an ordinary valve has only two or three positions (or possibly more in some cases), a servo-valve allows a continuous variation of flow with changing electrical signal. An electro-hydraulic servo-valve is thus defined as an electrical input servo-control valve capable of continuous control and it is commonly used to drive hydraulic actuators or motors in closed-loop servo-systems.

Typically, a flow control servo-valve consists of two stages: an electro-hydraulic pre-amplifier and a spool valve. The second stage is basically the same for the several types of servo-valve offered by various manufacturers; each manufacturer uses the same materials for both the sleeve and spool, the same heat treatment, the same diametral clearances and similar manufacturing and processing techniques. So, the only real basis for comparison and user selection of one type of valve over another is the first-stage configuration and the manner in which the inter-stage loop closure is accomplished.

Servo-Actuators

The actuator in practical servo-systems is a hydraulic cylinder, so a combination of a servo-valve and a conventional cylinder may be described as a servo-actuator. More specifically, however, the description is best reserved for cylinders which incorporate a suitable servo-valve or are designed for direct mounting of matching servo-valves. The advantage of such an approach (apart from making the system more integrated), is that it can provide unrestricted flow between servo-valve and actuator. Provision is also commonly made for mounting of the master reference — *eg* a linear potentiometer — on the actuator itself.

Types of Servo-Valve

Basically there are only three first-stage configurations which are commonly used in the design of servo-valves, each of which incorporates a 'functionless' open-centre arrangement. The variations are then:

(i) Half-bridge one-arm variable;

(ii) Full-bridge two-arms variable;

(iii) Full-bridge four-arms variable.

In all three types an 'elastic' feedback element can close a loop between the spool valve and the input torque motor.

A single nozzle flapper first stage is typical of the half-bridge one-arm variable configuration. The full-bridge two-arms variable is represented by the symmetrical double-nozzle flapper. Full-bridge four-arms variable is represented by the jet-pipe/receiver combination.

The single- and double-nozzle flapper arrangements normally have an operating clearance of about 0.025 mm (0.001 in) between the flapper and nozzle(s). Because, in the second case, the maximum flapper movement away from either nozzle is normally limited mechanically to about 0.05 mm (0.002 in) the contamination particle size which this double nozzle version can handle without malfunction is strictly limited. The movement of the flapper away from the single-nozzle is not tightly constrained; this type is, therefore, generally more tolerant of contamination.

A single-nozzle configuration, however, is inherently susceptible to null shift, *ie* to any change in the input current required to bring the valve to the condition where it supplies zero control flow at zero load pressure drop. Null shift may occur with changes in supply pressure, temperature or other operating conditions. The double-nozzle configuration is not inherently subject to null shift with temperature and supply pressure variations. A jet-pipe/receiver combination is most tolerant to contamination and is substantially insensitive to null shift arising from changes in temperature and supply pressure.

Flapper Valves

Both single- and double-flapper valves operate on the same principle. The system consists of a fixed and a variable orifice in series, the variable orifice being composed of a nozzle with a flapper plate very close to it. Flow is metered between the flapper and the circumference of the nozzle. The area between the fixed and variable restrictors is connected to one end of a spool valve. In the case of a single-flapper, the other end of the spool is connected, through a half area piston, to supply pressure (or to a spring). The spool forces are balanced when the control pressure is equal to half the supply pressure.

A current in the torque motor in one sense applies a force to the flapper to move it towards the nozzle. This increases the restriction and decreases the flow through the nozzle, causing a decrease in flow and pressure drop across the fixed orifice. The control pressure increases and causes the spool to move. The feedback spring between the spool and flapper applies a force to the flapper to return it towards the null position. When the feedback force and torque motor forces are equal, the flapper is restored to its null position, the spool forces are balanced and spool stops.

A double-flapper valve works on the same principle, except that the flapper operates between two nozzle/restrictor systems with the control pressures applied to either end of an equal area spool.

Jet-Pipe Valve

The first stage of a jet-pipe servo-valve consists of a torque motor, a jet-pipe and a receiver. An electric current in the torque motor coils develops a torque at the armature. The jet-pipe is rigidly attached to the armature and rotates with it. A very small flow of high-pressure oil is fed by a flexible tube to the jet-pipe. As the high-velocity oil flows out of the end of the jet pipe, it impinges upon the face of the receiver. Two small-diameter holes located side by side in the receiver are connected to either end of the valve spool. With the jet-pipe centred over the two holes, equal pressures are developed on either end of the spool, causing it to maintain its position.

When a signal is received by the torque motor, the resultant torque causes the jet-pipe to rotate off-centre and pressure unbalance occurs across the spool, causing it to move. As the spool displaces from null, it deflects the feedback spring, developing a force counter to the input torque. This force returns the jet-pipe to null and the spool comes to rest in its new position. Spool displacement is thus proportional to the torque input. For practical purposes, a linear relationship exists between torque and input current. This flow-control characteristic is fully reversible and a servo-valve can be used as a three-way or four-way type of valve. Continuous control of spool position can be applied from one extreme to the other.

Construction Considerations

Second Stage

All servo-valves use the same construction for the second stage, *ie* a hardened spool operating in a hardened sleeve or bore. This makes them resistant to erosion of the control edges of the spool lands. However, over long periods of operation, even in relatively clean fluid, erosion will occur. This erosion produces an effective underlap of the valve, increasing the flow gain and leakage around null. The effect of a given degree of erosion is dependent on the stroke of the valve: the longer the stroke of the spool, the smaller the effect of the erosion. This is because a longer stroke means a narrower port; hence, less leakage and less change in flow gain. A longer stroke also reduces the proportion of the valve's operating range over which erosion effects occur.

Torque Motor

The null stability of a servo-valve is controlled by the stability of its torque motor. Null shift can be caused by mechanical movement of components but, except in extremely high acceleration environments, this is very unusual. Null shifts are usually caused by small internal mechanical movements in the torque motor from stress changes and by the stress changes themselves in the magnetic circuit. It is important, therefore, to choose a torque motor construction which minimizes these effects.

When the torque motor is constructed, the magnets may be attached to the frame by screws or by welding. Both methods impose initial stresses. The stresses imparted by welding can be released by adequate temperature cycling. The same cannot be done completely for screwed parts, because they rely on stress and strain for retention. Over long periods and through thermal cycles and other strains, the stresses in bolted parts will change, causing changes in the magnetic circuit and some small movements of components. This causes null shift.

After thermal stabilization of a servo-valve, it is necessary to re-adjust it. It is important that the method of adjustment should not re-introduce the stresses just relieved. A biasing spring on the torque motor, physically isolated from the magnetic circuit, is a safe way of providing this feature.

There are two methods of closing the loop between torque motor and second stage spool. The two systems are: displacement follow-up; feedback of a force proportional to spool position. The follow-up system demands a torque motor or force motor displacement proportional to input current. This displacement moves a flapper device to a new fixed position. The resultant pressure unbalance moves the spool which, either directly or through a mechanism, moves the nozzles until the original null relationship is re-established. There are two principal short-comings to this method. It is difficult to make an electro-magnetic device with a completely linear displacement/current characteristic and the internal-loop gain is very poor because it is limited by the 1:1 follow-up system. This tends to increase hysteresis and threshold response.

The force-feedback system avoids these difficulties. It consists of a spring between the second-stage spool and the element driven by the armature – either jet-pipe or flapper. The receiver or nozzles are fixed, providing a first-stage null condition at a fixed geometrical position of the armature. Thus, with a linear spring between spool and armature, a linear-force output electrical signal is required from the torque motor. This is relatively easy to provide at one fixed position. The internal-loop gain can be readily adjusted by adjusting the change on the magnet. This has a large effect on the displacement-force gain of the torque motor, but only a small effect on the force/current characteristic. The principle is simple but requires some care in execution as the movements are small; a valve passing 22.5 lit/min (5 gal/min) has a spool displacement of 1.5 mm (0.06 in). One way is to attach one end of the feedback spring to the torque-motor member (jet-pipe or flapper) with a ball on the end fitting into a slot in the spool. This requires an extremely close match between ball and slot.

If a clearance or backlash does develop, there is an area where the spool can drift to and fro with no possible control over it, because there is no feedback intelligence between second and first stage. The forces on the ball can cause some friction which can degrade threshold, resolution and repeatability. A better way is to attach it rigidly at both ends and provide an elastic pivot to accommodate angular deflections.

Pilot Stage Efficiency

The first-stage pressure gain is directly related to the efficiency of the valve – *ie* the ratio of maximum available load power (to stroke the spool) to quiescent power loss. Flapper valves tend to have low efficiency due to inherent pressure feedback reducing the first-stage pressure gain. The lag in this feedback caused by control compliance is also a possible source of valve instability. Adjustment of the flapper-to-ground spring rate will stabilize the inner loop but this in turn limits the loop gain around the first- and second-stages. The jet-pipe design has no pressure feedback to the first stage, so this problem is avoided.

Again, because of its greater efficiency, for the same first-stage supply flow, the jet-pipe design can cover approximately twice as much useful flow as is possible with the double-nozzle flapper. In practice, the jet-pipe recovers up to 90% of the total first-stage supply flow, reducing the quiescent power demand and heat generation.

Values for ideal arrangements of the basic configurations are:

(i) Half-bridge one-arm variable : 0.192

(ii) Full-bridge two-arms variable : 0.135

(iii) Full-bridge four-arms variable : 0.185

In practice, with jet-pipe servo-valves, ratings as high as 0.25 have been attained.

See also chapter on *Electro-Modulated Hydraulics.*

Electro-Modulated Hydraulics

ELECTRO-MODULATED HYDRAULICS embraces both electro-hydraulic controls and complete servo systems operating as a closed-loop with feedback. Three basic 'blocks' are common to all such systems: (Fig 1).

(i) An electronic regulator.

(ii) An electro-hydraulic device.

(iii) A hydraulic actuator (*eg* cylinder).

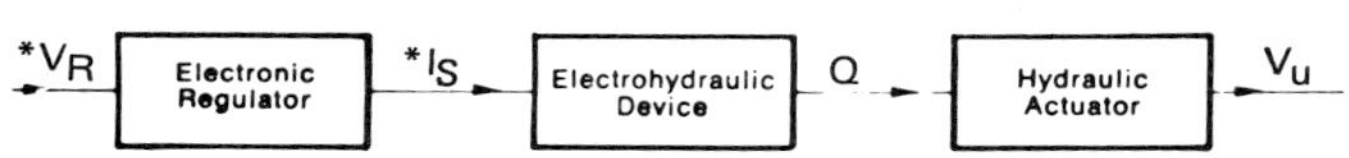

Fig 1 Basic 'blocks' for electro-modulated hydraulics.

*V_R – *reference voltage*
*I_S – *driving current*
Q – *hydraulic flow*
V_u – *controlled variable*

In an open-loop electro-modulated system input is provided by a reference voltage fed to an electrical regulator which passes on this information in the form of a driving current applied to an electro-hydraulic device (*eg* an electro-servo-valve). This in turn is responsible for the hydraulic actuator. The output thus corresponds with a value of the input signal. For good response and correlation it is important that pressure, load on actuator and other system parameters are kept constant since the system cannot sense or correct any discrepancy between input signal and resulting output.

In a closed-loop electro-modulated system a transducer is added and limited to the output to provide a feedback signal directed to the input side. This signal may be indicative of position velocity, force, or if necessary more than one output parameter (see Fig 2). The electronic regulator then compares the input signal with this feedback signal and varies its own output (driving current) to minimize or eliminate any discrepancy between input and output.

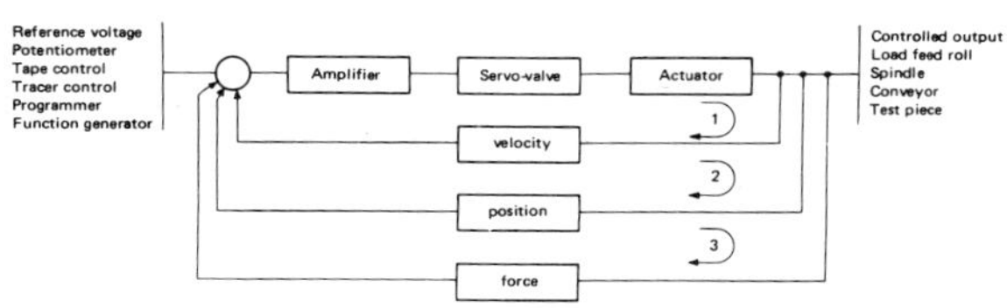

Fig 2 Three main forms of control using electro-hydraulic servo-actuators.

A further example of electro-modulation is shown in Fig 3. This is a conventional circuit, except for the fact that a main electro-modulating relief valve is included to control the pressure. The incorporation of such a valve allows the system to operate with pressure cycles of every complexity, with gradually increasing or decreasing levels rather than sharp pressure peaks.

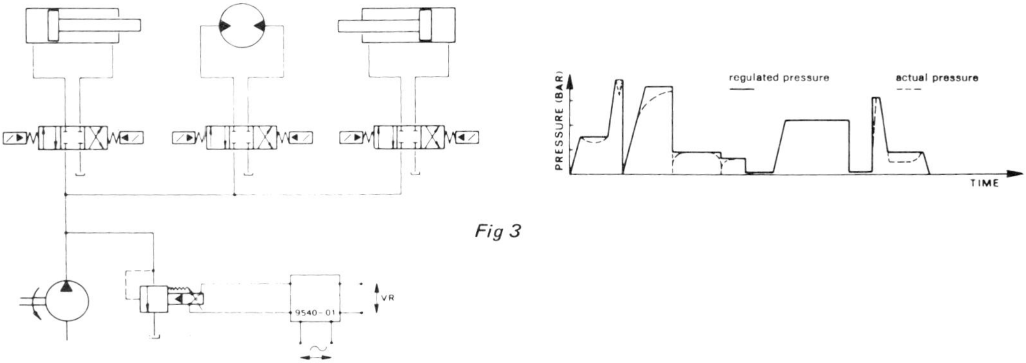

Fig 3

Fig 4 Flow plot for a servo-valve.

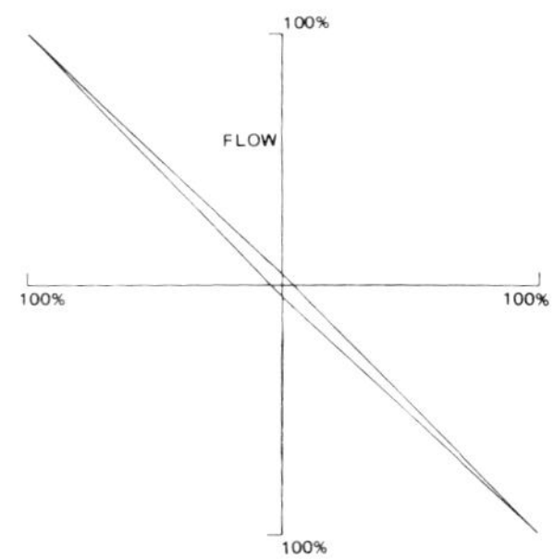

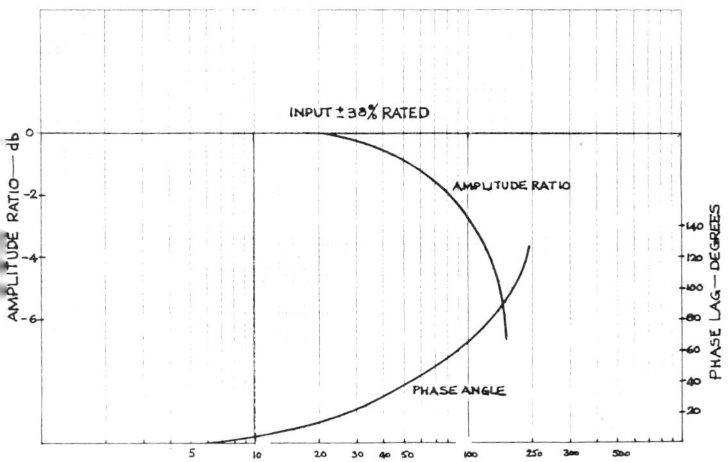

Fig 5 Frequency response plot for a servo-valve.

Valve Characteristics

The characteristics of a servo-valve can be described by a number of parameters. The rated flow, customarily defined at 70 bar pressure, can vary from 0.5 lit/min for a single-stage valve, up to 250 lit/min for a two-stage valve. The rated input at the rated flow can range from 4–500 mA, depending on valve and coil type. Linearity, a measure of proportionality between input and output, is usually better than 10%; a flow plot shows this (Fig 4). Separation of the two lines indicates the valve repeatability or hysteresis. The frequency response diagram (Fig 5) shows the ability of a valve to react to changes in input. This is an important characteristic and figures of 50—100 Hz are usual, while valves with a capability up to 400 Hz are available.

Valve Types

Control valves used in electro-modulated hydraulics are usually based on torque-motor operated servo-valves, or specially designed proportional pilot valves. Choice is largely governed by the degree of control required over flow and pressure. The main types of valves are:

(i) *Torque motor-operated electro-hydraulic servo-valves* — this kind of servo-valve is particularly recommended in applications where it is necessary to keep a bi-directional control on flow and pressure (controlled feeding of one of the two branches of a hydraulic actuator, motor, cylinder, *etc*). The features of the torque motor servo-valves allow them to give excellent performance and to meet the hardest requirements, that is: rapidity of response, accuracy, good null-point passing, *etc.* In order to avoid malfunction and to give long life without wear, efficient filtering system (5 to 10 μm) of the hydraulic fluid must be provided. The electric power required for operating the torque motor is low (0.1 to 0.2 W).

(ii) *Linear force motor-operated electro-hydraulic servo-valves* — this kind of servo-valve is particularly recommended in applications requiring unidirectional control of pressure and flow.
They are used as the first hydraulic operating stage of valves in which the output quantity (pressure and/or flow) can be directed by a pressure signal.

Solenoid-operated servo-valves have no critical filtering requirements (*ie* not higher than those of normal industrial components) and have the advantage that they can be used with normal type hydraulic components. Electrical power required to operate these valves is of the order of 10—25 watts.

TABLE I — SUMMARY OF TRANSDUCER TYPES

Main Type	Sub-Types	Remarks
Linear-displacement transducer	Capacitance	For small sizes only.
	Differential transformer	
	Potentiometer	
	Pulse	
Pressure transducer	Flat diaphragm	Insensitive to noise.
	Strain gauge	Good accuracy, wide range of response.
	Bellows	
	Bourdon tube	
Angular-measurement transducer	Potentiometer	
	Resolver	
	Pulse generator	Optical, magnetic, electronic, *etc.*
Velocity transducer	Linear	
	Pulse	Pulse frequency proportional to speed.
	Angular	Tacho-generators measuring rotational speed only.

Transducers

The accuracy achieved in a closed-loop electro-modulated servo-system depends primarily on the accuracy of the transducer. The main types employed are linear-displacement transducers, pressure transducers and angular-displacement transducers (see Table I). There are also separate types of transducers used for measuring velocity, *viz*

(i) *Linear-speed transducers* – obtained by transferring the linear motion into a rotary motion and measuring this motion, or by using the electric signal of a displacement transducer and determining the speed as a derivative of the displacement function.

(ii) *Pulse-speed transducers* – using a pulse transducer when the frequency of the pulses is proportional to speed. This can be used to measure either linear or angular speeds.

(iii) *Angular-speed transducers* – such as tacho-generators or alternators provide an electrical signal and a voltage proportional to the speed of rotation.

Electronic Regulators

Actual input signals may be derived from a reference voltage, potentiometer, tape control, tracer control, programmer or function generator. The transducer signal is rendered as a feedback voltage which may be a proportional signal (analogue) or a pulsed signal (digital). In either case the regulator is basically the same, except that with a digital error signal a digital-to-analogue converter has to be employed. In practice the electronics is considerably more complicated but digital control can be expected to provide for greater resolution; also it can utilize high-threshold logic elements for virtually complete immunity from noise.

A block diagram of a modern digital command and closed-loop control circuits for a precision feed drive is shown in Fig 6. The operating principle of the servo-system relies upon integration of command pulses generated from an internal crystal clock and feedback pulses from machine-driven digitizers. This is accomplished by using a multiple array of reversible counters. Errors

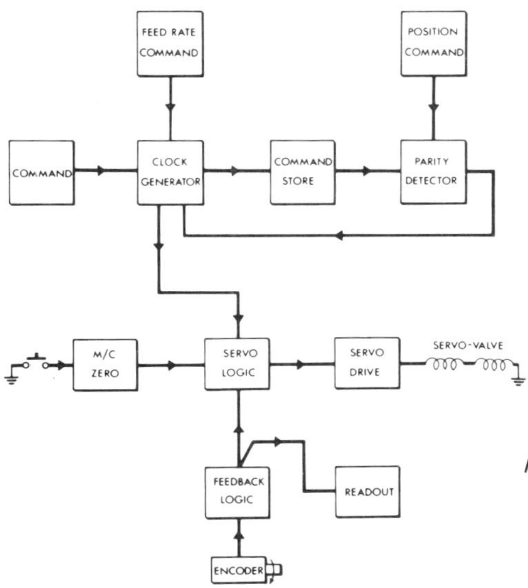

Fig 6 Block diagram of digital command and closed-loop control circuits for Keelavite precision feed drive.

derived from the feedback signals and the position and velocity commands are converted by a precision digital to analogue converted into an analogue voltage which is fed to a power amplifier controlling the input current to the electro-hydraulic servo-valve. Both proportional and derivative stabilizing terms are included for system optimization depending upon machines and a variable frequency filter oscillator is also provided to minimize stiction, hysteresis and dead band effects.

The manner in which control functions are performed is summarized under the following headings:

Position Control

The pre-selection of start, transition and stop positions on the panel edge switch causes a pulse count to be stored in the respective parity detectors. Subsequently when the command is activated, pulses from the clock generator are counted into the servo-logic unit and command store until parity exists with the pre-selected count from the position switches.

Feeding the pulses to a reversible counter in the servo-logic unit results in error signals which drive the servo-valve to feed oil to the actuator and move the machine slide. Feedback pulses from table/spindle motion are produced by a rotary incremental digitizer driven through a rack and pinion mechanism and by feeding this also into the servo-logic unit counter, a proportional error signal is produced to move the table until feedback counts equal command pulses.

Typically a 5 000 pulse/rev digitizer is mechanically and electronically coupled to achieve a single pulse count equivalent to 0.001 mm.

Velocity Control

The feed operation is also governed by the pulsed signals from the crystal clock; the output frequency of the clock is determined by the feed-rate setting registered on the feed-rate edge switches. When the clock pulses are received by the servo-logic unit counter and an error is produced, the digital-to-analogue converter generates a bias voltage which in turn causes the servo valve to monitor oil to the actuator and causes the digitizer to move. Without feedback into the counter a linear ramp wave form would be produced by the converter and the servo-valve opening and hence actuator velocity would progressively increase. However, in practice, since pulses from the digitizer are negatively summed with command pulses, the voltage error is exponentially modified to settle at some steady value when feedback frequency exactly equals input frequency.

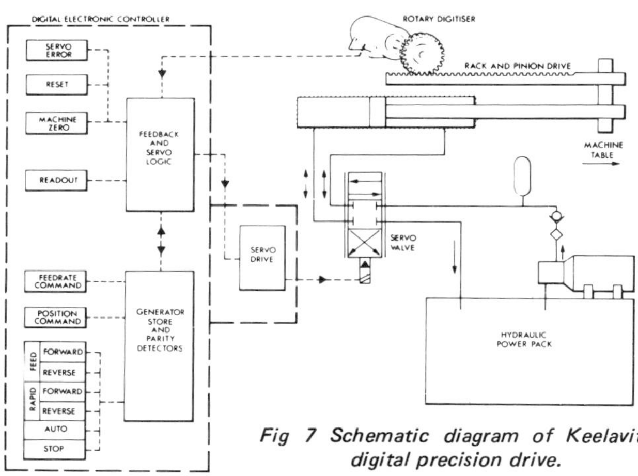

Fig 7 Schematic diagram of Keelavite digital precision drive.

Actuator velocity is therefore controlled in direct proportion to the input frequency. Increasing frequency produces a higher analogue voltage, the actual value of which is dependent upon valve amplifier gain and the number of bits in the counter.

By introducing clock or drive frequency into the detection system, command parity is registered prior to actual axis position parity from the feedback digitizer and proportional corrections to servo-valve feed are introduced to minimize final approach errors. A schematic diagram of this circuit is shown in Fig 7.

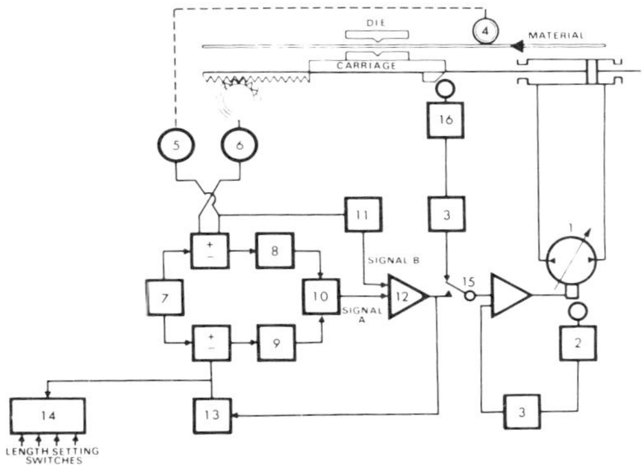

Fig 8 Digital position control. (Vickers).

1—Servo controlled pump.
2—Linear variable differential transformer — pump.
3—Demodulator.
4—Measuring wheel.
5—Material digitizer.
6—Carriage digitizer.
7—Clock.
8—Feedback counter.
9—Command counter.
10—Comparator.
11—Pulse rate tachometer.
12—Summing amplifier.
13—Voltage controlled oscillator.
14—Length counter.
15—Reed relay.
16—LVDT — carriage.

Another example of a digital-controlled system is shown in Fig 8. This is a flying cut-off drive. All electronic controls used are of the solid-state type, designed specifically for reliable operation in industrial environments. The digital controls are of a high-density integrated circuit type which has proved to be extremely reliable and maintenance free. Cut lengths and total number of pieces can be set on the operator's panel or can be fed into the system using punched tape or a computer output.

Pneumatic Logic Controls

BASIC LOGIC functions are NOT, OR and AND. For control purposes it is also necessary to have YES and MEMORY (to deal with sequential requirements). All these functions can be performed by simple valves or combinations of simple valves.

Simple 2-way valves can be used in combination to provide AND or OR logic response – Fig 1. Equally a 2-way valve can invert an output, to produce the NAND or NOR functions; or invert an input to provide INHIBITION. MEMORY, however, would require a combination of 3-way valves, or a 4-way valve (see also Table I).

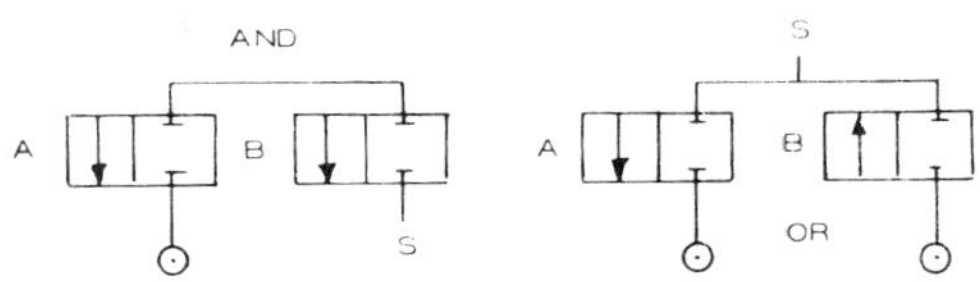

Fig 1

Interest in pneumatic logic circuits was originally stimulated by the development of pure fluidic devices capable of extreme miniaturization and also incorporating no moving parts. Whilst such a system can be shown to work it has several limitations in practice, notably the fact that it is essentially a low pressure system operating at air pressures of less than 1 bar, consequently needing amplification to power an actuator. Also such a system is critically dependent on air regulation and fine filtering of the air supply. It thus remains mainly of academic interest only.

Its counterpart in providing digital working – or logic circuitry – is derived from the miniaturization of more or less conventional valves *ie* moving part logic (MPL) elements operating at normal system pressure and capable of operating at high speeds. These now form the basis of virtually all practical pneumatic logic circuits, and have been adopted on a relatively large scale Properly designed MPL elements are reliable, positive in response, and suitable for use with both dry and lubricated compressed air supplies. Also the question of providing interconnections compatible with the size of these elements has been solved by the development of modular construction and assemblies.

Pneumatics *versus* Electronics

Pneumatic logic control circuits work on exactly the same principles as programmed electronic controls. Until comparatively recently a pneumatic logic control circuit of medium complexity generally proved cheaper than its electronic equivalent. With the advent of low cost microprocessors the position has changed considerably. Hybrid systems (electronic control of pneumatic systems) have many advantages to offer at comparable, or even lower cost, particularly as the complexity of the control requirement increases. Apart from compactness, the big advantage with

TABLE I – LOGIC CIRCUIT DEVICES

Logic	Hydraulic Symbol	Logic Symbol	Notes
NOT (3-way valve)	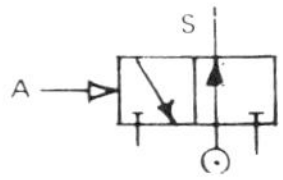	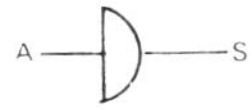	Output is present when there is NOT an input at A NOT A = S or $\overline{A} = S$ $A = \overline{S}$
AND (tandem 2-way valve)	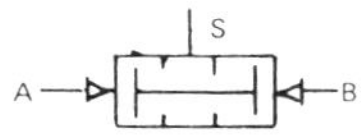	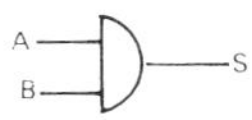	Output is present when there is input at A AND B A *and* B = S or A.B = S
OR (shuttle valve)	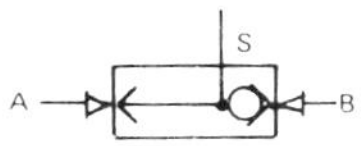	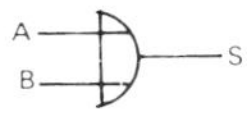	Output is present when there is input at A OR B. A *or* B = S or A + B = S
INHIBITION (3-way valve)	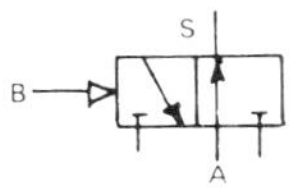	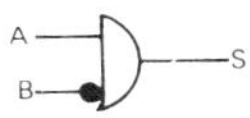	Output is present when there is input at B and there is no input at A. In other words, when a signal is applied to A it *inhibits* the signal path of B. B *and not* A = S, or $B.\overline{A} = S$ B.A = O
NAND (tandem 2-way valve and 3-way valve)	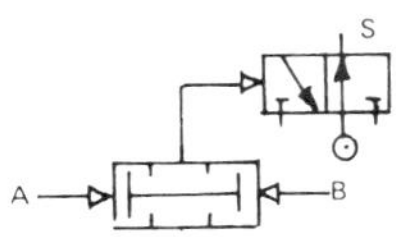	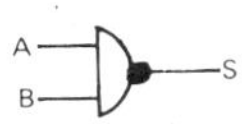	Output is present when there is *not* a signal input at A and *not* a signal input at B. *not* A *and not* B = S, or $\overline{A}.\overline{B}. = S$
NOR (shuttle valve and 3-way valve)	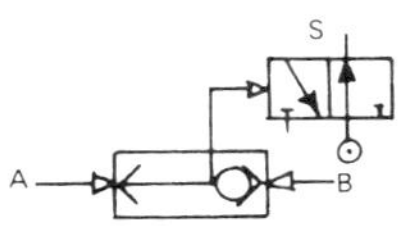	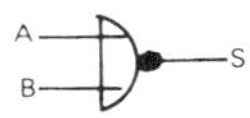	Output is present when there is *not* an input signal at A *or* an input signal at B. *not* A *or not* B = S, or $\overline{A} + \overline{B} = S$
YES (3-way valve)	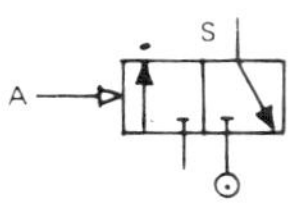	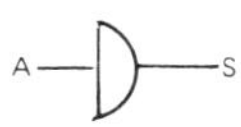	This is basically an amplifier. Line pressure is connected to B. This pressure is then available at the output when there is a signal input at A. (A = S) In basic terms, this is an inverted NOT with a second input.
MEMORY (4-way valve)	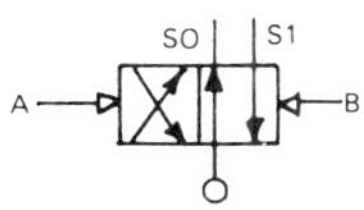	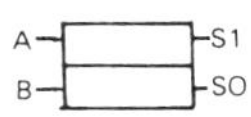	This is a flip-flop device. A signal input at A will give output state S1 and the memory will remain in this state, even when signal A is removed. A signal input at B then changes the output state to S2. It will remain in this state when signal B is removed, until receipt of the next signal A. In other words, the device 'remembers' the last input.

micro-processor is that the programme can be re-written and sequences changed without re-piping the control system.

At the present state of the art, therefore, it would be reasonable to state that where only a relatively few functions are required (say a maximum of ten), pneumatic logic control circuits still have much to offer. Above that number, the Programmable Logic Controller (micro-processor) offers increasing overall advantages. Where pneumatic controllers can continue to remain unchallenged is in high-risk fire areas, or hazardous areas – *eg* involving low and high flashpoint materials in processing.

Devices

The number of individual pneumatic devices required can be reduced considerably by working with a particular *type* of logic – *eg* AND logic, OR logic, NAND logic or NOR logic; or combinations of these. Certain functions will, however, remain necessarily common with any type of logic chosen – *eg* NOT, INHIBIT and MEMORY. As far as pneumatic logic circuit components are concerned the choice of type of logic depends largely on the *suitability* of miniaturized valves to perform the various functions.

If a multiplicity of elements is to be avoided, choice, basically, lies between NOR and NAND logic, or multi-functional devices. NOR and NAND can be made multiple-input devices when, by selecting either one or a number of inputs, the other logic functions can be derived. This reduces the number of physical elements required, but each is more expensive in terms of size and cost; and as deployed throughout the circuit, many will be operating with redundant features. Multi-functional devices can provide the basic functions of OR, AND, NOT and MEMORY in smaller (individual) packages; but using 4-way spool valves can present problems in miniaturization.

Circuit Design

The basic problem then remaining is how to design control circuits in terms of logic. The basic 'tools' are Boolean algebra, Karnaugh-Veitch maps and Cascade techniques primarily (there are others). Alternatively there are 'half-way' measures, such as the Step-counter method of electro-mechanical sequencing. All of the basic methods need considerable study to master and apply, and are skills not readily acquired by the intuitive designer.

For that reason, producers of modern MPL components have developed their own system methods to make logic design easier to understand and minimize circuit design time. This simply parallels computer technology where special, and simpler languages have been devised to make programming easier. In that case the language is tied specifically to a particular system, which in turn is based on a particular choice of logic elements providing all the logic functions necessary. This, like the language, can vary with different systems.

It is virtually impossible to generalize on this subject, different manufacturers of systems having different methods of approach and component construction. Thus only one system, which has proved particularly acceptable in Europe, will be described in more detail.

Pneumaid System

This system adopts the multi-element approach and is based on five functions – OR, AND, NOT, YES and MEMORY. These devices are shown in Fig 2. It is also allied to a sophisticated mounting system (Polylog) made up of standard modules which can be built up to form a complete mono-bloc control centre without inter-connecting piping.

The devices can be divided into two categories; active and passive. The active devices are basically 3-way, diaphragm-operated, air-return poppet valves, normally open for the NOT function and

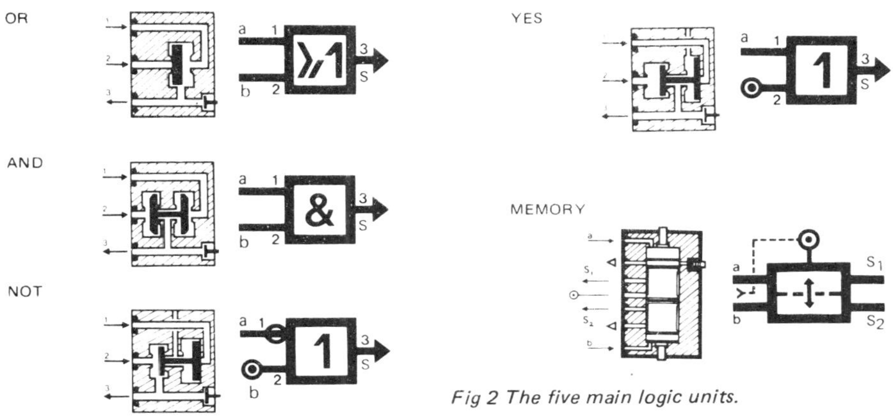

Fig 2 The five main logic units.

normally closed for the YES function. These units have separate supply and exhaust ports in addition to the input and output ports.

Passive devices are simple poppet valves actuated by the input signals only, and can perform the AND and OR functions. The MEMORY function is normally supplied by a miniature 4-way slide valve.

All MPL control systems require more than the five basic elements listed earlier. It is possible to produce a number of auxiliary devices from these basic gates.

With a fine restrictor valve having full flow in the reverse direction and a pressure reservoir (or even the capacity of the inter-connecting tubing), it is possible to construct simple RC time delays with good repeatability up to 30 μs. By combining these elements with the NOT and YES devices both normally open and normally closed delay functions can be achieved.

Similarly if a common signal is fed to the supply and input ports of a normally open delay arrangement, the result is a step-to-pulse converter which can be used to break the trapped signal in the pneumatic system. This is a problem with which most circuit designers are only too familiar. In a number of systems the logic device, reservoir and restrictor are stacked into integral time delay devices which occupy only one position of the mounting matrix.

By the addition of a simple diaphragm/nozzle pilot amplifier stage to the YES unit, a pressure amplifier can be developed which can accept signals from back-pressure air sensing units.

The diaphragm-operated NOT unit used in the multi-device logic system has been designed to carry out a function other than that of logic inversion. By setting the pilot pressure to reset at 10% (or less) of the supply pressure, the unit can be used to signal the end of the stroke of a pneumatic cylinder.

There is a refinement of this system in which the NOT unit supply pressure is derived from a port in the cylinder wall set just behind the piston when it has completed its stroke.

Polylog System Design

A circuit using the Polylog system can be designed and built after only a day or two's training. In fact it is possible to design and build the simpler circuit after only two or three hours' tuition. Therefore, no longer is it necessary to be a scholar of Boolean algebra and be able to calculate the resulting equations. Nor does one have to have the knowledge to find one's way round a Karnaugh map.

With the Polylog system there is a direct relationship between the machine cycle, the circuit and the way the elements are assembled. However, to make circuit design easier a simple form of diagram called Grafset is used. This is a European standard to show the graphical representation for expressions of automatic sequences. It is a step/command/transition control graph and a typical diagram is shown in Fig 3. This diagram relates to a circuit involving two cylinders A and B.

The right hand squares represent the movements of the cylinders. A+ refers to cylinder A out-stroking and A– the cylinder instroking *etc.* The squares on the left hand side represent each step in sequence.

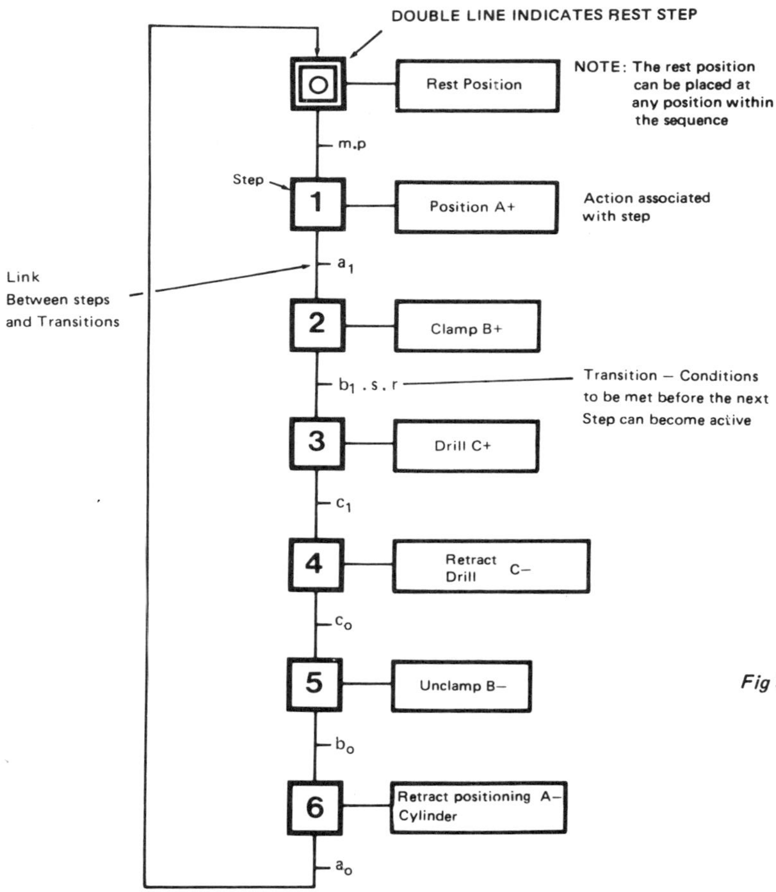

Fig 3

These are the principal steps. Each principal step requires a memory module – thus for this application we would require four memory modules. In between each step there are further letters; these represent interlocks confirming that the action of the principal step is complete or external signals setting the sequence in motion. These are called intermediate steps. This is basically all the circuit design that is necessary before beginning to assemble the system.

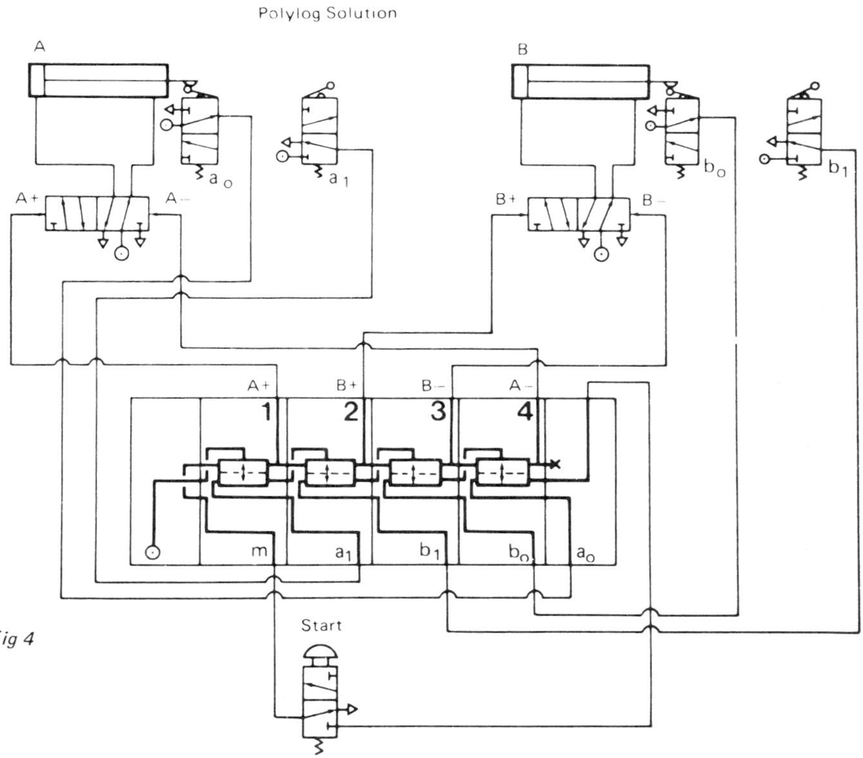

Fig 4

In order that the circuit can be quickly confirmed, special cards are available. When placed side by side these represent exactly the layout of the circuit. In this application the cards when put together would be as in Fig 4, shown together with the complete circuit. In this circuit four memory modules provide the four principal steps in the system. This example represents a very simple application.

Hydro-Pneumatics

THE CATEGORIZATION *hydro-pneumatics* applies to systems, machines or devices employing both a liquid working medium (or hydraulic component) and a gaseous working medium (or pneumatic component). These include air-hydraulic cylinders and various air-hydraulic systems, intensifiers, pulsation dampers, surge preventers and miscellaneous devices.

Air-Hydraulic Cylinders

The air-hydraulic system lends itself to integral construction, (*eg* the air-hydraulic cylinder – Fig 1). The rear or blank end of the combined cylinder is usually made the air cylinder. The two circuits are quite separate and are essentially the same in detail. The main difference is that since the hydraulic cylinder is through-rod only a nominal size oil reservoir is required, which is commonly of the spring loaded type. Reversing the positions of the air and oil cylinder would nullify this advantage.

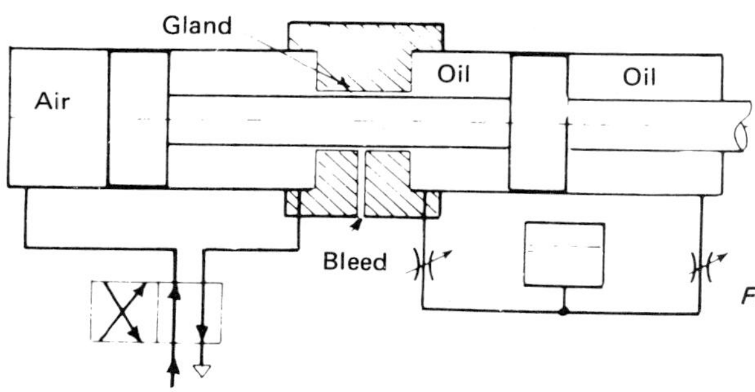

Fig 1

Throttle valves in the hydraulic lines again provide for adjustable speed settings, with the same possibility of introducing a mechanically operated bypass valve for fast motion over a particular travel. Also, of course, a plate valve can be used in the hydraulic piston to provide fast motion over a complete stroke in one direction.

Almost any desired sequence of speed control associated with linear motion can be obtained by suitable design of the hydraulic circuit. Independent valves can be inserted to provide a check at any particular point, or cam-operated valves can be used to provide skip feeds.

The hydraulic circuit affects only the speed of operation of the system. It has no effect on the air circuit and its controls, except to prohibit the use of unloading or pressure-relief valves sensitive to back pressures. These would be operated by damping produced by the hydraulic cylinder, resulting in loss of air pressure. The only effective way of unloading with hydraulic damping is by icroswitches or pilot valves. Apart from that, the design of the air circuit can follow conventional ractice and be as simple or complex as necessary.

Proprietary air-hydraulic cylinders are normally available with plain hydraulic pistons (for con-olled speed in either direction *via* regulator valves in the hydraulic circuit) or with a one-way alve in the hydraulic piston to give fast forward or return motion.

An example of the use of a plate valve in the hydraulic piston to provide undamped motion in ne direction is shown in Fig 2. Movement in one direction produces oil transfer through the rottle valve (and thus adjustable damping). Movement in the other direction provides flow rectly through the piston by means of the plate valve fitting, and thus basically undamped otion (*ie* viscous damping will be minimal).

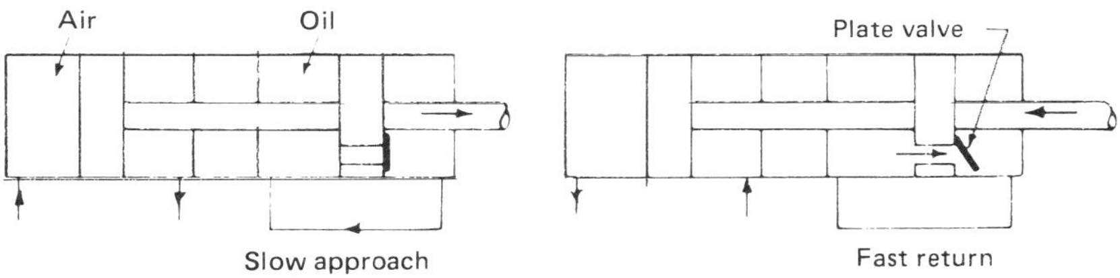

Fig 2

parate Air and Hydraulic Cylinders

milar solutions can be obtained mechanically, using separate air and hydraulic cylinders — *eg* the r cylinder rod can energize the hydraulic cylinder rod only on that point, or portion, of the roke where damping is required. Where only a proportion of the working stroke needs damping is has the advantage that the stroke of the hydraulic cylinder can be reduced to that actually quired, resulting in a smaller cylinder. It is important in such cases, however, that there should sufficient stroke length on the hydraulic cylinder to accommodate any mechanical adjust-ents required as otherwise it might be possible to adjust the working movement in such a way as 'bottom' the hydraulic cylinder.

More complex mechanical systems are also used, normally with air and hydraulic cylinders of ual stroke, the two locked and unlocked, as required, by mechanical couplings and latches. If cessary, free movement of the hydraulic cylinder can be taken up by springs, although this is not ual. Another solution is to use movements to manipulate regulating valves in the hydraulic cir-it itself, and this is preferred to mechanical systems where differential speed control is required.

Differential speed control can also be obtained by bypassing the appropriate flow regulator in e hydraulic circuit to give full bore flow for a suitable period of time or portion of the motion. ch a valve could be mechanically operated by hydraulic piston movement. The bypass valve can held open for an appropriate length of stroke either on the inward or outward stroke, or both, d then released. The remainder of the stroke is then speed controlled in the normal way by flow rough the regulating valve.

The combined air-hydraulic cylinder has the advantage of convenience — *eg* one unit only to be ounted; but it has two possible disadvantages compared with the use of separate air and draulic cylinders. One is that the longer length may make it unsuitable (or less suitable) for rtain installations. The other is that the through-rod places a premium on seal design to ensure at no air can bleed through into the oil circuit, as this could lead to jerky checking actions. The k is generally eliminated by incorporating a bleed hole in the central section between the two sets of rod seals. Any leakage from either cylinder will then be bled off to atmosphere, rather than into the other cylinder. The presence of air or oil at the bleed hole will also indicate seal or gland leakage.

A particular attraction of air-hydraulic cylinders (and air-hydraulic systems in general) is that where light output forces are required they can provide the rigidity and speed control normally associated with hydraulic systems without the cost of a hydraulic pump and driver. An example of a practical double-acting air-hydraulic circuit is shown in Fig 3.

Air-Hydraulic Systems

A basic air-hydraulic system is shown in Fig 4. The power unit is a hydraulic cylinder, but the fluid is pressurized from a compressed air supply feed to a second pistonless cylinder or supply vessel.

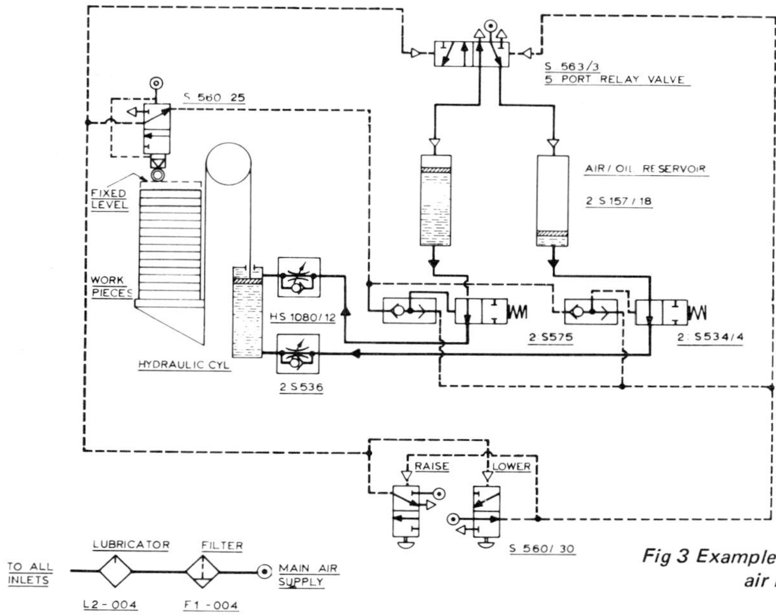

Fig 3 Example of a practical double-acting air hydraulic circuit. (Martonair).

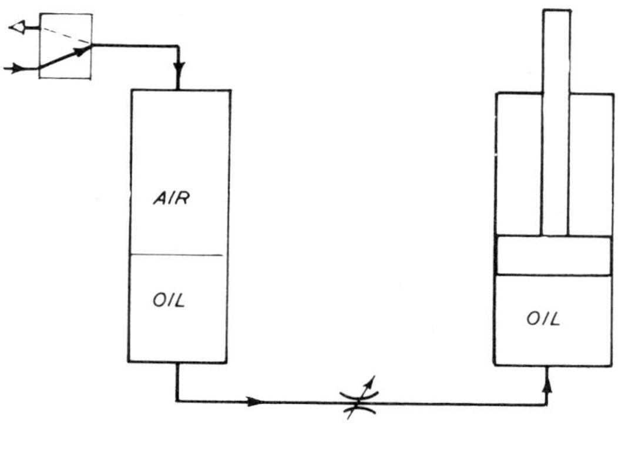

Fig 4

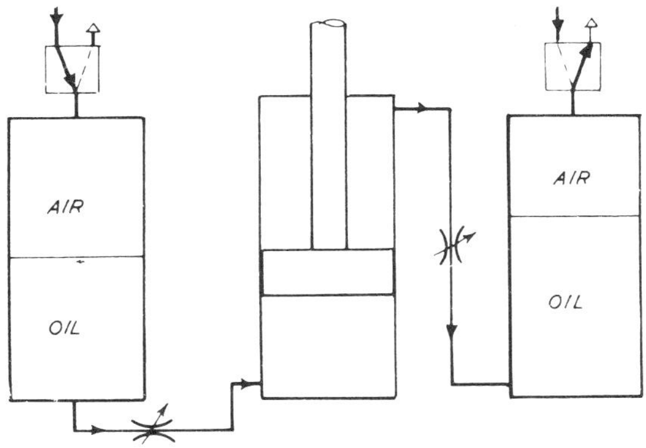

Fig 5

Movement of the hydraulic cylinder is initiated by opening the valve between the supply vessel and the cylinder, allowing pressurized fluid to flow into the hydraulic cylinder. At the end of the stroke the air supply is switched to exhaust and fluid is returned to the supply cylinder by reverse motion of the piston initiated by gravity, external load or spring force. If required, the same principle can be extended to double-acting working by adding a second supply vessel connected to the other end of the hydraulic cylinder which is now filled with fluid on both sides of the piston – Fig 5.

Such a system works at 'pneumatic' pressure – *ie* pressure for the working fluid is limited to that of the compressed air supply – hence large volume cylinders, the same size as an equivalent pneumatic cylinder, are required for high output forces. It does have the advantage, however, that the system is more rigid than a pneumatic cylinder, and also is capable of quite precise control *via* a throttle valve connecting the supply vessel and cylinder. A specific requirement is that the size of the supply vessel must be such that oil is never fully discharged from it, otherwise cavitation of the working fluid will occur with loss of rigidity and control. A sufficient residual volume of oil must always remain in the supply vessel to isolate the intermingled air/oil which may occur at the interface of the two fluids.

The more usual application of hydro-pneumatics is the use of a hydraulic cylinder as a check device for an air cylinder which provides the output movement required. In a basic configuration the two cylinders are mounted in tandem with a common rod (Fig 6). The hydraulic cylinder is usually smaller in bore than the pneumatic cylinder, but has the same length. This cylinder is connected end-to-end to form a closed loop for the hydraulic circuit, the rate of transfer of fluid from one side of the hydraulic piston to the other being controlled by a throttle valve.

In practice it will also be necessary to include a small reservoir in the hydraulic circuit to accommodate the rod volume extending into the cylinder of the 'design' stroke, and changes in

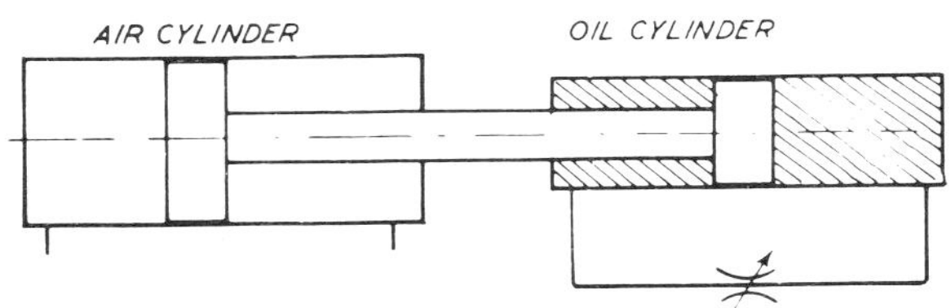

Fig 6

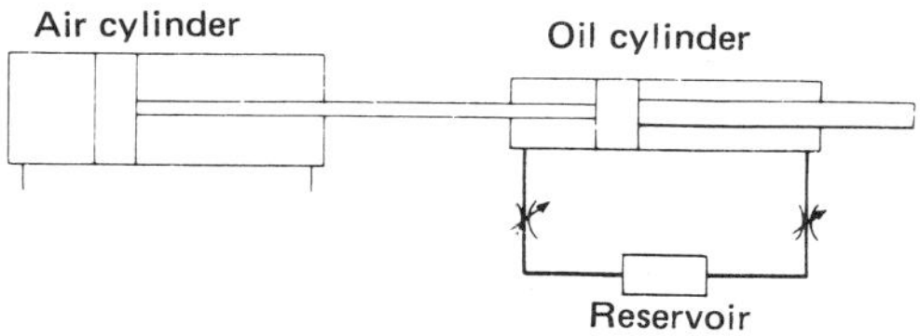

Fig 7

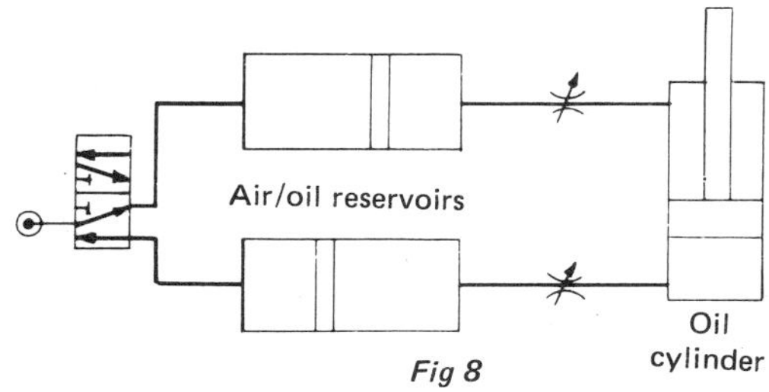

Fig 8

fluid volume due to thermal expansion and contraction. This also serves as a filling point for make-up oil to replace any leakages which occur in service through the rod seal. Only a small reservoir capacity is normally required – *eg* a little more than the full rod volume. Speed control is normally provided by throttle valves at each end of the hydraulic cylinder (Fig 7).

The need for a hydraulic reservoir in a double-acting hydro-pneumatic circuit can be eliminated by using two 'rodless' air cylinders (or effectively air/oil reservoirs) as shown in Fig 8. This is essentially the same circuit as Fig 5 but with the air/oil reservoirs controlled from a single selector valve. Control valve configuration and throttle valve positioning can be arranged to give a variety of different modes of working, *eg:*

(i) Controlled speed in either direction, or both directions;

(ii) Fast approach slow feed (as in Fig 8);

(iii) Automatic cylinder reciprocation, *etc*.

Check Units

The basic arrangement for using an air cylinder as a check unit for a hydraulic cylinder is shown in Fig 9. The hydraulic cylinder is the working cylinder the outward movement of which drives the piston of an air cylinder, increasing the pressure in the air cylinder. This will provide a cushioning action on the power stroke (*eg* to absorb impact loads), and on cessation of the stroke will maintain a backpressure on the workpiece. Back-up pressure is maintained over a full movement even if the stroke is reversed; alternatively, if necessary it could be relieved by opening the air cylinder to exhaust. The usual arrangement is for the air cylinder to be connected to an air bottle or pressure vessel, the cylinder size and pressure being selected to provide the required degree of cushioning and back-up pressure.

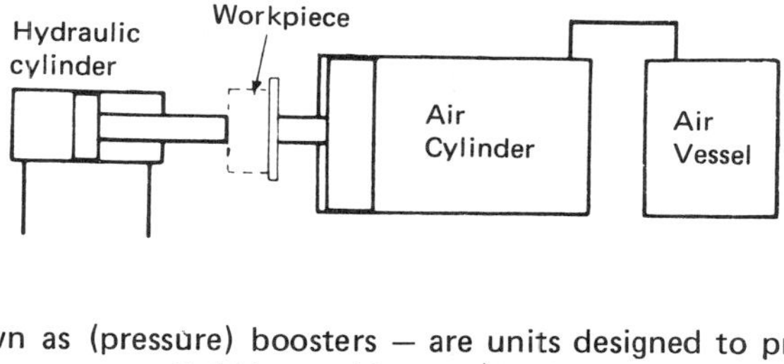

Fig 9

Intensifiers

Intensifiers – also known as (pressure) boosters – are units designed to provide a high pressure fluid output from a lower pressure fluid input. Most types of intensifiers provide complete separation of the low pressure and high pressure fluids, in which case different fluids may be used. Thus, compressed air (or gas), water or oil may be used on the low pressure side, with oil as the high pressure fluid (or *vice versa*). Water is favoured as the high pressure fluid for intensifiers designed

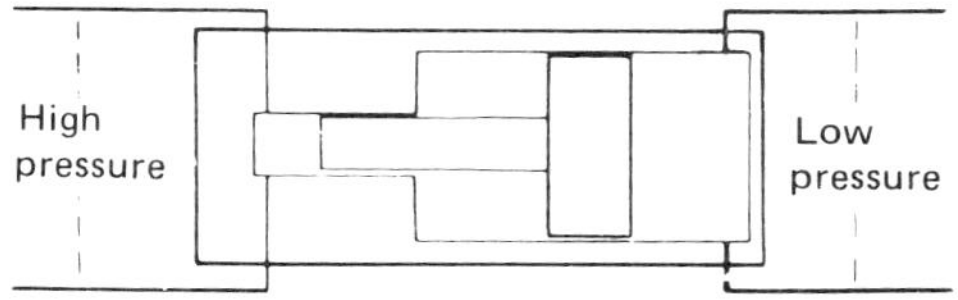

Fig 10

for high pressure test work because of its low compressibility (less than that of oil). For very high pressure testing, a fluid with even lower compressibility may be used.

Intensifiers with a hydraulic output can be all-hydraulic or air-hydraulic. The working principle is the same in both cases, the intensifier comprising two cylinders of different diameters mounted on a common rod – Fig 10. Low pressure is fed to the larger piston, generating high pressure *via* the smaller piston, with pressure multiplication directly proportional to the ratio of the piston areas, *viz*

The efficiency achieved

$$P_h \propto \frac{A_L}{A_h} \times P_L \quad \text{or} \quad P_h = \eta \frac{A_L}{A_h} \times P_L$$

where P_h = intensified output pressure
P_L = input low pressure
A_L = area of low pressure piston
A_h = area of high pressure piston
η = efficiency

The performance of intensifiers is commonly expressed in terms of pressure ratio, when:

$$\text{pressure ratio} = \frac{P_h}{P_L} = \mu \frac{A_L}{A_h}$$

(See also Tables 1A and 1B).

TABLE IA – PRESSURE INTENSIFICATION RATIO

Low Pressure Cylinder Dia. (ins)	Small Piston Diameter (inches)									
	1/4	3/8	1/2	5/8	3/4	7/8	1	1¼	1½	2
1	16	7.13	4	2.56	1.77	–	–	–	–	–
2	64	28.1	16	10.24	7.129	5.29	4	–	–	–
3	144	64.0	36	23.04	16.00	11.56	9	5.76	4.00	–
4	256	113.9	64	40.96	28.41	21.16	16	10.24	6.76	4.00
5	400	176.9	100	64.00	44.49	32.49	25	16.00	10.89	6.25
6	576	256.0	144	92.16	64.00	46.24	36	23.04	16.00	9.00
7	784	348.6	196	125.4	87.05	64.00	49	31.36	22.09	12.25
8	1 024	453.7	256	163.8	113.9	84.64	64	40.96	28.09	16.00
9	1 296	576.0	324	207.4	144.0	106.1	81	51.84	36.00	20.25
10	1 600	712.9	400	256.0	176.9	130.0	100	64.00	44.49	25.00
11	1 936	858.5	484	309.8	213.2	156.3	121	77.44	53.73	30.25
12	2 304	1 024	576	368.6	256.0	187.7	144	92.16	64.00	36.00
13	2 704	1 204	676	432.6	299.3	219.0	169	108.2	75.17	42.25
14	3 136	1 391	784	501.8	349.7	256.0	196	125.4	87.05	49.00
15	3 600	1 600	900	576.0	400.0	292.4	225	144.0	100.0	56.25
16	4 096	1 815	1 024	655.4	453.7	334.9	256	163.8	113.7	64.00
17	4 624	2 052	1 156	739.8	515.3	376.4	289	185.0	128.4	72.25
18	5 184	2 304	1 296	829.4	576.0	424.4	324	207.4	144.0	81.00
19	5 776	2 560	1 444	924.2	640.1	470.9	361	231.0	160.5	90.25
20	6 400	2 841	1 600	1 024.0	712.9	519.8	400	256.0	177.7	100.00

TABLE IB – PRESSURE INTENSIFICATION RATIO

Low Pressure Cylinder Diameter millimetres	Small Piston Diameter – millimetres									
	5	10	15	20	25	30	35	40	45	50
25	25	6.25	2.78	1.30	–	–	–	–	–	–
50	100	25.00	11.00	6.25	4	2.78	2.00	1.30	–	–
75	225	56.25	25.00	14.00	9	6.25	4.60	3.50	2.80	2.25
100	400	100.00	44.00	25.00	16	11.00	8.20	6.25	4.90	4.00
125	625	156.25	69.00	44.00	25	17.40	12.75	11.00	7.70	6.25
150	900	225.00	100.00	56.25	36	25.00	18.40	14.00	11.00	9.00
175	1225	306.25	136.00	76.50	49	34.00	25.00	19.00	15.00	12.25
200	1600	400.00	178.00	100.00	64	44.00	32.65	25.00	19.80	16.00
225	2025	506.25	225.00	126.00	81	55.00	41.30	31.50	25.00	20.25
250	2500	625.00	278.00	156.25	100	69.00	51.00	44.00	30.90	25.00

The efficiency achieved is dependent on friction and internal leakage, and also the amount of heating of the fluid. For simple 'one-shot' or single-stroke intensification such losses may be negligible, *ie* $\eta \approx 100\%$. Where the intensifier is operated to give a continuous high pressure output the actual working efficiency achieved may be reduced by up to 20%.

Continuous delivery can be provided by employing two intensifier cylinders operating alternately. They can be separate units or, more conveniently, co-axial or tandem units, where a single high pressure cylinder can be common to both 'working' cylinders. These are generally known as *continuous intensifiers.*

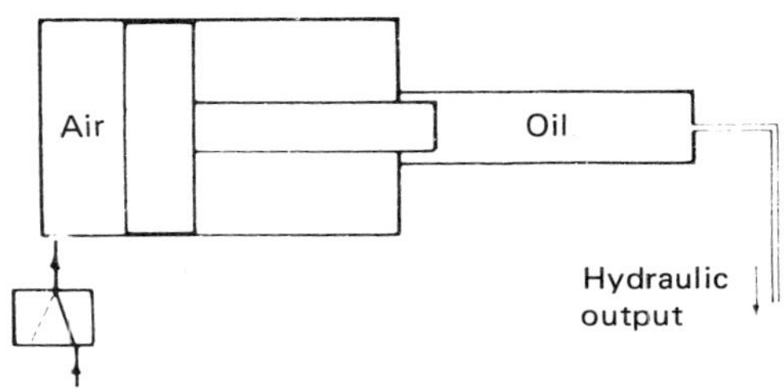

Fig 11

Simple cylinder construction can be used with the piston rod forming the high pressure piston – see Fig 11. For substantially higher pressure ratios the usual form of an intensifier is a double-acting air-cylinder controlled by a four-way valve operated at each end of the stroke to produce a continuous reciprocating action. This motion is used to drive a single-acting fluid pump incorporated in the cylinder construction which provides the supply of high pressure fluid. If a continuous supply of pressurized fluid is required, two single-acting pumps can be incorporated, worked alternately by the air cylinders.

Two examples of simple air-hydraulic intensifier circuits are shown in Figs 12 and 13. In Fig 12 control of the pressure intensifier is by 6-way valve. At the start D1 is in position A. Y1 and C1 are in the negative position. When D1 is brought into position B, O1 is pressurized and simultaneously O2 is vented. C1 moves in the positive direction under low pressure. In position C, Y1 moves in the positive direction and the pressure is intensified.

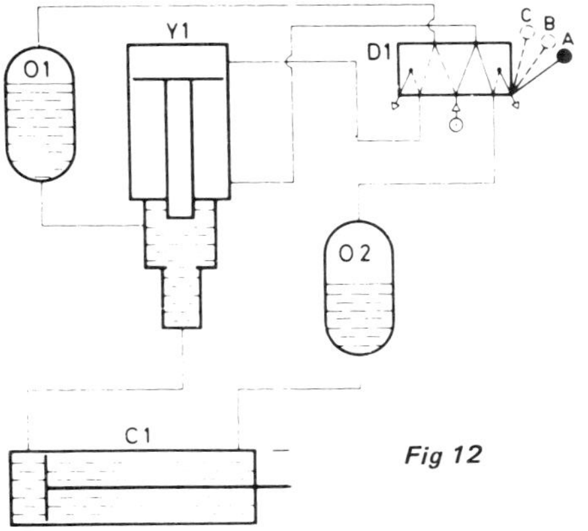

Fig 12

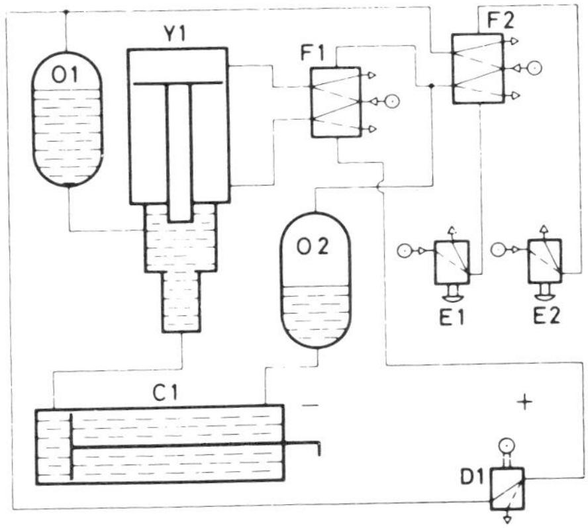

Fig 13

In Fig 13 control of the intensifier is semi-automatic. A signal *via* E1 results in positive movement at low pressure. When at C1, the piston rod actuates D1. Y1 is connected and the pressure is intensified. When E2 is actuated, C1 moves in the negative direction.

Figure 14 shows a simple air-hydraulic intensifier where the oil fluid reservoir is incorporated above the ram (air) cylinder. This is partly filled with fluid but also pressurized by the air supply. Hold capabilities are provided by shutting off the delivery flow with air pressure still applied on the appropriate side.

Figure 15 shows a double-acting intensifier designed for pressures up to 3.4 kilobar (50 000 lb/in^2). This embodies two high pressure cylinders with a central low pressure cylinder. At the end of each stroke a microswitch is triggered by plungers to give reversal through the solenoid valve and to supercharge the high pressure cylinder with low pressure fluid. This results in a considerable improvement in volumetric efficiency.

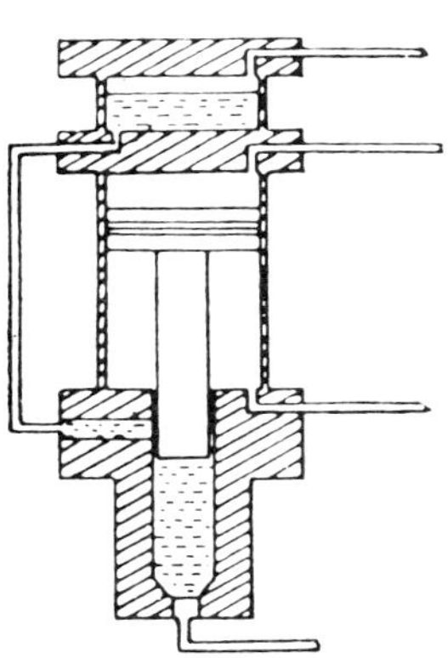

Fig 14

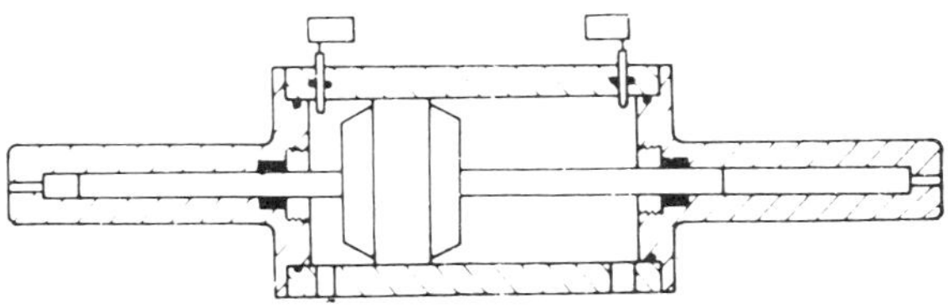

Double-acting high pressure intensifier arranged for continuous operation. The micro-switches are triggered at the end of each stroke by plungers passing through glands in the L.P. wall.

Fig 15

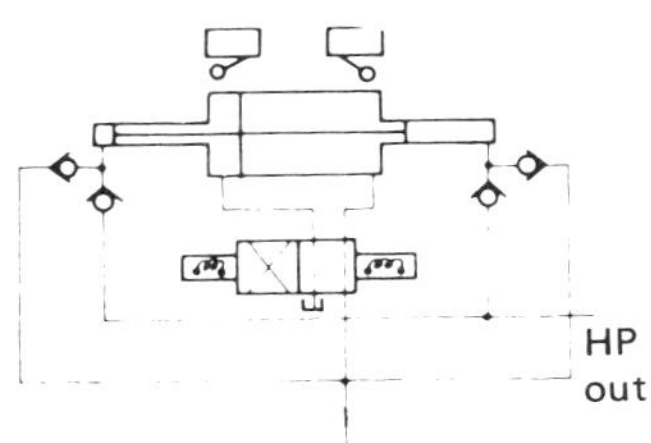

Continuous double-acting intensifier showing arrangement of valves.

All-hydraulic intensifiers for general industrial applications may be designed with quite moderate pressure multiplication ratios, *eg* 3:1 or less, to provide dual-pressure facilities in a system. These would normally be of the continuous delivery type. Very much higher ratios are employed for intensifiers used for high pressure test purposes with one-shot action.

Air-hydraulic intensifiers are normally designed to operate on shop air at a pressure of about 7 bar (100 lb/in^2) and may have pressure ratios as required.

One-shot Power Supplies

Since compressed air can readily be stored without loss (*eg* in a compressed air cylinder or accumulator), it can often be used as a source of energizing power for operation of a hydraulic system. In its most compact form this consists of an accumulator separate from the main hydraulic circuit, pre-charged to normal system pressure. In the event of hydraulic failure the system is switched to the energizing cylinder.

This is essentially a 'one-shot' energizing system since the available operating time with the energizing power supply is limited by the capacity of the accumulator.

Pulsation Dampers

Modern *pulsation dampers* can be divided into three main types:

(i) Low frequency units for use up to 10 Hz, also frequently used as suspension units when piped to a cylinder. Units in this category have a single connection to the system as this is adequate for the reciprocation of liquid in and out of it up to ten times per second. However, at higher frequencies progressively less effect on the removal of pulsation is achieved by a single ported unit because it requires very considerable energy to change the direction of flow so quickly and for this reason in-line or multi-ported units have been developed.

(ii) Multi-ported units which intercept the flow path, and therefore any variations in the flow itself in the form of pulsations, come directly into contact with the internal mechanism of the unit. This enables them to smooth out the flow by accepting momentary increases in volume and discharging those increases into the decreases in the rate of flow which immediately follow them.

(iii) When the cyclic frequency exceeds 500 to 1 000 cycles per second, the degree to which the capacitance of the unit is capable of responding to the fluctuations is severely decreased, and the mechanism by which these multi-ported units work is by wave cancelling and energy dissipation through heat. These, of course, are multi-ported or in-line units and they are normally constructed so that the incoming flow is split through a number of different length flow paths and is directed towards a sounding dome as remote from the incoming flow paths as possible.

The different lengths of the flow paths plus the bouncing of one wave against another in the sounding dome tend to wave cancel. The sounding dome is preferably a bladder (specifically a bladder as opposed to a bag as the liquid is inside the membrane) because then its excitation transmits to the surrounding precharge gas, the molecules of which become excited and create heat which is then dissipated from the pressure vessel shell which surrounds the gas and acts as a heat sink.

There is yet another category of pulsation damper which can be called flushability units; these are used, for example, in the food industry but can be used to advantage in any system which contains high levels of liquid borne contaminant. In order not to have a design which contains any crevices, it is normal for these units to be a variation on the original diaphragm accumulator. In this inlet and outlet port are provided in the bottom hemisphere and a unit looked at in section would show a 'V' shaped cross section to the diaphragm; this arrangement allows the minimum likelihood of particles remaining between the membrane and the pressure shell.

Shock Preventers

Units used for *shock prevention* or as *surge arresters* vary from accumulators or pulse dampers, generally, by having very large flow inlet apertures which are partially closed off by liquid trying to flow back out of them. The means by which a shock or surge is prevented is by allowing a period of time in which a liquid can decelerate itself against the increasing resistance caused by pre-charge gas being compressed.

They are not shock absorbers as the shock or surge has never been allowed to occur, neither do they attenuate a shock for the same reason. Were such devices ever used to deal with a standing wave already travelling down a pipe at the speed of sound in the liquid, it would probably shoot straight past or bounce straight off.

Shock Removers

This is a class of device which prevents a standing wave from passing further down a system or from bouncing back out of it. These are the most sensitive hydro-pneumatic devices and the only form which has come near to being adequately effective for this purpose is the tubular or sleeve type preventer.

Because of the length of these units, it is possible to open a membrane so that it is exposed to the increase in pressure of the wave and then to close it behind the wave which is thus completely contained.

See also chapters on *Accumulators* and *Accumulator-Type Devices.*

High Temperature Hydraulics

CONVENTIONAL HYDRAULIC fluids begin to break down at relatively moderate temperatures, and even below the break-down point will show accelerated deterioration. Mineral oils have a somewhat higher temperature rating than water-based fluids and can, in fact, give satisfactory service at fluid temperatures of up to about 140°C (280°F). Phosphate ester fluids can be used up to about 150°C (300°F); but neither type is particularly suited for working at high temperatures. Hydrolytic stability is often suspect at temperatures above normal working figures. Many otherwise satisfactory fluids may break under such conditions; others become excessively acid and promote corrosive attack or show an excessive rate of oxidation.

All fluids, therefore, can be given maximum temperature ratings for continuous duty consistent with a normal service life; and a higher short-term rating for intermittent duty. The latter will result in reduced life, depending on the severity of the over-rating. Typical values are:

	Continuous	**Short term maximum**
Water	38–50°C (100–120°F)	65°C (190°F)
Mineral oils	50–65°C (120–150°F)	120–140°C (250–280°F)
Water/oil emulsions	50–65°C (120–150°F)	65°C (150°F)
Water-glycol	50–65°C (120–150°F)	70°C (160°F)
Phosphate esters	65–82°C (150–180°F)	150°C (300°F)
Chlorinated aromatics	95°C (200°F)	150°C (300°F)
Silicones	up to 288°C (550°F)	316°C (600°F)

In general it is desirable to maintain fluid temperatures substantially below recommended maximum operating values as this will have a beneficial effect on fluid life. Also the use of synthetic fluids for higher temperature service may show unexpected limitations. It is a characteristic of synthetic lubricants that if they reach an excessively high temperature (as could occur at some localized point in the system) they can vaporize without leaving any residual surface film – *ie* leave surfaces dry and unlubricated.

Viscosity

Other characteristics of the high-temperature operation of fluids are poor stability and a loss of lubricity (due largely to a loss of viscosity). It is virtually essential that the fluid retains a viscosity of at least 2 centistokes at the operating temperature if efficient lubrication and sealing are to be maintained. This must also be accompanied by good shear resistance, otherwise the inherent shear losses in the system will still further reduce the effective viscosity. In any case the shear forces present in the higher-temperature system will probably be higher than in standard practice, due to the employment of closer-fitting components and finer tolerances to minimize leakage at low fluid viscosities.

Vapour Pressure

Another important factor to consider is the vapour pressure of the fluid at the operating temperature. If the vapour pressure is unduly high the fluid will tend to evaporate, which can lead to cavitation at the pump as well as seal failure, *etc.* It is necessary, therefore, that to be suitable for high-temperature working a fluid should have an acceptable vapour pressure figure at the system-working temperature. Provided these various requirements can be met, the ultimate choice of a high-temperature fluid can then be based primarily on thermal stability. Each fluid will have an absolute maximum service temperature above which it will dissociate or deteriorate to a degree where its properties are no longer acceptable, and a somewhat lower temperature above which it may still remain usable but with reduced life.

Table I shows the vapour pressure of typical fluids at elevated temperatures.

TABLE I – VAPOUR PRESSURE OF TYPICAL HYDRAULIC FLUIDS AT HIGH TEMPERATURES

Type of Fluid	Temperature °C	°F	Vapour Pressure mm Hg
High-quality mineral oil (DTD 585)	100 150 175	212 300 350	2.5 50 165
Phosphate ester	220	430	2.5–5.0
Silicone (Silicodyne H)	150 230	300 450	22 50

Cooling

In practice, where high service temperatures are involved, and consequently high fluid temperatures, it is generally more reliable, and certainly less expensive, to provide the system with a satisfactory method of cooling so that conventional fluids can be employed and maintained at acceptable fluid temperatures. This has the advantage of offering the widest choice of fluids and matching components, with known performance. It does, however, place an absolute premium on the cooling system being fully effective as any failure of the cooling system can lead to early breakdown, and possibly severe damage, to system components.

Various methods of cooling exist. The use of inter-coolers can be satisfactory where the environment temperature is not excessive and may be reduced to a single heat exchanger which ensures that the fluid temperature at the pump never exceeds a design maximum temperature consistent with the type of fluid employed. At higher environmental temperatures simple cooling will no longer be effective and individual components themselves may require cooling – for example, by water circulation through jackets or by water or alcohol sprays. This results in considerable complication of the system, and cooling systems of this type tend to be heavy and bulky and not always entirely reliable.

With a cooled system any of the conventional fluids may be employed (a fire-resistant fluid, for example, in place of a mineral oil, if required, to reduce fire hazard) and the design problem is simply one of providing adequate cooling. The choice of fluid may, however, be influenced by the specific heat and thermal conductivity of the fluid which govern the size of the heat exchanger needed to remove excess heat from the system.

The most satisfactory high-temperature fluids so far produced are of the silicone type, although methyl silicones are poor as lubricants unless suitably modified. Disiloxanes are also subject to hydrolytic instability. All silicones tend to have a low bulk modulus (that is they are fairly readily compressed) and are thus somewhat 'elastic' in very high pressure systems.

See also chapter on *System Cooling.*

'Tailored' Fluids

On the 'advantages' side silicones have a high degree of thermal and hydrolytic stability, retain high viscosity index and good resistance to shear breakdown. Lubricity can be obtained by 'tailoring' the molecule rather than using additives (which would be subject to thermal break-down).

A successful form of 'tailoring' is incorporating chlorophenyl groups to improve the lubricating properties of a basic silicone fluid and compounding with an oxygen inhibitor — characteristic of 'Silcodyne H' and 'Silcodyne M', respectively. Fluids of this type are suitable for continuous service temperatures up to 290°C (550°F) and have a life of 50 to 100 hours at 316°C (600°F) depending on service conditions. Above 316°C (600°F) there is a certain amount of degradation and molecular re-arrangement, the actual break-down rate being dependent on the temperature and of the order of 4% per hour at 370°C (700°F). The break-down products are low-molecular weight silicones which are volatile at the temperatures at which break-down can occur and soluble in the fluid at lower temperatures.

Compatibility

The main limitation of silicones is lack of compatibility with conventional elastomers and with certain metals at high temperatures. Aluminium, magnesium and copper are attacked at high temperatures and thus stainless steel is normally recommended for high-duty systems with silicone fluids. It is also recommended that all joints are welded in such cases. Stainless steel is difficult to flare and flareless compression joints or screwed joints may be prone to leakage. Conventional elastomers may be used with silicone fluids at low fluid temperatures (for example, up to 95°C (200°F), or 175°C (350°F) in some cases) but show some shrinkage. For higher temperatures Viton is applicable up to 260°C (500°F). Above this, no elastomeric seal is likely to be suitable in any case since all elastomers are limited in service temperature.

Seals

Where very high fluid temperatures are visualized, and a suitable fluid is available (for example a modified silicone), the design of suitable seals sets a specific problem. Not only is the service temperature of elastomeric seals limited but any lack of or marginal, compatibility is aggravated by high fluid temperatures. For very high temperature systems, therefore, and particularly in the case of dynamic seals, it may be necessary to design a 'sealless' system, or one which employs metallic seals only, in materials compatible with the fluid at the working temperature.

Silicone (and similar) high-temperature fluids are of fire-resistant or 'non-flam' type and thus provide an additional safety measure for high-temperature working. Their cost would, however, prohibit their selection as a 'non-flam' fluid in conventional systems operating at low or moderate temperatures. At such temperatures, too, their low bulk modulus would be a distinct disadvantage in high-pressure systems.

The use of components and fluids compatible with the (high) service temperature required represents a simpler, more direct solution, but at the expense of considerably reducing the choice of fluid and placing a premium on component specification. The problem in this case is two-fold — first the selection of a suitable fluid, and second the re-rating of components (or selection of special components) at the higher service temperature. In addition, the choice of suitable seals may be particularly restricted by the need to achieve compatibility with the special fluid at the working temperature involved.

All metallic components will suffer a certain loss of mechanical strength with increasing temperature, although this is seldom serious (and usually negligible) within usable temperature limits. Thus tungum alloy suffers a loss of some 20% at 400°C (750°F), compared with its strength at normal temperatures. Light alloys show a more marked loss and, in any case, are not suitable for lines in high-pressure systems. Steels are less temperature dependent and stainless steel can be considered to be unaffected by temperatures within the range likely to be encountered in practice for any system. Nickel alloy tubing is an alternative choice for very high temperature systems and has some advantage in manipulation.

Liquid-Metal Fluids

Liquid metals offer the possibility of being used as hydraulic fluids in highly specialized systems working at fluid temperatures up to 650°C (1 200°F). One of the most suitable metals so far evaluated is a eutectic alloy of sodium and potassium (23% sodium, 77% potassium) which has a melting point of −12°C (10°F) and a boiling point of approximately 705°C (1 500°F) under atmospheric pressure. The density of this liquid alloy is comparable to that of water. A further advantage is that the alloy shows no signs of degradation or deterioration with either time or temperature over its whole liquid range.

The problems of utilizing such a fluid in a practical system are, of course, considerable. Sodium-potassium alloy reacts violently with oxygen or moisture and so special techniques and extreme precaution have to be taken to prevent contamination of the fluid. Its lubricating properties are also poor and its tendency to provide fluxing action can promote localized welding although the high thermal conductivity is helpful in reducing high temperature in bearings and shaft seals.

Sealing presents a particular problem, demanding the use of metallic or ceramic rotating face seals, or metallic rings and reeds for sliding seals. Both vane and centrifugal type pumps with appropriate modifications and bellows-type shaft seals have shown themselves capable of handling the alloy at pressures up to 140.6 bar (2 000 lb/in^2) and temperatures in excess of 540°C (1 000°F). Bearings and seals lubricated by the fluid remain the most critical items.

With a liquid-metal fluid a unique opportunity is offered for the design of servo-valves utilizing the conductivity of the metal (fluid), such as the employment of a small electro-magnetic pump as a combination transducer and pilot stage.

Ultra-High Pressure Hydraulics

ABOVE PRESSURES of 300–350 bar (4 000–5 000 lb/in^2), the increasing elasticity of both fluids and system components, higher material stresses, and demand for closer working clearances and tolerances all begin to cancel out the advantages of higher pressure working. Thus for hydrostatic systems in general, maximum practical working pressures can be set at 560–700 bar (8 000–10 000 lb/in^2). The higher figure also represents the maximum normally adopted for rams.

Practical fluid systems can, however, be worked at very much higher pressures. Typical examples are liquid springs where the working fluid performs as a compressible medium, and chemical processing such as polythene production, where pressures of the order of 2 800–3 500 bar (40 000–50 000 lb/in^2) may be utilized. Ultra-high pressures are also used for metal extrusion, hydraulic forming, the pressing and compacting of metal and other powders and auto-fretting. Other fields include high-pressure testing and the further development of fluid-pressure devices in the still relatively unexplored fields of metal ductility, chemical reactions and changes of state under extreme pressure conditions.

The practical limit for fluid pressure devices yet developed is of the order of 400–500 kilobar (6–7 *million* lb/in^2). Within this limit, an arbitrary classification of pressure ranges is shown in Table I.

TABLE I – ARBITRARY PRESSURE RANGES

Application	PRESSURE		
	bar	kilobar	lb/in^2
Industrial hydraulics: low-pressure	20–35	–	250–500
medium-pressure	70	–	1 000
high-pressure	140	–	2 000
Aircraft and special duty hydraulics	210–280	–	3 000–4 000
optimum limit	350	0.35	5 000
Normal limit for linear actuators	420	0.4	6 000
Heavy-duty rams and presses	700	0.7	10 000
Limit for hydrostatic systems	700	0.7	10 000
Limit for mineral oils	3 500	3.5	50 000
Ultra-high pressures	over 3 500	3.5	over 50 000
Limit for fluid-pressure devices	14 000	14	200 000
Limit for pressure devices	–	400–500	–

Production of Ultra-High Pressures

Pumps of the mechanical type are generally unsuitable for generating pressures in excess of 3.4 kilobar (50 000 lb/in^2) and so hydraulic intensifiers are normally used as the source of fluid power. In the case of an all-hydraulic system the source of primary pressure is a pump, with pressure ranging up to about 1 125 bar (16 000 lb/in^2). Final pressures up to 14 kilobar (2 000 000 lb/in^2) can readily be achieved by such devices, although the majority are designed for lower pressure working. (See chapter on *Hydro-Pneumatics*).

Measurement of Ultra-High Pressure

The normal method of measuring pressure up to about 14 kilobar (2 000 000 lb/in^2) is the free piston gauge. The diameter of the piston is measured and the area calculated whilst the weight it supports can be compared very accurately with the standard. When pressures are low the area of the piston is fairly large and can be calculated with great accuracy. If a lubricating oil is used as the pressure medium the diameter of the cylinder is made 0.005 mm (0.0002 in) larger than the piston and the effective diameter is then 0.003 mm (0.0001 in) greater than the piston diameter.

With increasing pressure the piston diameter must be reduced, both to reduce the supported weight and limit the stresses in the cylinder. The relative accuracy to which the piston can be measured then becomes less, and also knowledge of the thickness of the oil film between piston and cylinder. As the pressure increases the piston and cylinder diameters also expand, and the true value at any pressure must be known.

One method of evaluation is to compare the behaviour of similar free-piston units made from materials with different elastic coefficients. Alternatively, the necessary data are obtained from theoretical considerations.

The hydraulic circuit for a typical proprietary deadweight tester is shown in Fig 1. The main pump supplies the intensifier through a directional control valve and a pressure-loaded relief valve. The springs of this valve give a relief pressure of *circa* 3.5 bar (50 lb/in^2), but when the needle valve is opened this pressure is applied to the spring end of the plunger, automatically increasing the relief pressure. Thus the operator need only turn this valve and the pressure rises automatically until the valve is shut. The intensifier discharges through non-return valves and the high pressure circuit is released through a special needle valve.

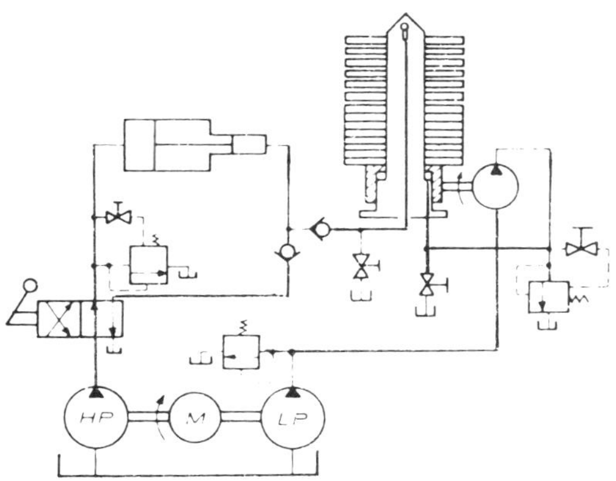

Fig 1 Hydraulic circuit for deadweight tester. The twin pumps supply the intensifier and weight manipulation respectively.

The low-pressure pump is connected directly to the hydraulic weight-revolving motor so that all the oil pumped passes through the motor, or if it is stationary, through the bypass valve. The motor exhaust is then led to a second pressure-loaded relief valve governing the opening of the loading valve, producing pressure to lift the weight off-loading cylinder. As each successive weight is picked up the pressure increases and is locked in the cylinder by the check valve. The weights are on-loaded by opening a release valve.

Where the bulk of a free-piston gauge prohibits its use a simple manganin gauge may be used. This comprises a coil of manganin wire mounted on a suitable holder and housed in a pressure chamber. Leads are taken through the walls of the chamber to a Wheatstone bridge on which the coil resistance is measured. As resistance is also dependent on temperature, the temperature must be maintained precisely constant when readings are taken.

For pressures up to 3 500 bar (50 000 lb/in^2) a secondary gauge is sometimes used, consisting of a pressure chamber to which is bonded a strain gauge. Expansion of the chamber under pressure changes the resistance of the strain gauge, which is measured on a Wheatstone bridge circuit – Fig 2.

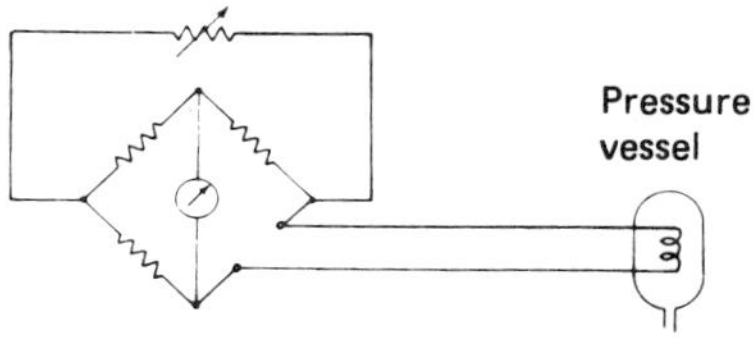

Fig 2

Dial gauges are also available for pressures up to 10.5 kilobar (150 000 lb/in^2), usually based on a Bourdon tube with a high-tensile steel tube.

Fluids for Ultra-High Pressures

The practical limit for working of conventional mineral oils is 3.5–4 kilobar (50 000–60 000 lb/in^2), above which they become so viscous as to approach 'solid' characteristics. The working pressure range for such fluids can, however, be extended up to about 6.9 kilobar (100 000 lb/in^2) by the addition of viscosity index improvers, provided the fluid itself is not subjected to continuous high-shear rates which cause a break-down of the additive and some loss of performance.

Water does not suffer from this effect, but due to its very low viscosity is extremely difficult to seal. It is thus less satisfactory as a working fluid for high-pressure systems, although it may be used as the fluid medium for ultra-high-pressure testing, particularly destruction testing where the fluid is lost.

TABLE II – PRACTICAL PRESSURE LIMITS FOR FLUIDS

Fluid	MAXIMUM PRESSURE	
	bar	lb/in^2
Straight mineral oils	3 500	50 000
Mineral oils with VI improvers*	up to 7 000	up to 100 000
Castor-oil	7 000	100 000
Castor-methanol	up to 11 250	up to 160 000
Glycerine	14 000	200 000
Water-glycols	up to 14 000	up to 200 000
Synthetic lubricants	up to 14 000	up to 200 000

*Depending on type

Castor oil has a superior performance to mineral oils as regards viscosity/pressure characteristics and can be worked up to about 6.9 kilobar (100 000 lb/in^2) without excessive thickening. Castor-alcohol mixtures, or similar castor-base fluids retain good fluidity up to very much higher pressures – see Table II.

Glycerine is one of the best fluids for ultra-high-pressure working, remains fluid up to the highest practical pressures and has a very high bulk modulus. It is more generally used in the form of water-glycol mixtures, however. The primary limitation of such fluids is their tendency to cause rusting on ferrous metal components.

Synthetic lubricants are an alternative choice, although they tend to be expensive. This can be partly offset by diluting the fluid with a compatible low-viscosity liquid. Many of the synthetic lubricants also have high bulk moduli, which are desirable to reduce compressibility effects, and low freezing points, which are essential to prevent solidification if the fluid is rapidly expanded on the high-pressure side.

In addition to governing the 'elasticity' of the ultra-high-pressure system, the bulk modulus of the fluid governs the efficiency of the intensifier. The more the fluid is compressed in developing the ultra-high pressure (*ie* the lower the bulk modulus of the fluid), the greater the wastage of power input in compressing the fluid. The bulk of this power loss is stored in the compressed fluid (a proportion being transformed into heat), and will be released when the pressure restraint is removed. Thus the higher the bulk modulus of the fluid the lower the 'explosive' release of energy in the event of a break in the system or a component, as in destruction tests. Typical bulk moduli for various fluids are given in Table III.

TABLE III

Fluid	Typical % Compressibility per atmosphere	Typical % Compressibility at 700 bar (10 000 lb/in^2)
Water	49.5×10^{-6}	3.3
Water-in-oil	51.8×10^{-6}	3.5
Water-glycol	38.5×10^{-6}	2.6
Phosphate ester	37.0×10^{-6}	2.5
Chlorinated hydrocarbon	35.5×10^{-6}	2.4
Mineral oil	50.4×10^{-6}	3.4
Glycerine	25.9×10^{-6}	1.75
Silicone	103.7×10^{-6}	7.0

Pipes and Fittings

All fluid devices and their connections on the ultra-high side are potentially hazardous. Special constructions may be required, and certainly special pipes and fittings. All designs of ultra-high pressure equipment try to keep dimensions as small as possible to minimize the bursting stresses involved. Tube bores are usually 3 mm or 1.5 mm (1/8 inch or 1/16 inch), with a suitable wall thickness to accommodate the maximum hoop stress within the maximum permissible working-stress for the material. Composite tubes may be used for maximum strength with minimum over-all diameter. Intensifier high-pressure cylinders may be as large as 25 mm (1 inch), but would normally be less for pressures above 7 kilobar (100 000 lb/in^2).

Standard sizes of tubing have been established in both alloy and Type 316 ST. The smallest size is 1.5 mm (1/16 inch) bore by 6 mm (1/4 inch) o.d. which is good for 6.9 kilobar (100 000 lb/in^2), whilst for 13.9 kilobar (200 000 lb/in^2) a composite tube having a stainless steel core and alloy steel envelope is preferred. The envelope is drawn down onto the core to pre-stress it.

Pipe joints are also standardized in principle if not in detail. The block is tapped with an appropriate thread – 9/16 inch x 20 tpi for 1/4 inch o.d. pipe and the inlet hole coned at an angle of 60° (Fig 3). The pipe is screwed with a left-hand thread on which a stop sleeve is screwed and coned at an angle of 59° so that the joint is first made at the small end of the cone. The force tending to burst open the joint is obviously dependent on the joint area and by this means it is kept to a minimum.

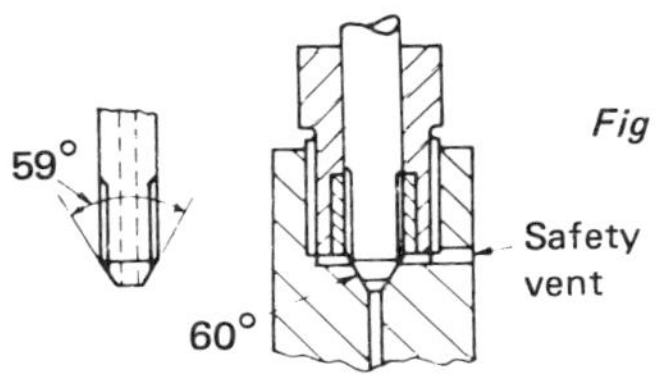

Fig 3 Typical high pressure pipe joint.

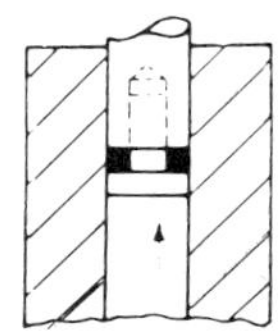

Fig 4 Bridgman 'unsupported area' plug seal.

Larger joints are more of a problem and the details may depend on the general design. The principle of unsupported areas first introduced by Bridgman is a useful guide; this joint, in the form of a plug or piston seal, is shown in Fig 4. The force on the end of the plug is transmitted to the packing – usually non-metallic – which, being annular, has a lesser area than the plug. The pressure tending to flatten it is therefore about 50% greater than if it had no hole and the packing is forced into the hole with correspondingly greater force. The force can be so great that the neck of the plug is pinched off.

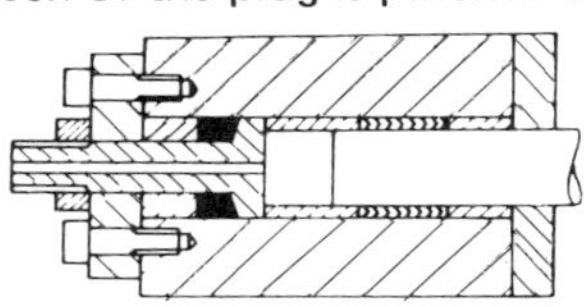

Fig 5 Sealing end of cylinders and outlet pipe.

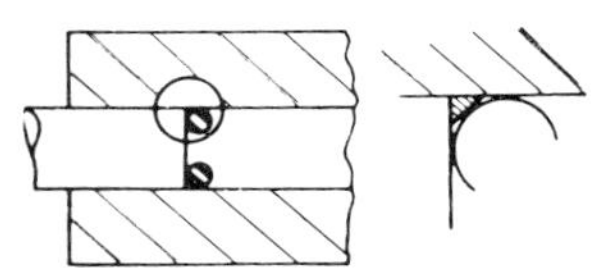

Fig 6 O-ring seal, showing an enlarged view method of preventing extrusion with metal backing ring.

The same principle is applicable to outlets for cylinder ends (Fig 5). As the fluid velocity is low the outlet bore can be small so that the spigot is not seriously weakened. The conical seat for the packing also tends to reduce the inward thrust as it is compressed under pressure.

Provided that extrusion can be prevented, O-rings have a greater sealing capacity. One method of doing this is to insert a triangular-section close-fitting ring behind the O-ring (Fig 6). The O-ring under pressure behaves as a very viscous fluid and it is reasonable to assume that, like all fluids, it becomes more viscous as the pressure increases.

Larger Components

Auto-frettage is widely used for treating cylinders which have to withstand high pressure although the highest pressure for which it can be applied is limited by the fact that auto-frettage requires application of a pressure appreciably higher than the working pressure.

Auto-frettage consists of pressurizing a cylinder until the yield point is exceeded; it is then given a low-temperature stress-relieving heat treatment. The effect is to produce high compressiv

:tresses in those parts of the cylinder which are subjected to tensile stresses when it is under ›ressure. The result is that the safe working pressure may be doubled, as the risk of fatigue, due to eversal of stress, is eliminated.

There is a limit to the pressures for which auto-frettage is applicable, simply because it is impos- ible to produce the over-pressure. One method of overcoming this, in suitable cases, is with taper :onstruction, as shown in Fig 7.

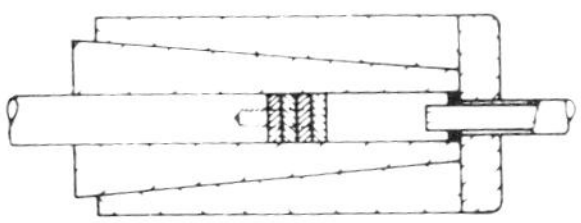

Fig 7 Taper construction for strengthening very high pressure cylinders. The piston packing of alternate washers of copper and leather is effective for high pressures.

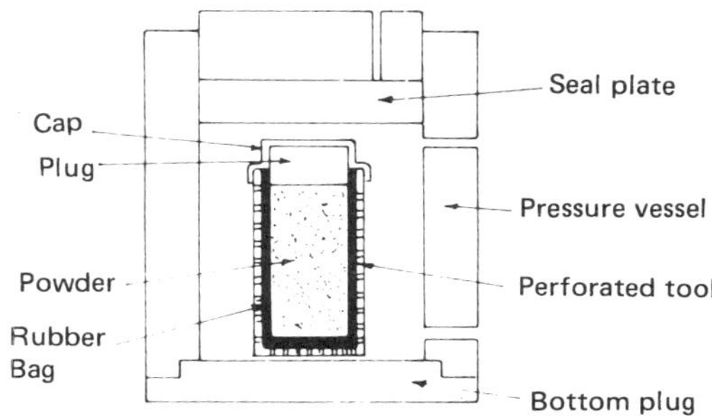

Fig 8

sostatic Pressing

sostatic pressing involves applying pressure uniformly over the whole surface of a subject under- 'oing compaction, and is applied particularly to powdered forms of solids. Various media may be ısed for applying the pressure, a high-pressure or ultra-high-pressure fluid generally being the most :onvenient for low-temperature working. A normal requirement in such cases is that the subject naterial and pressurizing fluid must be separated, usually by putting the material into some form f flexible container or flexible mould, which is then sealed to prevent entry of fluid. It is this rinciple of applying the compacting pressure uniformly over the whole surface being treated vhich distinguishes isostatic pressing from conventional die pressing and forming. A further ractical advantage is that the reactive force available can be very high compared with that given y a conventional press, with a consequent saving in bulk and weight of equipment.

A typical form of isostatic press is shown in Fig 8, comprising essentially a heavy-walled ressure vessel with a detachable top plug for access to the working space, with suitable seals. This ; allied to a pumping system, which must be capable of generating the required fluid pressure at a uitable flow rate to provide an economic pumping time, and in particular allow for the consider- ble reduction in displacement volume of the subject undergoing compaction.

The subject is loaded in the form of a 'tool' which comprises the flexible mould or bag isolating he material from the fluid, mounted as an assembly with the powdered material and a form of ıould for producing the required shape or form. This is vented as necessary to allow hydrostatic ressure to reach the outside of the flexible bag, compacting the powder to the pre-determined hape set by the rigid members of the assembly.

The pressure required depends primarily on the pressure-density characteristics of the powder eing compacted. The isostatic nature of the system means that the surface of the powder is ıbject to constant pressure, and thus implies constant powder density during compacting, lthough the pressure system within the piece will depend on the transmissibility of the powder. ı practice, very constant densities are normally achieved, certainly superior to those produced by ıechanical pressing processes.

Rate of pressure application, dwell time and rate of pressure release are also important para- ıeters affecting the quality of the finished product. Rate of pressure application is controlled by the rate of pumping, and rate of pressure release by releasing the pressurized fluid through a metering valve, optimum times for all three phases normally being determined by sample runs.

Isostatic pressing is also extended to sheet metal forming and other fabrication techniques, where the pressure systems within the subject may be further modified. The pressing of ready-formed solid materials, rather than compacting, is generally referred to as hydrostatic pressing.

Liquid Springs

The liquid spring consists essentially of a stout cylinder, stressed to withstand very high pressures, enclosing a piston and rod assembly adequately sealed at the rod end by high-pressure seals — Fig 9. The piston incorporates two orifices, one large enough to permit rapid flow and the other acting as a metering orifice or recoil orifice. Opening of the rapid-flow orifice is controlled by a suitable valve to give one-way opening. The metering orifice provides for flow in either direction.

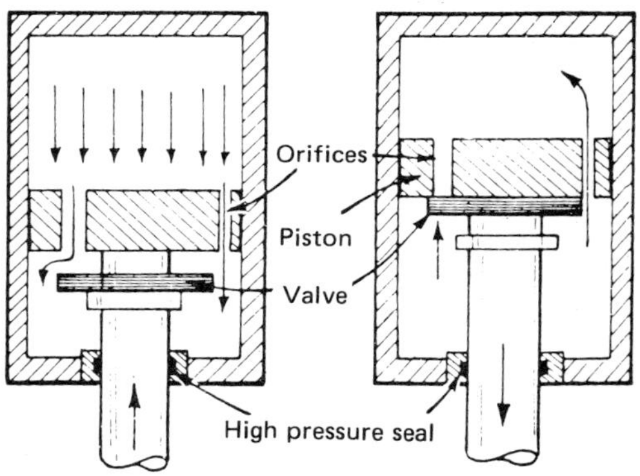

Fig 9 Principle of liquid spring (diagrammatic).

On the retraction stroke, the piston is driven up the cylinder and thus the volume available for the fluid is reduced by that volume of piston rod entering the cylinder, resulting in compression of the fluid. It is this compression to very high pressures (*circa* 3 500 bar or 50 000 lb/in^2) which provides 'spring' action due to the compressibility of the fluid and to a lesser (and usually negligible) extent, the elasticity of the cylinder itself. At the same time dashpot damping is provided by the main orifice being open and by flow through the recoil orifice.

On the extension or recoil stroke, piston movement is energized by the expansion of the fluid from its compressed state. The main orifice is closed by the valve, but flow from one side of the piston to the other can take place through the metering orifice. This flow provides damping on the recoil stroke.

Pressure build-up on the 'spring' stroke is progressive and directly proportional to the stroke (governing the length of rod entering the cylinder, and thus the physical reduction in fluid volume). Compressibility to be anticipated with a mineral oil fluid is of the order of 3.5% per 70 bar (1 000 lb/in^2), although the relationship will not be linear, particularly when higher pressures are reached towards the end of the stroke. Thus the spring action tends to become progressively 'stronger' with increasing stroke, and, if the rod section is particularly generous, may produce sufficient reduction in fluid volume to raise the fluid pressure to a point where it becomes virtually solid. Primarily, therefore, the spring reaction is determined by the diameter of the rod relative to the cylinder diameter, and small changes in rod size are more effective in controlling the spring 'rate' than the selection of fluids with different bulk moduli.

Hydraulic Stressed Bolts

Jltra-high hydraulic pressures of the order of 2 100 to 3 500 bar (30 000 to 50 000 lb/in^2) are used in special designs of high-duty bolts capable of being stretched longitudinally as a preliminary o fitting. The resulting reduction in diameter enables the bolt to slide easily into a specified size of hole. On release of hydraulic pressure, the bolt expands to provide the correct degree of fit to naintain a definite pre-determined fitting force between the bolt and hole. Similarly, such a bolt an easily be removed again, if necessary, by hydraulic stretching.

This form of bolt obviates the need for hammering bolts into position where particularly rigid or accurately aligned assemblies are required, and eliminates any possibility of scoring the bolt or bolt hole surfaces.

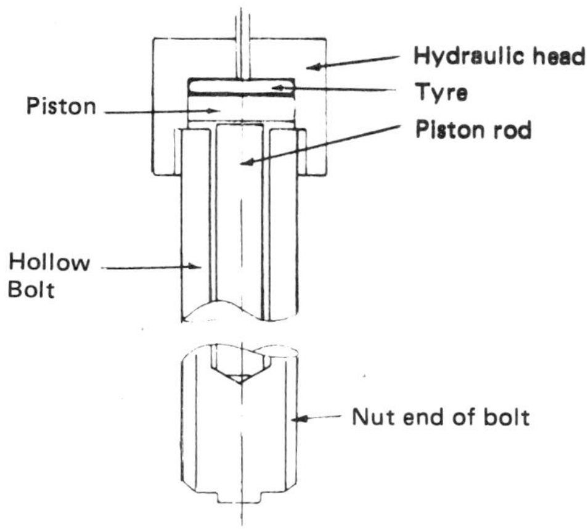

Fig 10

The design comprises a blind drilled bolt with a reduced diameter screwed head which can be ssembled as a unit with a loose piston rod, piston, rubber 'tyre' and a hydraulic head, in that rder – Fig 10. Hydraulic pressure is applied directly to the centre of the hydraulic head and is ansmitted *via* the tyre and piston to the piston rod, generating a pre-determined force sufficient o stretch the length of the bolt the required amount, but without exceeding the maximum ermissible stress in the bolt material. The reactive force is contained jointly by the hydraulic head nd the bolt head to which it is attached.

The total contraction which can be achieved on the bolt is of the order of D/2 500, where D is he bolt diameter. This is generally adequate to provide a radial gripping pressure of the order of 12 bar (1.4 tons/in^2) between the bolt and hole surface. Axial grip can also be provided, if equired, by controlling the run-up tightness of the nut.

Vibration and Noise

SOURCES OF noise in hydraulic systems are the pump and its driver, the distribution lines, control elements and actuators. Of these, the pump is normally the main source of noise, but this can usually be reduced to moderate levels by suitable acoustic treatment. The main aim at this end of the system should be to minimize pump-generated noise and vibration and any associated resonance. At the same time it is highly desirable to isolate the distribution line(s) so that vibration generated by the pump unit is not transmitted through the pipework, with the possibility of resonance occurring at other connecting points. This does not, however, eliminate the possibility of pressure pulsations being transmitted by the fluid to the pipework system and attached components, which phenomenon may need separate treatment.

It may be necessary to analyze the various possible sources of noise in the complete system in detail in order to arrive at satisfactory noise treatment. In this case the possible sources of noise generation are, in decreasing order of significance:

(i) Pump noise (where applicable)
(ii) Appliance noise
(iii) Control element noise
(iv) Water hammer
(v) Chatter
(vi) Cavitation
(vii) Resonance
(viii) Pipework noise
(ix) Thermal effects

Pump Noise and Vibration

Pump/motor vibration can be minimized by mounting the pump and motor on a common base (or mounting the motor integral with the pump) and isolating the complete unit on a resilient mount. A general recommendation is that the natural frequency of the isolated mount should not exceed one-quarter of the shaft speed (frequency), although it may be permissible to approach one-third of the shaft speed if a stiffer mount is required.

If further acoustic treatment is required the whole pump/motor unit can be fitted with a suitable enclosure. The majority of hydraulic pumps are driven by electric motors, so no special problems are involved other than ensuring an adequate airflow for cooling the electric motor. If necessary a forced draught ventilating system can be used with a completely sealed enclosure employing duct silencers of the absorptive type.

The reservoir or tank may be located in the same enclosure, or distant from it. Although inherently a source of fluid pressure damping, a reservoir does not necessarily attenuate noise. In fact, it may well reverberate as oil contained freely in a tank does not prevent it becoming an efficient sound radiator. The only modifying effect on noise in such a case will be a spectrum shift, unless the tank itself has high damping or can be suitably damped (*eg* with a damping/sound insulation covering).

A simple method of isolating or decoupling the pump from the delivery line is by a flexible hose connection. Isolation can be further improved (if necessary) by using two such hose lengths in close proximity and mounted at 90° to each other. Ideally, isolation by flexible pipe should include bends in two mutually perpendicular directions with equal distances between bends.

A further method of decoupling is the use of O-ring or similar elastomeric seals in a suitable coupling. This is particularly effective for decoupling high frequency vibrations but is less effective at lower frequencies. Couplings of a similar type are produced for semi-flexible connections, allowing for a limited degree of movement or misalignment. Such couplings are not effective isolators unless the resilient elements employed provide complete isolation between the components connected (*ie* do not allow direct metal-to-metal contact).

In general, isolation of the pump and motor from the tank by suitable mountings and decoupling from the pipework will free the rest of the system from the transmission of mechanical vibrations and the consequent possibility that these would be amplified. Care must be taken to ensure that there is no short-circuiting of the isolation or decoupling employed.

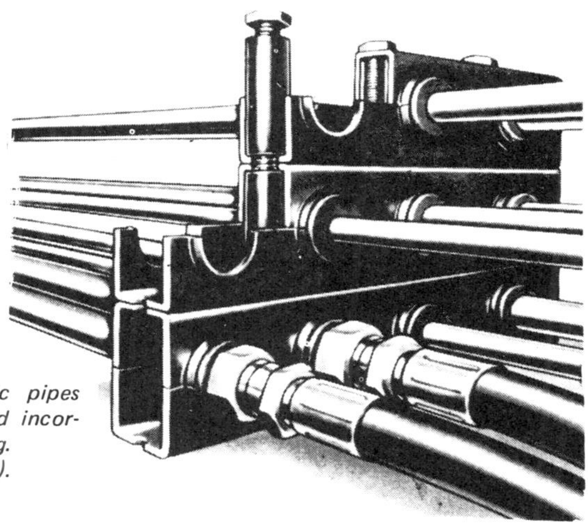

Multi-tube clamp for hydraulic pipes based on stacking modules and incorporating vibration damping. (Industrial Hydraulics Ltd).

Pipelines

Noise produced in hydraulic lines may be pump generated (changes in power and pressure, or varying amplitudes of pressure pulsations) or fluid generated (flow instability, turbulence or simple fluid friction).

Fluid-generated noise in small bore pipes with low to moderate flow rates is generally negligible, unless pressure pulsations are present, (*eg* due to valve cavitation). Thus pipe vibration, and consequent radiation of airborne noise, is usually due to the higher level of noise generated by fittings; pipe resonance is due to mechanical vibration or resonant noise generated in supporting systems.

Standard treatment for noise reduction is:

(i) Damping by means of suitable isolating pipe supports. This also provides decoupling for supporting structures.

(ii) Decoupling from other sources of noise or vibration in the system.

(iii) Soundproof 'lagging'.

For the majority of systems only (i), and to a lesser extent (ii), should be necessary. 'Lagging' is normally only required when there are pulsation vibrations present which cannot be damped or isolated by simple means. This is most likely to occur on pumped systems employing thin walled, large diameter piping, particularly on the suction side.

Sufficient damping for pipes is usually provided by suitable supports, or pipe clips spaced at regular intervals, the supports having resilient linings so that vibration in the pipe is not transmitted directly to the surface to which the supports are fixed.

Vibrating and isolating pipe clip. (James Walker & Co Ltd).

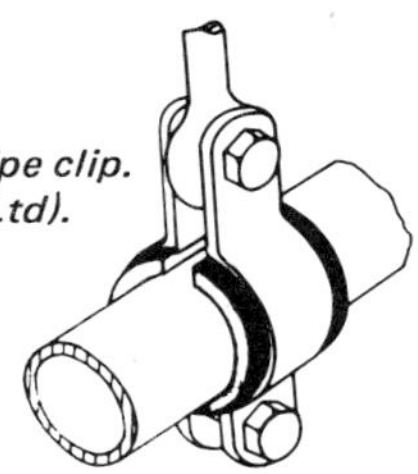

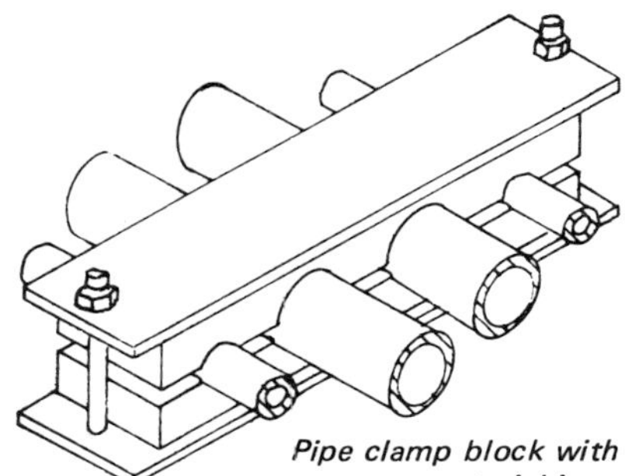

Pipe clamp block with vibrating isolating material incorporated. (James Walker & Co Ltd).

Standing Waves

Optimum pipe spacing can be analyzed in terms of standing wave phenomena, although this is seldom necessary. The case of axial standing wave is usually academic, for practical lengths are usually substantially lower than the critical length, which is defined by:-

$$L_a = \frac{8200n}{f} \quad \text{for steel pipes}$$

where L_a = resonant length of pipe in metres

n = number of half wavelengths in the vibrating pipe

f = frequency of any strong vibration

Theoretically, at least, the distance between pipe supports should always be less than this resonant or critical length.

Suction Line

The suction line is a first suspect in a hydraulic installation which proves noisy, and where the noise cannot be directly attributed to pump or components. Suction lines can generate noise if there is an excessive pressure drop when the pump is sucking at sub-atmospheric pressure and drawing air out of solution (hydraulic oils normally contain about 8% air in solution). The resulting formation of air bubbles, and their subsequent collapse, can cause 'mechanical' noise which is often erroneously diagnosed as pump noise.

Suction line noise can also be caused by a partially blocked or undersized suction filter, poor placement of the outlet pipe in the reservoir or entrained air.

Delivery Lines

Delivery lines can carry mechanical vibrations to distant parts of the circuit. These vibrations may be amplified at local points by the resonance of supporting structures or components directly connected to the pipework. Resonance can be eliminated by decoupling connections.

All pipework installations should be designed on the basis of avoiding abrupt changes of section which could lead to large flow velocity changes and generation of turbulence. As far as possible, too, generous bend radii should be employed for similar reasons.

Valves

Noise due to the operation of valves, regulators and control elements is transient and related to the degree of turbulence or cavitation produced, although in specific designs and certain circumstances individual elements may be subject to vibration and generate a continuous noise. So much depends on the design and finish of the flow passages involved that no general analysis can be attempted. The noise level of such devices is dependent on the design and the localized flow velocities produced and also on the response time, where applicable. The latter effect can be minimized by arranging that the response time is not shorter than that required by the system. This will result in minimum 'hammer'. 'Water hammer', in fact, depends on the switching velocity of the valve – *ie* on the spool-switching velocity in the case of spool valves. Valves operated by dry solenoids have in fact, uncontrolled response and so often produce 'hammer'. Wet solenoids are cushioned by the hydraulic fluid so move more smoothly and open the valve passages more gradually (but at the expense of some loss of solenoid power).

As a general recommendation, simple undamped ball-and-spring non-return and relief valves should not be used. On the design side, every effort should be made to ensure that the flow passages of valves are swept and free from sharp edges and corners as far as possible. Directional control valves must also be carefully designed to prevent flow instability occurring.

Cavitation in Valves

Cavitation is a breakdown in flow caused by the localized fluid falling below the vapour pressure of the fluid. Consequently, vapour bubbles are formed resulting in irregular and noisy flow. Such a reduction in pressure can occur in regions of localized high flow velocities, such as are caused by restrictions to the flow path. Thus, the onset of cavitation is marked by a critical pressure, which in turn is dependent on a critical velocity. The effective critical pressure is also a function of temperature, however, since this governs the vapour pressure of the fluid. Accurate prediction of cavitation conditions is most difficult, and usually impossible, in the design of valves and fittings, and problems have to be tackled on empirical lines. Although much has been done to design fittings which do not produce cavitation at normal or recommended flow rates, it does not follow that this will be maintained over the full range of operation. If the flow rate is sufficiently restricted, cavitation and noisy flow can be expected. Thus a partially closed tap or valve is nearly always noisier than when fully opened; also quite a small change in position, and thus flow rate, can cause a change from cavitating (and noisy) to non-cavitating (and relatively quiet) flow. It is also a characteristic of many valves, that for flow rates (valves openings) below that which produces cavitation, cavitation noise increases with increasing frequency; whilst for higher flow rates, where flow is non-cavitating, cavitation noise does not vary greatly with frequency. This also explains the considerable difference in interpreted sounds – higher frequencies being more readily radiated and sounding louder to the ear.

In the case of high pressure systems, or valves subject to high pressure drops, it is desirable to utilize flow paths designed to eliminate cavitation as this can cause physical damage to the valve components as well as excessive noise. The problem, basically, is one of preventing the pressure in the valve throat from falling below the fluid vapour pressure in order to prevent cavitation occurring. This requirement can be rendered in mathematical form:

$$\Delta P = F_L{}^2 (P_1 - F_F P_V)$$

where ΔP = allowable pressure drop
F_L = liquid pressure recovery factor
P_1 = upstream pressure
P_V = vapour pressure of liquid at temperature concerned
F_F = critical pressure ratio (normally between 0.68 and 0.92).

A simpler formula which may be used is:

$$\Delta P\ (\text{max}) = K_C (P_1 - P_V)$$

where K_C is the cavitation index of the valve (specifically an index of incipient cavitation)

The value of K_C for any particular type and design of valve can be determined experimentally and typically can be expected to be of the order of 0.6–0.7 for a simple single-stage valve.

To avoid cavitation in multi-stage valves, pressure drop in the first stage should be less than the maximum given by the above formula, with the pressure drop in the second stage less than that in the first, and so on. Ideally the pressure drop across each stage should be a constant percentage of the difference between the upstream or inlet pressure at that stage and the vapour pressure, *viz*

$$\Delta P_S/\Delta P_{ps} = 1 - K_C$$

where ΔP_S = pressure drop at stage
ΔP_{ps} = pressure drop at preceding stage

Reservoirs

Reservoirs should be designed to avoid the entrainment of air in the fluid, and the recommendations given in the following paragraphs are intended to assist in this

Return lines should enter and suction lines should leave the reservoir well below the surface of the fluid when at its lowest permissible level.

Return lines should be fitted with a pepper pot and suction lines with a bell mouth entry or a suction strainer where they enter the reservoir.

A baffle should be fitted between the suction and return lines. Variable capacity systems require special attention.

The reservoir should be fitted with a bubble separator. This may be a single 60 mesh wire gauze set between the return and suction tank openings. Maximum effect is obtained when the gauze lies at an angle of 30° to the horizontal.

Where a reservoir is pressurized with air, there should be a separator at the oil-air interface.

The pressure drop across the entire suction line should not exceed 0.3 bar regardless of reservoir pressure, nor should suction pressure at the pump inlet be less than 0.2 bar. Suction line velocities should not exceed 1.5 m/s. Working and return line velocities should not exceed 4.5 m/s.

Reservoirs should be flexibly mounted and isolated from surrounding structures.

Pulsation Dampers

Modern hydraulic pulsation dampers sub-divide into three categories:

(i) *Low frequency units* for damping pulsations up to about 10 Hz. These are invariably single ported units.

(ii) *Multi-ported units* which intercept the flow path and are more effective in removing pulsations of higher frequency. They can be effective up to 500–1 000 Hz.

(iii) *True silencers* which work on the principle of wave cancellation and heat dissipation, for pulsation damping of frequencies above 500 Hz. These are normally based on a multi-pass geometry (*ie* the incoming flow is split into a number of different flow path lengths) terminating in a 'sounding dome'. The preferred type of sounding dome is a bladder.

Recently considerable attention has been given to the development of hydraulic pulsation dampers capable of dealing with frequencies in the sonic range. They are designed to smooth small volumetric displacements and at the same time deal with the very high acceleration rates which can cause sound waves to be generated by the mechanical components of a system.

Shock Preventers

Shock preventers are pulsation dampers (or accumulators) characterized by having very large flow inlet apertures which are partially closed off by liquid trying to flow back out of them. They are not shock absorbers, as they prevent shock or surge occurring. For the same reason, they do not attenuate shock.

Shock Removers

These are sensitive hydro devices which prevent a standing wave from passing farther down a system or from bouncing back through them. They are normally of tubular or sleeve form with a flexible membrane. Because of their length it is possible to open a membrane so that it is exposed to the increased pressure of a wave and close behind the wave with it already shut in.

Acoustic Filters

Acoustic filters can be fitted to systems where pressure ripple is high. These are essentially tuned silencers which are critical in design and are usually effective over only very narrow frequency bands, although the attenuation achieved can be quite high. Untuned silencers simply comprise an expansion chamber with broader coverage but reduced attenuation. An accumulator is, in effect, an untuned hydraulic acoustic silencer and is most effective at lower frequencies. Dissipative-type silencers provide for dissipation of energy through viscous flow losses and, as a consequence, consume some fluid energy. They may be combined with an untuned silencer, although the attenuation will still be appreciably lower than that of the tuned type.

In general, wave cancelling filters are to be preferred since the frequencies involved are low. If the pressure transients are narrow band, a Quinke Tube and expansion chamber can be effective. A major disadvantage of this and other types of simple wave-cancelling filters, however, is the relatively high pressure drop produced. The more usual form of hydraulic silencer is the pressure-release type. This gives minimum pressure drop and broad band filtering, but is pressure sensitive and needs regular routine maintenance.

See also chapter on *Accumulator-Type Devices.*

SECTION 4

·lydraulic Motors

\S A general rule hydraulic motors differ from hydraulic pumps only in detail. Many types of ›ump will work as motors connected the other way round (particularly axial-piston pumps), when hey may be operated as combination pump-motors. Exceptions are those types fitted with ball or pring-loaded valves (*eg* inlet and outlet valves in piston pumps), or simple pumps with very low nechanical efficiency. In the former case, modification is needed for those machines to operate ffectively as hydraulic motors.

Particular differences which may be observed between some types designed as motors rather han pumps are modified port timing and the provision of case drains to protect shaft seals; also ane motors usually differ from vane pumps in having spring-loaded vanes to ensure good starting orque being generated.

Hydraulic motors work by accepting pressurized fluid at their inlet and converting pressure nergy into rotational energy (torque and speed). Basically torque is dependent on the operating ressure and rotational speed on the flow rate through the motor.

:lassification

lydraulic motors may be classified by type — *eg* piston motors, vane motors, gear motors, *etc*; or y their performance characteristics. A general classification for the latter is:

(i) *High-torque motors* — designed to provide high output torque with maximum torque (usually) available from starting up. Motors of this type are normally low-speed machines, generally of radial-piston or axial-piston configuration.

(ii) *High-speed motors* — designed for high operating speeds with low torque. Characteristics may vary widely with different types and individual designs. Chief type of high-speed motors are modified gear motors, axial-piston motors and vane motors.

(iii) *Medium-torque/medium-speed motors* — designed to provide good torque with higher operating speeds than high-torque motors. A widely differing range of performance may be offered by radial-piston, axial-piston, vane and gear motors in this category.

(iv) *High-moment motors* — designed specifically to provide good start-up torque with higher speed of operation than high-torque motors. They are usually of radial-piston type.

ston Motors

ere, unlike pumps, only the axial-piston configuration is used to any great extent for hydraulic otors — *eg* see Fig 1. These are usually true combination pump-motors, *ie* will operate equally

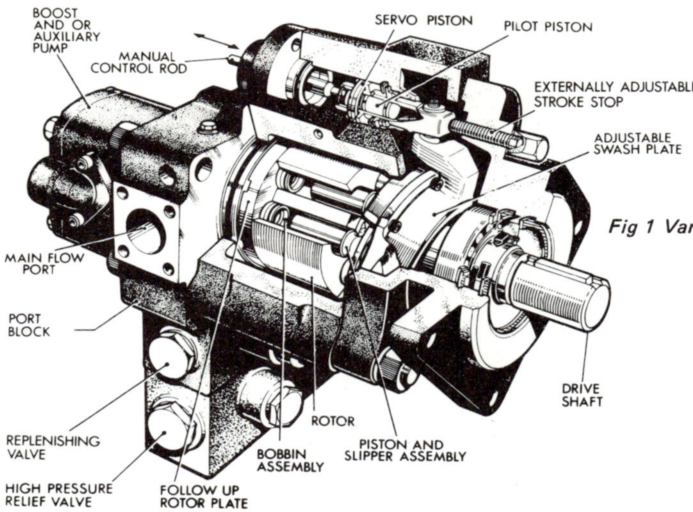

Fig 1 Variable-capacity axial-piston pump/motor.

well in either mode. Available models and sizes cover most requirements as regards speed, torque and power output, and suitability for working pressures up to 350 bar (5 000 lb/in^2), and the same specifications usually apply for pumps and motors. Some axial-piston motors are specially designed for high-torque/low-speed applications, or as specific matches to particular pumps for hydrostatic drives. Axial piston motors are also the main type used for infinitely-variable-speed drives, although similar variable-speed characteristics are available from any fixed displacement motor coupled to a variable-delivery pump. See also Table I.

TABLE I – AXIAL-PISTON MOTOR CHARACTERISTICS

Type	Speeds rev/min	Torque	Pressure	Remarks
In-line	Up to 3 000 or up to 5 000 for special duties	High to moderate	Up to 210 bar (3 000 lb/in^2) or up to 350 bar (5 000 lb/in^2) for special duties.	Widest variety of sizes and designs.
Bent-axis	Up to 3 000 or up to 5 000 for special duties	High to moderate	Up to 210 bar (3 000 lb/in^2) or up to 350 bar (5 000 lb/in^2) for special duties.	Wide variety of sizes and designs.

Vane Motors

Motors of this type are particularly suitable for medium-speed/medium-power applications, with moderate to low torque output. (See also Table II).

The chief limitation of a sliding-vane motor is poor starting characteristics due to high internal leakage past the vanes until these are fully extended by centrifugal force. This can be overcome by

TABLE II – VANE MOTOR CHARACTERISTICS

Type	Speeds rev/min	Torque	Pressure	Remarks
Simple sliding vane	100–3 600	Moderate to low	35–140 bar 500–2 000 lb/in^2	For constant speed moderate power drives. Efficiency up to 90%.
Sliding vane with spring or pressure-loaded vanes	100–3 000	Moderate to low	35–175 bar 500–2 500 lb/in^2	Improved low-speed performance. For constant-speed moderate-power drives.
Rolling vane	5–2 000	Moderate	35–140 bar 500–2 000 lb/in^2	Offers a very wide speed range with good efficiency.

using spring-loaded vanes, or providing additional porting to pressurize the bottom of each vane. In general, however, a vane motor even with such modifications is not suitable for low-speed running, although, again, loss of performance at low rotational speeds can be offset by increasing the number of vanes.

An alternative approach is provided by the so-called rolling vane or rotary abutment motor – Fig 2. The rotor housing the sliding vanes is concentric with the housing in this case, and pressure s applied continuously to two vanes to produce continuous rotation. The rotary abutment gears provide a seal between the rotor and casing, as well as timing of the rotor and rotary abutment.

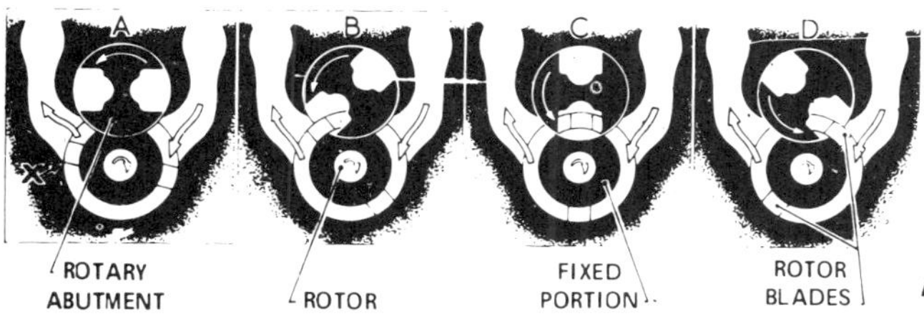

Fig 2 Keelavite rotary abutment motor.

Although considerably more complex than a simple sliding-vane unit, this form of vane motor perates with very low friction and has an excellent low-speed performance, and it can be operated ver a wide range of speeds.

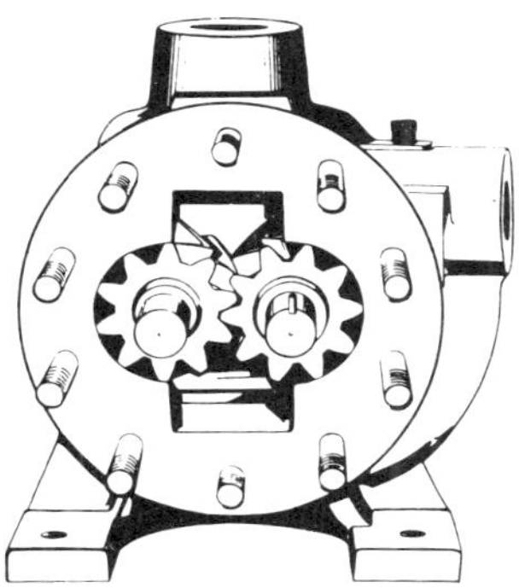

External gear motor

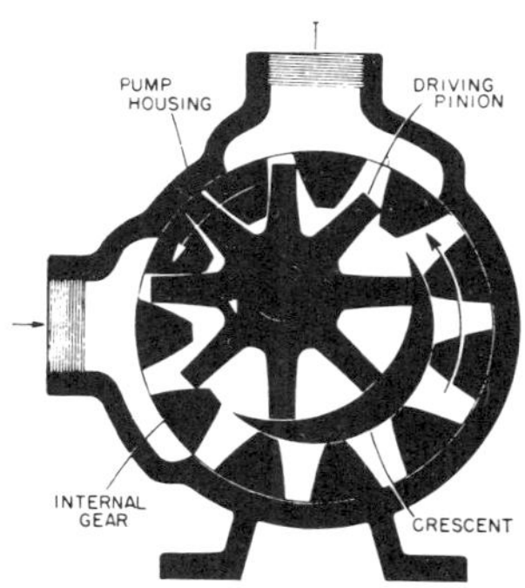

Internal gear motor.

Fig 3

Gear Motors (Fig 3)

The external gear pump can also work directly as a motor. However the necessity of providing a high volumetric efficiency places a premium on both design and construction to minimize internal leakage. Sometimes some mechanical efficiency is sacrificed in order to improve volumetric efficiency, but this is more applicable to a gear pump design than a gear motor design. Ideally, for easy starting up and smooth running, friction should be kept to a minimum with a gear motor and this normally calls for some degree of pressure balancing of the gears.

Conventional methods of pressure balancing applied to gear pumps to minimize internal leakage normally favour uni-directional running and may not always be applicable in the case of a motor suitable for general use. The construction of a high-performance gear motor with fixed side clearances involves a manufacturing cost penalty because of the greater precision required. Low-cost gear motors, therefore,may have certain limitations as regards overall efficiency and the maximum fluid pressures they can accept without generating excessive bearing loads and friction. On the other hand, a well-designed and precisely made gear motor can run at speeds of 5 000 to 10 000 rev/min and be capable of working at pressures up to 210 bar (3 000 lb/in^2) – Table III.

TABLE III – GEAR MOTOR CHARACTERISTICS

Type	Speeds rev/min	Torque	Pressure	Remarks
External	100–3 000	Medium to low (up to 2 500 lb-in)	Medium up to 140 bar (2 000 lb/in^2).	Good type for constant-speed moderate-power drives. Efficiency 90% or better.
External (modified)	5 000–10 000	Moderate	Up to 210 bar (3 000 lb/in^2)	More costly production for specialized applications, efficiency can approach 98%.
Internal	Up to 5 000	Medium	Medium	
Internal (lobe rotor)	25–1 000	High	Medium	Suitable for low-speed drives.

Internal gear motors are usually limited to low powers, but can offer specific advantages in such applications. The internal gear with (typically one tooth less in number) drives round the external gear and shaft, producing a shaft sheed higher than the rotational speed of the two gears. Thus a relatively high shaft speed can be achieved with relatively low gear-to-gear sliding speeds.

Lobe-Rotor Motors

The lobe-rotor pump-motor is a form of gear motor where internal and external gear 'teeth' are specially shaped to provide sealed 'pockets' of fluid in the same manner as a lobe-rotor pump. The pressurized fluid fed to the motor acts directly on the exposed internal gear 'tooth' *via* appropriate porting or a distributor valve. The inner gear is thus caused to rotate relative to the stationary outer gear. Various configurations are possible and the mechanical output can be derived through intermediate gearing, usually arranged to provide a drive reduction. Internal gears of this type are normally used for low-speed motors capable of generating high torques – see Fig 4.

Cam-Rotor Motors

The cam-rotor motor is a development of the vane motor, employing two sliding vanes housed in slots in the main casing and a rotor shaped in the form of a double-lobe cam. The vanes are spring

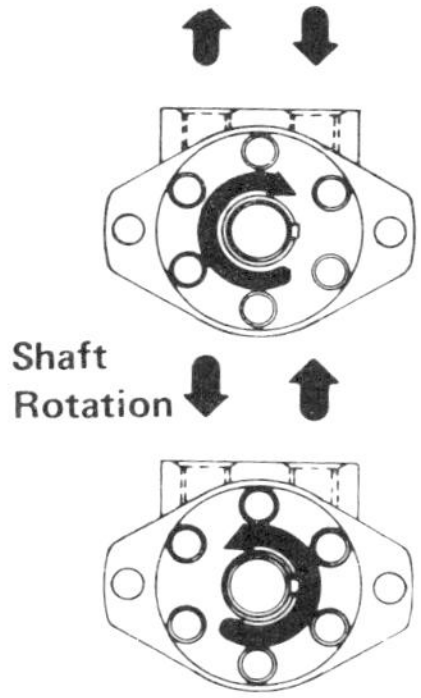

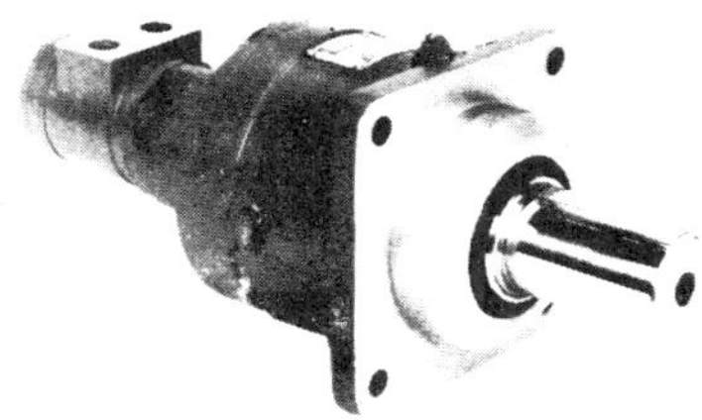

Fig 4 'Gerotor' high torque, low speed hydraulic motor. (Adan).

loaded to remain in contact with the rotor surface all the time — Fig 5. The double-cell configuration provides balance, with inlet and outlet ports diametrically opposed. Working is then similar to that of a conventional sliding-vane motor, but with a pulsed rather than a continuously generated torque. Two (or more) cells may be mounted in-line with the rotors on a common output shaft, each cell displaced angularly with respect to the other. Two such cells mounted at an angle of 90° will provide substantially constant output.

Fig 5 Denison Deri cam rotor motor.

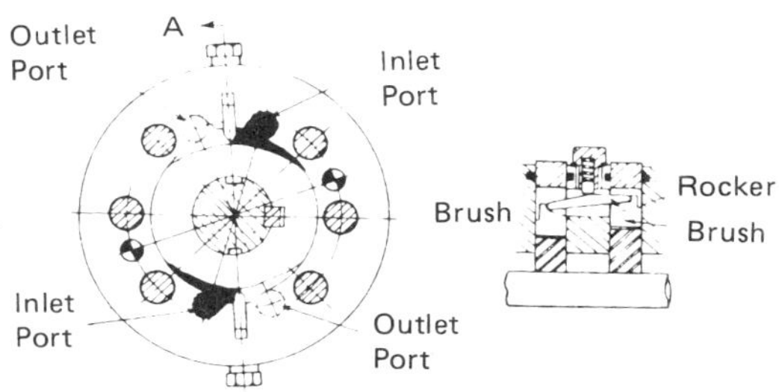

A further variant on the vane principle is the type where the vanes are housed in a stator with diametrically opposed pairs of vanes inter-connected by con-rods. The casing itself then becomes the rotor, the inner side of the casing being contoured to form the vane 'cells'. Such units are true combination pump-motors with low-speed, high-torque characteristics.

See also chapter on *Hydraulic Pumps.*

Hydraulic Couplings

A HYDRAULIC coupling, or *fluid coupling* as it is commonly called, is a three-component device incorporating an *impeller* mounted on an input shaft and a *turbine* mounted on an output shaft, with both enclosed in a suitably shaped *casing.* In most practical designs construction is simplified by attaching the casing itself to the input shaft, with the impeller blades cast-in or incorporated as an integral part of the casing.

The casing is filled with a pre-determined quantity of oil, when the coupling will work as a fluid clutch. When the driven shaft is rotated fluid is thrown outwards under centrifugal force and directed backwards and inwards on to the vanes of the runner, setting up a circular or vortex flow — Fig 1. At the same time the kinetic energy of the fluid pumped by the impeller is transferred to the turbine. Initially, the driver (driving the impeller) is virtually unloaded and rapidly increases in speed, with the torque increasing as the square of the speed. When the input torque is sufficient to overcome the load on the driven member the turbine accelerates and continues to do so until reaching a stable full-load speed with input and output torque always equal.

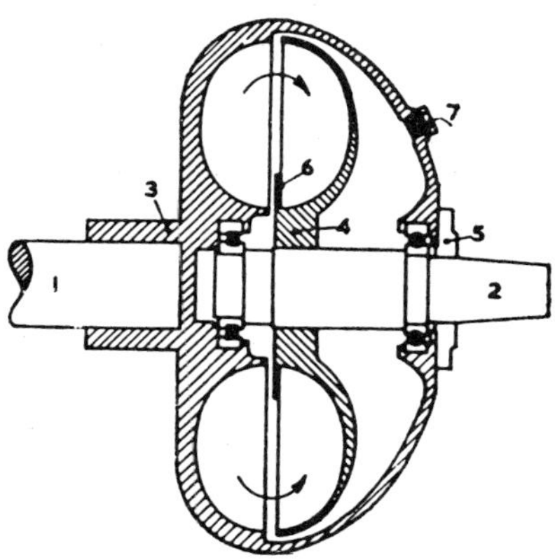

1. *Input shaft.*
2. *Output shaft.*
3. *Casing and impeller.*
4. *Turbine or runner.*
5. *Seal.*
6. *Baffle ring.*
7. *Filler plug.*

Fig 1 Fluid coupling.

The possible speed ratio between input and output can thus vary from zero (stalled) to 1.0 (no load). Over the same speed range slip varies from 100% (stalled) to zero respectively — see Fig 2. Slippage is an inevitable feature of the coupling operating under practical load conditions, with maximum efficiency usually realized with slip values of the order of 2.5% to 5%. A fluid coupling is, therefore, normally proportioned to operate at this corresponding speed under full design load conditions. Power loss at this design operating point is reduced to a minimum, and also the difference between input and output speeds.

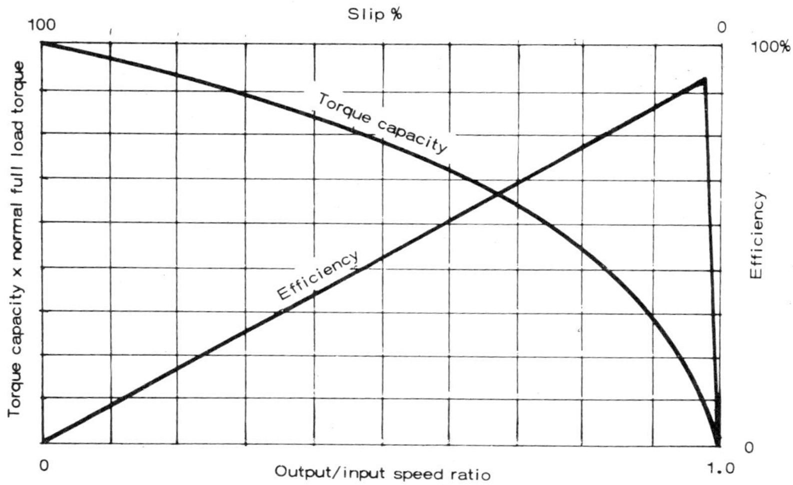

Fig 2

The several major advantages of this type of coupling are its ability to rotate in either direction with equal efficiency, to accommodate a reverse drive direction (that is, the turbine effectively driving the impeller), and the smoothing characteristics of the fluid link in absorbing shock, promoting smooth acceleration, *etc.* It can also continue to run stalled without harm, except that the input power is dissipated in the form of heat, increasing the temperature of the fluid. In simple fluid couplings of the constant-filling type, protection against over-heating of the oil can be given by a fusible plug screwed into the casing.

The transmission characteristics are related to both the quality and quantity of the fluid (almost invariably oil) in the coupling. It is therefore most important that the correct grade of thin mineral oil, as recommended by the manufacturers, is used.

The quantity of oil governs the coupling effect and overfilling or loss of fluid will detract from performance — *ie* any change in oil volume will alter the working characteristics of the coup ing. Design and performance of a typical *constant-filling fluid coupling* (traction coupling) is illustrated in Fig 3.

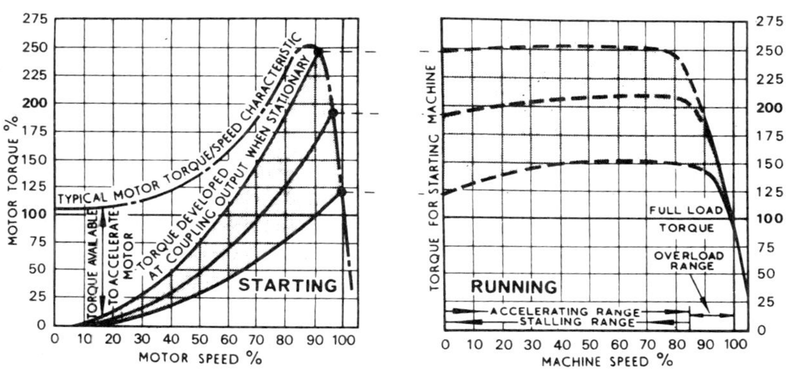

Fig 3

Traction type fluid coupling with constant filling.

The quantity of oil circulating for any period is identical for both the impeller and turbine, and speed changes in the direction of prime-mover rotation are identical to speed changes in the turbine. Therefore, torque conditions on both members are identical and there can be no torque conversion, that is, there can be no increase in torque on the turbine over and above that which the impeller receives from the prime mover.

The *scoop-control* coupling provides rather greater flexibility of operation at the expense of small mechanical complications. It is similar to a conventional fluid coupling except that the volume of fluid circulating can be changed whilst the coupling is running. This enables the torque capacity of the coupling to be varied so that the speed of the output shaft can be adjusted over a wide range. In effect it works as a form of infinitely variable speed 'gearbox' with fluid clutch characteristics.

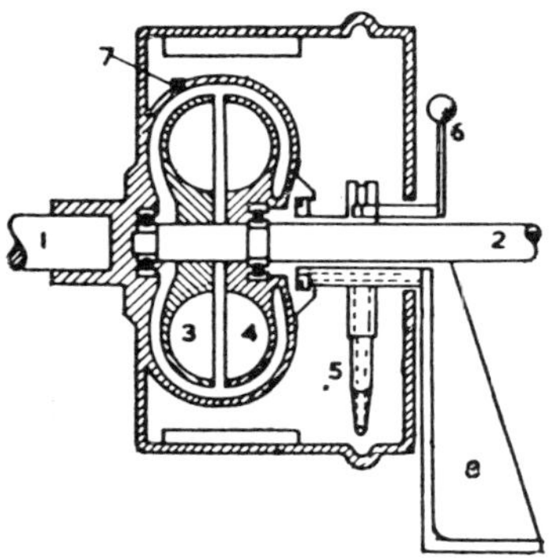

1. Input shaft.
2. Output shaft.
3. Impeller.
4. Turbine or runner.
5. Scoop.
6. Scoop depth control.
7. Leak off.
8. Scoop support.

Fig 4 Scoop control fluid coupling.

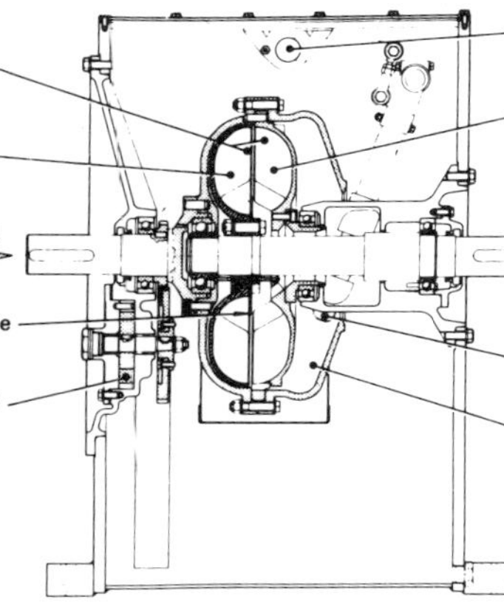

Fig 5 Variable speed scoop trimming fluid coupling.

A typical arrangement of a scoop-control coupling is shown in Fig 4. Impeller and turbine are completely surrounded by an outer casing, with a number of leak-off nozzles fitted to allow oil to flow from the inner to the outer casing under centrifugal force. The outer casing is capable of containing the whole charge of oil in an annular belt. A scoop tube dips into this belt and ducts oil back into the inner casing through a suitable axial channel. Provision is made to raise or lower the end of the scoop tube so that the quantity of oil circulating in the coupling itself can be varied – increasing the coupling volume to reduce slip, and *vice versa,* and thus providing variable speed characteristics. The fact that changing the volume of the operating oil will also affect the temperature of the oil (and thus its viscosity) is not normally significant since oil is constantly being bled off and circulated and thus experiencing a certain amount of cooling. If necessary, an oil cooler can be incorporated in the circuit to prevent over-heating.

A variation on the scoop-control coupling is the *scoop-trimming coupling,* wherein the major part of the oil is contained in a stationary casing below the coupling and a built-in pump is provided to circulate the oil from the tank through the cooler into the fluid coupling circuit – Fig 5. The purpose of the scoop tube is to adjust or 'trim' the oil level in the working circuit to the required value while the coupling is running. Such an arrangement is of particular value for use with high-speed electric motors.

A possible variation is a design of coupling which eliminates end thrust on the input and output shafts. The usual method of achieving this is to employ two mechanically connected turbines, back-to-back, each facing its respective impeller. The whole is then contained in a single casing (for example, carrying the impeller blades). The axial thrust generated by one half of the paired coupling is then exactly counter-balanced by that generated in the other.

See also chapter on *Torque Converters and Hydraulic Transmissions.*

Torque Converters and Hydraulic Transmissions

THE HYDRAULIC torque converter is similar in form to the hydraulic coupling but incorporates an additional reaction stage. This is produced by interposing one or more sets of stationary guide vanes between the impeller stage and the turbine stage – Fig 1. The effect of these vanes is to change the direction of the fluid flowing from impeller to turbine so that the resultant torque generated in the turbine is greater than that on the impeller, thus producing *torque multiplication.* With this mode of working the turbine is still dependent on the change in fluid speed but the impeller torque now depends on speed *difference.* The more the torque-converter output shaft is slowed down by the load the greater the amount of fluid which is forced through the reaction member to drive the impeller and maintain impeller speed. The resulting performance characteristics are shown in Fig 2. With the type of flow produced the blading shape becomes much more significant than in a fluid coupling and allows the blades to be shaped in different ways to produce a variety of operating characteristics. The design and construction of torque converter blading is, therefore, much more critical.

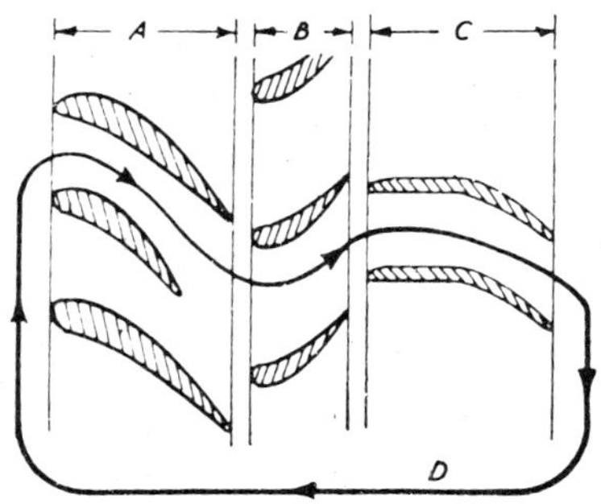

Fig 1 Elements of a torque-converter. A–impeller stage. B–reaction stage. C–turbine stage. D–fluid circulation.

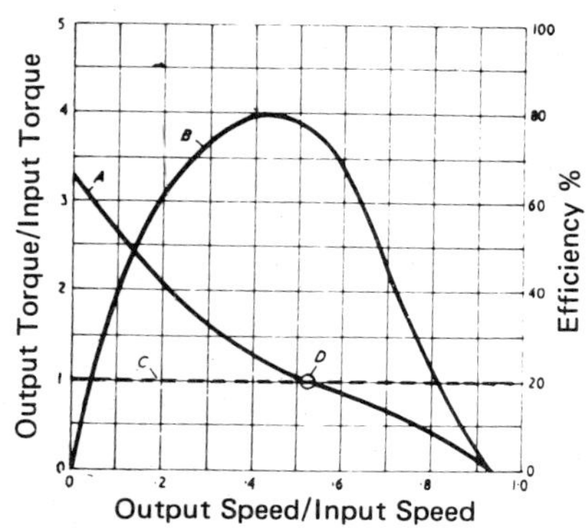

Fig 2 Typical characteristic curves of single-stage torque-converter. A–torque. B–efficiency. C–constant input torque. D–clutch point.

Any particular design will have a specific 'design point' related to the out/input speed ratio, at which maximum efficiency is obtained. Another important point on the characteristic curve for output torque is the *clutch point* or speed ratio at which the out/input torque ratio is unity. Operating above the clutch point (that is at higher speed ratio) both the torque ratio and efficiency fall rapidly. Under such conditions the performance of a torque converter is inferior to that of a fluid coupling. At any lower speed ratio (below the clutch point) torque multiplication is available. Maximum torque conversion occurs when the turbine is stationary (speed ratio zero), this value being referred to as the 'stall torque ratio'. For simple three-element torque converters the stall torque ratio obtainable is usually of the order of 3.0 to 3.5, and seldom exceeds 4.0 except in special designs. With multi-stage torque converters greater degrees of torque conversion may be obtained — for example, 6:1 with a three-stage unit. The clutch point in typical multi-stage torque converters usually occurs at much lower speed ratios however — Fig 3.

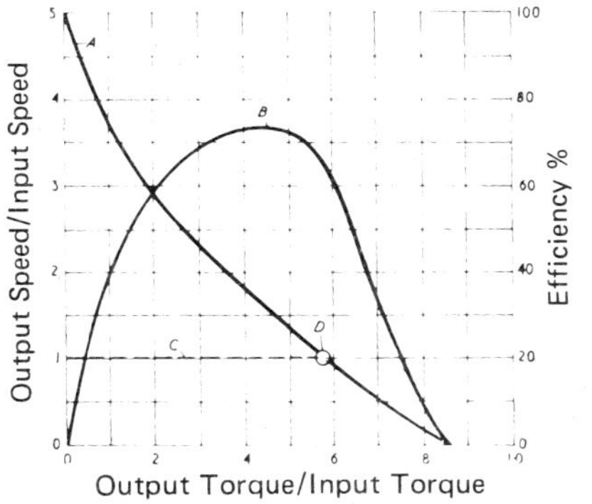

Fig 3 Typical characteristic curves for multi-stage torque-converter. A—torque. B—efficiency. C—constant input torque. D—clutch point.

Converter Couplings

The converter coupling combines all the advantages of a torque converter at speed ratios up to the clutch point, and at higher speeds continues to operate as a straightforward fluid coupling with equal input and output torques. Thus the 'racing' point is eliminated by maintaining a 1:1 torque ratio above the clutch point, whilst the efficiency also increases from the clutch point — Fig 4.

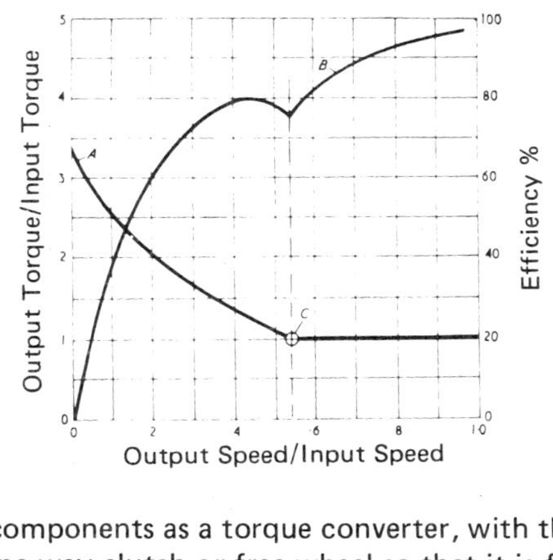

Fig 4 Typical characteristic curves for converter-coupling. A—torque. B—efficiency. C—coupling point.

The converter coupling utilizes the same components as a torque converter, with the exception that the reaction member is mounted on a one-way clutch or free-wheel so that it is free to rotate in the same direction as the impeller and turbine but locked to the frame against the opposite direction of rotation. Thus up to the clutch point the reaction member is locked in a stationary position and the device operates as a torque converter. Above the clutch point the reaction member free-wheels with the impeller and turbine and the whole unit acts as a straightforward fluid coupling.

Specific advantages associated with a torque transmission are:

(i) Automatic provision of high starting torque and tractive effort.

(ii) Exceptionally smooth take-up of load, with marked reduction in stack loading of following components (*eg* half shafts in the case of traction drives).

(iii) More efficient prime mover making sure (in the case of traction drives) the engine can be run at a substantially constant speed despite wide variations in translational speed (road speed).

(iv) The virtual impossibility of stalling the prime mover.

(v) Improved engine life by the cushioning effect between engine and load.

(vi) Simplified maintenance through elimination of a mechanical clutch and gearbox.

A comparison between the performance of a torque converter and a three-speed manual gearbox in a typical lift truck application is given in Fig 5. In the case of larger and heavier vehicles, however, engine power is generally limited by economic considerations and it usually becomes necessary to combine a torque converter with a multi-ratio gearbox.

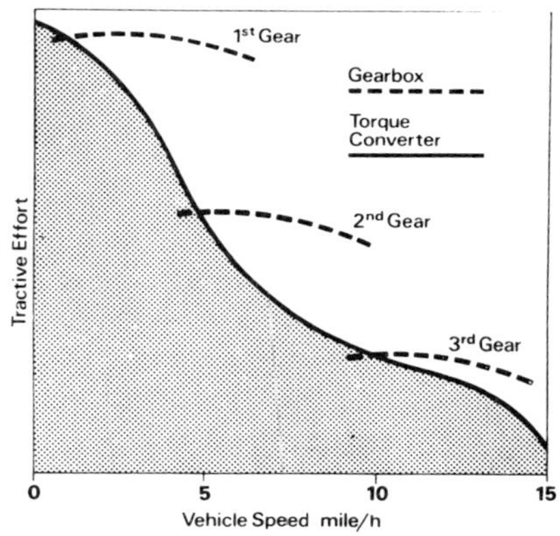

Fig 5 Graph comparing the typical performance curves of a three-speed manual gearbox and a torque-converter with single forward and reverse gears.

Such a combination may be employed in various ways, for example to modify the torque ratio available or the efficiency, or to provide reverse and neutral gear positions as well as forward speeds. With *series connection* mechanical gear ratios are inserted in the transmission before or after the converter in order to provide further torque multiplication as required. With *shunt connection* the input torque is split between the mechanical gearbox and the hydraulic converter. Series connection provides increased torque multiplication with some loss of efficiency, and shunt connection a loss of torque ratio with an increase in efficiency. Series-shunt connection can provide increased torque multiplication with an increase in efficiency. Epicyclic gearing is invaiably employed for shunt or series-shunt connection.

Hydro-Mechanical Drives

The limitation of an ordinary mechanical speed-change gearbox, where high powers are to be transmitted, is its inflexibility and an attractive solution is the provision of a hydraulic element in the form of a hydraulic coupling associated with some form of constant mesh gearbox. The only other suitable form of transmission for high powers is electric. Both involve heat losses and hence may appear less efficient than geared drives. By comparison, the majority of diesel-mechanical systems are limited to below 500 bhp.

Within the limitations of generalization, relative efficiencies for different forms of diesel power transmission are illustrated in Fig 6. The characteristic turbo converter efficiency curve is self-evident, and also the effect of changing to direct drive at the clutch point. Curves for similar transmission systems plotted as tractive effort against speed are shown in Fig 7.

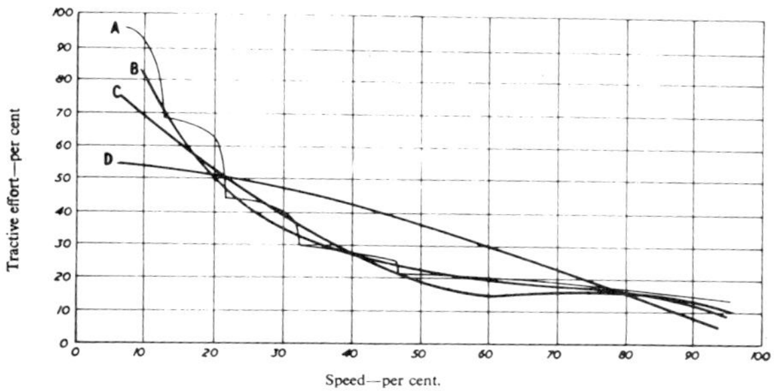

Fig 6 Relative efficiencies. A—5-speed mechanical gearbox drive. B—diesel-electric. C—partial-hydraulic drive through turbo-converters. D—partial-hydraulic-drive with direct drive in top.

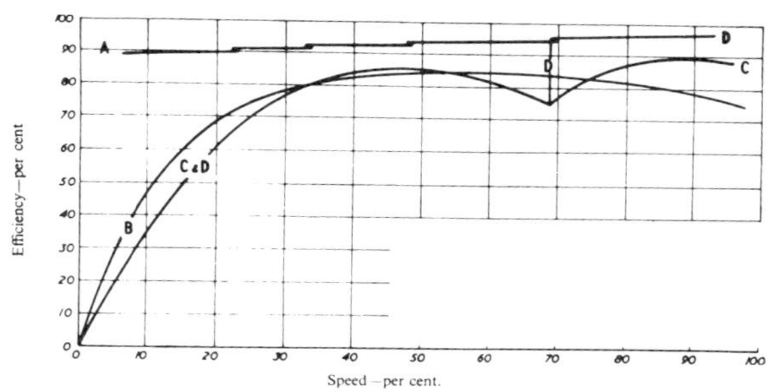

Fig 7 Relative tractive effort. A—5-speed mechanical drive. B—diesel-electric. C—partial hydraulic (turbo-converter). D—direct drive.

Hydraulic drives for locomotives are based on turbo converters plus a hydraulic coupling for direct drive or the Lysholm-Smith design with a torque converter and a friction clutch. Mechanical-hydraulic drives are exemplified by the traction type fluid coupling with Wilson gearbox and the scoop-control coupling with SSS Powerflow gearbox.

The purpose of the hydraulic coupling used in hydro-mechanical drives between the engine and gearbox is to provide a smooth take-up of drive — that is, to introduce flexibility into an otherwise inflexible system.

The Lysholm-Smith system employs a torque converter and a direct drive, the former used for accelerating up to speeds of 48 kph (30 mph), when the drive closes the throttle and reverses a clutch to unload the converter and bring into engagement a direct drive. This arrangement permits a reasonably simple control to be used for multiple unit working if required.

Hydrostatic Drives

IN A hydrostatic drive circulated fluid acts as a transmitter of pressure energy between a primary unit (a pump) and a secondary unit (a motor). Pressure energy is converted into torque and rotary motion by the motor. Practical power transmission is achieved by trapping, propelling, and finally utilizing the pressure energy of the fluid by hydrostatic reaction, with no significant change in the velocity of the fluid (hence its 'passive' character). This is in direct contrast to a hydrokinetic transmission (*ie via* a hydraulic coupling or torque converter) where it is the change in velocity of the hydraulic fluid in the circuit which is responsible for the transmission of energy. A further general distinction between the two is that, for a given power, the energy is transmitted by a small flow of fluid at high pressure in the case of hydrostatic transmission, and by large fluid flows at comparatively low pressure in the case of hydrokinetic systems.

Both pump and motor are displacement machines, the hydraulic power developed being a function of the flow rate and pressure drop. The overall efficiency of such a machine is expressed as the ratio of the mechanical output power produced to the input power required to drive the pump (usually multiplied by 100 and quoted as a percentage).

Specific advantages offered by hydrostatic drives are:

(i) Stepless speed regulation, making them particularly suitable for vehicle transmission.

(ii) Continuous transmission between motor and drive wheel or driven device regardless of speed variations.

(iii) Simple single-lever operation of speed, forward and reverse control.

(iv) Dynamic braking through reduction of pump delivery.

(v) Simply set limits to acceleration or deceleration.

(vi) Simple and reliable method of protection *via* pressure-relief valve(s).

(vii) Freedom to operate pump driver within its optimum speed and power range.

(viii) Considerable flexibility in the positioning of pump(s) and motor(s), adaptable to a wide variety of power-drive requirements.

Open-Circuit Hydrostatic Drive

With an open-circuit hydrostatic drive fluid is drawn from a reservoir by the pump and then pumped to the motor *via* a control system which regulates the motor speed and direction. Return flow from the motor is to the reservoir, Fig 1. The pump in such a system is always uni-directional (*ie* has specific inlet and outlet ports) and would normally be of the gear or vane type. It can be adapted to multi-rotor working, when the rotors are selected to run simultaneously or individually as required. Characteristics of an open-circuit transmission are that the direction of rotation of the

TRANSMISSION ACCOMPLISHED
Picture transmissions from outer space can be pretty spectacular. And there's something out of this world about Abex Denison transmissions too. Hydrostatic transmissions – so powerful for their compact size and weight, those space restriction problems soon diminish. Ideal for mobile and industrial applications. They may never orbit Saturn, but they make rings around the competition. For a complete briefing contact Transmission Control at Burgess Hill now.
INFINITELY VARIABLE~ INFINITELY SUPERIOR
HYDROSTATIC TRANSMISSION PUMPS & MOTORS
Abex DENISON
ABEX DENISON LTD
VICTORIA GARDENS,
BURGESS HILL, SUSSEX RH15 9ND
TEL: BURGESS HILL (04446) 5121 TELEX: 87168

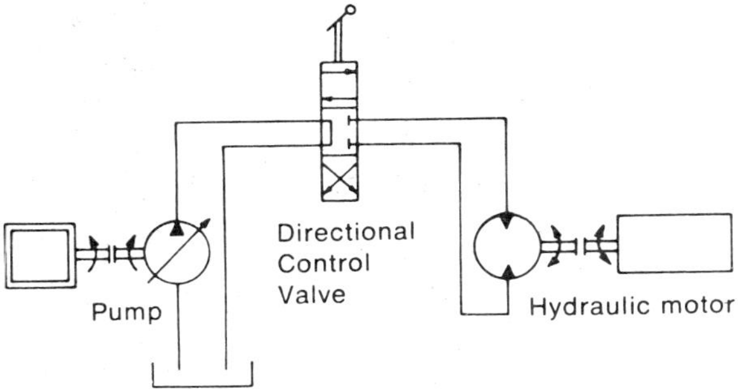

Fig 1a Open circuit hydrostatic transmission.

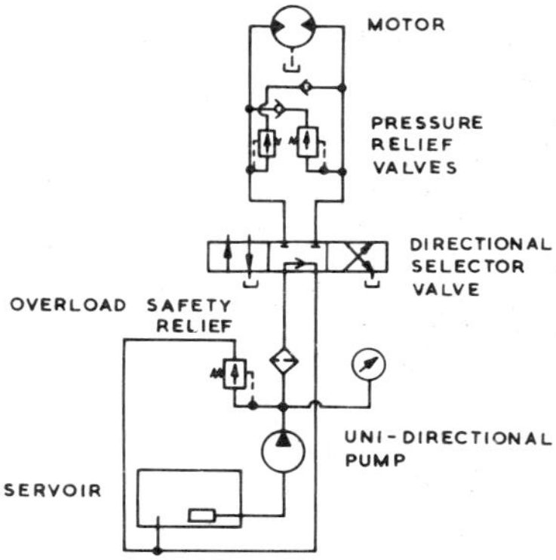

Fig 1b Basic open circuit.

motor can be changed by interposing a directional control valve between it and the pump, but braking is then only possible by restricting the return flow. Cooling is not normally necessary as adequate heat exchange is usually provided by a reservoir of suitable size.

The main limitation of an open-circuit transmission is that if the drive inertia is large, changes in speed at the pump end may cause over-running and thus cavitation in the pressure line with consequent noise and drive instabilities. If, however, the motor is driving something with a low inertia, say, a fan, then it is more likely this simple form of drive will be satisfactory. It is also possible, by connecting the pressure line between the pump and the motor through a non-return valve, to allow the motor with its drive inertia to freewheel either to rest or down to that speed indicated by the pump flow. Care must be taken to prevent surging in the drive, a possibility to which this form of transmission is prone.

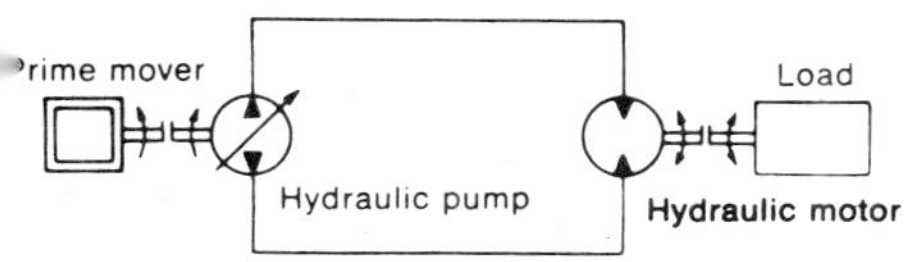

Fig 2a Closed circuit hydrostatic transmission.

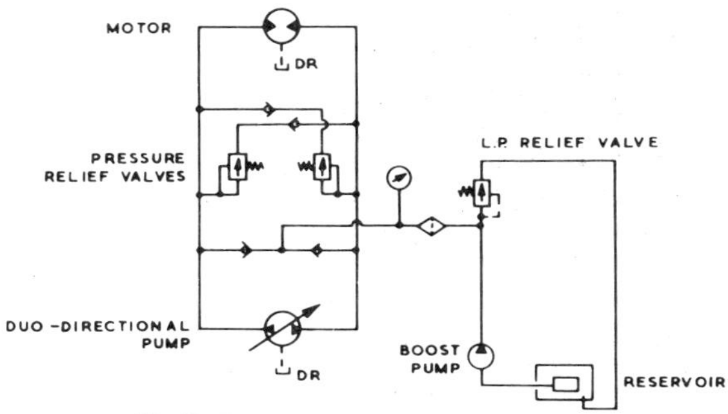

Fig 2b Basic closed loop circuit.

losed-Circuit Hydrostatic Drive

Vith a closed-circuit hydrostatic drive one pump is normally associated with one motor, the two eing directly connected in a closed hydraulic loop (Fig 2). The motor speed and direction of otation are controlled by varying the delivery and direction of the pump output, (calling for a ariable output); or by using a duo-directional pump (normally an axial-piston type). It is possible o operate a multi-motor system on this basis, although this will impose certain performance mitations. Thus with 'N' motors in parallel, and assuming an equal load on each, each motor peed should be proportional to (pump output — total system slip) divided by N. With series onnection of the motors, motor speeds would be proportional to pump output but there would be a successive loss of speed through each motor in the chain due to slip in the previous motors. Also the torques taken out of each motor would impose additional pressure load on the pump.

With a closed circuit operation is symmetrical. Thus high and low pressure rates are interchangeable and any speed between zero and maximum can be selected in both directions of rotation. It also follows that since the primary and secondary units can operate either as a pump or motor, dynamic braking is available. Thus the specific advantages of a closed-circuit transmission over an open-circuit transmission are that it has the features of reversibility and braking without the necessity of throttling (with subsequent generation of heat) – (Fig 3). It does, however, require additional valves and is thus more costly to construct.

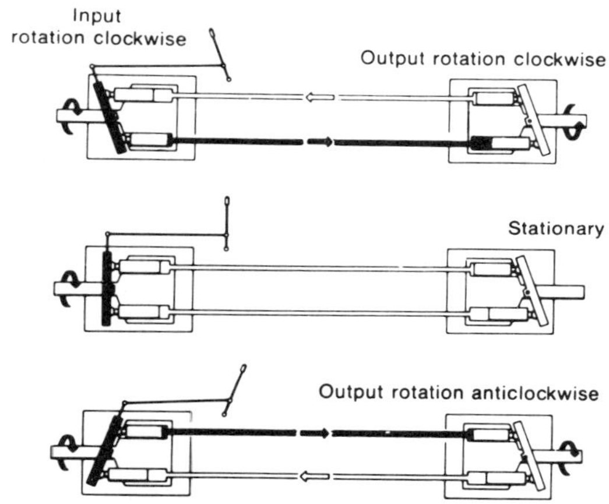

Fig 3 Reversibility and dynamic braking are features available in a closed circuit hydrostatic transmission.

Hydrostatic Transmissions

Closed-circuit systems are normally used for hydrostatic transmissions, the circuit being associated with a boost pump to maintain pressure in the system by compensating for slip in the pump and motor. The boost pump is commonly built into the main pump. It is also usually necessary to provide a heat exchanger to cool the oil returned, either from the leakage connections or spilled off from the circuit, depending on the power distribution in the hydrostatic transmission. If the circuit is generously designed then the temperature of the oil in circulation will rise slowly and may in some circuits effect a balance without the necessity for bleeding off and cooling separately. In this case most of the energy lost due to inefficiency will go into the oil which has leaked through the various clearances in the pump. Sometimes it is not possible to introduce generous dimensions in the hydrostatic circuit, in which case the friction of the oil circulating through the system is sufficient to raise the temperature to a point which, because of the effects of viscosity, might cause a falling off in performance. In this case a spill-off from the circuit, either temperature-controlled or through an orifice controlled by pressure, would be necessary. By suitable valving the full precharge flow can be passed through the transmission circulating lines. This usually gives ample circuit cooling.

Characteristics of Closed-Loop Hydrostatic Transmissions

Hydrostatic pumps and motors can be coupled together to form a transmission in the following basic combinations:

(a) A constant-capacity pump and a constant-capacity motor;

(b) A constant-capacity pump and a variable-capacity motor;

(c) A variable-capacity pump and a constant-capacity motor; and
(d) A variable-capacity pump and a variable-capacity motor.

(a) Fig 4 — a constant-capacity pump and constant-capacity motor; this combination provides fixed ratio drive. If the two units have equal displacements the ratio is unity and the combination analogous to a very flexible drive shaft. Unlike a drive shaft, however, there is a slight loss of ɔeed with increasing load.

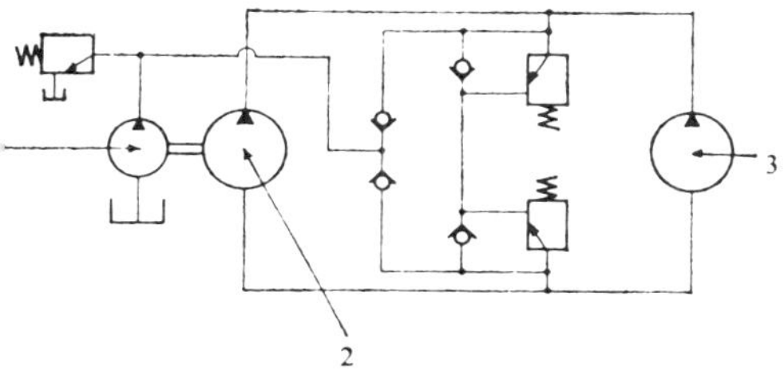

Fig 4
1 & 2 — constant capacity pump.
3 — constant capacity motor.

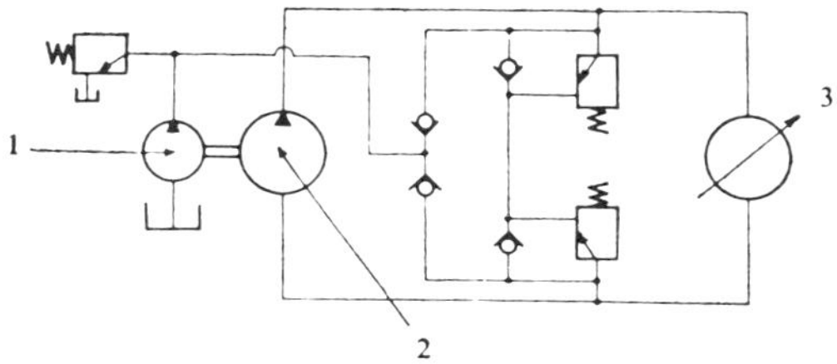

Fig 5
1 & 2 — constant capacity pump.
3 — variable capacity motor.

When the two constant-capacity units have different displacements the gear ratio is propor-ɔnal to the displacement of the two units.

(b) Fig 5 — constant-capacity pump and variable-capacity motor; this drive is similar to item (a) any given motor setting. It is usually designed with a larger capacity motor than pump, which eans that at maximum displacement the motor runs at a slower speed. Reduction in displace-ent of the motor increases the output speed but decreases the torque transmitted. The result is proximately constant horsepower over the designed speed range.

(c) Fig 6 — variable-capacity pump and constant-capacity motor; at zero displacement the ımp acts as a clutch. A stepless increase in the displacement of the pump causes the motor to tate, approximating constant torque up to the maximum designed speed. Reversal in the direc-n of rotation of the motor can be obtained by an 'over-centre' reversal of the pump displace-ent control. This ability to give variable speed, constant torque, clutching and reverse makes the nsmission suitable for a wide range of drives.

(d) Fig 7 — variable-capacity pump and variable-capacity motor; this transmission arrangement mbines the constant horsepower and the constant torque characteristics of items (a) and (b)

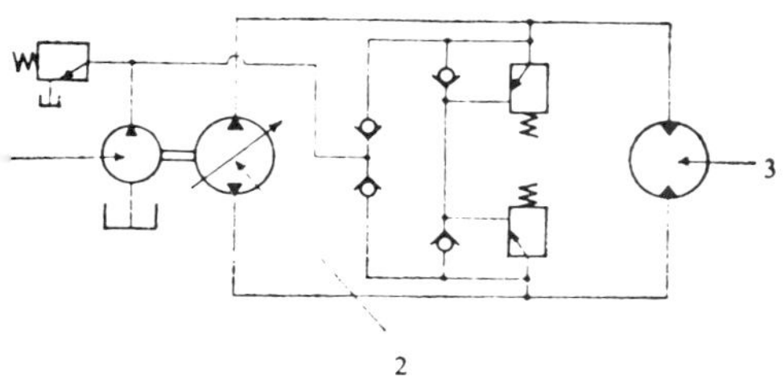

Fig 6
1 — constant capacity pump.
2 — variable capacity pump.
3 — constant capacity motor.

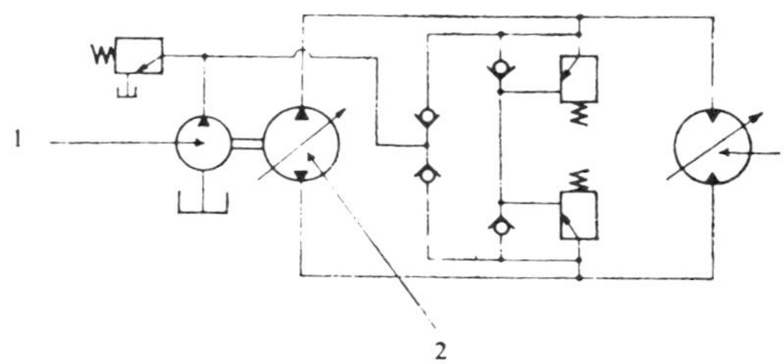

Fig 7
1 — constant capacity pump.
2 — variable capacity pump.
3 — variable capacity motor.

with the clutching and reversing characteristics of item (c). It is especially suitable for drives requiring a wide range of speed and torques.

The torque/speed range requirement, particularly in vehicle applications, largely determines whether a fixed pump/variable motor, variable pump/fixed motor or variable pump/variable motor provides the most economical system. For example, for a torque speed range of 2:1 the variable displacement motors give a substantial reduction in pump size, whilst for a range of 3:1 variable displacement motors have little advantage to offer either in maintaining good efficiency or in giving substantial savings in pump size.

Pumps used in hydrostatic transmissions are equipped with a variety of controls. These may provide either direct or remote mechanical control of the pump output and, in addition, servo-mechanisms to control pump output automatically in accordance with predetermined parameters. A manufacturer's catalogue will normally list any or all of the following options:

(1) Manual control of output — by mechanism directly attached to the pump casing.

(2) Pilot-operated control of output — this would allow the rate of flow from the pump to be remotely controlled by a directional control valve.

(3) Constant horsepower control — this method regulates the pressure flow-rate relationship of the pump so that any power may be obtained from the motor up to the maximum horsepower setting, *ie* not exceeding the horsepower output of the prime mover less power losses in the transmission.

(4) Constant pressure control — this would permit any desired pressure to be pre-set, the pump then automatically adjusting its output flow-rate to maintain this pressure.

In addition, combinations of these controls are possible so that, for example, a constant-horsepower control can be obtained with a maximum output limit, or a constant-horsepower characteristic with over-riding manual control.

Control systems adopted may be direct, hydraulic servo or electro-hydraulic servo. Direct controls may be manual, hydraulic or electric in operation, but would not utilize a follow-up system. Hydraulic servo-systems may be manual, hydraulic, or electrically operated and are usually dependent on a follow-up servo-valve built into the pump. Electro-hydraulic controls represent a far more sophisticated system with high response, making them most suitable for fully automated drives (*eg* machine tool drives).

High-Speed and Low-Speed Motors

In many cases, the final drive requirements for speed and torque will determine the type of motor required. In applications calling for high torques at slow speeds, however, there is often a choice between a slow-speed high-torque motor and a high-speed motor driving through a reduction gear. Low-cost gear boxes sometimes render the latter financially attractive.

Generally, however, it is more economical to employ slow-speed high-torque piston motors and confine high-speed motors to high-speed applications. The main reasons are higher overall efficiency, more effective break-away torque and less space taken up.

For extremely low rotation speeds, a slow-speed motor and gearbox unit is often used.

The range of hydraulic pumps and motors provided by the hydraulic manufacturing companies today gives industry the ability to meet almost any demand for rotary power transmission. These include split systems capable of providing both high pressure for a vehicle hydraulic system and the hydrostatic transmission.

SECTION 5

Pressure Gauges

THE MOST common types of pressure gauges used on hydraulic systems are the Bourdon gauge and the spring-loaded plunger type gauge. Portable electronic pressure gauges are now becoming more readily available, replacing the Bourdon gauge for 'spot test' measurements (*ie* plugged into various points in turn to obtain pressure readings), and may also be employed to give continuous monitoring of pressure in critical circuits. These are normally of resistance bridge type, employing a pressure transducer. Pressure may be indicated by a meter or digital read-out *via* light-emitting diodes. The accuracy of such gauges is determined by the accuracy of the transducer and is generally comparable to, or slightly better than, a Bourdon pressure gauge. The pressure range covered by such instruments is from 0–700 bar (0–10 000 lb/in^2); but even higher pressure can be accommodated in some designs.

Alternative instruments with a higher accuracy, or higher range, than that offered by a Bourdon gauge are the precision Bourdon gauge or Budenberg gauge, but both are rather delicate. They would normally only be used as standard or test gauges, although their future in this respect is undoubtedly rivalled by the electronic pressure gauge.

A typical Bourdon gauge is shown in Fig 1. The accuracy obtained is reasonably good, although inclined to drift with use. British Standard BS 1780 specifies that the accuracy of such a gauge when new should be within plus or minus 1% of the maximum graduation. In practice it is safer to assume that the likely accuracy of a Bourdon gauge is of the order of plus or minus 2½% of full scale.

The gauge used should have a suitable full-scale deflection consistent with the pressure which is to be measured, *ie* the normal readings taken should lie at the upper end of the scale. If not, the

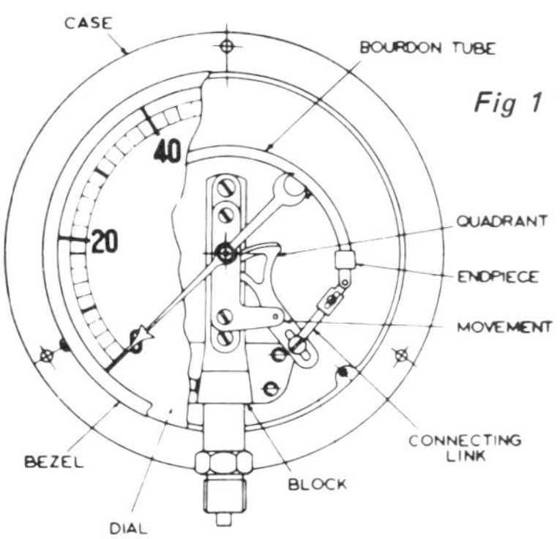

Fig 1 Typical form of Bourdon pressure gauge.

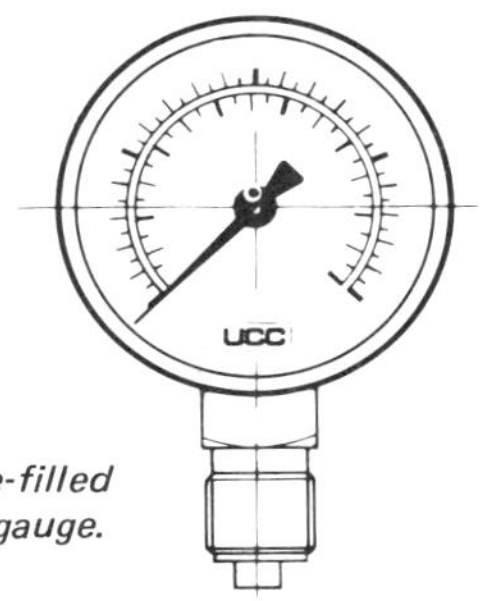

Glycerine-filled pressure gauge.

possible error will be proportionately greater. Also, gauges should be of generous size. A minimum dial diameter of 50 mm (2 in) is usually recommended, with a preference for 75 mm (3 in) or even 100 mm (4 in) diameter.

The accuracy of the gauge will also depend on how it is installed, the conditions to which it is subjected in service, and, particularly, how well it is protected against pressure surges and mechanical shock. The latter can seriously 'shift' the calibration.

For continuous 'in circuit' use the gauge should ideally be mounted on a rigid surface isolated from mechanical vibration produced by the pump, valves or machine, which could affect the delicate mechanical movement. Mounting on a separate instrument panel is a logical solution, but there is no basic objection to mounting the gauge on the machine, or even on a rigid pipe, provided the mounting point is free from vibration. Complete isolation, however, is preferable.

Normally Bourdon gauges are mounted with 'snubbers' or 'gauge savers' in the gauge connection line. These are essentially throttling devices which restrict flow and thus dampen pressure surges under conditions of fluctuating pressure. They may be pre-set or adjustable by means of a needle valve or fixed orifice.

For complete protection against pressure surges a pressure gauge can be isolated from the circuit *via* an isolating valve which holds the gauge at zero pressure. The valve is then operated manually to bring the gauge into circuit when a reading is required. On release, the valve returns to its isolating position.

Other types of valve are also produced where the gauge is normally left continuously in circuit but automatically isolated by the valve on the appearance of surge pressure peaks.

An alternative approach is to fit self-sealing couplings at points in the system where pressure readings are required. A single gauge can then serve a complete system, being itself connected to a matching coupling and simply plugged in at various points in turn to obtain pressure readings. No protection is provided should a pressure surge occur when the gauge is actually plugged in, but if necessary, a 'snubber' can be incorporated in the gauge line.

The spring-loaded piston gauge is more rugged and largely shock-proof — *ie* it will withstand pressure surges and shock loads. Also such types can readily be designed for pressure up to 170 bar (12 000 lb/in^2) and can be used anywhere for continuous measurement of pressure. The accuracy of this type of pressure gauge is not as good as that of a Bourdon gauge, however, nor is the ultimate pressure range so high.

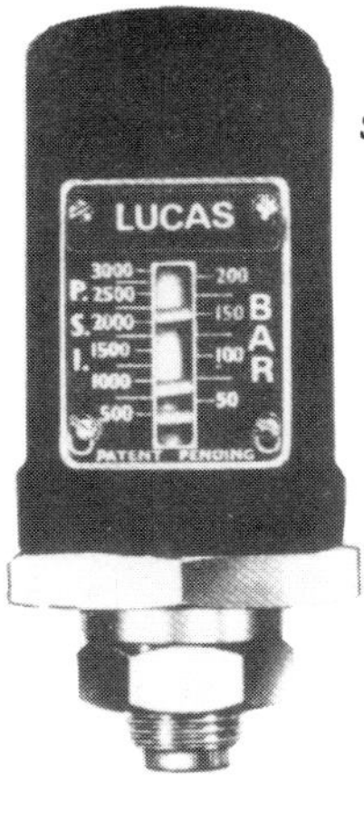

Shockproof spring-loaded plunger type pressure gauge. (Lucas Fluid Power).

Pressure gauge for brake testing. (Foundrymeters Ltd).

Uses of Pressure Gauges

Pressure gauges may be fitted purely as a nominal check that the system pressure is of the required order, relying on the gauges to indicate the presence of some faults should the readings fall unduly. In a properly designed system, however, pressure gauges should be incorporated (or at least provision made to plug them in) at all points where pressure measurement can be of real value for system tuning or fault finding.

Pressure monitoring is particularly useful in the case of complex circuits. In this case it will usually be found that the most important 'pressure points' are in the pilot system, as, if a main valve fails to operate, they will immediately show the possible reason.

Pressure tappings can be taken from the line or passage between the pilot and main valve or the main valve may have a special port which, when the valve operates, opens to the line leading to the pressure gauge.

To save having a separate pressure gauge for each point a multi-way valve can be used which connects them in turn to a single gauge.

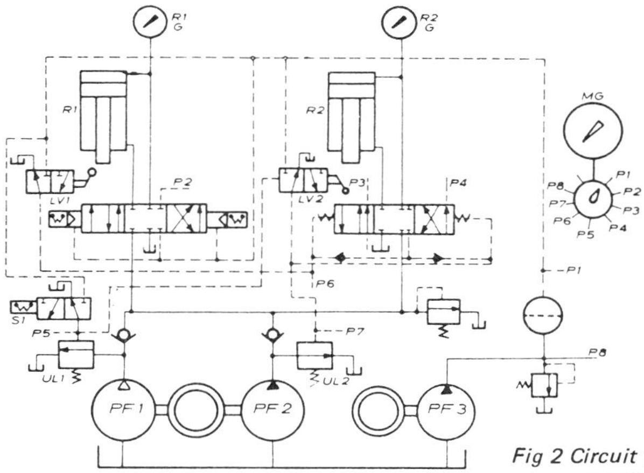

Fig 2 Circuit to show location of monitoring points.

A simple circuit illustrating this is shown in Fig 2. In the circuit a solenoid pilot valve controls a ram R1 and a pilot valve ram R2. R2 is operated first and pump PF1 is loaded by the solenoid valve S1 and unloading valve UL1.

The limit valve LV1 applies pilot pressure to the valve for ram R2. At the bottom of the stroke of R2 limit valve LV2 reverses R2 and at the same time closes unloading valve UL2. Finally LV2 is reset by pressure from S1. The remaining controls are electrical.

Pressure tappings P1, P2, *etc,* are taken from every point in the pilot circuit where pressure change occurs when a valve operates. For pilot valves a tapping is taken immediately after the valve, but for the main valves it is necessary to use a separate port. The pressure at the rams is not covered by this scheme which is purely concerned with valve operation. If the ram pressure is important it must be registered by separate gauges for each ram.

For rapid pressure monitoring a multi-way valve which can be rotated to connect the pressure gauge to each point, P1, P2, *etc,* in turn is very convenient.

Besides monitoring the valves it is also possible to check the state of the filter by taking readings first on the inlet and then on the outlet sides through tappings P1 and P2. With a pilot pressure of 28 bar (400 lb/in^2) and a pressure gauge graduated to 42 bar (600 lb/in^2), a pressure drop of about 0.2 bar (3 lb/in^2) can easily be detected on a 152.4 mm (6 in) or larger dial gauge.

Calibration and Checking

To ensure continuing accuracy pressure gauges should have their calibration checked periodically. The general recommendation is an annual check. Gauges should be checked monthly, or at least every two months, if readings are to be relied on. This applies particularly to gauges which are used for circuit 'tuning', whether or not they are in continuous use.

The simplest method of checking is to compare readings with those given by a more accurate test gauge (preferably a Budenberg type), although where a number of gauges are involved a dead weight tester is a good investment for the instrument or test shop. This is a standard form of pressure gauge test apparatus comprising a screwed ram and a piston weight platform assembly, as shown in diagrammatic form in Fig 3. The instrument is filled with oil, which is then pressurized to the required level by placing weights on the platform. The screwed ram is rotated until sufficient additional pressure is applied to lift the weights, and the weights then spun to ensure freedom of error from friction. With the weights spinning the pressure in the system is at the desired value and the gauge can be read. Comparative readings can be made between a test gauge and the gauge being checked.

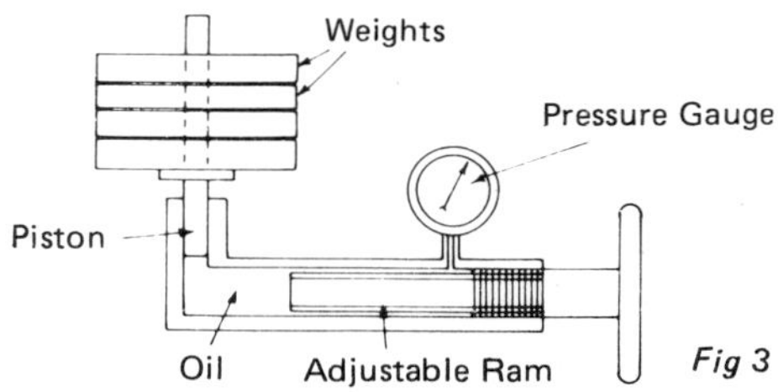

Fig 3

If a gauge while still on the machine does not return to zero when pressure is released, it should first be checked by undoing the pipe connections at the gauge itself. If the pointer then drops to zero the trouble is due to air trapped in the gauge tube and will probably disappear in a short time. The readings at higher pressures are not affected. If the pointer does not return to zero, the gauge should be corrected or re-calibrated. A common cause of zero errors on hydraulic gauges is the sudden release of pressure which strains the mechanism. If it occurs, one of the protective measures mentioned earlier can be adopted.

Testing and Test Equipment

BASIC TEST measurements on hydraulic systems can be made with a pressure gauge to check system pressure; a thermometer inserted in the reservoir to check fluid temperature; and a flow meter (or volumetric tank) to check flow rate. More conveniently, a portable hydraulic test unit can be employed to give all the information likely to be required for routine work. An example of this type of unit is shown in Fig 1. Here pressure up to 420 bar (6 000 lb/in^2) is indicated by a twin-scale glycerine-filled pressure gauge; flow is measured by a displacement turbine, the speed of which is electrically maintained; and temperature by a thermometer in the turbine housing. Fig 2 shows connections for three different types of tests.

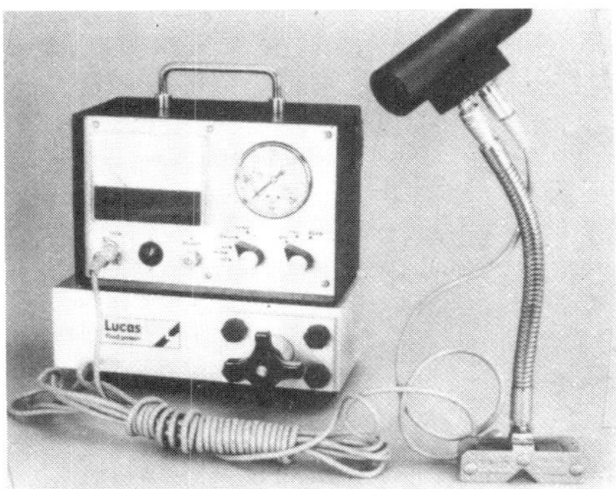

Fig 1 Lucas portable hydraulic test unit incorporates combined flow, tachometer and temperature gauge; pressure gauge; and loading valve.

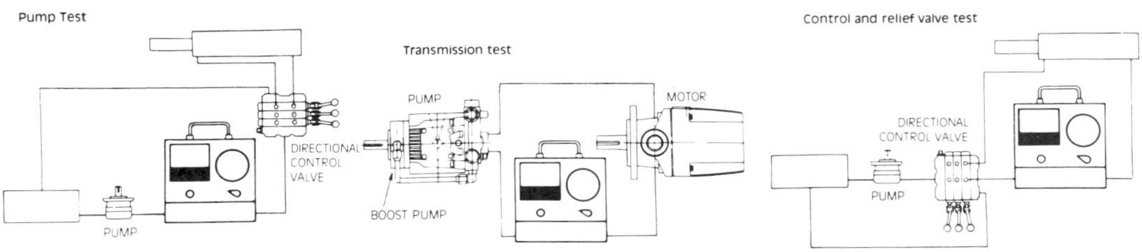

Fig 2

Rig Testing

Detailed testing may be needed to evaluate components or to check complete systems. In the latter case this may be done on a rig duplicating the actual working conditions, or on an installed system using its own or an external source of hydraulic power.

A rig is specific to the individual system design and thus both time consuming and costly to construct. It can, however, be fully justified for prototype testing where (i) the system is critical either in terms of design or performance requirements, or both; (ii) the cost can readily be recovered by standardizing a proven design for economic production of a sufficient number if it is not likely that further changes will be needed in the light of service experience. In the latter case, besides proving the system design, the rig may also be useful for the evaluation of the prototype design of individual components.

Besides providing performance-proving data, such testing can also establish wear and life characteristics of individual components, as well as ultimate life of critical systems. In many cases the latter can be governed by a single small component rather than by the fatigue limits of the main structure or structural components.

A further use for such rigs, after prototype proving, is for the running-in of individual components, although this can equally well be done on much simpler rigs for production components. This has been found desirable in the case of many high-performance components manufactured to very close tolerances which can sustain a high degree of wear during the first few hours of use. Such components can be fully run-in, then, if necessary, stripped down and examined before being assembled in the system proper.

Testing Pumps

A simple set-up for testing the performance of pumps is shown in Fig 3. A particular point to note is that the tester is inserted between the pump and its relief valve and so is no longer protected. The pump should not be started unless the tester pressure control valve is open and pump pressure (as indicated on the tester) must not be allowed to exceed the maximum setting of the relief valve. This test mode enables pump delivery to be measured on no-load and maximum (pressure) load. The difference between the two is an indication of the state of the pump.

Tester positions for testing the relief valve are shown in Fig 4. The tester load valve should be in the no-load position before starting up. After a run up to system temperature the load valve is closed until zero flow is indicated, when the pressure gauge will show the system relief valve setting required. Irregularity in indicated pressure reading will show a leak in the intake side of the

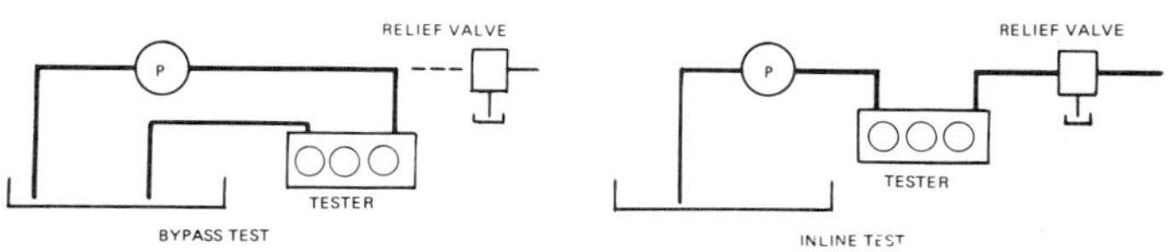

Fig 3

Fig 4

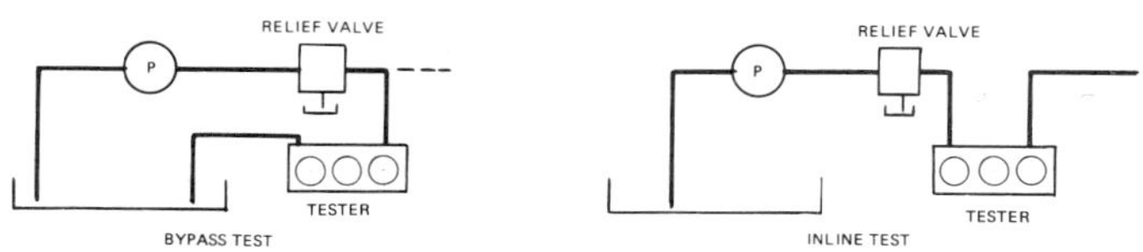

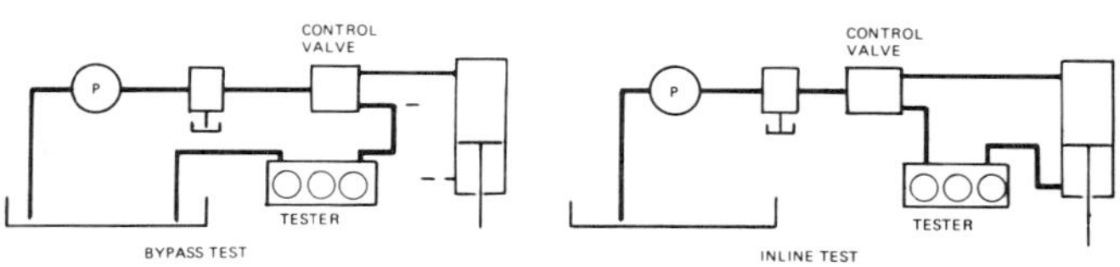

Fig 5

pump, or cavitation. Flow rate indicated on no-load should approximate to pump rated delivery. The flow rate at 50% pressure setting should be the same – any marked difference indicating a relief valve fault.

Testing Valves

Testing facilities for checking the directional control valve (selector) in the circuit are shown in Fig 5. The flow rate with no-load should be equal to the rated pump delivery. The same flow rate should be indicated when the pressure is adjusted to near maximum system pressure. Any marked loss in flow rate means that there is a leaking selector.

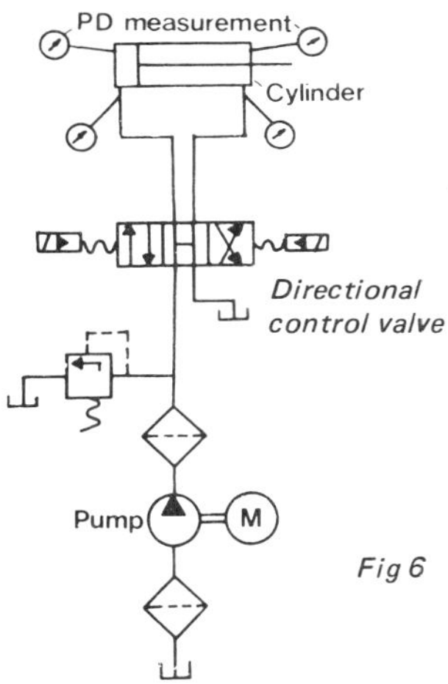

Fig 6

Testing Cylinders

A relatively simple rig can be constructed to test hydraulic cylinders (see Fig 6). Components required are a motor-driven hydraulic pump drawing fluid through a reservoir, a directional control valve, filters, and a pressure control valve located in the circuit between the discharge filter and the directional control valve. For testing, the direction of working of the cylinder is repeatedly reversed by the directional control valve using any suitable switching arrangement. Testing is then continued for the desired number of operating cycles to be evaluated.

Similar considerations apply for the determination of actual operating rates, a series of tests being taken at the different flow rates required. Practical measurement will determine realistically the effects of friction and back-pressure on performance and will establish the true flow rate required. Further, a test rig provides scope for investigating the value, or necessity, of fitting restrictors in forward or return lines, or both, in order to arrive at optimum movement times, or to assist in synchronization.

Testing Pipes and Tubes

Until comparatively recently the strength of pipes and tubes was evaluated on purely empirical lines – working pressure ratings were based on pressure test figures with a suitable factor of safety applied. This method of 'proof' rating is still widely used for non-ferrous metal tubes, non-metallic tubes, and non-seamless tubes.

A distinction can be drawn between a pressure test to establish an ultimate bursting pressure for a particular tube and a true 'proof' test which is employed merely to 'prove' that the tube can withstand a suitable higher working pressure than the normal working pressure rating. In the former case the tube is tested to destruction and figures for a number of different specimens will tend to show appreciable scatter. This is not necessarily significant since once a mean bursting pressure figure can be established the maximum working pressure rating will normally incorporate a safety factor of at least three, and usually very much higher in the case of cast or fabricated tubulars.

'Proof' pressure testing implies that a particular specimen will be tested at some figure above its maximum working pressure rating to establish that the safety factor value is at least that of the ratio 'proof' pressure/maximum rated working pressure. Theoretically, the proof pressure selected could be anything between the maximum working rating and the burst pressure, but if the component is to be used in the system after proof testing, it is important that the proof pressure employed should be less than the yield point of the material, otherwise permanent weakening may result. Yield point is clearly established in the case of ferrous metals as the point where the stress/strain curve departs from linearity. Non-ferrous metals may not exhibit a definite yield point, in which case the limit is best set by the 0.1% proof stress for the material. In the case of non-metallic hoses where neither of the aforementioned apply, the 'proof' or testing pressure is normally set at a specific fraction of the maximum rated working pressure, the actual factor employed depending on the construction involved.

Testing Aircraft Hydraulics

A complete hydraulic test system for an aircraft overhaul base as shown in Fig 7, typically comprises a static power installation with a 150 hp power unit (1) supplying a motor/pump test unit (2) with an electronic control bench (3) which automatically displays and prints out test parameters and a component test bench (4) supplemented by mobile hydraulic test console (5) which are coupled to an aircraft for system tests. Mobile petrol diesel, or electrically powered trolleys (6) and (7) provide complementary test facilities inside or outside the hangar.

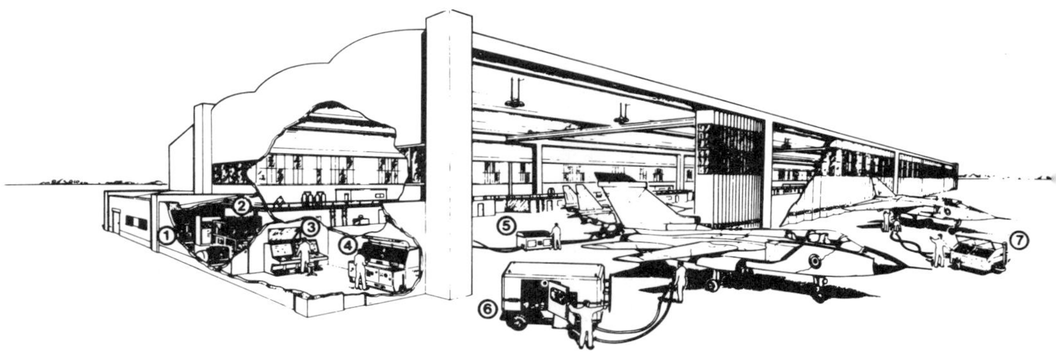

Fig 7 Hydraulic test system for aircraft.

All connections from the pumps to the rest of the trolley system are made by means of flexible hoses to minimize the effects of pulsation and noise throughout the rest of the circuit. This arrangement together with a pulse attenuator downstream of the main pump, dampens amplitude of pulsation by about 90% with a reduction in noise of up to 19 dB. Filters in use are rated 1 μm absolute to 3 μm absolute to conform with the high cleanliness requirement of fluid used in modern aircraft hydraulic systems; they are of the non-bypass full strength element type fitted with differential pressure switches to give visual indication at the control panel that the elements require changing, and to shut down the trolley, automatically, when the elements become blocked.

Pressure control over the trolley delivery outlets is achieved by pilot-operated pressure-compensated reducing valves which enable the pressure at each outlet to be adjusted within the pressure range of the trolley. Return line back-pressure in each line is adjustable by means of pneumatic operated pilot back-pressure valves within a range of 2–17 bar, the setting being governed by pressure requirements of the aircraft reservoir. A pre-charged nitrogen cylinder supplies the pneumatic pilot pressure.

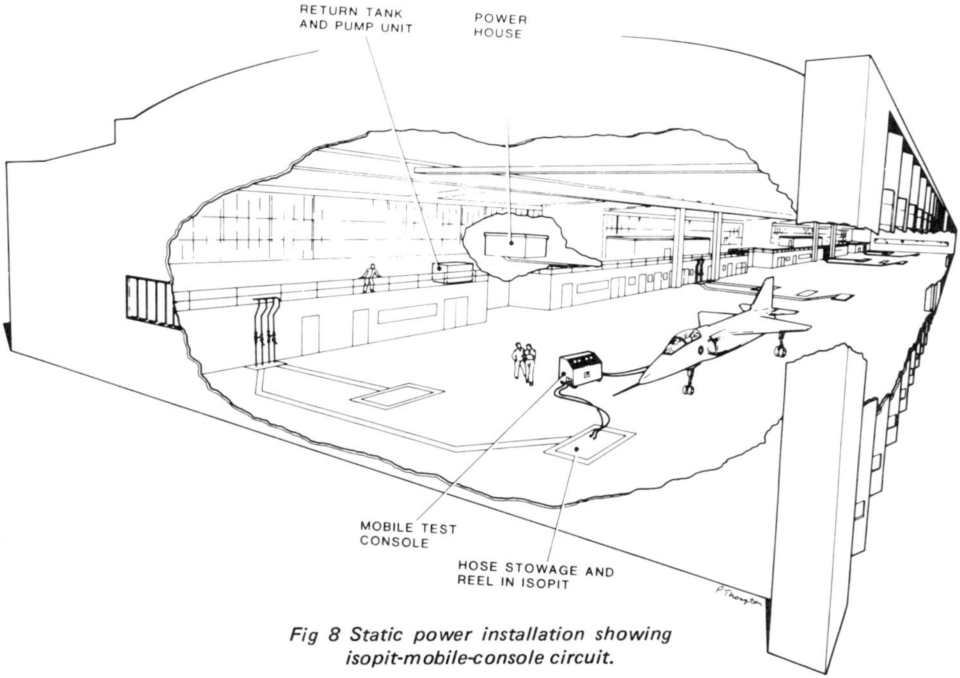

Fig 8 Static power installation showing isopit-mobile-console circuit.

Static Power Installation

An example of a static power installation is shown in Fig 8. The power unit, the size of which is unrestricted by the requirements for portability, is housed in an isolated, sound-proofed power-house which contains filtering, cooling, and de-aerating units. Remote control is provided in the hangar, and with the addition of automatic protective devices, direct supervision is not required during operation.

A typical power-house would accommodate the following items of equipment:

A motor (diesel or electric) driving two axial-piston, constant-delivery, high-pressure pumps.

A large main hydraulic fluid reservoir open to the atmosphere.

A smaller sealed tank to supply filtered, de-aerated fluid to the main pumps.

Hydraulic filters (1–10 μm).

Electrically driven pumps to circulate oil and water through an oil/water heat-exchanger.

A fan-cooled water-cooling unit.

A main-control unit containing motor switchgear, electrical controls and system indicating lamps.

Connection points for an auxiliary power source.

Hangar equipment would comprise:

A coupling point with self-sealing couplings installed in each of several isopits (Industrial Services Outlet pits).

An additional auxiliary supply point mounted on the hangar wall, with self-sealing couplings but without an accumulator boost unit.

Return tank and pump units which pump the return fluid back to the power unit.

An electrical control panel carrying a duplicate set of system controls and indicating lamps.

Emergency stop buttons.

Mobile consoles which connect to the aircraft and the isopit connections and which enable the test supplies to the aircraft to be individually controlled.

Fluid from the main reservoir is cooled and filtered by the oil-circulating-and-cooling unit and then passed to the sealed reservoir of the power unit. Water for the cooling unit is fed from a water-mains cistern which can be heated in extremely cold conditions by a 1 kW heater, but is normally cooled in a cascade type cooler with an electrically driven fan for additional cooling.

The power provides a high-pressure, high-flow supply to the wall-mounted connection and the isopit, to which mobile consoles are connected by flexible hoses to interface with the aircraft. If required, two consoles may be operated from one isopit by using a simple flow-dividing manifold which has three pressure and three return self-sealing quick-release connections. The consoles provide control, filtration and sampling of the fluid to and from the aircraft's system. Fluid returning from the aircraft to the isopit is pumped back to the main reservoir by two return tank-and-pump units.

A wall-mounted junction box consisting of two stop-valves and two connection manifolds may be used to connect with a servicing trolley, such as that used to service helicopters, using a different type of fluid from that of the SPI. The helicopter servicing trolley is powered by a hydraulic motor driven by pressure fluid supplied by the SPI. Component test benches may be similarly powered.

Leakage

WHEREAS LEAKS were often commonplace on hydraulic systems a decade ago they should not be tolerated on any modern system.

The most common cause of leaks is faulty joints or couplings, when disassembly and retightening may provide a cure. The use of gasket compounds to seal a leaky coupling is generally to be avoided, as sealing compound may be introduced into the pipework as a contaminant. If a compression-type coupling cannot be sealed adequately by tightening, then the olive probably needs replacing, or possibly the pipe end has been damaged during initial assembly and needs refinishing. In most cases leaks developing at joints or couplings are due to a bad initial fitting, such as overtightening of joints.

Minor leaks may not be very much in evidence, except for a fall in oil level in the reservoir calling for unexpectedly frequent topping up, or a deposit of oil on floor or machine surfaces. In the latter case the source is also indicated. Where the source is obscure, it may be necessary to examine the whole run of the system logically starting at the pump and following through the high pressure circuit and then the low pressure circuit, back to the reservoir.

Massive leaks are generally instantly recognizable by the oil spray under pressure, or the considerable pool of residual oil collecting under the system. If the system is working, a large leak will also immediately indicate its presence by loss of system pressure and a slowing down of the actuators. The pump will also become starved and cavitate as soon as the oil level in the reservoir has fallen below its normal 'low' level. This will call for immediate shut-down, as continued operation will only empty the system and could cause mechanical damage to the pump.

Seal Damage

Leakage from components is usually due to damaged seals or packings, although slight leakage may be normal from some components. If seal failure is premature, every effort should be made to diagnose the cause of failure, as an alternative type of seal may be indicated for improved life. Also when a seal has failed, necessitating disassembly of the component, the condition of the rubbing surfaces on the component should also be examined, *eg* cylinder bores. These could have been damaged as a result of seal failure or perhaps been the primary cause of seal failure, due to deterioration or rusting. In either case, such surfaces need re-finishing before new seals are fitted or, as an expedient to avoid excessive shutdown time, an alternative seal material, more suitable for the degraded rubbing surface, could be tried. Normally, however, seals should always be replaced with similar devices and recommended materials.

External leaks are obvious and their source is easy to determine and deal with. Internal leaks are often more difficult to detect and may involve tackling one component at a time. In most cases it is possible to determine from the circuit diagram which ports should be blanked off from the pressurized supply at a given condition. If such a port is opened by disconnecting the line to it, and flow is evident from the port, then there must be internal leakage to that port. The circuit diagram will also indicate which components are suspect in providing a leakage path direct from the pressure line to the return line flow to the tank.

The usual cause of internal leaks is seal failure. They may also be due to scored surfaces against which the seals rub, or oil viscosity being too low. In the former case the cause of scoring should be determined if possible, and action taken to prevent a repetition (*eg* fitting a gaiter to a cylinder rod to prevent ingress of abrasive contaminants if working in a dirty atmosphere). If low oil viscosity is the cause, this could be due to an excessive fluid temperature.

Seal Failure

Seal failure may be due to lack of compatibility of the elastomer, although this is unlikely if specified seal materials are used. The most common causes of seal failure are:

(i) *Extrusion* — caused by excessive pressure, lack of support and back-up for the seal, excessive clearances, or faulty groove or gland geometry. Seals should always be used within recommended limits for pressure (with due regard to intermittent peak pressures). Extrusion or wedging can normally be eliminated by reducing clearances and/or providing the seal with back-up rings.

(ii) *Cracking* — may be due to age-hardening, physical deterioration, thermal hardening at very low temperatures, excessive heating due to high friction (lack of lubrication or too tight a fit) or abrasive wear. Age-hardening can occur during long idle periods, particularly at low ambient temperatures. Abrasive wear is primarily caused by the mating metal surface finish being too rough. Elastomeric seals normally demand a surface finish of 0.4 microns (16 micro-inches) or better for good life.

(iii) *Spiral twisting* — normally limited to O-rings and usually caused by side loading (*eg* reciprocal motions). This condition can often be relieved by the use of a back-up or gland ring.

(iv) *Surface damage* — abrasion and wear caused by a rough mating surface or sharp edges on grooves or back-up rings.

Where there is definite evidence of surface damage or roughening, the condition of the surface on which the seal runs should be checked. Surface finish conditions for satisfactory seal life are summarized in Table I.

TABLE I — SURFACE FINISH REQUIREMENTS FOR SATISFACTORY SEAL LIFE

Type of Seal	Light Duty Applications		Normal		Heavy Duty with Limited Life	
	microns	microinches	microns	microinches	microns	microinches
O-rings	0.4	16	0.4	16	0.4	16
Moulded rubber seals	0.4	16	0.4	16	0.4	16
Semi-reinforced rings	0.4	16	0.4	16	0.4	16
Rubber-impregnated fabric rings	0.8	32	0.4	16	0.8	32
Fabric packings	—	—	0.4–0.8	16–32	0.8–1.2	32–48
Leather rings	1.2	48	0.8	32	1.2	48
Compression packings	—	—	0.4–0.8	16–32	0.8–1.2	32–48

Static Seals

Static seals are generally less troublesome than dynamic seals as regards development of leakage, although couplings can be a source of trouble if piping is subject to marked vibration. Couplings incorporating O-rings or toroidal rings have vibration damping properties, but elastomeric rings are likely to deteriorate if subject to high temperatures (above 100°C) and can fail as a consequence. Where such temperatures are likely to be realized elastomeric static seals can be replaced with metal wedge seals.

Bonded seals and bonded washers provide excellent, long-lasting seals in themselves but centralizing can sometimes be a problem if there is no centralizing shoulder on the coupling itself. Alternatively an internal washer is sometimes used, in which case it should be in a resilient material, *eg* PTFE. A plain steel washer used between a union head and a bonded seal will introduce a leakage path. There are other reasons why a bonded seal may not seat correctly and thus lead to leakage. Such leakage problems are attributable to other items than the seals or washers themselves.

See also chapter on *Hydraulic Seals.*

Maintenance and Trouble-Shooting

REGULAR ROUTINE inspection should be regarded as an essential part of the working of any hydraulic system, both to ensure that the system is working at normal efficiency and to instigate preventative maintenance as necessary. Suitable schedules for routine maintenance should be laid down (and adhered to), based on intervals determined by experience or specific recommendations of pump and component manufacturers. There is no realistic saving in exceeding such intervals since although no actual fault may occur there is always the possibility of a progressive loss of working efficiency which could be more expensive than the labour-cost of routine maintenance.

Filters

Full flow filters do not necessarily provide protection from contaminated fluid since they are normally provided with a bypass which operates when the filter element becomes clogged. This is a necessary precaution to prevent further increase in differential pressure across the filter element which could cause it to fail and break up, distributing particles downstream. Thus, whilst a clogged filter is normally 'safe' as regards element migration, the flow is now fully bypassed and any contaminants in the fluid will be circulated through the system. This presents some difficulty since the 'life' of a filter element depends primarily on the system itself and the degree of contamination inherent in the working of the system. This emphasizes the importance of establishing intervals for preventative maintenance for filter checking on an empirical basis.

In the absence of specific recommendations from filter manufacturers, a programme of filter maintenance can be based on the following:

(i) Estimate element 'life' on an hourly basis and change filters at such regular periods as are found necessary.

(ii) Use filters with clogging indicators, with periodic examination of the indicators at intervals shorter than the anticipated 'life'. Here it should be noted that many types of clogging indicator can be triggered by the higher fluid viscosities normal when a system is started up from cold.

(iii) Use a pressure gauge across the filter to measure the differential pressure and indicate when a change is required. This is only as reliable as the accuracy of the gauge, and also depends on the gauge being read regularly.

(iv) Couple a clogging indicator to give a visual and/or audible warning when the element becomes clogged and the bypass is opened.

The filter 'life', or clogging time, will decrease with degradation of the fluid and with any unanticipated increase in the amount of contaminant in the system. Any marked reduction in filter life, therefore, should call for an analysis of the fluid. It is particularly important that any fluid sample withdrawn from the system should be taken from a suitable point on the system when working and at normal running temperature — preferably through a sampling valve. Withdrawing a fluid sample from a static system, particularly from inside the reservoir, is virtually use-ess, except as a check for possible water contamination. In cases of doubt, separate samples can be withdrawn from different parts of the system.

Fluid Life

The life of the oil is dependent on the grade employed and the system operating characteristics particularly the service temperature). With any high-quality hydraulic oil, properly matched to he system requirements, life should normally exceed 15 000 hours.

With more severe service conditions life will be shortened and periodic sampling within the imes specified in Table I should be included in routine maintenance. It is also generally desirable o check and record the working temperature of the fluid at each routine maintenance interval as n excessive temperature at any stage can indicate potential trouble and accelerated degradation of he fluid properties.

TABLE I — PARAMETERS AFFECTING FLUID LIFE

Type of System	Normal working Temp. of fluid	Oil Life (hours)	Remarks
Industrial hydraulics	Not exceeding 40°C (105°F)	Over 20,000	Straight mineral oils may prove satisfactory provided filtration is effective.
	Not exceeding 45°C (130°F)	15–20,000	Hydraulic quality oils should be used.
	Not exceeding 65°C (150°F)	10,000 or less	Premium on quality of oil used.
	Not exceeding 90°C (175°F)	Depends on degree of oil cooling present.	Premium on oil quality — straight mineral oils unsatisfactory without cooling.
Hydrostatic drives	Not exceeding 80°C (175°F)	Possibly up to 10,000	Using light viscosity oils.
Precision hydraulics	Not exceeding 55°C (130°F)	10–15,000	Premium on adequate filtration for maximum oil life.
	Above 55°C (130°F)	Drastically reduced unless oil-cooling is incorporated.	

'Workshop' examination of an oil sample is not a very accurate method of assessment. The eneral state of the oil as regards sludging, *etc*, will be indicated by the amount of build-up of ontaminants on the filters. Thus any oil removed from a filtered system should be clear at any age of its life. Cloudiness will indicate possible water contamination. If this does not disappear on anding, with the water separating out at the bottom, it is a sign that the demulsifying agent is xhausted.

A 'workshop' test for water contamination is to place a drop on a hot-plate. A normal oil will noke or burn. An oil drop contaminated with water will 'hiss'. If water contamination is ispected, of course, the sample can be left to settle, when the water should separate out at the ottom. This will not necessarily happen, however, if other contaminants in the fluid are present the extent that they act as effective emulsifying agents. Under such conditions the sample will ot clear and water will not separate out.

A lightening of the oil colour, relative to the original, may mean that the oil has become contaminated or diluted with a thinner oil and its viscosity lowered as a consequence. The admixed oil may or may not have similar resistance to oxidation, and so could affect overall oil life as well as viscosity. A colour change (lighter or darker) may also mean that a different oil has been used for topping-up the system. Again this may or may not affect the useful life of the oil, depending on the quality and viscosity of the added oil. The best safeguard with a colour change, if the oil sample is otherwise clear and smells 'sweet', is a viscosity check.

Oil samples may be submitted to an independent laboratory for test, or to the oil supplier for his laboratory report (where the latter provides this service). In the first case it is necessary to know the original specification (or preferably a laboratory report from the same source on the new oil) in order to interpret the test data. In the latter case the state of the oil will normally be evaluated as part of the test report.

An electronic microscope and X-ray detection equipment form a sophisticated microprobe assembly for visual and X-ray examination of contaminant. Investigations into the nature and composition of fine silts as well as wear products are possible.
(Pall Industrial Hydraulics Ltd)

Pumps

Pumps should be serviced at regular intervals, as recommended by the manufacturer(s). If pump performance is at all suspect, this should be checked by an individual test. Apart from deterioration of the pump itself, parameters which can affect pump performance are:

(i) Low oil level in reservoir;
(ii) Filters clogged, or restrictions in intake line;
(iii) Air leaks in intake line;
(iv) Entrained air drawn from reservoir;
(v) Oil viscosity too high;
(vi) Oil viscosity too low;
(vii) Pump speed too low.

Items (iii), (iv) and (v) will normally be accompanied by noisy pump operation.

Fault Finding

Faults fall into three main categories:

(i) Leakage in the system;
(ii) Loss of system pressure;
(iii) Pump not working properly.

Loss of System Pressure

In a working system there will inevitably be pressure variations, such as those due to variations in the effective compression ratio of an accumulator and pressure surges, produced by back-pressure effects, in the case of double-acting cylinders. Thus the actual pressure level at a particular point in a system can only be established empirically, when ideally pressure variations should not be more than about plus or minus 3% throughout a working cycle. A figure of plus or minus 5% is more realistic, and about the limit of accuracy for the majority of less expensive pressure gauges which may be fitted in the circuit. Such pressure gauges are usually protected against surge pressures and thus will not indicate momentary surge pressures. They will, however, give a reliable indication of any marked change in average pressure, indicating a potential or actual fault, provided the gauges are checked for calibration at regular intervals. The error which can develop on a gauge which is not periodically checked can be well outside the 5% limit and present a false indication of a 'fault'.

Loss of pressure is a general indication of a leak or partial pump failure, or the lesser likelihood of the failure of a bypass or relief valve. A poppet type valve opening to a bypass can also 'starve' a particular line by lifting at a pre-determined pressure and then failing to reseat until a very much lower pressure is reached, because of the tendency for such a valve head to remain 'floating' in the fluid stream. Valves with a degree of inherent modulation are thus preferred in such circumstances.

Pressure Variations in the System

The causes of marked pressure variations in the system are most likely to be found under the following headings:

(i) *Undamped bypass valve* — The cure in such cases is to replace with a modulated or constant pressure type.

(ii) *Clogged valves* — Calling for cleaning or, if desirable, replacement by a more suitable type with a 'self-cleaning' or non-clogging motion.

(iii) *Pump pressure pulsations* — Which are more marked with some types of pump than others. Undesirable pressure fluctuations of this type can usually be damped by an accumulator; alternatively, a different type of pump could be used with more constant pressure delivery characteristics.

(iv) *Excessive foaming in the reservoir, caused by air entrainment* — This could be caused by poor reservoir design, an air leak in the system, or similar design fault, allowing air to become entrained with the fluid.

(v) *Air inclusion* — Such as produced by air leaks allowing pockets of air to become trapped in the system, or unsatisfactory bleeding when the system is filled. Air inclusion will be marked by lack of rigidity in the operation of the system. Additional bleed points may be needed in the system, or the circuit design revised.

Excessive System Pressure

Excessively high system pressures may be caused by partial blockages, although these would normally be relieved by relief valves in the system. Excessive flow through relief valves, however, may starve the flow to the actuators, resulting in slower motions. An excessive system pressure will also increase the fluid heating and may lead to excessive fluid temperatures.

TABLE II – OIL WORKING TEMPERATURES

Type of System	Working Conditions	Working Temperature		Oil Life
		°C	°F	
Industrial	Ideal	under 43	under 110	Very long (over 20 000 hours).
	Normal	43–55	110–130	15 000–20 000 hours.
	Acceptable	55–66	130–150	Reduced oil life; premium on oil quality.
	Acceptable with inter-cooling	66–80	150–175	Reduced oil life; premium on oil quality.
	Maximum without intercooling	94	200	Very much reduced life; straight mineral oils unsatisfactory.
High-temperature		up to 140	up to 280	Short life; straight mineral oils unsatisfactory.
Hydrostatic	Normal	up to 80	up to 175	Light viscosity oils used (10–30 centistokes at working temperature).

Excessive Fluid Temperatures

Recommended working temperatures for oil fluids are given in Table II. Basically, the higher the oil temperature the more rapid the onset of oxidization and the development of degradation products, as the oxidization inhibitors are used up. For maximum oil life the fluid working temperature should not exceed about 50°C (120°F), although working temperatures of up to 80°C (175°F) may be acceptable with mineral oil fluids.

Possible causes of excessive oil temperature are:

(i) *Use of oil with too high a viscosity* – This causes an excessive load on the pump, with lowered pump efficiency and an excess proportion of the input power being transformed into heat. The fluid viscosity should be selected according to the pump manufacturer's specification. Note, however, that this may be modified if there are other factors present, causing an excessive rise in fluid temperature in the working system.

(ii) *Pump bypass valve set too high* – This allows the pump to overwork and generate excess heat, which is transferred directly to the fluid.

(iii) *Pump not suitably off-loaded, allowing the full input power to be developed in the form of fluid heating.*

(iv) *Insufficient fluid in the reservoir.*

(v) *Internal leakage* – This may be due to wear on the pump. Equally, a fluid with too low a viscosity will promote excessive slippage, but not necessarily over-heating unless lubricity is low and excess mechanical friction results.

(vi) *Unloading valve blocked* – The unloading valve to the tank or accumulator may be blocked, not allowing the pump to unload. Check and rectify as necessary.

(vii) *Restrictions in lines* – Damaged, kinked or partially blocked lines may introduce excessive load on the pump and consequent over-heating.

(viii) *Output motions obstructed or stalled giving excessive flow through relief valves, with high heating through these restricted flow paths.*

(ix) *Excessive flow velocities* – Check flow velocities against recommended figures, *eg* not exceeding 4.6 m/sec (15 ft/sec) on pressure lines. High flow velocities promote high frictional losses and over heating.

(x) *Reservoir too small* – Oil volume is insufficient to allow normal cooling during cycling, hence there is a gradual build-up of temperature. Check reservoir capacity against system demand.

(xi) *Insufficient normal cooling* – Check that lines, *etc* are not located in high ambient temperatures where they may be subject to heating rather than cooling through normal radiation. If impossible to re-arrange, inter cooling will probably be necessary.

(xii) *Inadequate inter cooler* – Check performance of the inter-cooler against system requirements. Inter-cooler may be faulty, too small or badly positioned.

Malfunctioning Pump

A suspected mechanical fault in a pump can be checked by disconnecting the pump from its driver and turning over by hand. Any excessive tightness or non-smooth rotation is a certain indication of internal mechanical damage to the pump or its bearings. If the pump motion is smooth, but noisy when connected and run, this is most likely due to cavitation, caused by starvation on the inlet side. This could be due to low oil level in the reservoir, clogging of the inlet strainer, or air in the suction line.

A diagnostic summary of pump faults is given in Table II.

TABLE III – CHECK LIST FOR PUMP MAINTENANCE

Cause	Action
Low oil level	Check oil level in reservoir (recommended level should be marked, and intake line must be below oil level).
Restriction in intake line	(i) Check filters and strainers in reservoir. (ii) Check flow through line if necessary (line may be kinked or damaged).
Air leak in intake (noisy pump operation)	(i) Check oil level in reservoir. (ii) Check intake line for leaks.
Pump speed too low	Check actual speed against manufacturer's specification; if low, driver may be faulty.
Oil viscosity too high (noisy pump operation)	(i) Check oil viscosity, or replace with new oil if suspect. (ii) Oil in reservoir may be too cold on starting up, causing lack of prime.
Pump out of line	Check under normal working conditions, running alignment may differ from static alignment.

System Malfunction

The basic problem here is to determine whether the malfunctioning is due to one of the more obvious faults described earlier, or to the failure of one or more of the individual components controlling the operation of the system. The greater the complexity of the circuit the more the inter-dependence of the control and behaviour of the various elements, and the greater the number of individual loops or circuits which may be involved. Diagnosis then relies on isolating the fault within a specific group, and from there determining the basic fault responsible for the malfunctioning of that group. Treatment is thus specific to the circuit and requires close study and attention to the circuit diagram as well as full appreciation of the function of each individual component. The most straightforward method is usually to 'read back' from the actuator involved to establish the point at which the lack of behaviour or control is initiated, ignoring those components which are not directly concerned with the function which is at fault.

In the case of simpler circuits more direct and straightforward action is usually possible, particularly if the system pressure is first checked.

System Pressure Normal, Lack of Actuator Movement

- (i) Absence of signal;
- (ii) Solenoid failure in the solenoid valve accepting the signal;
- (iii) Mechanical obstruction.

System Pressure Normal, Loss of Actuator Speed

- (i) External oil leak;
- (ii) Internal oil leak in actuator or valves in actuator pressure line circuit;
- (iii) Faulty or badly-adjusted relief valve;
- (iv) Partially-blocked control valve;
- (v) Pump fault or blockage in delivery line reducing delivery, but not necessarily pressure;
- (vi) Overheating of fluid causing loss of viscosity;
- (vii) Excessive wear on actuator;
- (viii) Excessive loading of actuator, caused by unduly high external load or eccentric loading;
- (ix) Increase in actuator friction, *eg* due to maladjustment of compression packings, distortion through applied bending loads, *etc.*

System Pressure Low, Loss of Actuator Speed

- (i) Loss of delivery through a pump fault or leak;
- (ii) Accumulator failure, or loss of gas pressure requiring a recharge;
- (iii) Faulty pump drive;
- (iv) Incorrect valve settings;
- (v) Faulty relief or bypass valves;
- (vi) Maladjustment of metering valves in the case of linear actuators, or faults or dirt in valve;
- (vii) Dirt under cushion ball valve (linear actuators).

System Pressure Variable, Variable Actuator Speed

- (i) Worn pump, or pump faulty, with failure to produce required maximum demand;
- (ii) Under-sized accumulator, or loss of compression ratio (accumulator may need a gas recharge);
- (iii) Internal leakage in part of the system;

(iv) Partially-blocked control or sequencing valve;
(v) Faulty pressure-relief valve;
(vi) Out of order cushion ball valve (linear actuators).

Irregular Action

Irregular or erratic action is commonly caused by air entrainment. Check as aforementioned for 'Variable Fluid Pressure'. Other possible causes are:

(i) *Excessive friction* — Caused by seals or packings being too tight or incorrectly fitted so that 'wedging' results. Extrusion of seals and 'wedging' may occur at high pressures with O-ring seals, unless provided with back-up washers. Lack of lubrication or binding on slides, *etc,* can be another cause of excessive friction.

(ii) *Misalignment of actuators* — Misalignment of actuators, tables, slides, *etc,* can cause irregular action.

(iii) *Compressibility effects* — The compressibility of the fluid under high pressure can affect precise movement and control. This is a normal fluid characteristic. A synthetic fluid is better in this respect. Air entrainment also increases elasticity of the fluid.

(iv) *Lack of synchronization* — Exact synchronization is difficult to achieve within straightforward hydraulic systems and it may be necessary to apply synchronization *via* mechanical linkages or the use of accumulators.

System Pressure Remains High with System Unloaded

(i) Faulty non-return valve isolating the unloaded part of the system;
(ii) Badly-adjusted unloading valve;
(iii) A combination of (i) and (ii).

Note: If adjustment of the unloading valve produces no change in system pressure, then the fault lies with the non-return valve.

Maintenance of Hydraulic Cylinders

Particular points to check in cylinder maintenance are summarized under separate headings:

Rod End Fittings — Wear on these components can be the result of poor or even a total lack of lubrication on moving parts. A misaligned cylinder can also cause damage, particularly to the mating threaded parts if excessive bending moments are involved.

Gland End Cap — Thread pick-up and damage here can very often be traced back to insufficient cleaning-down of the whole cylinder from a previous strip-down. It is imperative that all dirt and contaminants be thoroughly cleaned off the cylinder's external surfaces before any work is commenced. All cylinder oil should be drained off and parts cleaned using a lint-free cloth. A note should be made of the order of component disassembly to ensure that all items are replaced in the right order. The correct spanners should always be used as damage caused by poor tools can mean that end-cap removal can only be achieved by destructive means.

Wiper Seal Housing — Can often be overlooked as a possible source of cylinder failure. Wiper seal seating should be examined carefully to ensure that there is no surface deterioration which would allow ingress of abrasive or corrosive materials behind the wiper seal.

Wiper Seal — Because this is not a pressure seal its importance can be overlooked. All seals should be renewed throughout as a matter of good maintenance policy. Re-fitting a used wiper seal is uneconomical and, although not immediately obvious on re-instatement, it would permit ingress of dirt leading to an early cylinder strip-down.

Gland Seal – This is the main pressure-retaining seal to prevent oil leakage to atmosphere. Once again, it is important to check the seal cavity housing in the gland bearing assembly, as any seal is only as good as the surface on which it works. Care should be taken when fitting new seals, which must be well lubricated with system fluid, to ensure the correct orientation to mating parts. Incorrectly fitted seals do not effectively maintain the hydraulic fluid pressure.

Static O-Ring Seals – These can imbed themselves into the surface of mating parts leading to seal deterioration and possible corrosion in the seal housing and on the cylinder tube wall. These material surfaces may need treatment or attention before seal replacement.

Bearings – The gland and piston bearings are usually bronze and wear can be recognized by burnishing or high-spots. Excessive wear at these points can destroy the strut principle of the cylinder, leading to 'knuckling'. Score marks on the gland bearing could indicate possible shaft damage. Scoring on the piston head indicates oil contamination within the cylinder, which can also lead to damage on the internal bore of the cylinder casing. Oil contamination can only be cured by flushing out the complete hydraulic circuit, with special attention to filters, *etc,* and re-filling with the correct grade of clean oil.

Casing – Care should be taken to flush out thoroughly before examination. The internal bore should be examined with a high intensity light but the polished bore can make damaged areas very difficult to recognize. Casing damage must be removed by honing, where practicable within seal manufacturers' tolerances. External casing examination is visual and damage may be obvious, as it is usually mechanical damage at the bearing points caused by lack of lubrication or cylinder misalignment. Excessive damage will mean complete casing replacement. Indentations on the outer casing wall can transfer through to the inner wall and are recognized as concentrated burnishing on the piston head. However, this is not readily seen on examination, owing to the reflected light. Piston head wear on the cylinder bore can usually be recognized by extruded seals. Any external pipework damage, such as indentations, can reduce oil flow and piston speed.

Shaft – Damage can come in many forms including indentations, surface abrasions or even bending. Slight bending may not be obvious visually but shows as high burnishing on the gland bearing, burning or discolouration on the shaft high-spot, and sometimes wearing through of the shaft plating. Excessive bending is obvious and usually requires complete shaft replacement. Bending is caused by over-stressing or similar maltreatment. General wear on the shaft-plated surface must be dimensionally checked for acceptability. Corrosive conditions can be indicated by the extent of the wear and shaft re-plating to manufacturers' tolerances may be necessary.

Cylinder Re-assembly – Absolute cleanliness is essential on re-assembly and the subsequent need for re-furbishing can often be traced back to a lack of care in previous cleaning and re-assembly procedures. The correct tools should always be used and only soft hammers to avoid damage to plated or polished surfaces. All components should be freely lubricated with system fluid and care taken with the correct order, alignment and fitting of all items on re-assembly. Couplings for oil feeds should be tightened to prevent oil leaks and cylinder malfunction. The incorrect coupling of oil feeds can cause burst feed tubes or cylinder casings due to hydraulic lock-off within cylinders. Pressure intensification of 4:1 or 5:1 can be experienced due to full bore pressure acting upon a locked-off annulus.

Hydraulic Hose

Maintenance problems are usually, but not necessarily, matched to the severity of the application.

The system design needs to take this into account to ensure that likely maintenance points are readily accessible. Equally the maintenance schedule and frequency should be matched to the severity of the application.

Any of the following conditions will require hose and/or fittings replacement:

(i) Leaks at fittings or in the hose;

(ii) Damaged, cut or abraided cover;

(iii) Kinked, crushed, flattened or twisted hose; hard, stiff, heat-cracked or charred hose; blistered, soft-degraded or loose cover; cracked, damaged or badly corroded fittings; and fitting slippage on the hose.

Visual inspections will reveal when clamps, guards and shields need to be tightened or replaced.

Handling Flexible Hose

For in-plant hydraulic hose service to continue on an uninterrupted basis, the engineer must initiate a suitable maintenance plan. Well-planned maintenance helps ensure the safety of personnel and enhance hose performance.

When hose is received at the plant, some applied common sense at the outset, in the unpacking procedure, for example, can preserve hose life. The use of sharp or heavy tools should be avoided as they have the potential to damage a hose, and, therefore, create an unknown safety hazard. The hose should be stored in its original container until installed, particularly pertinent if the carton or crate is durable. Also, protective wrapping should be used both when handling hose and when placing it in storage.

Storage

Hydraulic hose stored for a long time should be in a coiled-flat plane, not hung haphazardly on hooks or spike-like protusions. The storage area should be cool and dark and free from dust, dirt, dampness and mildew. Things that attack hydraulic hose in storage include temperature extremes, humidity, sunlight, oils, solvents, corrosive liquids and radio-active materials. Also, insects and rodents are attracted by rubber and proper protection should be provided.

Ozone oxidizes rubber, so the hose must be stored away from electric motors and other ozone generators. Hydraulic hose needs to be kept away from heated surfaces such as steam pipes and radiators because excessively high temperatures can harden and crack rubber surfaces. Recommended storage temperatures are +10°C (50°F) minimum, to +38°C (100°F) maximum.

See also chapter on *Leakage.*

SECTION 6

Mechanical Handling

THE HIGH conversion efficiency of hydraulic pumps — and in particular the more recent development of gear pumps with high volumetric efficiencies — has resulted in the extensive application of hydraulics in mechanical handling. The other great advantage of hydraulics in this field is its versatility.

The most widely used source of hydraulic power is the gear pump embodying pressure-balanced bushes or side plates to minimize clearance gaps without the necessity for high side-plate forces. At the same time lubricating systems have been developed for these pumps circulating low-pressure oil through the bearings with negligible loss of volumetric efficiency.

In the interests of power saving and the elimination of hydraulic oil coolers some of the larger diesel-powered mechanical handling vehicles employ multiple-pump systems, often of the split-circuit type. Commonly a tandem gear pump comprising a large and a small element has its combined capacity matched to the total power available for hydraulic services at full system pressure, whilst the delivery from the individual pump sections is sized to suit the services being supplied.

A further refinement, though more frequently used in earth-moving equipment, is the dual pump offloading system. In this case the combined capacity of the pump is matched to the engine power available but at a pressure lower than maximum system pressure — say 140 bar (2 000 lb/in^2) in a 200 bar (3 000 lb/in^2) maximum-pressure system. Then, the capacity of one of the individual elements is matched to absorb the same total power but at maximum pressure. An unloading valve dumps one of the pump deliveries when the pressure level in the system reaches the predetermined setting, thus giving high-speed operation of the hydraulic services when at moderate load and a lower speed at full load.

The advantages of multiple fixed-delivery pump systems are heightened by the pressure capabilities of some of today's more advanced gear pumps. Continuous operating pressures of 250 bar (3 600 lb/in^2) with a 20% allowance for pressure transients are now available in pumps with individual displacements up to 92 cm^3/rev (5.6 in^3/rev), whilst in the smaller capacities the continuous operating pressure can be as high as 275 bar (4 000 lb/in^2). Fig 1 shows a typical operating duty envelope for a high performance pump of this type. In tandem or multiple pump arrangements the combined sections can operate at main system pressures of around 207 bar (3 000 lb/in^2), whilst individual sections can operate up to the higher pressure levels where necessary to provide, for example, high clamping loads with small, economical actuators.

Whilst the future of the efficient, low-cost, fixed-displacement pump is assured, even the versatility offered by multiple fixed units cannot match that of the variable-delivery pump. Its fundamental advantage is the reduction of power wastage by its ability to reduce delivery flow in

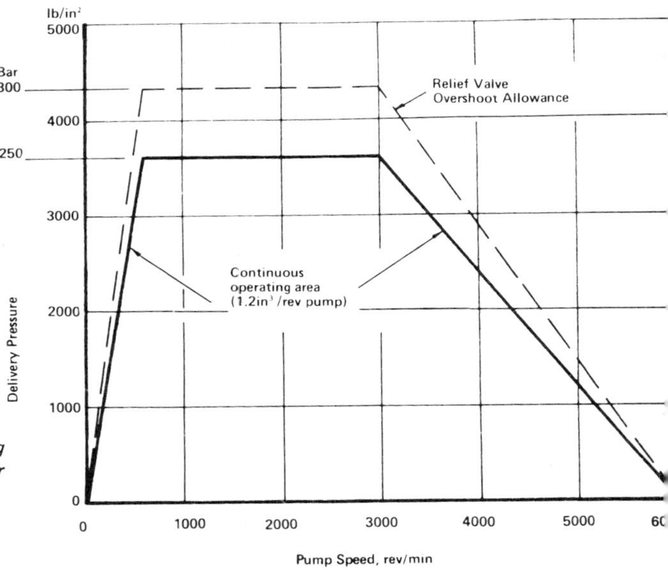

Fig 1 Typical pressure/speed operating envelope for Dowty high pressure gear pump.

accordance with the system requirement. This facility is achieved by a swash control servo whic can be arranged to respond to various circuit parameters. Pressure-compensating control change pump swash to maintain a predetermined constant pressure level in the system. Constant-flow controls limit pump stroke to give a pre-set constant delivery irrespective of pump speed chang (ideal for large vehicle steering duties). Power-compensating controls give constant multiples c flow and pressure along a constant horsepower line drawn between the limits of maximum syster pressure and the pressure at maximum pump stroke. Manual control of swash gives an ideal mear of speed control for, say, hydraulic motors without spilling flow.

Hydraulic Motors

Hydraulic motors are particularly suitable for conveyors and machines which need constant spee and very frequent start, stop and reverse actions. The moving parts of a hydraulic motor are fr quently so small and light for the torque rating that very high rates of acceleration and deceler tion can be obtained. All hydraulic motors can run as pumps, and in some cases, the same un may be used for either purpose.

A characteristic of the hydraulic motor which can never be matched by the electric motor is th stall condition. In this state the hydraulic motor can be held stationary at full torque witho energy being wasted or heat building up. Additionally there are several basic types of semi-rota actuator. One of the most popular has a sealed vane in a semi-circular chamber. Pressure admitted to either side of the vane and movement is thus produced through a limited angle up about 300°.

Static Hydraulic Systems

Static hydraulic systems are usually designed and built for one specific purpose. This makes necessary to have a very comprehensive range of valves to satisfy the requirements of the mc problematical situation. This range will include a variety of pressure-control valves, that is, reli reducing, off-loading, balancing, dividing and check valves. Most of these may be controlled pilot pressure. Directional valves have a great variety of spool configurations for obtaining the characteristics required. Control may be by mechanical means, pilot pressure or solenoid.

In some cases a static system may be based on an accumulator. Thus, when the maximum pressure is reached the pump is unloaded or the motor stopped, then when any circuit demand causes a sufficient decrease in pressure, the pump will automatically reconstitute the full pressure. The fluid pressure may be converted back into mechanical energy by using cylinders, semi-rotary actuators or rotary hydraulic motors. All these may be used in static mechanical-handling installations.

Static mechanical handling systems are finding an increasing number of applications in industry, civil engineering, commerce, *etc.* Examples include:

Scissor lifts – These come in a variety of types and sizes. The advantage is that a load may be lifted from a very low minimum height to several times that height. Sometimes automatic control is used so that sheet material may be loaded or off-loaded a sheet at a time, and the top of the stack will remain at a constant height. In the simplest form a feeler or finger is used to sense the top of the stack, and this is connected to a valve which controls the flow of oil into or out of the supporting hydraulic cylinder. Such a device is ideally suited for feeding printing machines or other paper-converting machines.

Passenger or goods lifts – Hydraulic power has many advantages, such as low initial cost when used only as far as the first or second floors. In such applications the single-acting displacement ram is accommodated in a borehole directly under the centre of the floor of the lift cage. For greater heights a hydraulic ram is mounted in the side of the lift shaft and operates a jigger mechanism which, with a cable, doubles the movement. This hoisting system has the added advantage that there is no mechanism at the top of the lift shaft. The motor, pump and control equipment are usually housed in the basement, where any noise is not a nuisance.

Warehousing – Storage systems are in use from which any quantity of many hundreds of items may be extracted automatically. In this way the central store of a chain of supermarket stores can be operated by a relatively small staff using a computer. Individual orders are computer processed in such a way that the automatic mechanism can produce exactly the goods required.

Electric Fork Lift Trucks

Hydraulic power for most battery electric fork lift trucks is supplied by gear pumps directly driven by compound wound motors which offer a light-load speed 50% to 60% greater than when fully loaded. Maximum light-load speeds reach 4 000–5 000 rev/min, and the only type of pump capable of efficient operation at speeds of this order is probably the gear type. Motor speeds can, of course, be reduced, but this will add considerably to the cost, size and weight of the truck at the design stage.

Most lift trucks feature a simple hydraulic circuit which consists of a reservoir, constant-displacement pump and open-centre control valve. Standard trucks are fitted with a single-acting spool valve controlling the hoist cylinder and a double-acting spool valve controlling the tilt function. Valves with up to six services are used on trucks fitted with attachments such as fork-sideshift device, clamps, or a rotating head.

An additional double-acting spool may be necessary for extending and retracting the mast where the truck is designed to be used in gangways of varying width.

In order to effect a high degree of accuracy in load spotting, the control valve in all lift trucks must have first-class metering characteristics. It is also imperative that both sides of the tilt cylinder are fully charged with oil at all times, since any voids which are allowed to form will result in loss of control and consequent instability of the load or the truck itself. Exhaust charging valves or pilot-operated check valves are normally fitted to prevent such an occurrence.

The forks of a lift truck are usually lowered by gravity. The pressure generated by the weight of unloaded forks is not high and the control valve and pipework must have a low pressure loss if an acceptable lowering speed is to be achieved. Lowering at dangerously high speeds, on the other hand, can result from unskilled use of the control valve with the forks fully loaded. A burst pipe or hose between the lift cylinder and the control valve would give rise to the same problem. To remedy this a one-way flow-regulating valve can be fitted directly to the lift cylinder in order to maintain approximately constant exhaust flow at all pressures between that equivalent to light-loaded forks and fully loaded forks.

Modern gear pumps operate with high overall efficiencies but fixed displacement. Any metering of the pump supply, such as occurs when stopping the load, tilting or side-shifting, involves considerable power wastage. Variable-displacement piston pumps offer the possibility of improved average operating efficiency. Service speed would be a function of lever travel, so that all spotting operations would be carried out with a low pump flow and consequently absorb a minimum of power.

A further demand made by the battery electric truck application on pump design arises from one of the truck's own main attributes. Its cleanliness, freedom from fumes and quietness make it ideal for indoor operation but in coldrooms, warehouses, stores, *etc,* the noise made by the truck's hydraulic pumps can be the greatest source of annoyance in the building. However, pump designers have made significant advances in minimizing both mechanical and hydraulically-generated noise by careful attention to manufacturing standards on gears and the detail design of fluid relief and associated porting in the pump. Equally, vehicle designers can minimize noise amplification by careful attention to pump mounting and the location and attachment of pipework.

Battery electric trucks are limited in capacity by the present economics of battery design, and find their greatest use in factories and warehouses where their silence and absence of noxious exhaust gases are obvious advantages. They are not suitable for duties involving long travelling distances, because of their relatively low speeds and the problem of battery capacity.

Engine-Powered Lift Trucks

Larger fork lift trucks are powered by diesel engines (or sometimes by petrol engines) when the hydraulic supply can be provided by an engine-driven pump.

Increased lift speeds are achieved as a result of the more powerful pumps having higher flows. Because adequate engine power is necessary for traction purposes and is not used when the hydraulic system is operational, hydraulic power consumption does not usually become a major factor in trucks below 5 000 kg (5 tons) capacity. For trucks of larger capacity it is necessary to reduce hydraulic power consumption at high loads. A variable-displacement pressure-compensated piston pump would achieve this.

Current systems use a two-speed lift approach which can be achieved by a double-acting lift cylinder in which the annulus and head sides can be inter-connected by a valve. Low speed is achieved by supplying only the head side of the cylinder and thus exhausting the annulus side, whilst high speed is attained by inter-connecting the head and annulus sides of the cylinder and therefore pressurizing the annulus and adding its exhaust to the supply at the head side. The lifting capacity is, of course, reduced in the proportion of the rod area to the piston area and the lifting speed increased inversely. The ratios of high to low speed are obtained at the design stage by the choice of the appropriate rod size. In some designs, the valve inter-connecting the annulus and head sides of the cylinder is automatically pressure-operated. Normally the valve is biased to the high-speed position, the maixmum load which can be lifted being a function of the system pressure and rod area. Loads in excess of this result in a higher system pressure which shifts the valve, blocking the inter-connection and diverting the annulus side of the cylinder to tank. Loads equivalent to system pressure and the piston area can thus be lifted.

LIFT TRUCK CIRCUIT WITH PUMP UNLOADER

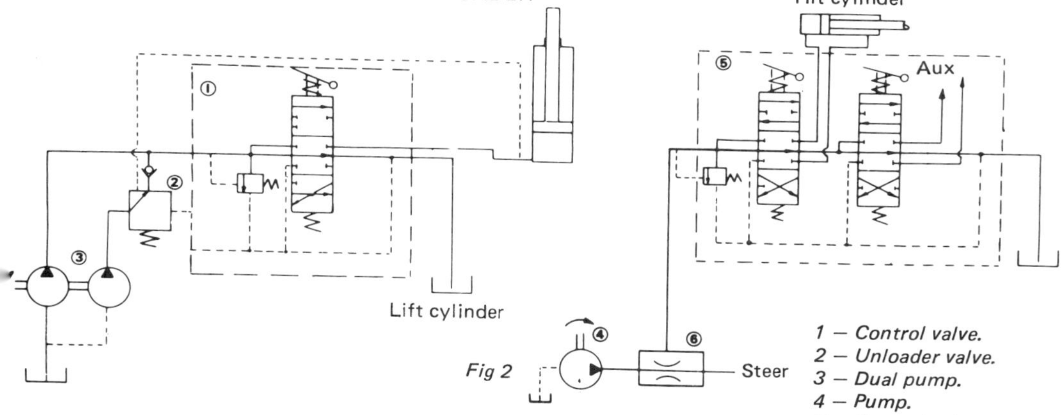

Fig 2

1 — Control valve.
2 — Unloader valve.
3 — Dual pump.
4 — Pump.
5 — Control valve.
6 — Bypass flow regulator.

The circuit diagram of an alternative two-speed system is shown in Fig 2. Two lift circuit pumps re used, one of which is unloaded automatically and dumped to tank when the system pressure eaches a value equivalent to the maximum allowed input power to the hydraulic pumps. The emaining pump can thus operate up to a higher pressure before input power again reaches its naximum allowed level. This system has the advantage that a normal single-acting lift cylinder can e used. In conjunction with a bypass flow regulator, a third and smaller pump is fitted, the conrolled outlet of which supplies oil for power-assisted steering purposes. The bypass outlet supplies separate valve for controlling tilt and other auxiliary services where full system flow would be an mbarrassment and would require a permanent throttling device in a single-pump system. A multiump system allows freer selection of pump capacities for the differing needs of the services.

Power-assisted steering is necessary for all but the smallest lift trucks to ensure rapid low-effort nanoeuvrability in confined areas. Hydraulic power for this purpose is sometimes provided by a eparate engine-driven pump incorporating its own small reservoir, flow regulator and system relief alve. Some manufacturers utilize the main hydraulic system.

ide-Loading Trucks

ide-loading lift trucks are designed to carry long loads and complete containers on larger size ehicles. The mast is mounted on one side of the chassis and moves across to pick up or deposit he load outside the wheelbase of the vehicle. The load is moved in and lowered on to the deck for ansportation. To give stability during lifting and lowering, hydraulically-operated stabilizer jacks re used. The problems confronting the hydraulics designer are very similar to those posed by ont-loading lift trucks. Mast extension and retraction is a standard requirement. Stabilizers can e powered by an auxiliary circuit. Power steering is standard and the necessary pressure oil supply can be made available from either a separate pump or bled from one of the main circuits by a flow-regulating valve as already described. A typical hydraulic circuit for a side-loading truck is shown in Fig 3.

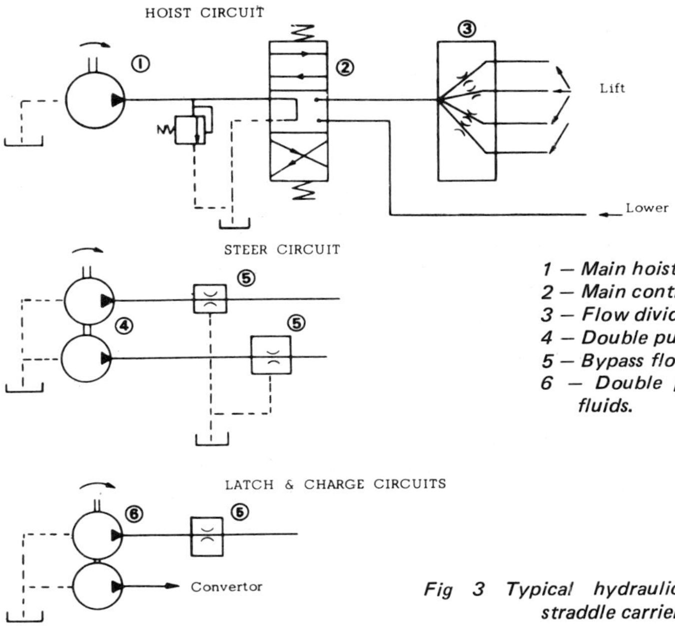

1 — Main hoist pump.
2 — Main control valve.
3 — Flow divider.
4 — Double pump.
5 — Bypass flow regulators.
6 — Double pump sealed for separate fluids.

Fig 3 Typical hydraulic circuit for straddle carrier.

Straddle-Carriers

The straddle-type lifting vehicle grew out of the need to move standardized containers on and off trucks and flat cars for marshalling purposes. To ensure that the rows of containers can be marshalled closely together it is necessary to keep the supporting legs of the vehicle as narrow as possible, which leads to the use of traction chain drive from the upper deck to the wheels. The deck itself houses the driver's cab, engine, transmission and all hydraulic power units.

Straddle-carriers are now more or less standard equipment at marinas for lifting and launching craft up to about 10 000 kg (10 tons) deadweight, although here the design is usually less sophisticated with mechanical lift and hydraulics involved only for power-assisted steering. Engine and cab (or steering position) are side-mounted. Straddle-carriers of this type are generally known as travel-hoists.

Drives

Nearly all modern engine-driven lift trucks, side-loaders and container handlers are mechanically driven through a torque converter, 'hot shift' gearbox and conventional differential axle. All are ideal applications for hydrostatic transmissions. The speed range over which maximum power must be transmitted is not great — very high speeds are neither necessary nor practicable on load carriers of this type — but the greatest precision of control (*ie* the ability to start and stop smoothly and to position the vehicle with great accuracy) is absolutely vital. Hydrostatic transmissions can meet the control requirements better than any other form of transmission.

Three types of output drive are possible:

(i) High-speed hydraulic motor or motors connected through reduction gears to conventional axles with differential units;

(ii) Low-speed wheel motors;

(iii) High-speed wheel motors with integral planetary reduction gears.

In theory, it would be desirable to dispense with axles, which are expensive, and impose certain constraints on the layout of the vehicle. Low-speed wheel motors or high-speed wheel motors with reduction gears obviate the use of axles. However, the wheel-motor units have to be attached to the vehicle frame and one pair at least must be steered. Provision for oscillation about the longitudinal axis of the vehicle may be necessary to confer stability on uneven ground. Possibly the greatest drawback to the use of wheel motors is that the most suitably shaped radial piston type is difficult to design as a variable-displacement unit, while the variable-displacement axial piston unit — either as a low-speed direct drive unit or as a high-speed/planetary gearing combination — has poor physical shape characteristics for wheel mounting. Stepped displacement characteristics can be designed into radial piston units by allowing a proportion of the pistons to be disengaged when low displacement is required. Alternatively, series/parallel flow to a number of constant-displacement wheel motors can be arranged to give steps of effective displacement and, therefore, several speed ranges. The necessary valving is complicated however, and although technically possible does not commend itself commercially. For any vehicles whose required speed/torque characteristics are such as to need a variable-displacement motor, it would seem that a high-speed motor or motors driving through conventional differential axles is the most practical solution.

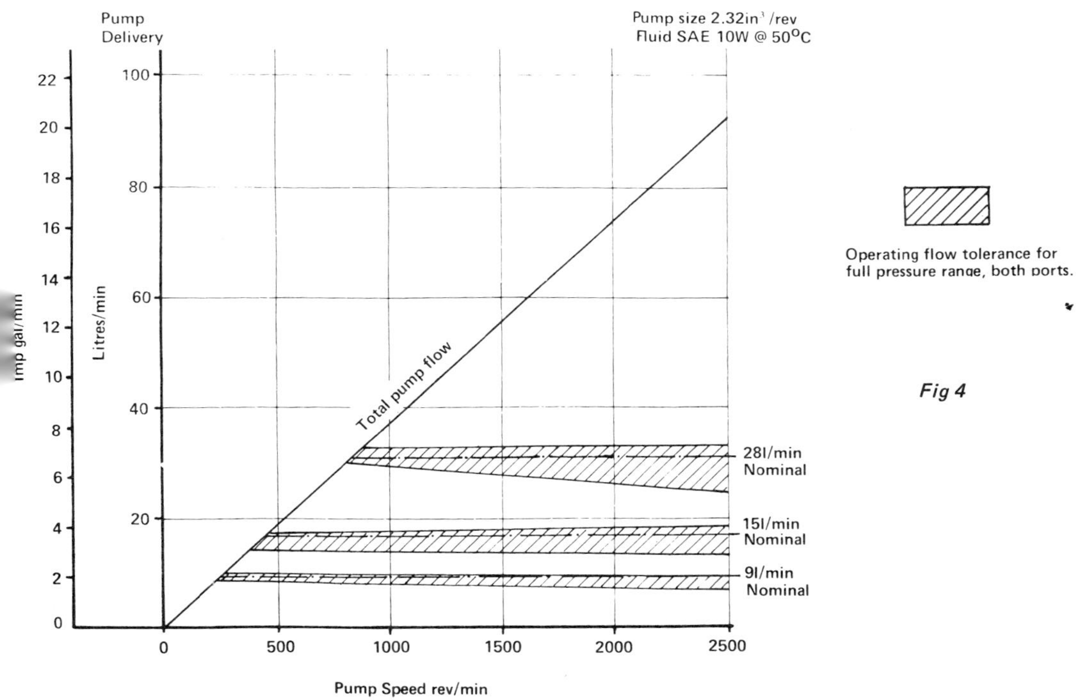

Fig 4

Power Steering

Most mobile mechanical handling equipment in use today employs power steering, generally of the hydrostatic type without mechanical linkage. The metering units are normally of the geared rotor or vane type and are used in conjunction with power steering pumps having integral flow-control valves so that the steering circuit is supplied with virtually constant flow irrespective of engine speed. In the simple spill type valve any flow which exceeds the pre-set requirement of the steering circuit is dumped back to tank or to pump inlet; circuit protection is usually provided by a pilot relief valve which, on cracking, signals the main metering spool to dump flow to tank. Such systems are, of course, power wasting, but having the advantage of simplicity and low cost are nevertheless very widely used.

The true priority-flow valve is similar in function except that the excess flow from the priority service, usually a large proportion of the total pump output, can be supplied to other hydraulic services. These services can be used at entirely different pressure levels from that of the priority service. Typical pressure levels are 103–140 bar (1 500–2 000 lb/in^2) for the priority (steering) circuit and 175–207 bar (2 500–3 000 lb/in^2) for the 'secondary' services. It is normal practice for this type of valve to include an integral full-flow relief valve, with its own return line, to protect the priority service; the alternative of dumping flow into the secondary line can be undesirable when other services are in use. Fig 4 shows typical control characteristics obtained from a modern flow-control valve.

A further facility offered by some flow-control valves is the provision of a variable adjustment for priority-flow settings, useful on applications like augers where differences in material density may require a range of settings rather than one fixed flow level.

Control Developments

Problems of size and siting of directional-control valves have led to the development of remote operation of the main valves by various means. Hydrostatic systems have been developed and used on applications such as manned work platforms, but more commonly low-power pilot valves at the operator's hand servo operates the main control spools. Pilot operation can also give advantages of control; manual operating loads are very light giving accurate 'feel' and the servo-actuators can be designed to give a non-linear relationship to the main valve spools to improve inching control of services.

A further variety of remote operation is the recently introduced electro-hydraulic controller. These small units, connected to the main control valve only by electric cable, offer proportional control (*ie* the control valve spool position, hence the load speed, changes in proportion to the movement of the remote hand control), giving a high degree of controllability and positioning accuracy. The freedom given by the cable connection enables the manual-control units to be incorporated in a portable box, which can be worn on a neck strap; machines can thus be operated in hazardous areas whilst the operator is in complete safety.

A recent important development in directional-control valves, particularly in the USA, has added yet a further dimension to the versatility of variable pumps. These valves, now known as 'load sensing', change the delivery of the pump in relation to movement of the control lever. In this system the pump servo is arranged to give a basic constant-flow control function and the spool valve opening is the variable orifice used to set the pump flow level. In maintaining a constant (low) pressure drop across the spool orifice the pump changes its flow output in relation to the orifice size and this principle gives the following important advantages:

> For set spool selections the pump flow produced is independent of system pressure and pump speed ensuring absolutely minimum power loss. Spool deadband is minimized and spool flow forces are small, due to the small metering pressure drop, giving rise to precise control and feel with service load speed proportional to valve movement up to maximum pump flow.

Mobile Cranes

Most mobile cranes in the 15 000 kg (15 ton) class are hydraulically actuated and considerable inroads are currently being made into the 30 000 and 60 000 kg (30 and 60 ton) and even larger classes. Hydraulics can provide infinitely variable speed actuation at least as effectively as any other form of power transmission. It affords an extremely high power to weight ratio and is now more competitively priced in relation to the alternatives.

Cranes demand greater control than any other materials-handling vehicle. Safety, as in other handling equipment, is of paramount importance. Lift trucks, side-loaders and container handlers use linear power output actuators (cylinders or rams) for most functions. In addition, they require rotary actuators or hydraulic motors for some functions, such as winch drive and slewing operation. Telescopic mobile cranes normally have derrick cylinders, a telescopic cylinder to extend and retract the jib and hydraulic motors for the hoist winch and slewing. Lattice jib or strut cranes will normally have two or three hydraulically-driven winches controlling the jib and hook through wire rope connections. Both types require stabilizer jacks or outriggers to confer the stability necessary for working at extended radii round the machine.

Most hydraulic cranes are powered by constant-displacement gear pumps. Winches are driven by gear motors through reduction gears or, in some larger sizes, directly by slow-speed radial piston motors. Starting, stopping and speed regulation through the entire speed range is handled by manual directional-control valves. Fine metering control is essential. Starting a suspended load requires that the hydraulic motor develops a high stalled torque efficiency (low torque or a 'running' start is not possible). Although modern gear motors have excellent stalled torque efficiencies, there is nevertheless a considerable variation as each gear tooth passes through its engagement cycle. Many times there is sufficient 'slack' in the drive for the variation to be unnoticed — backlash in splines, reduction gears, flexible couplings, *etc.* However, in a crane with a suspended load all 'slack' is taken up as the load must be supported by the motor immediately the brake is released. A dual motor in which the tooth engagement in each section is phased so that one section is at its most favourable torque position while the other is at its least favourable, offers a considerable improvement, giving consistent stalled torque efficiencies in excess of 90% of theoretical maximum torque output. A dual motor can also be arranged for series-parallel operation, offering the possibility of a two-speed drive which can be invaluable for crane winch operation where high, light-load hook speeds are desirable. Piston motors also exhibit starting torque variations, depending on the number of pistons used, for which allowance must be made when choosing the displacement of the motor for the particular job.

Leakage across the internal elements of a fluid motor, whether of the piston or rotary type, demands the installation of a friction brake on crane winch applications. The use of a high-speed motor with reduction gears allows a relatively small brake to be mounted on the high-speed input shaft. The load is generally lowered against hydraulic pressure (imposed by a valve) sufficient to support it. Consequently the brake is not used dynamically. Many crane winch motors now incorporate an inbuilt disc brake which is spring-loaded to the 'on' position and provided with an automatic hydraulically-powered release mechanism.

Most modern hydraulic cranes which utilize constant-displacement pumps are equipped with two, three or even four pumps. Multi-pump combinations permit the incorporation of stepped speed ranges providing, in combination with control valve metering, infinitely variable speeds at the output units without excessive power wastage. A typical mobile crane hydraulic circuit such as would be used on a 15 000 kg (15 ton) machine is shown in Fig 5.

Mobile cranes with extremely long lattice jibs are commonly used during the construction of high rise buildings, bridges and other civil engineering work. Considerable skill is needed to position loads which may be more than 30 m (100 ft) above the operator's cab. High lifting speeds are necessary to keep cycle time as low as possible. However, high lowering speeds with heavy

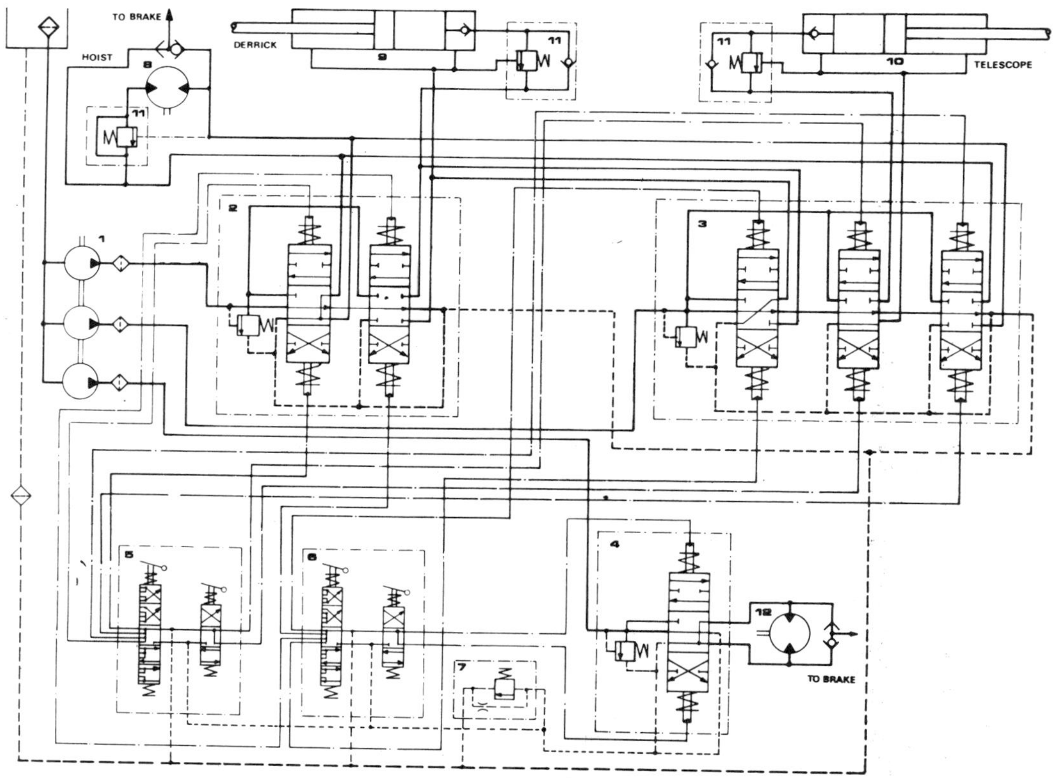

Fig 5 Typical circuit diagram for mobile crane with remote power control.

1 – Triple section pump.
2 – Control valve. 1st speed hoist. 2nd speed derrick.
3 – Control valve. 1st speed derrick, telescope and 2nd speed hoist.
4 – Control valve. Slew.
5 – Remote control valve 1st and 2nd speed hoist and telescope.
6 – Remote control valve 1st and 2nd speed derrick and slew.
7 – Exhuast loading valve.
8 – Hoist motor.
9 – Derrick ram with lock valve.
10 – Telescope ram with lock valve.
11 – Counterbalance valve.
12 – Slew motor.

loads are undesirable and dangerous. Hydraulic control gives the possibility of improved performance in all these respects.

Pressure-operated interlocks can be arranged to prevent high speeds automatically when high loads are lowered. The possibility of providing feed-back servo-controls to allow predetermined lifting or lowering speeds to be 'dialled in' by the operator is also worth investigating. Close control of a load so far away as to be almost 'out of sight' could be ensured and many breakages eliminated.

There is no doubt that the use of closed-circuit variable-displacement hydrostatic transmissions for winch drives is now economically feasible. Some large capacity crane winches already incorporate this type of drive in preference to the more rudimentary open-circuit gear pump-motor combinations. An extension of the use of true variable-displacement hydrostatic drives can be envisaged, particularly to large power outputs where the gains in average operating efficiency could be considerable.

The problems of providing effective manual operation of the control valves on many materials-handling machines can be difficult to solve. Cabs are small with limited accommodation for direct lever-operated control valves, especially those of large capacity. Hydraulic lines of sufficient capacity for the large flows which are now common are difficult to run to a central location, particularly if a considerable number of services is involved. There is considerable virtue and, indeed, necessity in the case of large units in siting the control valves in positions affording maximum accessibility and minimum pipe runs. By intelligent siting, the length of pipes and hoses and the numbers of connections and bends can be considerably reduced, offering a reduction in cost not only in pipe and fittings, but also in installation, which may be of even greater significance.

Mechanical links between the operating levers and valve spools are possible in some installations but may require excessive ingenuity and cost in installations with turntables or elevating cabs. A solution utilizing remote hydraulic power controls is finding increased acceptance. Small pressure-control servo-valves can be hydraulically inter-connected to the spools of the main control valves and so arranged that the displacement of the main spool is a function of the movement of the servo-valve lever. A linear relationship equivalent to that which would be obtained by direct actuation of the main spool is possible and can be improved on. The servo-valve can be designed so as to give a large increment of controlled outlet pressure for a small spool travel and thereafter small increments of controlled pressure for the remainder of the spool travel. The main valve spool can be centred by multiple spring packs giving a discontinuity of spring rate and therefore force increment at various points in its travel. The combination of these two features enables a large proportion of the servo-spool travel to be utilized while the main spool is traversing only a relatively short metering band. This results in better metering characteristics than if the main spool were directly operated. The servo-operating pressure is of the order of 15–20 bar (217–290 lb/in^2) so that small bore plastic hose with low-pressure fittings can be used for inter-connecting the main and servo-valves. Supply lines to and from the servo-valve would be similar. Instantaneous flow rates of up to about 14 lit/min (3 gal/min) from the servo-valve are all that is required to ensure a maximum response time of 0.2 secs in the largest size of main control valve. The servo-valve is arranged for console mounting and designed for a low level of operating force consistent with providing adequate 'feel'. Hold detents, if required, are provided on the servo-valve rather than on the main valve and are arranged for pneumatic, hydraulic or electrical release as may be appropriate. In addition to providing remote control, the system also ensures extremely low-effort operation of even the largest control valve. Operator fatigue is markedly lower with a consequent improvement in machine effectiveness and the main control valve is totally enclosed and does not require any dynamic seals.

Industrial Robots

THERE ARE many different types of industrial robots which can be categorized in various ways. The most common classification is by degree of sophistication, *viz*

(i) *Manual manipulators* — which are basically simple tools, handling devices, *etc* worked manually from a remote position.

(ii) *Pick-and-place robots* — low-technology robots limited to a range of simple movements.

(iii) *Fixed sequence robots* — capable of performing movements in a pre-determined sequence, and where the set information cannot readily be changed.

(iv) *Variable-sequence-robots* — where the programmed performance can readily be changed for another.

(v) *Play-back robots* — which repeat a programme initially 'taught' to them by a human operator.

(vi) *NC Robots* — programmed by numerical control data, *ie* basically the same control mode as a numerically controlled machine.

(vii) *High-technology robots* — which in addition to being programmable have a built-in capacity to detect changes in the work environment or work condition and correct accordingly.

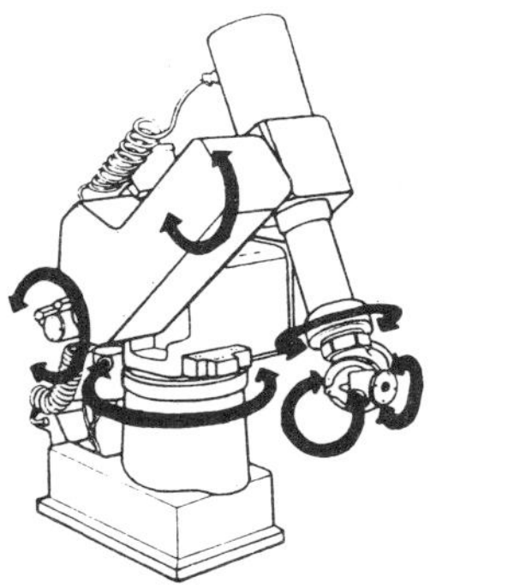

The six degrees-of-freedom commonly employed in industrial robots.

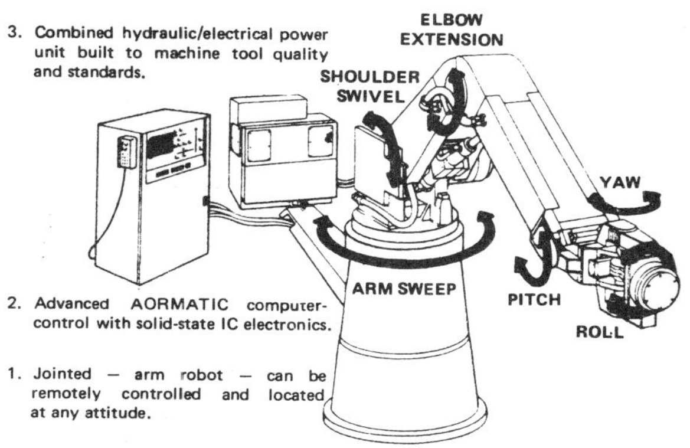

Cincinnati Milacron T3 robot system.

Pick-and-place, and a proportion of fixed sequence robots may have relatively straightforward movements which can be mechanically, pneumatically or hydraulically operated. Other categories with computer-type control are invariably based on servo-systems for movements. On an overall basis non-servo-types are decreasing in proportion to servo-types; and whilst pick-and-place robots represent the greatest number of robots in use, a high proportion are now servo-types. Non-servo-types now account for less than 25% of all industrial robots.

The fundamental components of an industrial robot are the mechanical system, the power-drive system, the sensor and/or servo-system, and the functional control system.

Four different kinds of mechanical systems have evolved, each with a readily identifiable pattern of movement.

Cartesian or Rectilinear — the gripper moves along by three perpendicular tracks to achieve the required height, width and depth.

Cylindrical — here an extendable arm is mounted on a central slide which can go up or down and swivel on its mounting.

Polar — an extendable arm is mounted on a central pivot.

Anthropomorphic — based on the human form, the mechanical arm can bend at an 'elbow' and swivel at a 'shoulder'.

As to which is the best geometry is dependent on the application. Whilst the anthropomorphical type is the most flexible, it is the most difficult to control. Flexibility and dexterity both increase with the number of joints and the quality of control.

Three types of drive system are in wide use:

(i) Pneumatics — requires a clean compressed air supply.

(ii) D.C. Electrics.

(iii) Hydraulics — usually with a built-in power pack.

Unimate 2000 presents the cast rotors to the stripping press which removes the mandrels at Brook Crompton Parkinson Motors.

The Unimate (R) 2000 industrial robot removes the brass casting from the rear of the press at J.W. Singer and presents it to an infra-red eye which immediately detects any faults or defects in the shot.

Pneumatics gives a cheap and simple power system but does not allow easy control of either speed or position. D.C. Electrics are convenient and precise but there is a power limitation. Hydraulics is highly reliable and there are a number of positive advantages of this system over the others, the most important being the greater overall power and response capability. Also very useful is the availability of particular hydraulic components, such as the cylinder for linear actuation and accumulator for energy storage and meeting peak demands.

Finally, for situations where there is a flammable atmosphere hazard (*ie* in paint spraying) intrinsically safe electro-hydraulic systems can be used to ensure that electrical power levels are kept so low as to eliminate the chance of ignition. This offers a very considerable weight and space saving over flameproof equipment.

In general, electric motors are particularly suitable for high-speed light-duty movements. Pneumatics is also well suited for simpler, high speed units. Hydraulics becomes almost an automatic choice for larger robots. Currently something like 50% of all industrial robots use hydraulics; 25–30% pneumatics; and about 20% d.c. electrics. Some industrial robots use a combination of two different power sources. A very large proportion of the more sophisticated robots are hydraulic powered.

A Unimate 2000B robot at Metal Castings (Worcester) Ltd presents a casting to a photo-cell scanning device whilst the machine has recycled to make another shot.

The body shells of the Metro being spot welded together by 28 Unimate robots operating on two body framing lines at British Leyland's Longbridge plant.

Controls

The sensor and servo-control system can be either quite straightforward or as sophisticated as required. Simple pick-and-place robots use on/off controls to drive against mechanical stops or limit switches with springs and dampers to stabilize motion. Movement may, of course, be jerky: if a smoothly controlled action is wanted then a servo-system is necessary. The power supplied to each axis can be modulated by a d.c. power amplifier or by an electro-hydraulic control valve. This allows the arm to be driven in a continuous and pre-determined manner using, for example, a signal recorded in the functional control unit or from a control handle. Combining the sensor system and servo-system allows feedback control in which the control signal to the servo-system becomes the displacement of the gripper as sensed by the instrumentation; this allows increased position accuracies to be attained. Continuous path control can be provided which is really an extension of the position control just described. The path is broken down into a sufficient number of intermediate positions with point-to-point control between each. This leads to a teach mode emerging in which, as the robot is moved through a trajectory, a recording is made of the sensory position feedback. This recording can then be played back, as quickly or slowly as required, as a drive signal to achieve the desired path.

The functional control system can be as basic as a programmable plug board, consisting of a matrix where the rows correspond to each axis and a stepping switch energizing each row in turn to operate the robot through its cycle. Electronic computers are clearly the most powerful and attractive means of control. Here the remarkable reduction in cost brought about by the more recent development of micro-processors opens up many possibilities. Robots can be programmed to carry out specific tasks, take alternative actions dependent on results from sensory feedback and, through communication links, interact with other robots.

Effect of Robot Size

A large, powerful robot normally works in noisier and less pleasant surroundings than its smaller counterpart. A smaller robot will have an easier environment and probably a softer job. Also, because of its location, it will be treated with greater care. All of these factors make it possible to design the small robot with a smaller safety factor than that required for a larger robot. As a result, the cost advantage of a hydraulically driven robot diminishes with size.

With installation, maintenance and other operational costs to be taken into account, the most cost-effective small robot is usually electric driven. The exact crossover point between hydraulic and electric drives may vary with robot configuration and the robot's intended use.

A main limitation which may be noticeable with electric-driven robots is that drives are under-powered, necessitating the use of high gear ratios to obtain the necessary dynamic performance. As a consequence they can be fast acting for small movements, but prove very slow for large transfer moves.

A particular advantage of hydraulics in this respect is that energy can easily be stored in an accumulator and released when an extra burst of robot activity is required. Thus it is not uncommon to find some hydraulic robots which momentarily require a delivery of say 60 gal/min being supplied by a pump of only a quarter of this capacity operating part time in loading an accumulator.

Other particular advantages are the simplicity and reliability of a hydraulic cylinder for providing linear motions; and the efficiency of a hydraulic motor in providing rotary motions. Part-rotary motions are also readily provided by suitable designs of semi-rotary actuators (*ie* vane-type actuators).

In paint spraying and other applications the environment may present an explosion hazard and the robot must either be explosion proof or intrinsically safe so as not to ignite the combustible environment. Here, the hydraulically driven robot has a great advantage over the electrical since the electric energy from feedback devices and the energy to drive servo-valves can be small enough not to ignite the explosive fuel/air mixture.

Similar considerations apply to pneumatic driven robots, with the advantage of faster movements. However electric drives are widely favoured for paint spraying robots because of their suitability for generating fast, small movements.

Machine Tools and Automation

PARTICULAR ADVANTAGES of hydraulics applied to machine tools are:

(i) It can provide a complete system for operating all machine tool movements from a single power source (*eg* a single electric motor driving a single pump or pump group).

(ii) The elimination of costly lead screws and the attendant anti-backlash equipment always associated with mechanical drives.

(iii) Extremely smooth movement under infinitely variable speed control.

(iv) The availability of proportional control response *via* proportional control valves and electric signalling.

(v) Ready adaptation to automation through manual control; or sequential or combination control *via* logic control circuits.

(vi) Simple and easily replaceable power cylinders or hydraulic motors, and accurate positioning using servo-control systems.

(vii) The possibility of using the hydraulic power for auxiliary functions as well as primary controls, and even for spindle drives.

Basic Hydraulics

A basic hydraulic system for a machine tool consists of a motor driving a pump, which circulates oil from a reservoir to the various control valves. Filters in the system are essential to maintain a high standard of cleanliness in the oil, removing all types of solid contaminant, *ie* to provide effective silt control. Judicious positioning of the control valves can save piping complications and it is often convenient to locate them in groups fed from a common manifold.

Motion control can be by cams, limit switches and stops; from numerical or tape control systems, working through solenoid-operated or hydraulically piloted valves; or sequential/combinational logic circuit designs. A particular advantage is the range of speeds possible, which would be very difficult to achieve with a standard a.c. motor without the complication of sophisticated electronics.

In some cases, the actuators are simply cylinders with the piston mechanically connected to the moving part, saddle, slide or table; in other cases, the cylinders are combined with the slide itself, with the piston and rod remaining stationary. Normally, they are double acting, but in a few instances it is possible to have a spring return, *eg* in work-holding clamps. For milling or multi-drill heads, where torque requirements and speeds are high, hydraulic motors are used to advantage.

The actuator rod diameter is simply determined from the designed maximum thrust to be transmitted. The desired lowest slide feed rate (cm/min) is compared to the smallest repeatable flow rate (cm^3/min) of the flow control valve to be used (typically around 30 cm^3/min) and an effective piston area calculated ($cm^3/min \div cm/min = cm^2$). Knowing the effective area and the maximum thrust, the maximum oil pressure at the actuator can be computed and, by taking control and other circuit element pressure drops into consideration, the maximum working pressure at the pump is determined. Flow rates are dependent on actuator size and speed of movement and the number of actuators operating simultaneously, *ie* the machine sequence of operations. This information is usually combined on a machine sequence diagram, from which the circuit elements (pumps, accumulators, control valves, filters, pipe diameters, *etc*) are selected.

The control elements, such as directional, pressure control and flow control valves, can be mounted on individual subplates and piped together, or they can be mounted on a module block with internal connections.

In the case of elements that are piped together, common problems are that the control requires too much space, that the pressure drops through the resulting long pipe runs are too great and that these are expensive and not always reliable. The result is a high price to pay for the control. When the individual elements are mounted on module blocks and these are then put together to form control systems, the expense of pipework is reduced.

However, if the pipe connections to the actuators and to the oil reservoir go out from a particular side of the control, this can lead to awkward runs for the connecting pipework. Moreover, any subsequent additions to the controls, for either of the aforementioned types, can only be carried out with difficulty. It would be better in these cases, therefore, on standard machine tools and standard machine-tool elements (*eg* unit heads), for designers to consider proven module

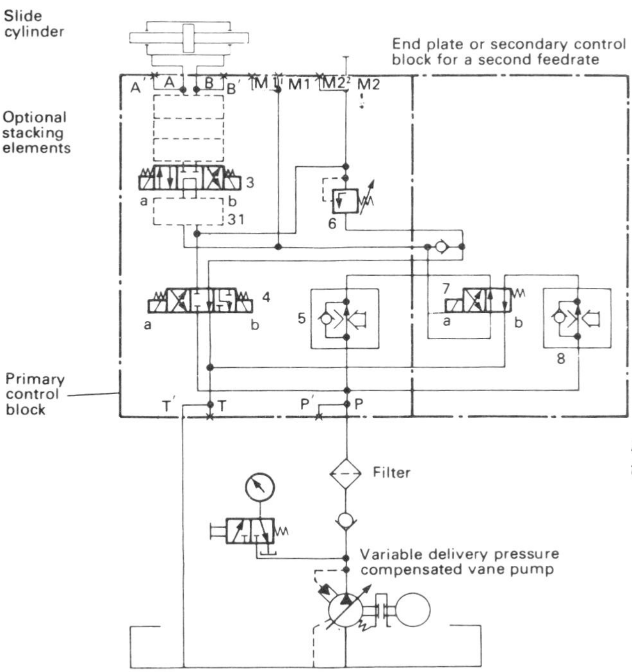

Fig 1 Simplified sketch of a machine tool hydraulic control system showing internally connected modules. (G. L. Rexorth Ltd).

Depending on the type of work-holding, a sandwich pressure-reducing valve and double-solenoid directional valve are fitted to the clamp or chuck cylinder. Displaced oil is returned to tank *via* the directional valve and the T port of the block. Reverse operation is effected by energizing the second solenoid. The check valve gives the clamping system a measure of protection against any drop in the supply pressure, especially when an accumulator is fitted.

The versatility of mounting is similar to the feed-control blocks, having ports suitable for either manifolding or pipe connections on the underside and duplicate threaded ports on the sides. Applications of these controls to machine tools are diverse, but include as examples: fine boring with boring units, drilling machines, radial turning with boring heads, turning and forming with single slides and parting-off on lathes, and the operation of checks and clamping systems of all types on various types of machine tools.

Automation

There are various degrees of automation ranging from semi-automation to a fully programmed system, *viz*

(i) *Semi-automation* — typified by sequential control where an operator is responsible for start/stop and also inspection and supervision.

(ii) *Automation* — typified by sequential control with 'feedback' where an operator is only necessary for stop/start.

(iii) *Full automation* — or a programmed sequential control system using punched tape or a computer.

(iv) *Fully programmed automation* — where the control is fully programmed and is capable of self-analysis, correction, *etc.*

Control Circuit Design

Control system design must start with a precise analysis of the movements which must be performed, normally using a distance-time diagram. It is then necessary to allocate power components (*eg* cylinders) to achieve these movements and select suitable control components (*eg* robots). The major problem is then to plot the circuit to achieve correct working with no possibility of lock-up and, preferably, elimination of all redundancies. This can demand considerable skill and experience in circuit design and is normally best done on a logic basis.

See also chapters on *Pneumatic Logic Controls* and *Industrial Robots.*

The return stroke is usually hydraulically operated but the system is simplified if a spring return is fitted. Pumps and motors vary with the individual makers but plunger pumps, giving a pressure of 280 to 420 bar (4 000 to 6 000 lb/in^2) are preferred for applying the working pressure and the speed can be varied by unloading one or more plungers. To avoid over-loading the motor the maximum pressure may be reduced as the speed is increased.

If the main force has only to be applied for the last fraction of the stroke, a dual pump unit has advantages, a gear or vane pump giving a rapid approach. The high-pressure pump can then be comparatively small.

Medium-Sized General-Purpose Presses

Both upstroking and downstroking presses are made but the preferred layout, except where operational considerations conflict, is the downstroking press with all possible hydraulic gear mounted on top of it (Fig 2). This arrangement keeps floor space required to a minimum, although it does pre-suppose adequate roof height; the tank should be immediately above the cylinder, so simplifying the prefill valve. The pump can then be mounted inside, above or alongside the tank.

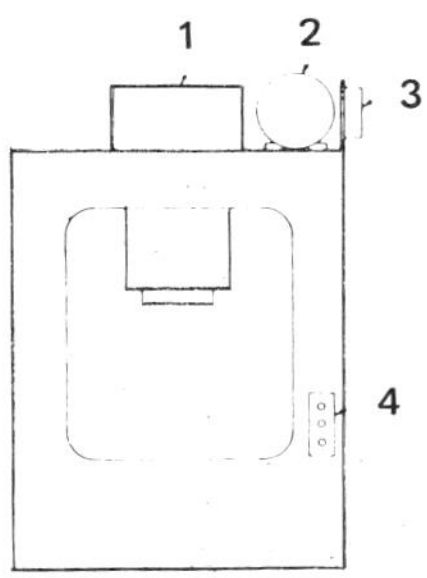

Fig 2 Medium sized hydraulic press with pump and control gear mounted on top. 1—tank. 2—pump. 3—valves. 4—push buttons.

Fig 3 Circuit for hand controlled press. 1—main cylinder. 2—tank. 3—prefill valve. 4—slow-down valve. 5—reflux valve. 6—main valve.

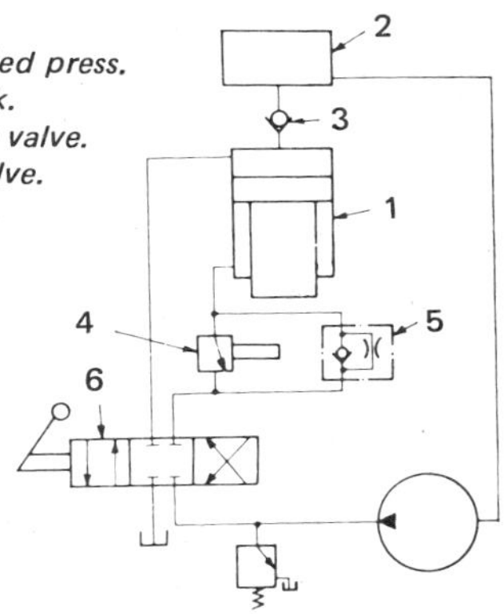

Double-acting rams avoid the complication of having additional return cylinders. Metal piston rings may reduce maintenance costs, although piston head packings have been omitted altogether and the consequent slight leakage on the return stroke tolerated.

If a double-acting ram is used gravity has to be relied upon to speed up the approach; the weight of the suspended parts is normally sufficient to ensure this, and indeed the speed may even need regulating.

A typical manual-control circuit for a medium-size press is shown in Fig 3. On the down stroke the main cylinder (1) draws oil from the tank (2) through the prefill valve (3). Oil from the annulus passes at first through valve (4), but as the ram reaches the end of its troke, valve (4) is closed by a cam and the oil must then pass through the restriction of the reflux valve (5), and is slowed down. When the ram meets the resistance the prefill valve closes and pressure builds up until the relief valve blows. On the return stroke the oil from the cylinder passes through the main valve (6) back to the tank, and the oil to the annulus passes freely through the check on the reflux valve (5).

Electrical controls are now normally preferred and can result in some simplification of the hydraulics on more complex working cycles. A typical electrical-control circuit is shown in the neutral position in Fig 4. Oil is circulating back to the tank but restriction (8) causes sufficient back-pressure to operate the reducing valve (6) which supplies the pilot system for the main directional valve (5). In moving valve (5) to the 'down' position, pressure is applied to both ends of the pull-back rams, the net result being a downward force, the speed of descent being controlled by the reflux valve (3). The main ram draws in oil from the tank (1) through the prefill valve (2). When the ram encounters resistance pressure builds up, and when it reaches about 14 bar (200 lb/in^2), valve (4) is opened, connecting the pump directly to the main cylinder. The main relief valve (7) also protects the pull-back circuit if, for example, reflux valve (3) is set too closely.

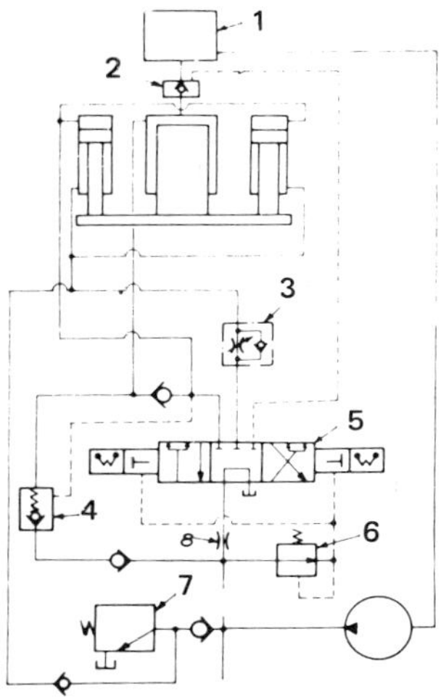

Fig 4 Hydraulic circuit for electrically controlled press.
1—tank. 2—prefill valve. 3—reflux valve. 4—pressure operated valve. 5—main valve. 6—reducing valve for pilot system. 7—main relief valve.

For the up stroke, pressure is applied to the annulus of the pull-back rams and also the pilot of the prefill valve, so that the main cylinder discharges directly back to the tank and the top of the pull-back cylinder discharges through the main valve.

A limit switch would return the main valve to the neutral position at the top of the stroke.

Larger presses are usually made for specific duties, of which representative types are:

Drawing Presses

The superiority of hydraulic presses for deep drawing is well established. Presses equipped for single-, double- and triple-action drawing are now made in a variety of sizes.

The conventional method of double-action drawing (Fig 5) uses a pressure plate controlled by rams, usually four in number, clustered around the main ram. These are arranged to apply the pressure plate before the main punch makes contact. The force on the pressure plate can be varied as pressing proceeds and made proportional to the resistance to the punch ram. Suitably shaped pressings can be passed right through the tool, making ejectors unnecessary.

It is often more convenient, however, to invert the punch and die (Fig 6) and the blank is laid on the pressure plate which is carried down against the resistance of the bottom ram. The bottom cylinder can be connected to an accumulator or to the main hydraulic system.

With the conventional stationary bottom cylinder the main ram and press frame must be suitable for a load equal to the sum of the forces required for drawing and for the pressure plate. The size of press can be appreciably reduced by fitting the pressure plate cylinders to the crosshead

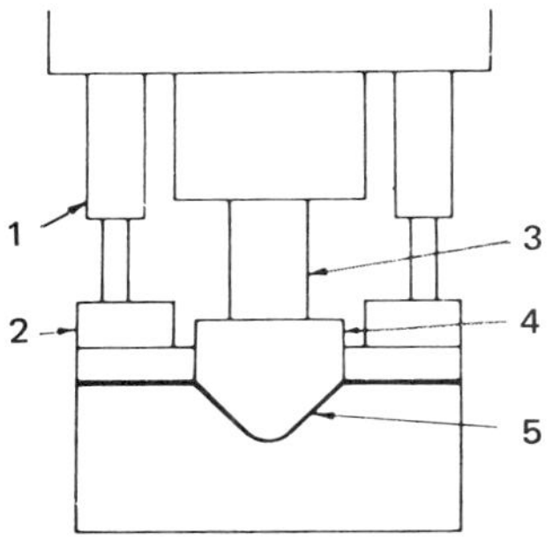

*Fig 5 Double action drawing.
The pressure plate grips the blank whilst the punch forces the metal into the die, so preventing puckering.
1—pressure plate rams. 2—pressure plate. 3—main ram. 4—punch. 5—die.*

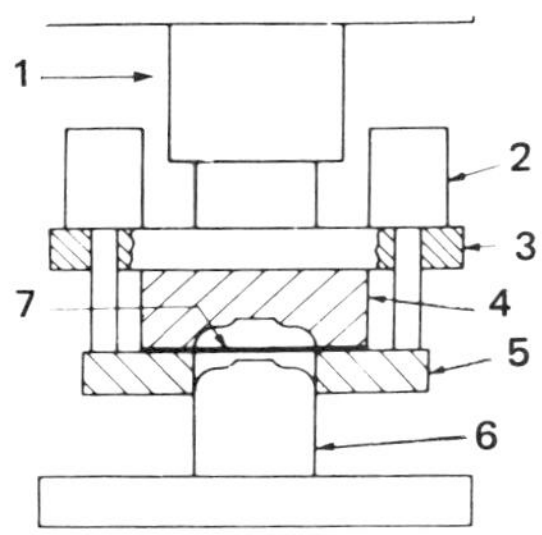

*Fig 7 Double action drawing with gripper cylinder carried on crosshead. This arrangement enables a lighter press to be used.
1—main ram. 2—pressure plate cylinders. 3—moving crosshead. 4—die. 5—pressure plate. 6—punch. 7—blank.*

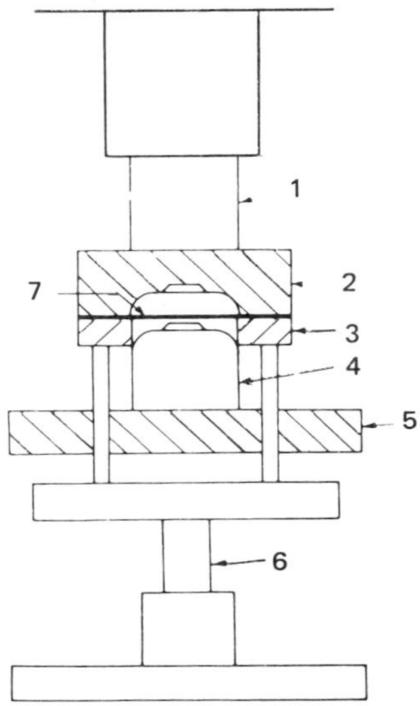

*Fig 6 Double acting press with die cushion acting from below. The die is forced over the punch and carries the pressure plate with it. The main ram exerts sufficient force to overcome the resistance of the material and the upward thrust of the bottom ram.
1—main ram. 2—die. 3—pressure plate. 4—punch. 5—press bed. 6—bottom ram. 7—blank.*

carried by the main ram (Fig 7). The pressure plate force is now completely self-contained and the main ram need only be large enough to give sufficient force for the drawing tool. With this method it is necessary to provide a separate pressure supply to the pressure plate rams and they cannot be connected directly to an accumulator. An alternative construction, giving the same effect, has a single ram below the press tool supported from the crosshead by tie-bars.

Triple-Action Drawing

Triple-action drawing tools enable a wide variety of intricate shapes to be drawn including those with portions indented in the main surface. In one method the metal is first gripped between the outer die and pressure plate (Fig 8) by rams clustered round the main ram, and the drawing operation is done by the punch pushing the metal into the bottom die in which a counter punch recedes against the thrust from the bottom rams.

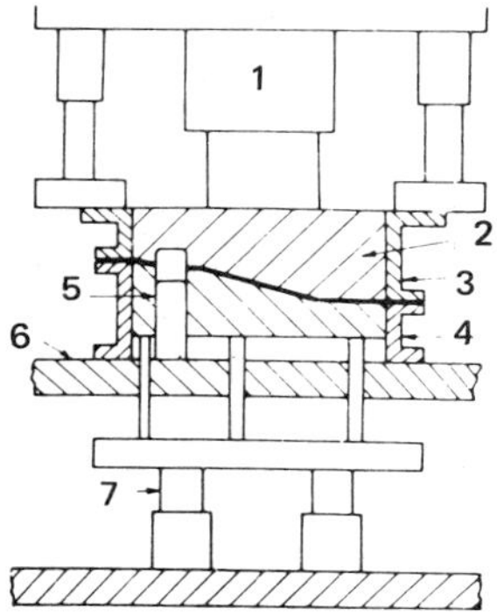

Fig 8 Triple action drawing used for production of difficult pressings. 1—main ram. 2—punch. 3—pressure plate. 4—die. 5—ejector. 6—main bed. 7—ejector rams.

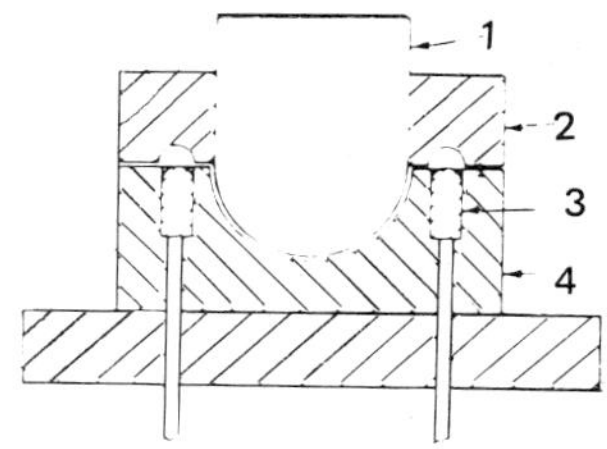

Fig 9 Triple action drawing tools for pressing the metal in the reverse direction to the main draw. 1—punch. 2—pressure plate. 3—reverse punch. 4—die.

Another method draws the metal in the usual way and afterwards it is drawn in the reverse direction by a reverse punch which is forced upwards by the lower ram into a shaped cavity in the pressure plate (Fig 9).

Forging Presses

The efficiency of a forging press is measured by the speed at which it can produce a forging of exactly the right size — the higher the speed the less the number of re-heats required. This calls for automatic stroke control, reversing the hammer at the exact point necessary to achieve the desired thickness.

Stroke control can be achieved mechanically or electronically. In a typical mechanical system the position of the tool relative to the press frame is continuously measured by a counting device, initially set to register the number of counter units representing the final tool position. The counter then counts down as the tools converge, until zero count is reached. At this point the return valve is triggered and the ram is reversed. Correction must be incorporated for ram speed (usually by tacho-generator) and frame stretch (as a figure proportional to pressure).

A typical hydraulic circuit, using electro-hydraulic servo-valves, is shown in Fig 10. Main valve (A) is operated by the two servo-valves (B) and (C). For bottom reversal both valves apply a 'blip' of pressure to the main valve which is held in place by a detent. If either valve (B) or (C) fails, the other valve is sufficient to operate valve (A). At the top of the stroke the pressure held in the main valve is normally exhausted through valves (B) and (C) in series. If one has failed, the passage is blocked and the press is stopped.

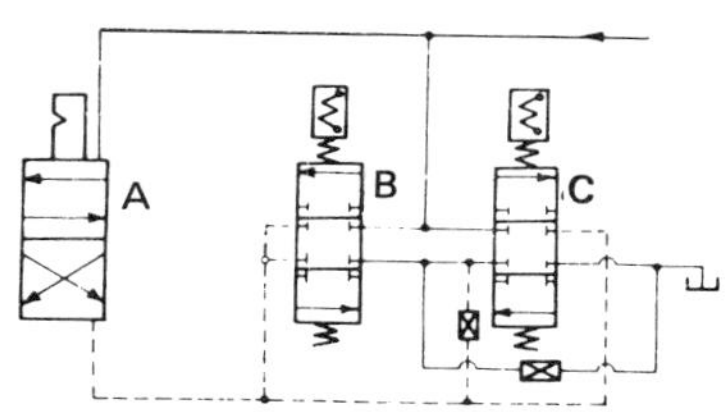

Fig 10 Arrangement of two servo valves for reversing at top and bottom of stroke with automatic stopping at top of stroke should either valve fail.

Electrical controls can be analogue or digital. The best accuracy which can be obtained with an analogue system under industrial conditions is about 1 in 500. There is no limit to the accuracy of the digital system, but for a forging press, the accuracy corresponding to 0.79 mm in 3 251 mm (1/32 inch in 128 inches) or 1 in 4096 is sufficient.

A simplified layout of a digital control system is shown in Fig 11. Here there is the top (moving) tool (1), the workpiece (2) and the bottom tool (3). The position of the moving tool is sensed by the counter (or digitizer) (5) which is sensitive to every 0.79 mm (1/32 inch) change. The final position of the top tool, which is determined by the forging thickness, is set by the operator on the counter (4). The feeds from both (4) and (5) are taken to the subtractor (6) from which, when the difference is zero, an impulse is sent to the 'up' servo-valve (9) through the amplifier (8). At the top of the stroke the signal to the 'down' valve (10) is given when the signal from the subtractor (6) is the same as that fed from the top of stroke positioner (11), the two being compared by the comparator (7). The simplified block diagram (Fig 12) shows more precisely how the electronic units are connected.

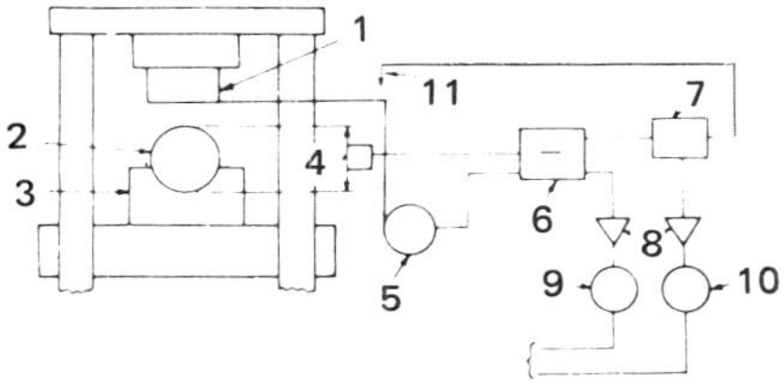

Fig 11 Simplified layout of digital control system for forging press thickness control.
1—top tool. 2—workpiece. 3—bottom tool. 4—counter. 5—digitizer. 6—subtractor. 7—comparator. 8—amplifier. 9—'up' servo valve. 10—'down servo' valve. 11—stroke positioner.

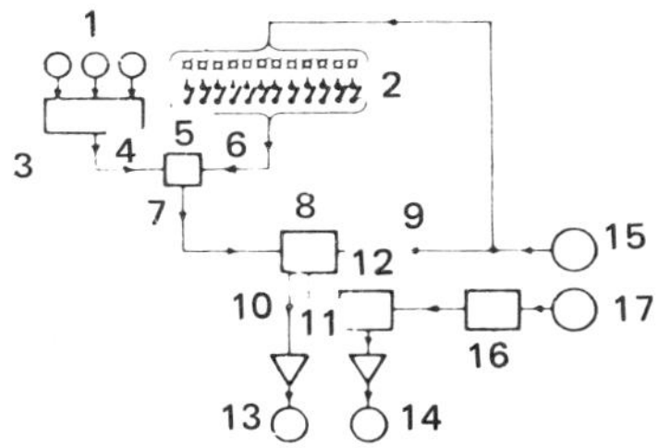

Fig 12 Block diagram of circuit for stroke control. 1—setting switches. 2—zeroing equipment. 3—decimal-binary matrixes. 4—forging height. 5—add. 6—bottom tool height. 7—required stop position. 8—subtract. 9—press position. 10—zero error signal. 11—error. 12—compare. 13—up valve. 14—down valve. 15—digitizer. 16—matrix. 17—return stroke setting switch.

The setting switches (1) allow the operator to set the forging thickness by three dials to the nearest 0.79 mm (1/32 inch), and this is converted into binary signals by a simple decimal-binary matrix, these being taken to the adder where they are combined with the zero height signals. The adder is a transistorized logic unit. The function of the digitizer has already been explained. The zeroing equipment works as follows — the press is brought tool-to-tool, without a workpiece and the digitizer output is taken to 12 lamps, one for each power of 2. A switch below each lamp is then set 'on' if the lamp is illuminated and 'off' if it is not. The signals are taken to the logic adder so that the signals from the adder to the subtractor now represent the bottom tool height plus work thickness, or the reversing position of the tool. The remaining operations are as described for Fig 11.

Signals representing hammer speed and frame stretch are fed into the electronic system and a variable dwell of 0–5 seconds is provided at the top of the stroke to give time to manipulate the ingot. It is also possible to decelerate the press, which may be falling at three times the forging speed, at about 25 mm (1 inch) from the bottom position. This fast approach saves time but better control is given if the press is slowed down gradually. The deceleration signal is also obtained from the digitizer.

High-speed forging presses can be operated automatically if co-ordinated with suitably controlled manipulators and a system is shown in Fig 13.

Digitizers, similar to those fitted to the press, measure the height and forward position. The manipulator is traversed by a hydraulic motor or cylinder, whilst it rotates the head in pulses of a pre-determined number of degrees.

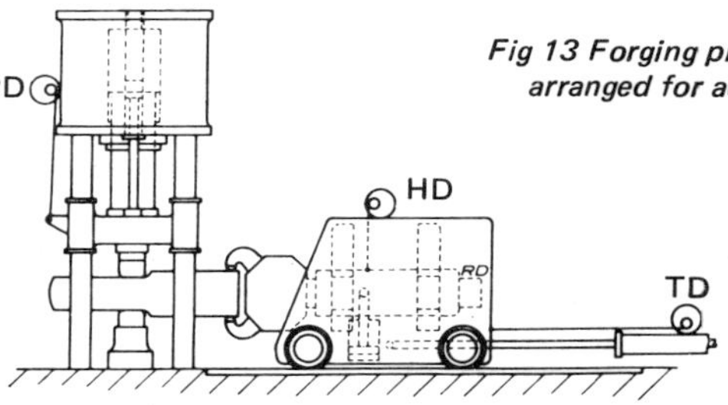

Fig 13 Forging press and manipulator arranged for automatic control.

The programme for the production of the forging is prepared, giving the final dimensions of each phase, the reduction for, and the forward travel between, each squeeze. Such information can be recorded on punched cards.

Coining Presses

Coining presses are similar to forging presses in requirements, except that the piece is cold-worked by stamping dies designed so that the area of the faces to be flattened is reduced to a minimum. Forces between 300 and 1 500 tons may be needed, with precise pressure-control and fast cycling times. Such machines are commonly fitted with an automatic feeding device and hydraulic ejectors.

Extrusion Presses

The size of extrusion presses varies with the type of product and the extrusion ram may exert a force of from several hundred tons for the smaller presses to several thousand tons for the larger. As the efficiency of the press depends to a large extent on the speed with which it can be fed with heated billets and the dies changed, these items require special consideration and are often operated pneumatically, or, on the larger presses, hydraulically.

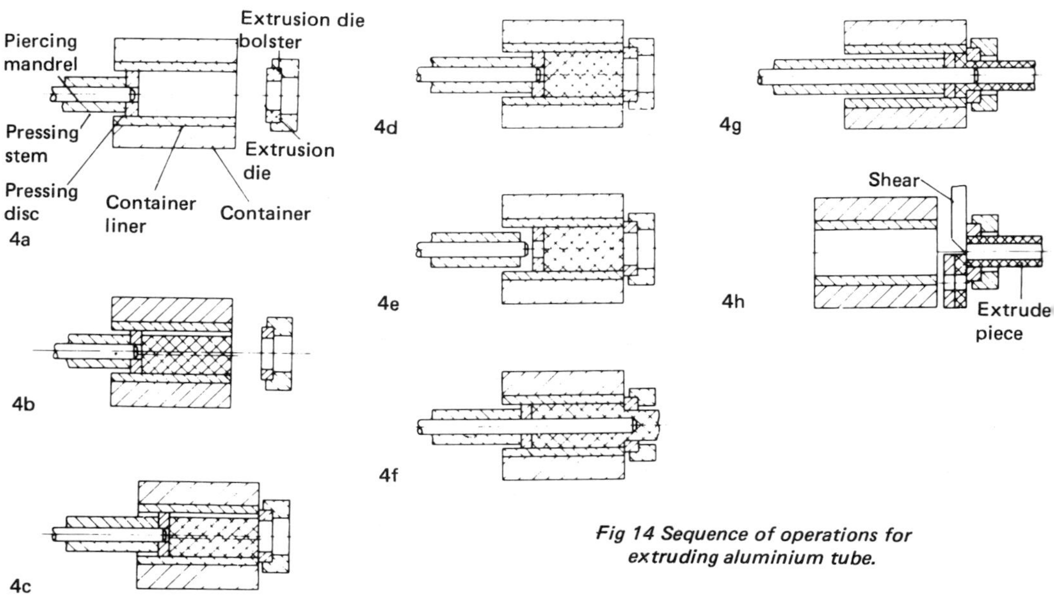

Fig 14 Sequence of operations for extruding aluminium tube.

Hydrostatic Extrusion

Hydrostatic extrusion is done cold and involves the introduction of a liquid filling between the ram and the billet so that the liquid supports the billet over its whole area and also acts as a lubricant. Pressure as high as 300 kbar (200 tons/in^2) may be involved. Use is also being made of a technique which employs both direct force and liquid pressure.

Fig 14 shows a billet with a high length/diameter ratio being extruded. The method is equally suitable for tubes, using a travelling mandrel, clad materials and wires. The latter are coiled before placing in the liquid cavity. Brittle materials are extruded into a back pressure as this is found to improve ductility.

Stretching Machines

Many types of rolled and drawn non-ferrous sheet and sections are treated by stretching beyond the yield point. The material is straightened and its physical properties are improved. Stretching machines are made in a wide range of sizes and are able to deal with every size of drawn and rolled non-ferrous section.

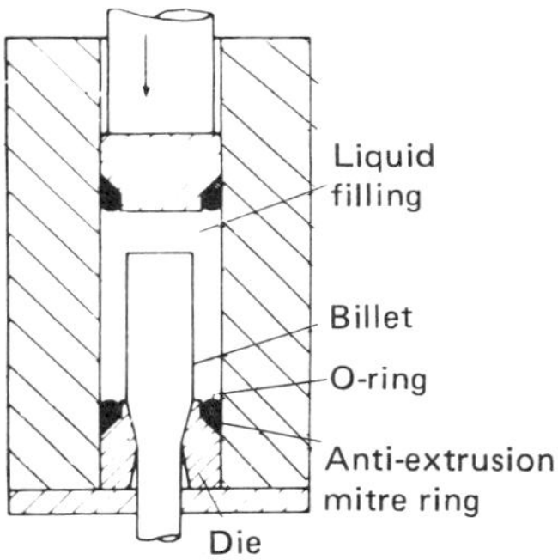

Fig 15 High pressure chamber on hydrostatic extrusion press.

The circuit shown in Fig 15 is for a 152 t (150 ton) machine with a 254 mm (10 inch) diameter working ram and 89 mm (3½ inch) diameter return ram. Main interest in this circuit lies in the provision for preventing damage should the work break.

As it stretches the material, the main ram pushes the return ram back. The oil escaping from the return cylinder is at first forced through the subsidiary relief valve RV2. As the main ram pressure increases, the servo-pressure on valve SV is sufficient to operate it, so that it bypasses the relief valve and oil passes back to the tank at a fairly low pressure. If the material breaks, the pressure in the main ram will fall, valve SV will close, and the oil forced out of the return cylinder will have to pass again through the relief valve, so providing a powerful brake. The operator closes the main valve immediately.

Some other types of special-purpose hydraulic presses are:

Body Panel Presses — large table-area presses for forming sheet metal panels. Typical maximum pressure 300–350 tons.

Rubber Die-Forming Presses — used throughout the aircraft industry for the production of wing and fuselage components.

Gap-Type Presses — for wheel forming, forging, straightening and bending. Typical size 80–85 tons.

Cold-Forging Presses — with capacities up to 1 500 tons for making components such as gear shafts with minimum material usage, improved grain structure and enhanced UTS.

Wheel Rim Presses – producing pressed steel wheel rims. Typical size range 220–500 tons.

Gramophone Record Presses – fully self-contained presses with steam heating and water cooling for pressing gramophone records.

Plastic Moulding Presses – for either continuous or single-cycle automatic control, incorporating automatic platen heating and cooling control.

Laminating Hot-Plate Presses – for making plastic laminates. Typical maximum sizes 2 000–2 500 tons; steam-heated and water-cooled.

Board Presses – with variable pressure and dwell for the production of insulating and building boards. Typical size 3 000 tons.

Baling Presses – for baling cotton, coir fibre, sisal and wool.

Brake Lining Presses – typical capacity 220 tons.

Dry Ice Press – for making blocks from carbon dioxide gas.

Hobbing Presses – for the die manufacturing industry.

Grinding Wheel Presses – hot and cold platen presses for the manufacture of grinding wheels.

Powder Compacting Presses – for powder metallurgy production, *etc.*

Special Hydraulic Requirements

Special valves are required in hydraulic presses for dealing with the high pressures and high flow rates encountered, and also for dissipating the energy stored in press frames and cylinders and in the fluid itself due to compressibility effects. If such energy was released through a conventional spool valve there would be severe shock forces generated. Also the heat generated and the high oil velocity would lead to damage to the valve seat. A special decompression valve must therefore be used and this is often incorporated in the main directional-control valve. It is of the 4-way type, with ports for high-pressure supply, exhaust to reservoir main cylinder, and return annulus or subsidiary cylinders. To operate the press the control lever is pulled outwards; this connects the pressure port to the cylinder port *via* the non-return valve and the return oil passes through the hollow piston to exhaust.

For releasing the pressure and allowing the ram to return, the decompression valve is shown in detail in Fig 16. It contains a subsidiary ball valve and a push-rod. When the control lever is pushed inwards, the end of the valve piston contacts the push-rod X, which lifts the ball valve off its seat against pressure. This allows the oil to escape through the small holes Y past the ball valve and out to exhaust through holes Z. The relatively small passages restrict the speed of the oil flow to a reasonable figure. As long as there is an appreciable pressure in the cylinder, it keeps the exhaust valve seated, but if force is maintained on the control lever the valve piston will push it open when the pressure has dropped sufficiently. When the piston valve is pushed right over the return cylinders are connected to pressure and the oil is forced out of the main cylinder. It should be noted that the main cylinder and reservoir connections are appreciably larger than the other ports so as to cope with the considerably larger flow when the ram is retracting.

The Use of Unloading Valves

An alternative method employs a direction-control valve which is somewhat similar to that just described, but an additional pilot-operated valve vents the pump in the neutral position. The circuit is shown in Fig 17.

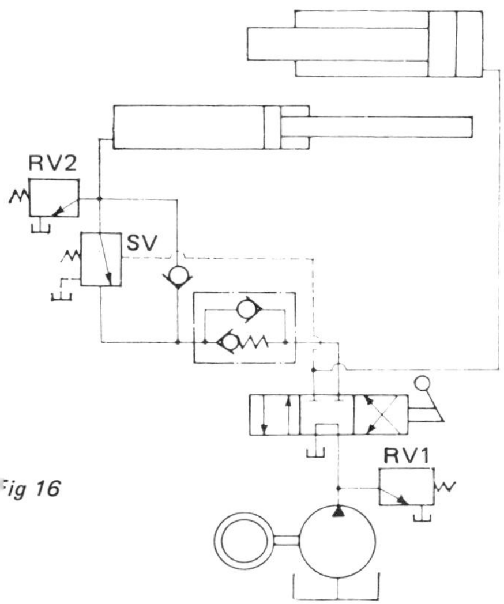

ig 16

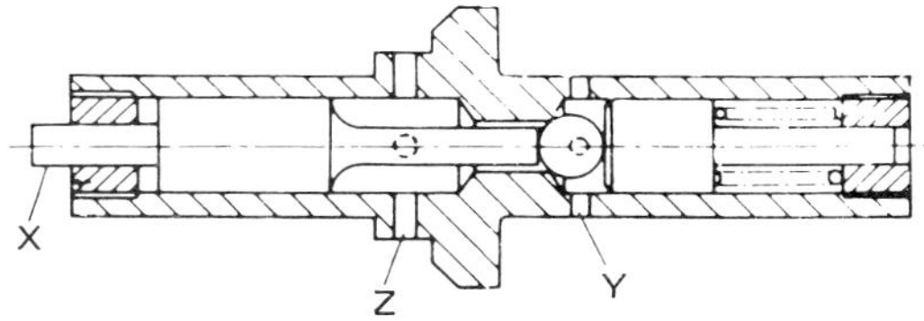

Fig 17 Detail of decompression valve fitted in spool of main valve.

The directional-control valve (Fig 18) has a pressure port A, exhaust port B, main cylinder port ; and return cylinder port D. The valve isolates all these ports in the neutral position so that the am is locked in position. For the working stroke of the ram the piston is moved to the right (by ushing the lever inwards) to connect pressure port A to the main cylinder port C through the assage E. The return port D is connected directly to the exhaust cavity.

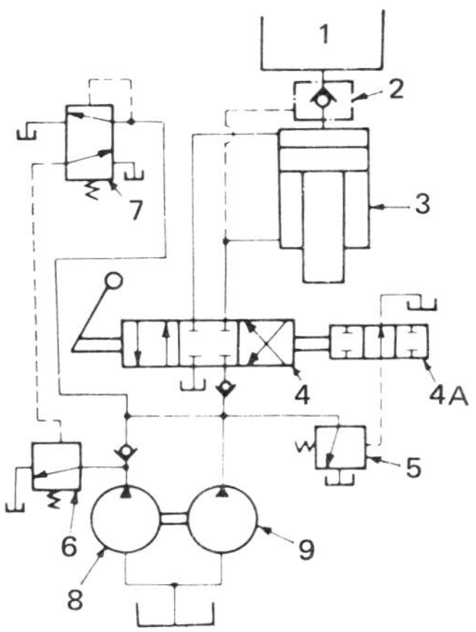

Fig 18 Press circuit showing functions of different types of valve. 1—tank. 2—pilot-operated prefill valve. 3—main double-acting ram. 4—main valve. 4A—pilot section of main valve. 5—unloading or amplifier valve. 6—unloading or amplifier valve (L.P. pump). 7—main relief valve cutting out L.P. pump before H.P. pump. 8—lower pressure pump. 9—higher pressure pump.

For the return stroke the piston is moved to the left, when it is prevented from making the full troke by striking the closed non-return valve. The holes F are, however, opposite the recess and ecompression takes place as the oil passes through these holes into the exhaust chamber. When he pressure has dropped sufficiently, further pressure on the lever opens the non-return valve and he cylinder contents flow freely to exhaust. The same movement connects the high-pressure port o the return cylinder port.

On the right of the lever is the unloading valve G which controls the amplifier valve. In the eutral position the two ports H are connected and the amplifier valve is unloaded by connecting the pilot circuit to drain. When the press ram is to be operated the unloading valve is closed by the piston moving across the gap between the lines. There is a certain amount of free motion between the main piston and the unloading piston, so that the pump is not loaded until the valve has been moved right home. The movement of the unloading piston is controlled by the washers K, which are unloaded by the spring in both directions.

The passage L is intended for draining the cavity M. The lever spindle is fitted with oil seals.

Relief Valves

Relief valves are normally pilot-operated for high-pressure operation on oil. One type has been developed for use with a two-delivery pump, giving a large volume and small volume high-pressure supply, one of which it unloads at zero pressure just before the valve reaches the pressure to which it is set. The other source is then bypassed to exhaust at the valve-setting pressure.

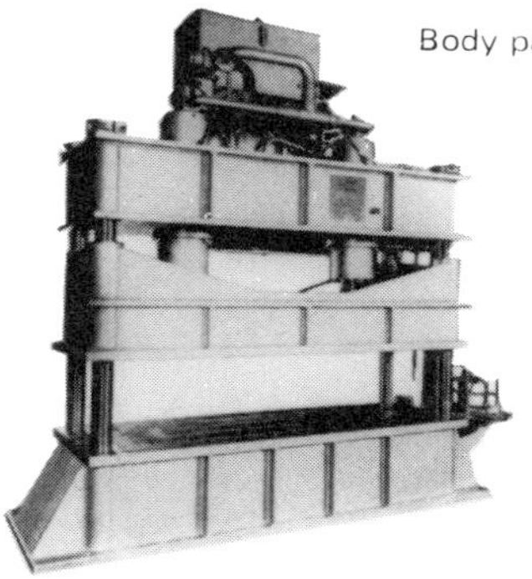

Body panel

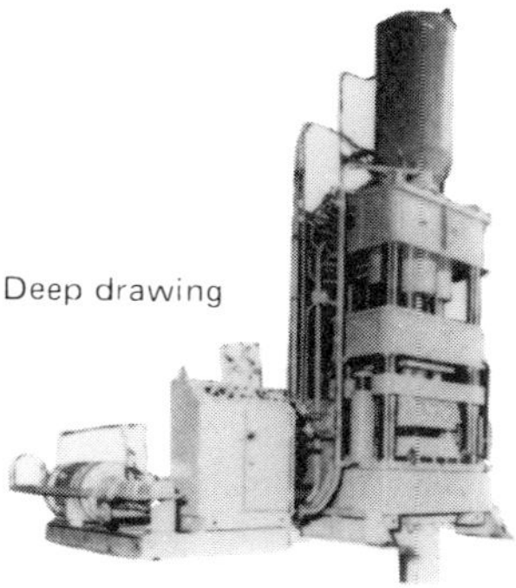

Deep drawing

Cold forging

Standard form

Wheel rim

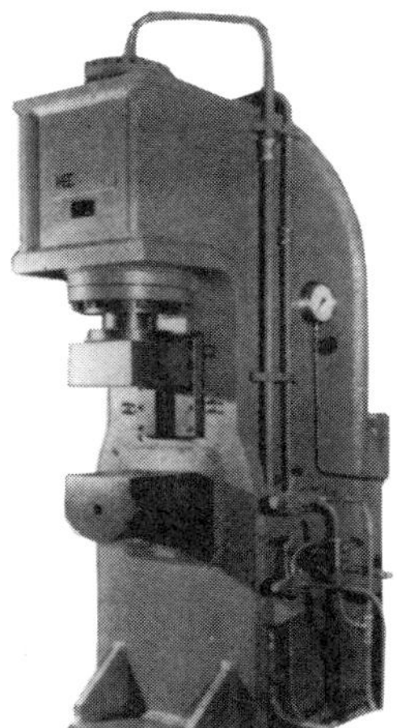

Gap type

Examples of hydraulic presses.
(The Hydraulic Engineering Co. Ltd).

Relief Valves for Large Presses

In large presses certain conditions can cause a rapid pressure rise, which can become dangerous, before the ordinary relief valve has time to open.

A relief valve which is responsive to change of pressure obviates this difficulty, as it will open if the rate of rise is rapid enough at a pressure below the ordinary relief valve setting. Basically the valve is similar to a balanced pilot-operated valve, but with spindle balanced by pressure on its ends (Fig 19).

The capacity above the spindle is connected to the main pressure line through a restriction and normally the pressures are balanced with the valve held closed by a light spring. A rapid rise of pressure is delayed by the restriction from building up in the capacity chamber, and, if the difference is sufficient the spindle will open and in doing so will compress the oil which will act as a 'liquid spring'. This valve can function as a conventional relief valve by connecting a pilot relief valve to the capacity.

Fig 19 Hydrostatic extrusion press. (Fielding & Platt Ltd).

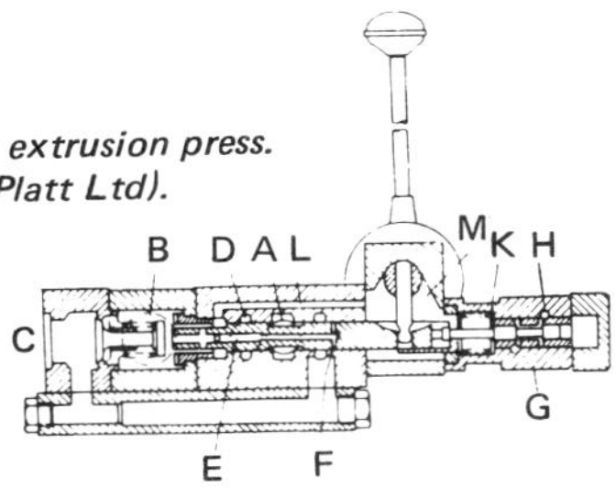

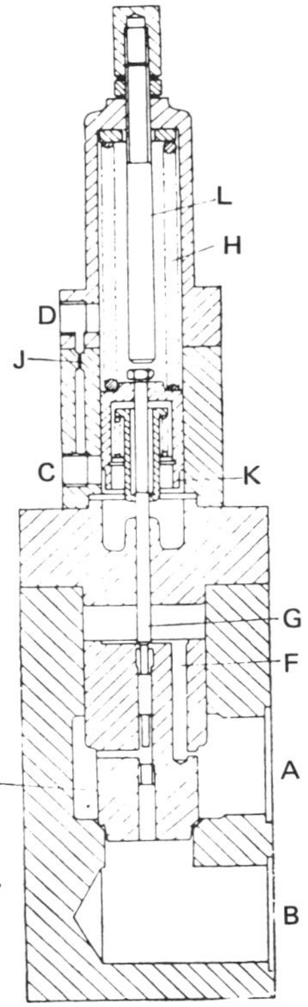

Fig 20 Amplifier or unloading valve. operated by pilot pressure.

Amplifier Valves

An amplifier valve which can be opened by the application of pressure is useful in some circuits, and an example of this is shown in Fig 20. The pressure is admitted at port A and port B leads to the reservoir. The pilot connections are made to ports C and D. Normally, the valve is closed by piston E, which is kept seated by the pressure from connection A acting on the differential area of the valve; it will be noted that the upper portion of the valve is larger in diameter than the valve

Fielding 16MN extension press for the aluminium industry.

Scrap press.
(Kammerich-Reisholz GmbH).

Hydraulic press
(Dunkes).

seat, the pressure being equalized through the internal passage F. The centre hole in the valve is sealed by the pilot valve spindle. The valve can act as an ordinary relief valve, as the pressure acts on the spindle C, and if it becomes too high the pilot valve is pushed off its seat against the load of the large spring H. The oil above the main valve can now escape past the pilot valve seat *via* a drilled passage, since the restriction to the internal passage in the main valve is greater than the restriction through the central hole and drilled passage between the top of the main valve and the exhaust port B. The absence of pressure on the upper side of the main valve now allows the differential area to act in the opposite direction and the main valve lifts, so allowing excess pressure to flow to drain past the tapered end of the main valve.

If it is desired to open the pilot valve by the application of a pilot pressure this is applied to port C and it acts on the underside of the piston K and lifts the pilot valve to unload the main valve as before. As the upper part of the housing is filled with oil, port D must be connected to exhaust to allow the displaced oil to escape and also to drain the flow through the choke J. The pressure difference between the connections A and B need only be 0.14 bar (2 lb/in^2) to keep the main valve open. The actual maximum valve opening can be set by the adjusting screw L.

When the pressure to the pilot connection C is interrupted the pilot valve will close as the oil flows through choke J, and by closing the central hole in the main valve, will cause it to close, the two valves lifting together.

If the choke is replaced by a solid plug, then the valve can be opened and closed by applying pressure to connection C and then connecting it to exhaust. If more convenient, connection C can be permanently connected to pressure and connection D connected alternately to pressure and drain.

Prefill Valves

For a given size of pump the approach speed of a press can be increased considerably by fitting a prefill valve which enables the oil to flow in and out of the cylinder without having to pass through the pump. Usually these valves are used with down-stroking rams, the valve being mounted on top of the cylinder, preferably in an oil reservoir.

Prefill valves are also made with an integral decompression valve, which takes the place of a decompression valve in the main control valve.

On small presses the pilot opening feature can be dispensed with and the cylinder exhausted through the main valve.

Hydraulic Workshop Tools

THE ORIGINAL hydraulic 'workshop' tool was the hydraulic hand jack incorporating a lifting cylinder and a pumping section in one combined unit. This basic design remains largely unaltered. Jacks for direct lifting actions are normally based on straightforward, rugged cylinders with short rods or plungers (for maximum capacity). They are 'sized' by lifting capacity in tonnes (tons), and stroke. Tonnage is determined directly by the cylinder bore and system pressure available. Strokes may vary from less than 25 mm (1 in) upwards. The majority of such jacking cylinders are single-acting, with load/gravity return, although if used for horizontal thrusting they may incorporate spring-return. Alternatively, a double-acting cylinder can be used in such applications.

Other operations which can be performed by simple hydraulic tools include the following:

Pushing	Pulling	Lifting	Forming
Bending	Straightening	Spreading	Pressing
Cutting	Clamping	Work-holding	

Other operations which can be performed by simple hydraulic tools include the following:

All the motions necessary for such modes of tool working can be derived from mechanical systems powered by a hydraulic cylinder. Tools may be individually designed for the job, in which case the mechanical system will be included as an integral part of the cylinder construction; or may be in the form of attachments designed to fit a single standard work cylinder. In the latter case a single cylinder can serve a variety of work duties, although, of course, only one attachment (tool) can be used at a time.

The simplest arrangement is where an individual tool can be connected to a hand pump *via* a flexible hose. This gives a basic pump/cylinder set. Equally the hand pump can be replaced by an electric or engine-driven pump set – Fig 1.

Fig 1 Hand pump and cylinder (left); and pump set and cylinder (right).

Where the use of more than one tool is contemplated a complete system of individual implements can be fed from a motorized pump connected to a manifold which acts as the distribution point for individual take-off lines. This system can be quite simple (Fig 2) or incorporate additional controls for pressure and flow regulation, an accumulator and additional valves in

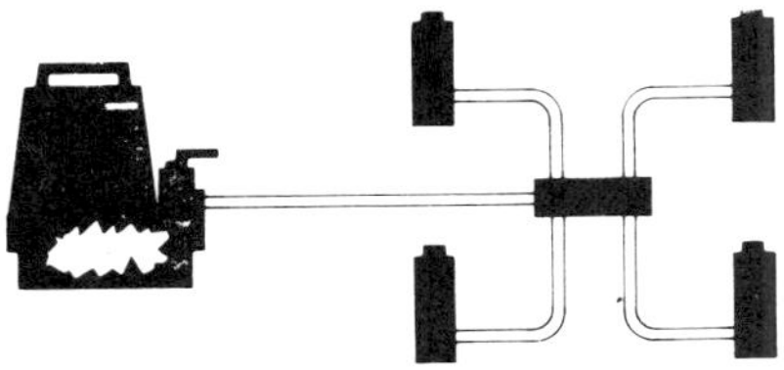

Fig 2 Basic pump and cylinder set, using an electric pump and additional cylinders.

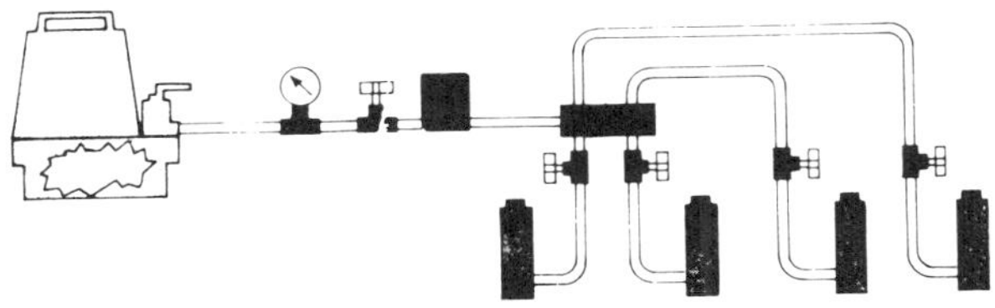

Fig 3 Basic pump and cylinder set, using additional control valves and cylinders for a more sophisticated circuit.

individual lines (Fig 3). The degree of control will depend on the particular system envisaged. Valves may be necessary to control flow rate, direction, pressure relief, shut-off and load-holding. Electric controls may be contemplated to limit, govern or stop cylinder motions, or to provide automatic pressure control. Pressure gauges may also be incorporated to provide a visual indication of pressure generated by the pump at individual work points as a guard against over-loading the tool or workpiece.

Cylinders

The design and construction of the work cylinder needs to be more robust than that of conventional hydraulic cylinders. In particular, the rods must be strong, with adequate bearings to resist bending and eccentric loads without damage to the cylinder. Since workshop tools are often used in unfavourable conditions, extra attention should be given to the design of seals to prevent the ingress of moisture, dirt or anything adhesive.

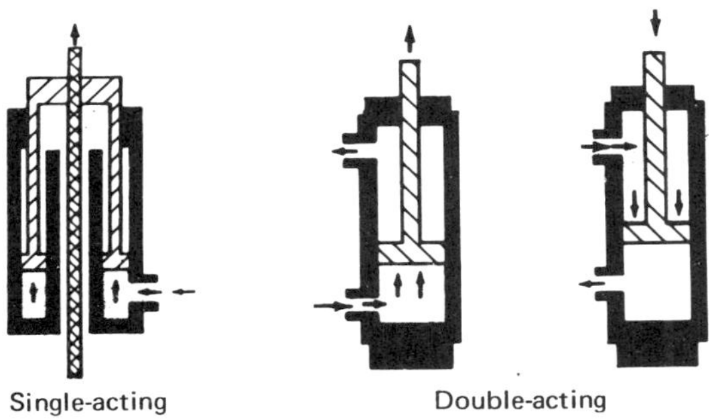

Fig 4 Hollow plunger cylinders. (Enerpac Ltd).

A single-acting cylinder can also be designed to provide both push and pull outputs — *eg* by making it a through-rod type, or by using a hollow plunger design as shown in Fig 4.

Double-acting cylinders have specific advantages for pushing, pulling and lifting applications, but have the disadvantage of requiring two lines with a selector valve for control. It is good practice to incorporate an integral safety valve in the cylinder to protect it from over-pressurization in the event of a return line being blocked.

In the case of plunger-type cylinders the output force is usually provided directly by movement of the plunger, or from a saddle attached to the end of the plunger. If the surface to which the force is applied is not exactly horizontal, the saddle can be of the tilting type to relieve the cylinder of excessive side-loading.

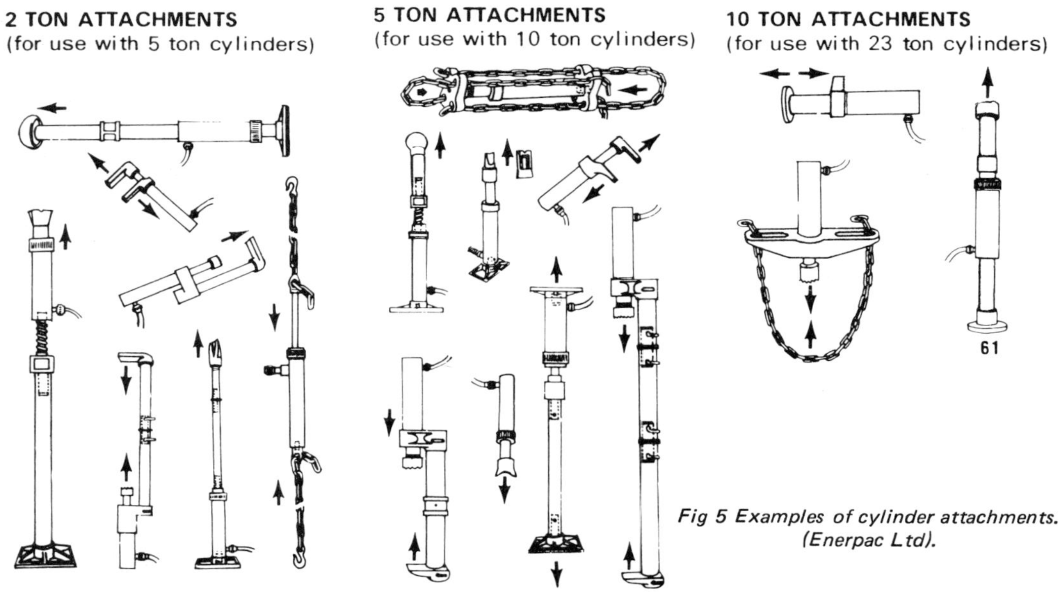

Fig 5 Examples of cylinder attachments. (Enerpac Ltd).

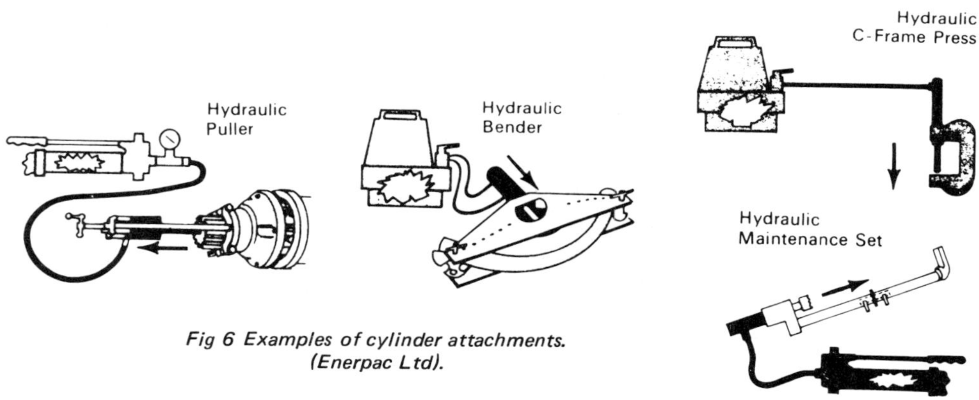

Fig 6 Examples of cylinder attachments. (Enerpac Ltd).

Similar saddles may be used on the end of cylinder rods, although the variety of attachments which can be used with 'force' or 'jacking' cylinders is considerable and largely dependent on the ingenuity of individual manufacturers. Some examples are shown in Fig 5.

Clamps, Benders, *etc*

The same principle can be extended to clamps, benders, pullers, *etc,* where the basic tool is designed to accommodate a standard cylinder as a source of power – some examples are shown in Fig 6.

For work-holding, cylinders can be fitted with an end-flange for rigid mounting with the clamping device attached to the end of the rod, or the cylinder can have a threaded body for mounting (the latter arrangement normally being restricted to miniature clamp cylinders, although work

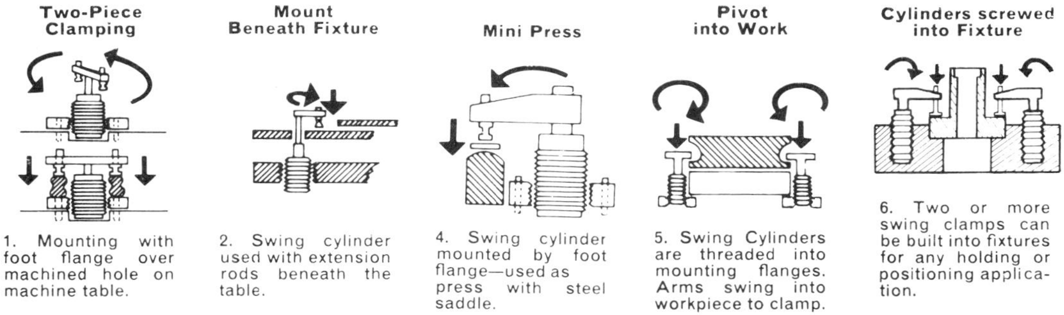

Fig 7 Examples of work-holding cylinders.

capacities may extend up to 4-tonnes (4 tons). Alternatively, a valve cylinder can be used to power a suitable design of mechanical clamp. Clamp cylinders are normally single-acting and fitted with spring-return. Some examples of swing-type work-holding cylinders are shown in Fig 7.

The hydraulic vice is another useful workshop tool which can provide controlled force of up to 4 tonnes (4 tons) or more in a very compact unit. This can be designed as a bench unit or for fitting machine tables. Another particularly useful tool in this category is the hydraulic collet chuck, used for holding individual workpieces for milling, drilling, slotting, *etc*.

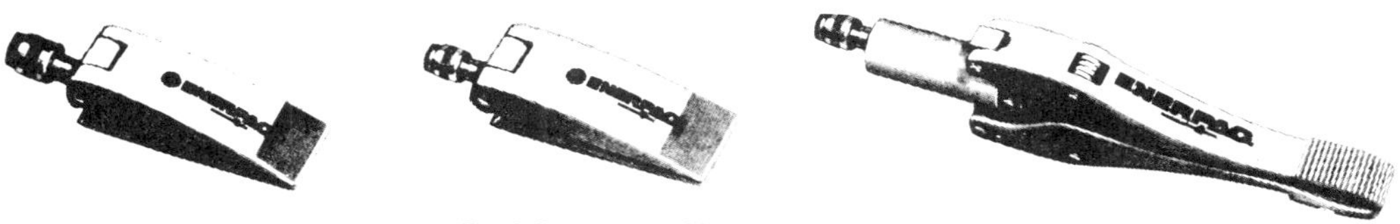

Fig 8 Examples of hydraulic wedges.
(Applied Power Ltd).

Examples of hydraulic wedges are shown in Fig 8. These may be designed as attachments to thread directly on to a standard work cylinder, or with integral cylinder. Wedge jaws and linkage can be proportioned to provide a considerable range of reaches and tip lift. Jaws are normally spring-loaded for automatic return. Since the maximum capacity at the tip is dependent on the mechanical system it is desirable to have a relief valve in the hydraulic supply, or at least a pressure gauge, to ensure that the maximum specified capacity of the wedge is not exceeded.

Hydraulics in Vehicles

HYDRAULIC BRAKES are standard in automobiles, the basic system employed having changed little over the past two or three decades. A master cylinder of the order of 25 mm (1 in) diameter is operated by a foot pedal through mechanical linkage so that a maximum pedal pressure of the order of 45 kgf (100 lbf) will produce a fluid pressure of the order of 40–60 bar (600–800 lb/in^2). This master cylinder supplies fluid to four slave cylinders simultaneously, Fig 1. The slave cylinders are normally of smaller diameter than the master cylinder, and their stroke is quite small so that the stroke requirements of the master cylinder are also quite small. The smaller size of slave cylinder also means that in the hydraulic system there is force reduction rather than force multiplication.

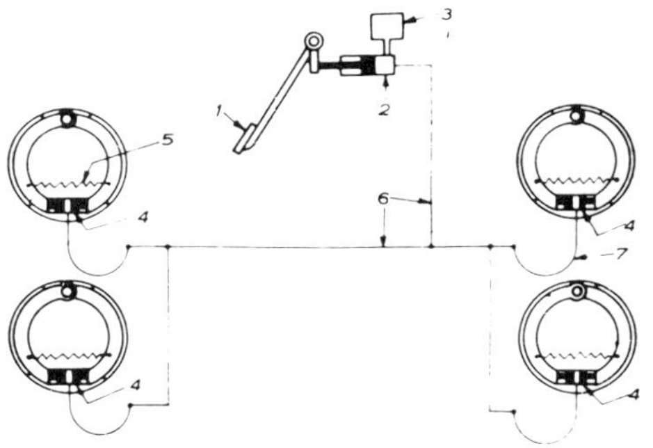

Fig 1
1—Brake pedal. 2—Master cylinder. 3—Reservoir. 4—Slave cylinders. 5—Return springs. 6—Metal pipes. 7—Flexible hoses.

The hydraulic circuit is a simple one-way type with spring return of the slave cylinders. No control valves are necessary since fluid flow is controlled by pedal pressure providing master piston movement in one direction and by spring return of the slave pistons when pedal pressure is released. A small reservoir is, however, necessary in the circuit both to ensure that the system always remains filled and to provide a compensation chamber to accommodate changes of fluid volume with temperature. This reservoir is also the filling and topping up point for the system.

Forces produced at slave cylinders on hydraulic braking systems can be readily altered by adjusting the size of the cylinders, but this is essentially a design factor. Thus to increase brake actuating force at, say, the rear wheels, the diameter of the slave cylinders for these wheels can be increased. Further redistribution of effective braking can also be achieved by modification of the actual mechanical units involved, *eg* to provide mechanical 'servo-action'.

Mechanically, the braking effort achieved will be directly proportional to the pedal pressure but this proportional effect will occur at different points of input movement as wear takes place. This effect is exaggerated by the fact that the mechanical movement is a force multiplier.

The efficiency, (or effectiveness) of the braking system is mainly dependent on the mechanics of the braking elements themselves, together with vehicle characteristics such as weight distribution and road surface condition, rather than the efficiency of the hydraulic system. The efficiency of force transfer (from pedal to wheel brakes) should, however, approach 100% provided the system remains full and free from entrained air. The main requirements for the fluid are that it should have a viscosity capable of providing sealing at all likely ambient temperatures and sufficient lubricity to lubricate the piston seals adequately. It also needs to have good chemical and physical stability since it is in contact with air in the reservoir. Seal requirements are compatibility with the fluid, low break-out friction (low 'stiction') and high resistance to ageing and cold setting. setting.

Main developments have been in the widespread adoption of disc brakes rather than drum brakes, and the diagonal splitting of the hydraulic circuit so that in the event of failure in one circuit braking is still available on the other (diagonally opposed) wheels. The even more desirable anti-locking system on individual wheel brakes remains slow in being adopted, mainly because of the increased cost of such systems.

Vacuum-Hydraulic Brakes

Where the brake-actuating forces required are greater than can readily be obtained through a manipulation of the various factors involved it is desirable to apply boost to the manual input. This normally involves the employment of a booster cylinder which supplements the pedal pressure above a pre-determined level of input effort. Higher input pressures are then accompanied by 'boost' pressure, resulting in a substantial increase in the force applied to the master cylinder and a consequent increase in system pressure without requiring excessively high pedal loads. Such a system also retains substantially the same 'feel' as a simple system, with braking effect directly proportional to pedal pressure. A typical vacuum booster unit is shown in Fig 2 which, used in conjunction with an otherwise conventional hydraulic circuit, is normally referred to as a vacuum-hydraulic servo-system.

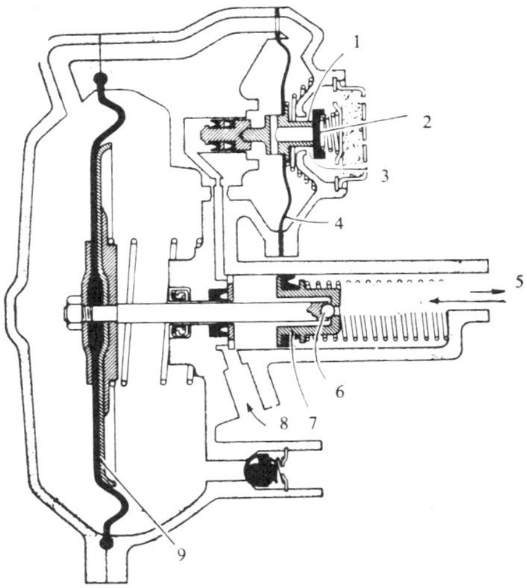

Fig 2 Mot-A-Vac vacuum-hydraulic servo in applied position.
1—vacuum valve seated. 2—valve disc. 3—atmosphere valve open. 4—plunger and diaphragm moved forward. 5—fluid displaced to wheel cylinder. 6—ball-end seated in piston. 7—piston moved forward. 8—from master cylinder. 9—main diaphragm moved forward.

Servo-power action can usefully be extended to assist manual operation of mechanical clutches. Since the engine is invariably flexibly mounted and the engine unit and clutch actuating linkage can have differential movements, a hydraulic link is also effective in both eliminating wear on all mechanical linkages and providing more precise response. It is becoming increasingly common to connect the clutch pedal and mechanical clutch *via* a simple hydraulic system of master and slave cylinders which is unaffected by relative motion of the engine unit and chassis. Force multiplication, (that is, 'servo-assistance') may or may not be provided, according to the size of cylinder and operating force required.

Servo-Operated Brakes

For heavier vehicles a complete servo-power system is generally preferable for brake actuation, utilizing the delivery from an engine-driven pump. A circuit of this type is shown in Fig 3, employing a tandem power valve and two accumulators. The pump draws hydraulic fluid from the tank and delivers it to a cut-out valve on the first accumulator and a non-return valve on the second accumulator. Each accumulator is connected to a unit on the tandem power valve which in turn connects to the appropriate brake lines. One accumulator feeds the front and the other the rear brake lines.

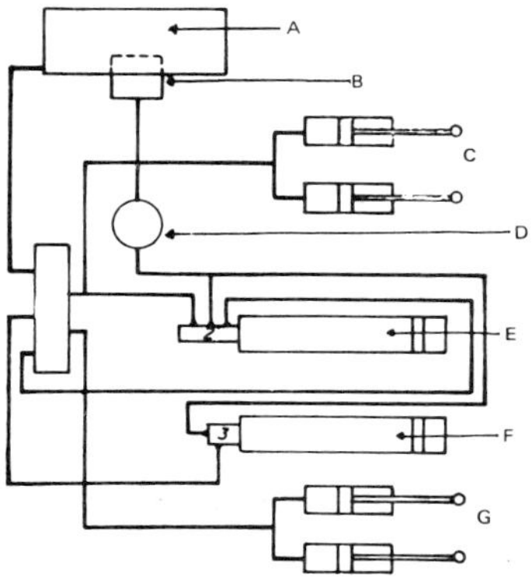

Fig 3 Example of power braking system circuit.
A-tank. B—filter. C—frame cylinders. D—pump. E and F—accumulators. G—frame cylinders. 2—cut-out. 3—non-return valve.

When the brake pedal is depressed this operates the power valve admitting pressure from the pressure lines into the brake lines, and actuating the mechanical braking system in the usual manner. Release of brake pressure allows fluid to flow from the brake lines to the return connections.

In power brake systems it is usual to arrange the lines so that in the event of a pipe breakage or similar failure power braking is still available on at least one axle. Manual reversion may or may not be made available in the event of complete power failure. If required, pressure switches can readily be incorporated into the system to indicate loss of pressure in the system or in either circuit of a dual system.

Power Steering

Power-assisted steering, in its basic form, comprises a hydraulic booster cylinder with one end fixed and the other linked to the steering arm, fed by an individual hydraulic circuit with its own pump and control valves. A typical system comprises a hydraulic pump (commonly of the vane type) driven by the engine, an oil reservoir (usually integral with the pump), an actuating cylinder incorporating a directional control valve, and flow-regulating and pressure-relief valves. These two valves may be incorporated in the same block which may be integral with the pump/reservoir unit. The actual arrangement of components can vary with the type of vehicle and space available. Some alternative methods of mounting are:

(i) Pump, valve and reservoir in a single block, requiring only two lines to connect to the actuating cylinder. The cylinder directional-control valve is then operated by mechanical linkage from the bottom of the steering column.

(ii) Separate pump and reservoir. This has the advantage of allowing the pump to be mounted at the front of the engine on an adjustable bracket. The reservoir unit may incorporate the valve block (Fig 4a); or the safety relief valve may be incorporated in the cylinder – (Fig 4b).

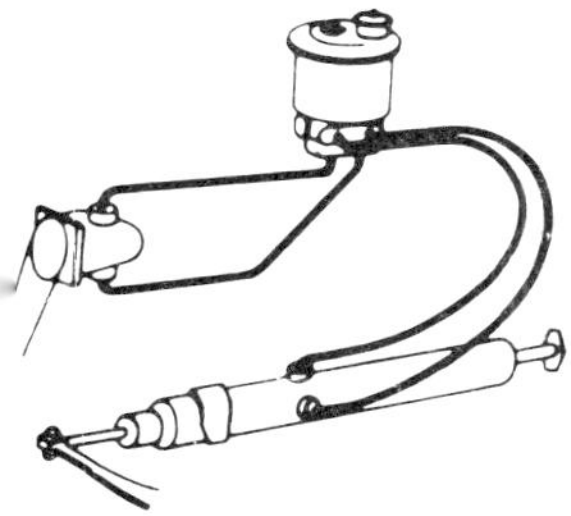

Fig 4(A)

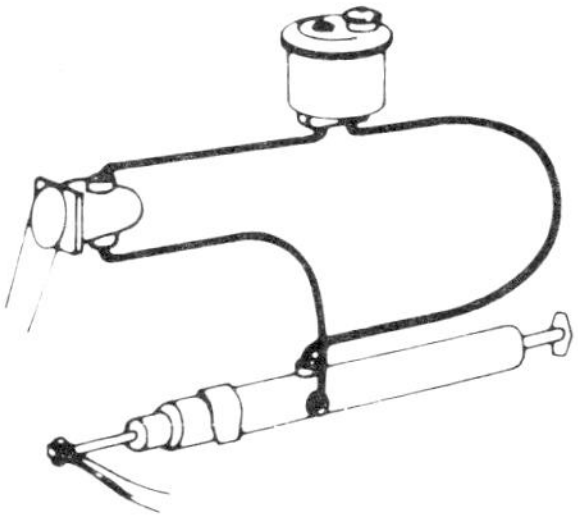

Fig 4(b)

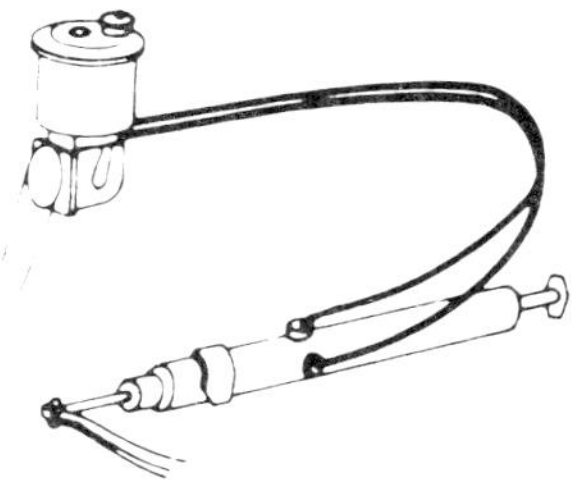

Fig 5

(iii) Relief and directional-control valves incorporated in the cylinder and the flow-regulating valve omitted. The pump is then associated with a simple reservoir, connected to the cylinder by two lines. This simplified system is suitable for use on slow moving vehicles – Fig 5.

(iv) Power steering worked off an existing hydraulic circuit in the vehicle. In this case the existing reservoir can be used although the power steering needs its own pump, which can be separately mounted at the front of the engine. The valve block is then separately mounted (Fig 6a). On slow vehicles the flow-regulating valve can be omitted and both the safety-relief valve and directional-control valve mounted on the cylinder (Fig 6b).

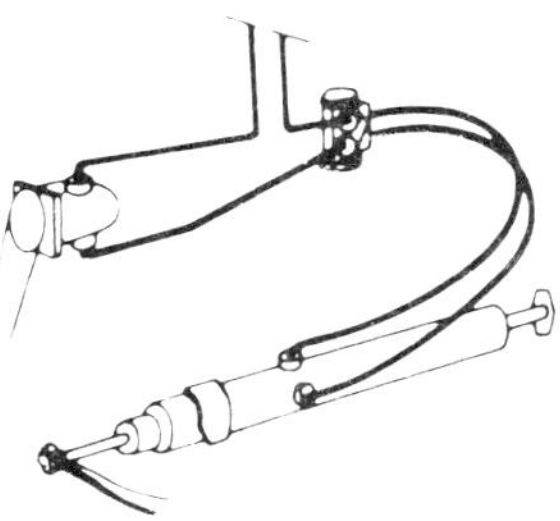

Fig 6(a)

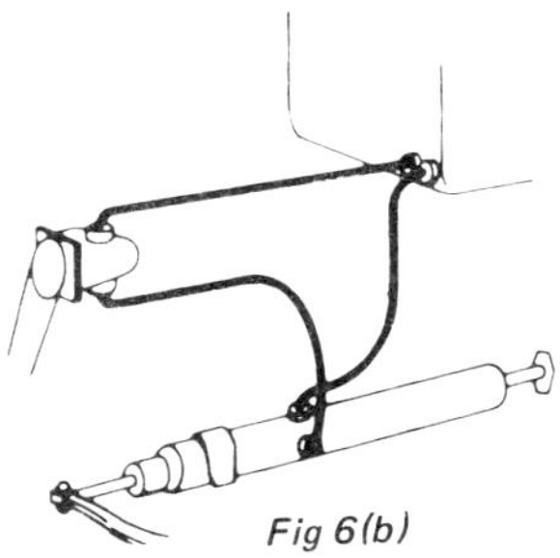

Fig 6(b)

There are also more sophisticated systems used, *eg* to provide proportional response and/or vary the degree of feedback force or 'feel' on the steering wheel itself. Typically, using a spool valve as a control valve, the amount of hydraulic assistance is then dependent on the movement of the spool, which is in proportion to the effort applied by the driver to the steering wheel. If the steering resistance at the road wheels is small, as on an icy road or when making small corrections at speed, the amount of effort and valve movement will be small, as will also the degree of power applied, thus avoiding over-violent response.

To obtain maximum advantage from power-assisted steering the ratio can be made to vary progressively from a moderate ratio at the straight-ahead position to a numerically lower ratio towards each lock. Thus the overall number of steering wheel turns between locks can be reduced without having gearing too direct for the straight-ahead high-speed position. It is then possible to turn the steering wheel from lock to lock with the car stationary or when maneouvring at low speeds with a very moderate effort at the steering wheel.

Secondary Systems

Secondary systems which may be operated by an engine-driven hydraulic pump include door opening, hood raising, built-in jacking systems, *etc.* For jacking purposes, however, a manual pump is more usual since there is a relatively limited demand for this type of system. Ancillary systems of this type commonly employ a gear pump which actuates one or more miniature cylinders through a control valve. Rotary actuators are occasionally used. Hydraulic motor drives are also used to a limited extent on primary services in some designs, for example, driving the cooling fan. In this latter example the flow rate to the fan motor, and thus the speed of the fan, can be thermostatically controlled so that the degree of cooling is matched to the engine temperature.

Auxiliary Services

On certain classes of commercial and other heavy vehicles, hydraulic power may be used for ancillary services, such as tipping motions on trucks, or the operation of special attachments. Where the demand on the hydraulics is light or intermittent such services may well be powered by the same hydraulic motor as the power-steering circuit, and an accumulator may possibly be included in the circuit. Normally, however, such demands are best met by a separate pump powering an entirely separate circuit.

Since 'live' hydraulics is essential for power steering, the pump for such a circuit is normally driven off the engine crankshaft and typically mounted at the front of the engine. If a second pump is mounted so as to be driven off the gearbox, then this circuit will be live only when the clutch is disengaged; this can limit its use unless a special type of clutch is used.

A more satisfactory solution, therefore, is to mount the second pump in tandem with the first to be driven by the engine crankshaft, or alternatively to employ a double pump which yields the same results. In the latter case, one half of the pump feeds the power steering circuit whilst the other feeds the circuit powering the auxiliary service or services.

Where power steering is not fitted then a single crankshaft-driven pump can obviously provide the necessary 'live' hydraulics for ancillary services. Certain vehicles (such as tractors) may, however, be fitted with a hydraulic lift or some other pump-driven service as standard whose circuit it may be practical to tap for additional services.

Engine Starters

The high power necessary to start large diesel engines (up to 2 000 hp) can readily be provided by a hydraulic motor and hand-primed hydraulic system. This has the advantage of making the starting system completely independent of battery or other external sources. An example of such a starter system is shown in Fig 7.

Power is derived from a hydraulic accumulator pre-charged with nitrogen to 105 bar (1 500 lb/in^2). When oil is pumped into the accumulator the gas is compressed to 140 bar (2 000 lb/in^2). This stored energy is released by opening the hand-operated starter valve allowing a small quantity of oil to pass through the hydraulic starter motor giving slow rotation of the Bendix unit to achieve full engagement with the flywheel. The valve then opens fully allowing full flow to the hydraulic starter motor, thus turning the flywheel through nine complete revolutions at 375 rev/min and starting the engine. The oil returns to the reservoir where it is pumped back to the accumulator.

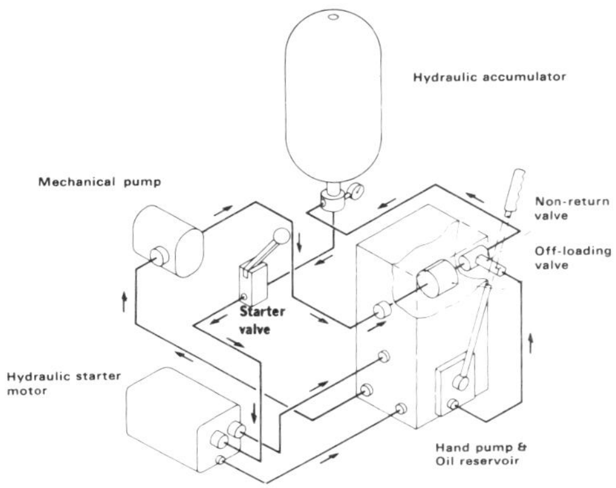

Fig 7 'Startorque' hydraulic starter for diesel engines.

The system can be re-charged by an engine-driven pump in only thirty seconds after starting. The system is protected against over-charging by an off-loading valve which, when the pressure reaches 140 bar (2 000 lb/in^2) re-directs the oil in an open circuit from the re-charging pump back to the reservoir, where it will continue to circulate at zero pressure.

Specialized Hydraulic System

Automatic transmissions are another major application of hydraulics to vehicles. These are highly-developed individual designs based on hydraulic torque converters, planetary gearing and hydraulically-operated multi-disc clutches and band brakes. Power flow in each gear ratio is achieved by locking various elements of the planetary gear set. The size of each transmission is determined by the torque characteristics of the engine(s) they are designed to match.

Hydraulics may also be applied to suspension design, particularly response to dynamic attitude changes of the vehicle body in cornering, braking and acceleration as shown in Fig 8.

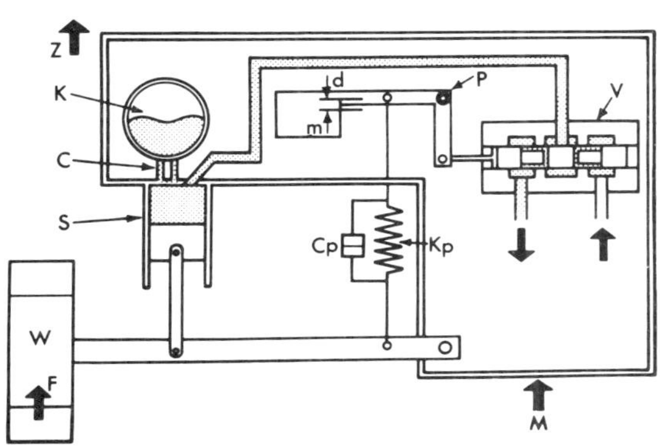

Fig 8 Schematic arrangement of active ride control applied to a single wheel.

A hydraulic strut S, with tandem gas spring K and damper C, supports the body mass M relative to the wheel W. A pendulous mass m is pivoted on the body at p and is supported approximately horizontally by a small secondary spring/damper Kp and Cp connected to the suspension linkage. Any swing of the pendulum is transmitted by a bell-crank lever to a body-mounted spool valve V which, fed by a hydraulic pump, regulates the flow of fluid to or from the strut S. If the system

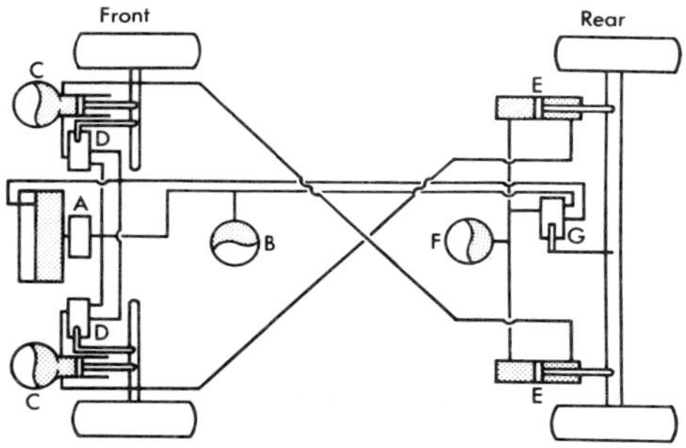

Fig 9 Active ride control for a car.

A–pump
B–accumulator
C–front strut and gas spring
D–control valve
E–rear strut
F–gas spring
G–control valve

parameters are correctly chosen there will be no actual movement of the pendulum, regardless of the magnitude of the input force, so the spool in V does not move either.

The effect of roll, induced pitch or a change of carried load is to vary the mass m carried by the suspension. If this mass is increased, the body and the pendulum pivot p move downwards towards the road. Since the resistance from Kp and Cp is unchanged, the pendulum is rotated clockwise. The spool therefore moves to the left and allows fluid to be admitted to the strut. This additional fluid compensates for that flowing into the spring K because of the mass increase, so the overall length of the strut and spring is unchanged. Once the body regains its original position, the spool valve closes again to complete the correction.

A typical car system is shown in Fig 9.

A requirement for the hydraulic pump (engine-driven by means of a toothed belt) is that its delivery, even at low speeds, must be sufficient to cope with the high lateral forces that can be applied to the car in violent manoeuvres. For all the installations so far developed an axial piston unit has been the choice; with constant pressure control, the swash-plate angle is varied automatically, according to the speed and fluid flow demand. This arrangement minimizes power consumption and therefore heat input to the fluid. The hydraulic circuit incorporates an accumulator to damp peak pressures from the pump and to help provide the initial flow surge necessary when lateral (or longitudinal) forces are suddenly generated. Fluid pressure is applied to a control valve at each front suspension and to a single valve at the rear axle.

It will be noted that, in the diagram, the front struts are integral with their gas spring/damper units, whereas the rear ones share a centrally mounted spring with its own damper. At the front, the struts are single-acting but at the rear they are double-acting, the smaller areas beneath their pistons being connected to the tops of the diagonally opposite front struts. Since the rear control valve is insensitive to lateral weight transfer, the pressure above both pistons remains constant during cornering, but the underneath pressure in each is modified to that of the inter-connected front strut; this modification opposes the lateral weight transfer at the rear wheels by a couple proportional to that exerted at the front wheels. The result is a constant ratio of front/ rear roll stiffness which ensures consistent handling qualities throughout the range of lateral acceleration. Appropriate proportioning of the effective areas of the front and rear pistons enables any desired degree of under-steer to be achieved in the interest of stability.

See also chapter on *Mobile Hydraulics*.

Mobile Hydraulics

THE PRINCIPAL advantages of hydraulics as the choice for powering mounted or remote systems on working schedules are:

(i) High power output from a compact actuator;

(ii) High power conversion efficiency;

(iii) High power/weight ratio;

(iv) Extreme flexibility of approach and control.

These characteristics are largely responsible for the fact that the hydraulic cylinder or ram has become the more or less standard choice for lifting, slewing, shoving, digging, tipping, *etc* on chassis-mounted equipment, including fork lift trucks.

Such equipment can be towed, or be self-propelled. In either case, the engine necessary for movement can be utilized to drive the hydraulic pump with a considerably higher efficiency than that achieved by mechanical power take off.

Pump Power Sources

Mobile hydraulics implies a mobile 'platform' on which is mounted the hydraulically-operated working motors, the two together representing the *total* demand for power. Normally the greater demand is from the (mobility) drive motor, but there may be cases where it is more economic, or more convenient, to provide the hydraulic system with its own driver so that it becomes, in effect, a self-contained power pack. There is, of course, also the possibility of using an all-hydraulic system where mobility is also provided by a hydrostatic drive.

Where full engine power is used for the hydraulics, the hydraulic pump is driven direct from the engine crankshaft. Equipment carried on a self-propelled chassis commonly has the hydraulic pump driven from the engine gearbox or crankshaft. Where the prime mover already incorporates a power take-off, as in the case of tractors, a pump/gearbox unit may be fitted directly to the power take-off, or it may even be shaft driven if the pump is an integral part of separate trailer-mounted equipment.

Linear Actuators

Normally a cylinder is used for deriving all linear motions required, operating as directly as possible (*ie* with minimal mechanical linkage). The lower the system pressure, the larger the size of cylinder required to produce a given output force and the greater the pump delivery required to achieve a given rate of working. Except in special cases, however, the work done is seldom significant. It is the force output which counts, provided this is achieved at a 'reasonable' or acceptable rate. This applies to most of the motions provided by mobile equipment for mechanical handling, *etc.* The input horsepower required by the pump will, however, be directly related to its delivery (governing work rate of the actuator) and pressure (governing force output of the actuator).

The highest working efficiencies can usually be achieved with higher working pressures. Thus 350 bar (5 000 lb/in^2) is about an optimum figure for 'hydraulic conversion', although this is used only in specialized systems such as aircraft hydraulics. For mobile equipment, 140 bar (2 000 lb/in^2) is a more realistic optimum taking into account practical considerations and cost, and even this demands careful attention to cylinder design. In particular, even though friction and cost are higher, seal sets of chevron type are favoured for piston seals instead of single lip seals as they have a longer life and lower leakage. A 140 bar (2 000 lb/in^2) system is capable of providing a force output of the order of 1 tonne (1 ton) from a cylinder only 31 mm (1¼ in) in diameter, or 10 tonnes (10 tons) from a 93 mm (3¾ in) cylinder.

Even more significant is the fact that such large output forces are available with the minimum of exposed parts (and minimum environmental hazard) and with very light, literally 'finger tip', control, since this involves only movement of a small valve spool. It is also very easy to develop more sophisticated systems, if necessary, such as servo-controls for positive positioning, whilst proportional movements can be obtained *via* proportional control valves.

See also chapter on *Hydraulic Cylinders.*

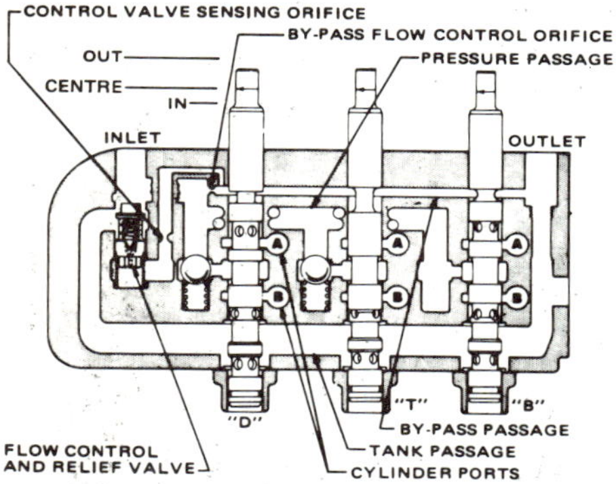

Multi-section directional control valve. (Sperry Vickers)

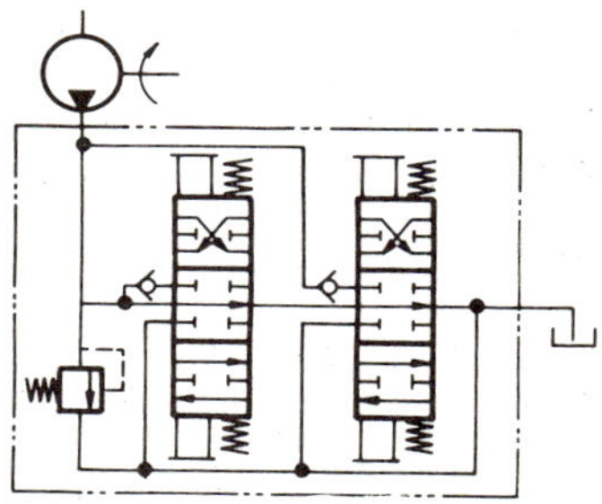

Fig 1 Parallel circuit.

Control Valves and Selectors

Control valves and selectors for mobile hydraulics are invariably designed for stacked assembly, with internal circuit configurations for parallel circuits, series circuits or tandem circuits (series-parallel). Some offer a combination of these circuits in any one valve block to adapt directly to a specific application (instead of employing a combination of individual valve types). For example, a tractor shovel requires a separate digging operation, with dipper arm and swing services simultaneously. The digging service would have a series-parallel circuit while a series circuit would be used for the other two. The combined pressures of the two services, simultaneously operated, are generally less than the pressure required on the digging service, so that the pressure limitation of the series circuit would be of no significance in this instance.

Parallel Circuit Configuration (Fig 1)

In a parallel circuit, the valves have a common internal pressure gallery, which allows simultaneous flow to any number of service ports. In theory, it is possible to select any number of spools and

JOHN DEERE

thus provide full pressure to as many services as are required. In fact, splitting the flow between a number of services means that the operational speed will be lower than the maximum available. In practice it is rare that each service requires the same pressure. The oil will go to the lowest pressure first, so that this will move before the other services. This can be counteracted by throttling the flow on that particular spool, and thus creating sufficient pressure in the valve internal gallery to operate other services at higher pressures. But the effectiveness of this is dependent upon the metering or inching characteristics of the valve.

Series Circuit Configuration (Fig 2)

The series circuit also allows simultaneous operation of more than one service, but in this case the exhaust from the spool nearest the inlet is the supply for the next downstream spool operated. The disadvantage is that the operating pressure of the last service is additional to the pressure required by the preceding upstream service, which again is additional to that required by the valve spool further upstream. Although the flow to each service is approximately the same, the total main relief-valve pressure available is split between the number of services selected. Consequently, although simultaneous operation is available, the usefulness is restricted by the limitations of total pressure.

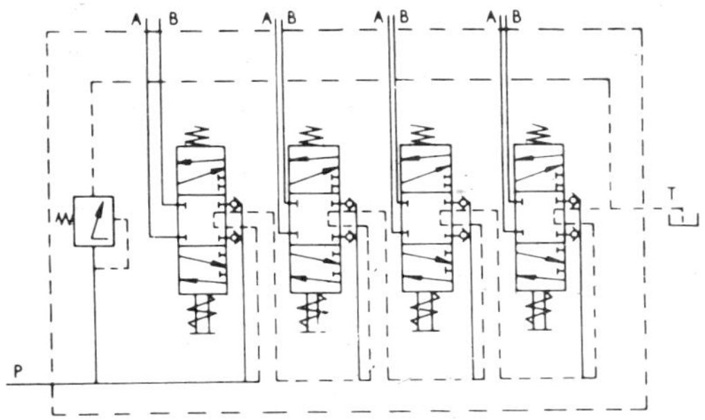

Fig 2 Series circuit.

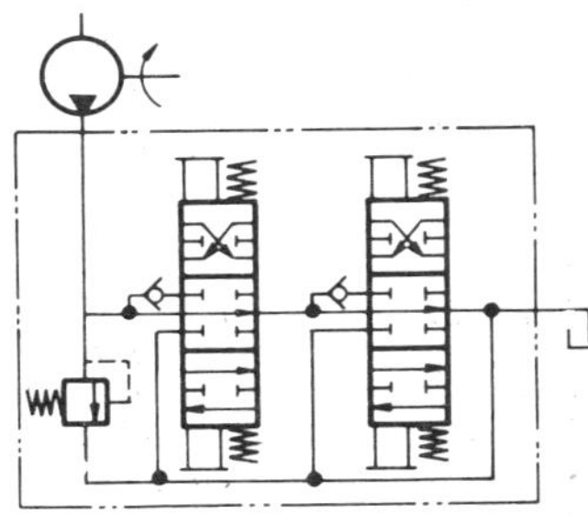

Fig 3 Series-parallel circuit.

Series-Parallel or Tandem Circuit Configuration (Fig 3)

In this type the spool nearest the inlet has first priority. If two spools are operated at the same time, the upstream spool closes off the through-bypass circuit. The flow will then automatically go to the next spool downstream. It is therefore possible, by throttling, to operate the spools simultaneously, but normally this circuit is used where it is required to feed the whole of the pump supply to only one service at a time.

The most usual method of mounting spool valves has been to bolt them together in banks, sandwiching gaskets between them, but there is a developing tendency to buy them as complete units containing encapsulated relief and check valves.

Hydraulic Motors

The application of hydraulic motors and hydrostatic drives for both linear, and more obviously, rotary movements is a field which is still developing. Hydraulic motors may well be considered for linear applications in certain fields where cylinders have limitations – particularly long-stroke movements where side loads may complicate cylinder design requirements and add to costs. A motor-driven rack and pinion output, or a hydraulic motor-driven rack and pinion output, or a

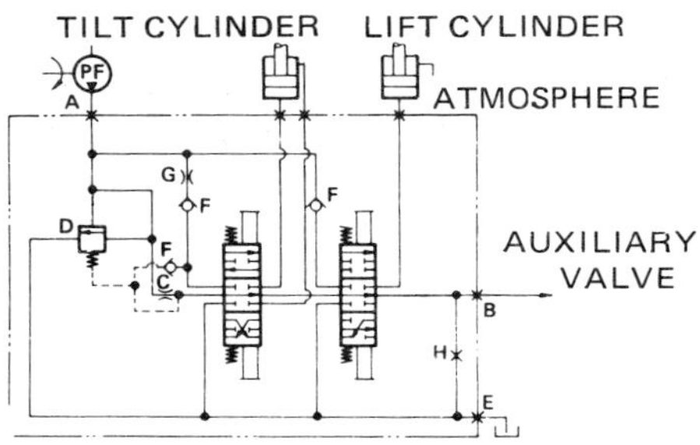

Typical hydraulic circuit for a lift truck.

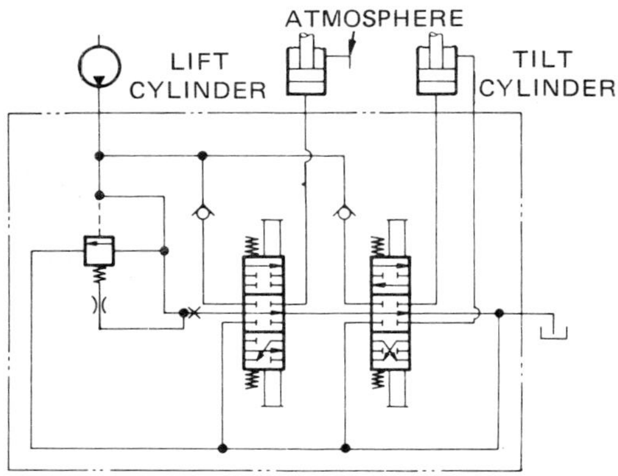

Typical closed centre circuit.

hydraulic motor driving mechanical linkage, at a comparable overall conversion efficiency, may prove a reliable alternative.

A particular advantage of the hydraulic motor is that it can give a starting torque approaching that of its running torque and run very smoothly at very low speeds. Inertia is also much lower than with most other forms of drive, so that starting and stopping can be almost instantaneous. The use of hydraulic motors, in fact, has revolutionized the powering of slow-speed devices in many industries, notably winches, conveyors and cranes. And 80% conversion of energy is a very conservative estimate (*eg* the hydraulic pump required to drive a motor of P horsepower would need a power output of the order of 1.25 P horsepower).

Like the cylinder, the hydraulic motor can be mounted at the working point, requiring only fluid-line connections and giving the same sort of flexibility of operation as that of an electric motor but with far greater flexibility of performance. Hydrostatic drives are, in effect, merely an extension of this principle. And both hydraulic motors and hydrostatic drives provide yet another field of further application for mobile equipment anywhere, or in fact, for any service. The advantages are there to be exploited – together with the undoubted, and proven, reliability of hydraulics with its almost complete absence of wear, exposed working parts, and mechanical transmission components.

Earth-Moving Vehicles

The heavy duty demands of earth-moving vehicles have favoured the adoption of hydrostatic transmission as well as hydraulic machinery. The latter most commonly involves linear motions which can be provided by a cylinder or cylinders through a linear mechanism. Where rotary power is required, or can conveniently be utilized, it can be provided by high-speed low-displacement hydraulic motors or low-speed high-torque piston motors. Torques in the latter case are considerably in excess of those provided by any alternative machinery of comparable size.

Power for hydraulic machinery is provided by a separate pump driven by the vehicle (diesel) engine. In the majority of cases this is of fixed-displacement type. The first problem is coping with the power developed when all the services are inoperative: boom, digger, slew motor and track drives. In addition, there is the problem when the actuator reaches the limit of its stroke of limiting the pressure by some pressure-limiting device, such as a relief valve, which can also serve as a spill for the unused flow from the pump.

There are two main solutions to the problem of 'off-loading' or 'idling'. One relies on the use of a variable-delivery pump where the lever setting the pump delivery control is acted upon by the pressure. As the pressure increases beyond some set value up to the maximum the lever moves in such a way as to reduce the pump output gradually to zero.

Alternatively, a fixed-delivery pump can be used with some form of unloading valve which, behaving in a similar manner, dumps the flow back to the reservoir at a relatively low pressure. The system is maintained at its high pressure by means of a small delivery pump protected by a further relief valve. Both these systems maintain a live supply.

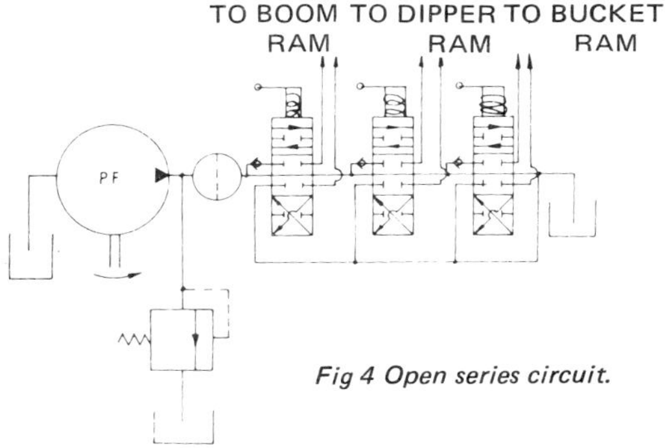

Fig 4 Open series circuit.

'Dozers, Loaders and Diggers

The second type of system, widely used on 'dozers, loaders and diggers, is the so called 'open' system. Here the particular pump feeds a set of selectors in series as shown in Fig 4. There are at least three positions and in the mid setting the output from the pump is fed directly through them back to the reservoir. Moving any one then cuts off the flow and allows the pressure to build up and operate the particular service. If the selection is complete then the selectors further downstream are inoperative but if the selection is only partial then it is possible to operate services simultaneously.

With all line supply systems enough heat may be generated by the 'off-loaded' pump to necessitate a small radiator or other heat exchanger. Again, in the neutral position the service connections to the actuator can either be blanked, maintaining it in any desired position for booms and buckets on loaders; or it can connect the service connections together (and incidentally to the reservoir) on some track motor controls. It is also common to incorporate check valves to prevent the inadvertent release of pressure on the operation of a subsequent service.

Where multiple fixed-delivery pumps are used, for example three on a front loading shovel, they are usually arranged so that one is normally directed to the power steering whilst the others feed the bucket and boom rams. The power available from the engine is limited and if one pump running at its maximum absorbs all the power, a special valve has to be incorporated to sense this and limit the combined power of the pumps within the engine capacity.

Excavators

Excavators have many varieties of pump combinations and, in addition to the digger operations of the boom, dipper, bucket and slew actuators, there is also the problem of providing power for the track motors which normally drive through a gear pair to produce the necessary low revolutions. In order to increase the speed of rams when extending, or to transfer fluid under following load conditions, use is made of a differential circuit.

Here the annular area fluid is directed into the full area to help the pump which now increases its speed as only the piston rod volume has to be filled, although the effort is reduced as the pressure acts only on this area. Units known as demand valves may be incorporated so that the necessary services are given priority.

Pressurization of the track motors also releases their brakes hydraulically and the slew motor is operated by an open neutral selector which allows the assembly to rotate freely to the desired position at which the operator applies a holding brake.

With variable-delivery pumps, in particular, it is possible in a very simple manner to produce a constant horsepower control to avoid stalling an engine.

On crawlers of all sorts it is now quite common to drive the tracks by means of hydraulic motors, using either high-speed units and reduction boxes, or low-speed piston motors and a single pair of gears. Each track is then controlled separately by the operator to steer the vehicle and the application of pressure to the motor releases the brake holding it.

When it is realized that these machines can climb gradients of 1:3 it is obvious that the torque applied at the drives is of considerable magnitude and can really only be obtained from a hydraulic transmission or torque converter, the former giving a much larger range of speed control.

Another use for hydraulic motors is in the slewing mechanism, used on excavators, which rotates the cabin, engine and digging assembly on a large bearing at high speeds and requires high accelerations only obtainable from fluid power.

Tractors

The basic method on which the draught and position-controls of most modern tractors are based still follows the original Ferguson system where implements were mounted directly on the tractor, lifted and lowered by hydraulic rams, with the further refinement that hydraulics also provided a means of controlling the manner in which the implement moved through the ground. By means of linkage and levers the draught of the implement was used to signal to the control valve handling power to the lift, so that weight could be transferred to the rear wheels of the tractor to increase its pulling power.

The increasing mechanization of agriculture led to the development of larger and more powerful tractors with larger matched implements which could cope with tasks beyond the capabilities of manual effort and earlier forms of mechanical lift. Thus for some years hydraulically powered rear linkages were offered as optional equipment, but today most farm tractors now incorporate hydraulic power lift (HPTO) as standard, and may have quite sophisticated systems for draught and position-control.

The built-in hydraulic system, with the increasing use of the tractor as a mobile power source, has enabled designers to provide power for cylinders used on tractor-trailed equipment such as tipping trailers, and has led to the introduction of many services for which hydraulic power has replaced human muscle power.

Front-end loaders and ditch digging and cleaning attachments are commonplace, and there is an increasing use of hydraulics to power external and auxiliary equipment such as hedge and verge trimmers, power saws and even the pumps used on machines for a portable milking bail.

While the availability of the in-built hydraulic system for tractors has led to the idea of powering external applications such as these, the power available is often insufficient. This, coupled with the need for more sophisticated control systems, has led to the provision of an additional source of power, so that an auxiliary hydraulic system has been added to meet the larger capacity or control refinement demanded.

Though the demand for auxiliary power is probably not yet such that fitting increased capacity systems would be justified economically on all tractors, many manufacturers now offer machines with such a system.

As the benefits to tractors from hydraulics became apparent, so their use in other farm machines developed and the combine harvester, on which a single hydraulic service replaced the electrical lowering and lifting of the cutting table some twenty years ago, now has the pick-up reel raised and lowered as well as rotated hydraulically, and also incorporates other hydraulically powered functions including power steering.

Independent hydraulic power-assisted steering systems have long been available for fitting to tractors and combine harvesters. The systems in use are powered by an independent hydraulic pump, driven directly from the engine, which provides power assistance through servo-controlled hydraulic cylinders connected mechanically to the steerable wheels.

There is also the simpler form of powered steering, which allows the machine designer greater flexibility, and which is referred to as hydrostatic steering. This system consists of an engine-driven hydraulic pump, a circuit-relief valve, a steering wheel and column, a pump/meter rotary unit with control valve and an operating power cylinder.

The pump provides the power to supply the pump/meter and valve unit, which is controlled by the driver through the steering wheel, so that turning the steering wheel to the left or right causes the control valve to direct oil from the engine-driven pump to the pump/meter unit, which displaces oil to the appropriate side of the power cylinder.

The use of this system, in addition to reducing operator fatigue, enables the designer to eliminate the usual mechanical steering linkage and to site the steering unit in the position best suited to the driving requirements.

See also chapter on *Mechanical Handling.*

Hydraulic Power Packs

HYDRAULIC POWER packs – also known as hydraulic power units – are self-contained hydraulic supplies normally comprising an integrally mounted electric motor/pump unit with associated tank together with necessary valves.

The pump/motor unit may be mounted on the tank, or separately, and packs are also usually available in either horizontal or vertical configuration. In some cases with vertical configuration the electric motor may be mounted on top of the tank with the hydraulic pump itself within the tank.

Relief and check valves may be mounted on a manifold (*eg* on the motor end of the pump), or on the tank. The basic unit may then be piped to cylinder(s) or actuator(s) through a suitable control valve; or more tidily mounted in a console with line connecting point(s), and control valves and any required instrumentation (*eg* pressure gauge(s)) mounted on the fascia. Hose assemblies are generally preferred to rigid piping for connecting the power pack to actuator(s), but this depends on the circumstances. With hose assemblies, for example, it is a simple matter to disconnect the power from one machine and transfer it to another, should this be required. Some power packs are specifically designed to be portable, *ie* fitted with lifting handles or skid- or trolley-mounted in the case of larger sizes, whereas others are intended for more or less permanent or console-type mounting.

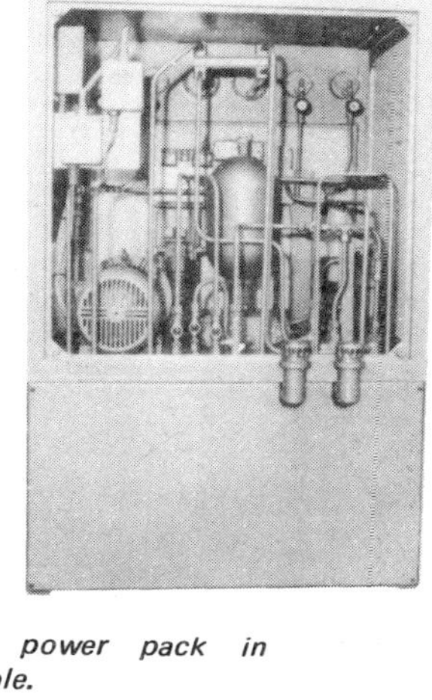

Brecknell hydraulic power pack in console.

Servotel hydraulic power pack with integral 70 gallon tank.

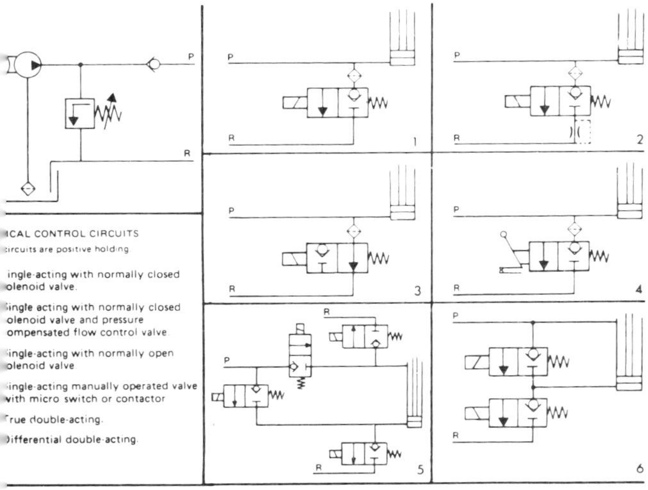

Typical circuits for hydraulic power packs. (Smiths Industries).

Power packs produced by individual manufacturers are normally based on one or other of their andard sizes of hydraulic pumps driven by a matching fan-cooled electric motor or motor otions. A range of tank sizes may also be available, giving the customer the choice of a range of elivery capacities. The majority of power packs fall into the category of smaller (hydraulic) ower units with pump deliveries ranging up to about 6 litres/minute, tank capacities up to 9 res, and single- or three-phase motors up to 2 or 3 kilowatts. Maximum pressure rating is usually the order of 210 bar. Larger power packs are also produced but these are less readily available standard productions and are usually produced for more specialized applications.

At the other end of the scale miniature hydraulic power packs are also available driven by 12 24 volt d.c. electric motors; or even smaller sizes for special applications.

pplications

representative list of applications is: fork lifts; scissor lifts; access ladders; tailgate lifts; lorry-ounted cranes; dump bodies; snow plough; power steering; paraplegic lifts; hose swaging equip-ent; and rubbish compactors.

Fenner electric motor powered hydraulic power pack.

A.C. power units are used in a wide range of machine tool applications and provide hydraulic power and pressure for lubrication of bearings, and for the operation of clutches, clamps and shuttles of all kinds. They provide the power to move and feed cutting equipment such as drill heads, boring heads and other hydraulic-fed cutting-tool holders. An especially important application in the machine-tool field is in the retrofit area. Users who wish to retrofit tracer equipment, special clamp devices, special tooling, or other typical machine tool features can power them with a separate hydraulic power unit without having to tap into the main hydraulic system of the machine tool in question.

Fenner manual hydraulic power pack.

Hydraulic power units are widely used for rubbish compactors, and in the supply of hydraulic pressure to thrust cylinders. These power units usually incorporate a Duplex pump providing two fixed deliveries to achieve maximum output of oil at low pressures for fast traverse and quick return of the cylinder. It also provides a high pressure low volume supply for the compaction work stroke. This arrangement not only reduces heat but also permits the use of smaller electric motors at full power output at both of these volume and pressure requirements.

All the components for this operation can be incorporated in a Duplex pump located within the reservoir. When the compaction cylinder meets the load resistance, the pressure in the system rises until it overcomes the spring force setting on the unloading valve stem in the high-volume pumping section. The valve stem moves, allowing the high volume pumping section's flow to bypass at low pressure into the reservoir. An internal check valve between the two pump section's prevents the high-pressure flow from leaking back during the unload phase. When the pressure in the system is reduced below the unloading-valve setting, the combined flow of the two pump sections is made available.

An externally adjustable relief valve is fitted in the hydraulic circuit to prevent excessive pressure build-up. The electric circuit incorporates a differential pressure switch taking pressure from the thrust delivery. When resistance is met and the pressure builds up to give maximum thrust, the pressure switch operates. This provides an electrical signal to a solenoid-operated four-way spool valve to return the cylinder to the 'load' position.

See also chapter on *Workshop Tools.*

Marine Hydraulics

ONE OF the chief applications of hydraulics on ships is for the operation of steering gears on all sizes of commercial vessels and the larger sizes of pleasure-craft. Systems range from small manually-powered units where system pressure for operating a steering jack is derived directly from manual movement of the ship's wheel to complete servo-systems with their own close-coupled pump or pumps. Hydraulic actuation is also employed with steering devices other than rudders, *eg* Schottel rudder-propellers and Voith-Schneider cycloidal propellers.

On large commercial ships many functions originally operated by steam or electric motors are now also commonly powered by hydraulics. These functions include:

Valve control — VLCC, ULCC and Product Tankers, OBO, Bulk Carriers, LNG/LPG;
Hatch covers — OBO, Bulk Carriers;
Loading ramps — Ferries, Ro-Ro;
Deck machinery — All classes;
Cranes — Bulk Carriers and Specialized Ships;
Fire pumps — All classes;
Lifts — VLCC, ULCC, OBO;
Pumps — Specialist and refined Product Carriers;
Deepwell pumps — LNG, LPG, clean Product Carriers.

Marine System Pressures

Choice of system pressure is largely arbitrary (subject to availability of suitable components). It seems generally accepted that the most cost-effective pressure range for marine hydraulics is 100—140 bar (1 500—2 000 lb/in^2) consistent with the ready availability of components meeting the regulations of the Classification Societies. There are, however, advantages in using low-pressure pumps and motors on certain types of ships since the components can be made more rugged and running and maintenance costs are lower.

Pumps

A wide range of gear pumps, piston pumps, screw pumps and main pumps are used in marine hydraulics ranging from very small sizes (*eg* to operate small cable winches) to quite large units for working deck machinery. Vane pumps appear to predominate in the smaller sizes. The main hydraulic pump(s) are commonly driven from the ship's engine. Other pumps may be driven by an auxiliary engine or electric motor.

Actuators

Hydraulic motors are widely used as actuators in marine hydraulic systems and provide a direct solution where a continuous rotary output is required. Semi-rotary actuators are used in preference to mechanical linkages where limited rotary output movements are required (*eg* in steering units). Linear output motions are normally derived directly from hydraulic cylinders.

In the latter case the choice of seals is particularly important. Sometimes the actuator is mounted directly on to a valve which does not have a top seal, *ie* the actuator bottom seal is the valve top seal. This seal must then be compatible with whatever cargo or ballast may be carried. In refined product carriers this can present a problem. Thus, it is always wise to mount, in this instance, an actuator having a twin seal arrangement and a vent between seals. This will ensure that should one or other seal fail at any time there is no fear of contamination of either product or hydraulic fluid.

Choice of actuator depends on the type of equipment to be operated, but in the case of large valves used on modern tankers, choice may not always be obvious. The following is a general guide:

Valve Type		*Actuator*
Wedge gate	:	Linear and rotary
Bulkhead	:	Linear and rotary
Butterfly	:	Semi-rotary and rotary
Ball	:	Semi-rotary and rotary
Non-return	:	Linear and rotary

In general it requires a greater effort to open a valve than to close it. This is particularly true of the wedge gate where the wedge is driven into its seat until it is sealed. As the sides of the seat grip the gate a high degree of force is required to break it out again.

Butterfly, ball and bulkhead valves show this characteristic to a lesser extent. In the closing operation it is essential that the moving element is able to travel far enough to ensure tight sealing. The wedge gate travels until the valve resists the closing thrust exerted by the actuator. In the butterfly valve there is normally some positive mechanical limit to travel, set so that the paddle is then in its fully-sealed position. Before reaching this point, the seat resistance may have been allowed to stall the actuator.

The ball and bulkhead valves rely primarily on accurate positioning of the moving element to sel tightly and this is achieved with mechanical stops.

To stop an actuator in the fully-open position of the valve, either a positive mechanical stop is used or the hydraulic power supply is cut off at a pre-determined position in the valve travel. The mechanical stops may be incorporated in the valve itself, in a gearbox mounted on the valve, or in the actuator. Some degree of adjustment during commissioning is generally required.

Pipes and Fittings

For marine hydraulics the use of carbon steel (and even some types of stainless steel) pipes is ruled out because of corrosion problems. Any such systems will normally fail in from three to five years, if not earlier. In many combined systems, where cupro-nickel, tungum or corrosion-resistant stainless steel pipe (properly painted) has been used, correctly protected heavy-duty carbon steel accessories have been shown to be good for about 10 years but generally require replacement at that time, the pipe tolerances usually remaining within reasonable limits. The most successful piping systems are those which use cupro-nickel or tungum (an alpha-brass alloy), and where possible all pipe to pipe connections are made by brazing or welding using sleeve couplings.

Wherever the piping has to be connected to equipment, then the use of pipe accessories and couplings using the compressible O-ring concept show the best results, since no deformation of the base pipe material is caused and all hydraulic sealing is by means of a flexible oil-compatible rubber ring. Such fittings in carbon steel, stainless steel, cupro-nickel and tungum are available, all but the first one being expensive on a first-cost basis, but necessary if very costly delays are to be avoided due to pipe system or connection malfunction.

Given the simplicity of actuators mounted directly on the cargo or ballast valve, or hydraulic motors on pumps, submerged in the tanks of the vessel, many vessels have extensive amounts of hydraulic pipe in the tanks and here cupro-nickel or tungum is very acceptable. Even with the inerting of tanks and ullage spaces, which should reduce the corrosion rate (by reduction of the amount of oxygen in the atmosphere), the sulphurous vapours of many cargoes attack the exposed piping underneath the deck and it is recommended that such pipes be properly sheathed for 4 or 5 m below the deck penetration. A very satisfactory alternative for valve control is to use reach rod drive.

A simple method of providing a more reliable hydraulic system is to design the complete system so that an absolute minimum number of hydraulic fittings and lengths of hydraulic pipe are required by integrating all the hydraulic equipment. This can be achieved by 'blocking' or 'manifolding' hydraulic equipment so that all the control functions are simply carried out inside the block with only a pressure and return line being required at the control station.

Deterioration

Many hydraulic systems and their associated equipment and components deteriorate in the marine environment simply out of neglect, since the operators are very often ignorant of the requirements of the systems for regular servicing and maintenance.

A general rule applicable to the aforementioned is that it is better to replace rather than repair on board ship, since the conditions when most repairs are carried out, often in drydock, are extremely unhygienic in the hydraulic sense. The replaced unit may be removed from the vessel at a convenient point and either repaired there or preferably returned to a proper service centre so that when repaired the component concerned may be adequately tested and, where necessary, re-calibrated, prior to return to the vessel or to the owner's store.

Corrosion is, of course, the chief cause of deterioration.

The corrosive effects of sea water are well known. Carbon steel is eaten away almost visibly. Chrome plating, which gives a hard surface resistant to seal wear, is porous and permits corrosion of the base metal and resultant blistering and flaking of the chrome. This in turn causes damage to the seals as the blisters foul the seal lips and hence allows oil leakage from the actuator. The porosity of chrome plating can be reduced by the use of a suitable intermediate plating treatment before the chrome is applied. Unfortunately the best of these intermediate materials are rapidly attacked by hydrogen sulphide which is present in many crude oils. An obvious material to resist corrosion by sea water, crude oil and petroleum products, is stainless steel. Several different types are available. None is completely corrosion resistant but the most resistant is 18/10/3 (BS 316 S16 or AISI Type 316).

External corrosion will attack all outside and exposed surfaces, so all exposed moving parts should be in non-ferrous materials or non-corrosive stainless steel.

Where shafts are exposed, all sealing should be double and adequate arrangements must be made to permit protection by painting or greasing. External trim should be as for exposed moving parts.

External surfaces of equipment should use cast iron in preference to mild or carbon steel. Admiralty bronze may also be used. All exposed surfaces should be painted at the factory of origin with the recognized marine primer paint, usually requiring shotblast to base metal and immediate coating before oxidation of the metal surface can commence. If not factory supplied, equipment may deteriorate severely whilst awaiting outfitting in the shipyard, or on board the vessel.

Internal corrosion will affect the internal parts of hydraulic equipment such as springs, spools, valve surfaces *etc* and the only real way to prevent internal corrosion is to ensure that a high quality hydraulic system oil is used and that all the hydraulic piping and accessories are tight, particularly those in exposed deck positions and in tanks where the pressure of a tank of sea water may be considerable.

All equipment designed to go to sea must be arranged with effective sealing against the ingress of water or cargo. It must be remembered that on deck, wave pressure may be substantial, and that pump rooms are liable to flood, so that very careful attention should be paid to gasketing and the basic design of the component. Whenever possible, equipment should be so arranged as to prevent the components from retaining sea water, cargo, or rain if exposed – in effect to be self-draining

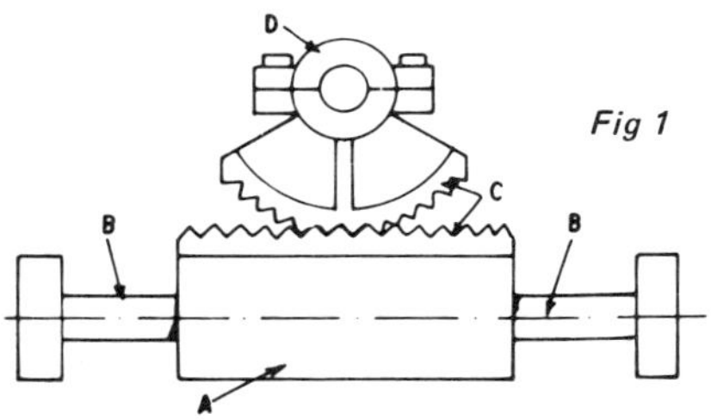

Fig 1

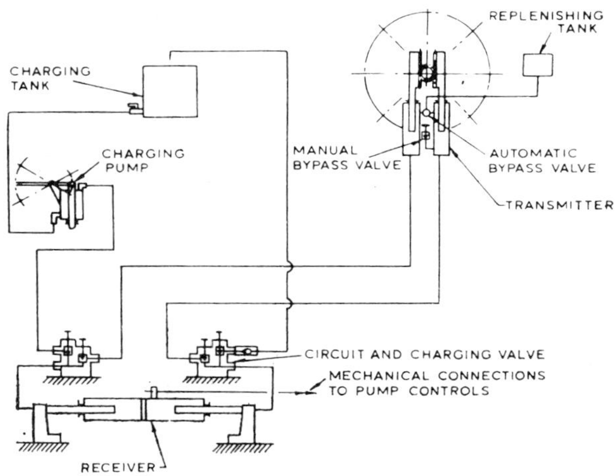

Fig 2 Hydraulic telemotor system.

A simple manual hydraulic steering unit is shown in Fig 1. The actuator is in the form of a cylinder (A) sliding over a fixed ram (B). The external face of the cylinder is fitted with a rack engaging a pinion (C) which is fixed to the rudder quadrant (D) clamped to the rudder stock. Sliding movement of the cylinder over the fixed ram – induced by admitting hydraulic fluid under pressure from one side or the other – is thus transformed into rotary movement of the rudder stock. Pressurized fluid is supplied usually by a rotary pump driven by movement of the ship's wheel. In effect, this is similar to a mechanical system but with fluid columns replacing the mechanical linkage between wheel and rudder quadrant. Relief valves are normally incorporated to protect the system against shock loads or excessive back pressure which may be generated by 'blowback' (when the actuator itself is momentarily working as a pump).

Such an arrangement is generally called a telemotor system, although true telemotors normally employ cylinders as transmitters rather than a rotary pump. Such a system is shown in Fig 2 where it will be noted that the system is continuously pressurized, being charged by a manual pump as required.

Telemotor systems are, however, largely superseded by electrical transmitters. A typical modern electrical control system for hydraulic steering is shown in Fig 3. This is a high-pressure system with hydraulic pressure supplied by two electric motor-driven pumps. Two completely independent electric control systems are present. The main system provides synchronized movement in the manner of a proportional servo-system. The standby system is push-button operated – one button for port and one for starboard. Movement in either direction depends on the time a particular button is kept depressed. The actual resultant rudder angle is displayed on a rudder position indicator.

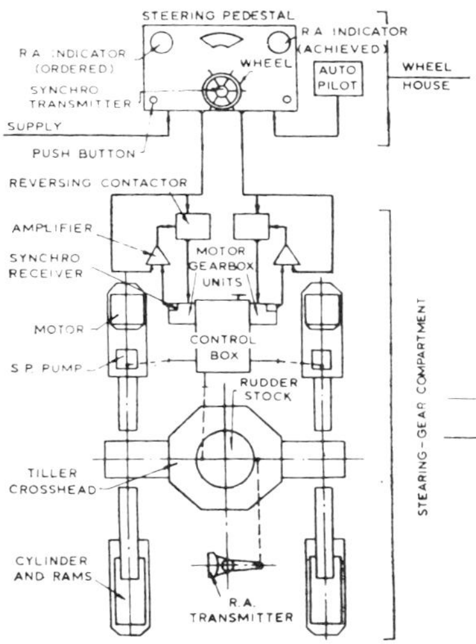

Fig 3 Electrical control system.

A modern servo-system is shown in Fig 4. Again, this is a high-pressure system supplied by a constant delivery pump which runs continuously. The pump delivery, and thus the rudder transit time, is determined by the setting of a flow control valve with both coarse and fine adjustments.

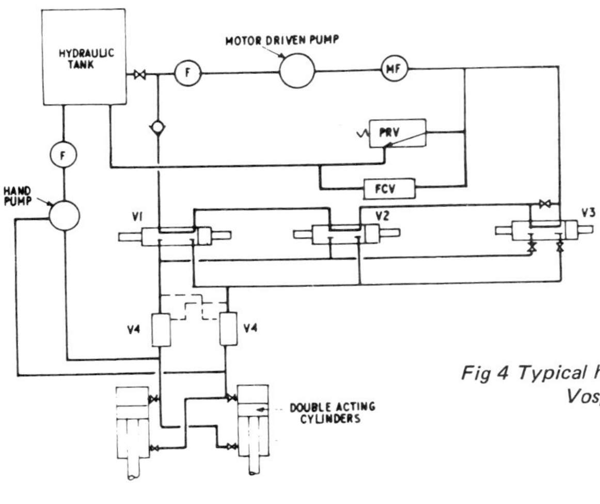

F – filter
MF – micro-filter
PRV – pressure relief valve
FCV – flow control valve
V1 – manual power control valve
V2 – secondary control valve
V3 – master control valve
V4 – pilot-operated check valve.

Fig 4 Typical hydraulic circuit for Vosper gear.

The three-position open-centre control valves are connected in parallel to a pair of double-acting cylinders through pilot-operated check valves. The master and standby control valves are solenoid operated from the electrical system. If the master valve fails it is automatically isolated and the standby valve brought into the circuit. If the electrical system fails the third valve (VI) becomes the control valve, which can be manually operated.

Feed-back is generated by potentiometers at the wheel and rudder head, generating signals which are fed to the switching unit from which the error signal is derived. Wheel steering is normal but push-button controls may also be provided.

Marine Gearboxes

Hydraulic power is widely used for clutch operation with all sizes of marine gearboxes. In such cases the hydraulic system commonly uses lubricating oil tapped from the gearbox itself, when the unit is normally referred to as an oil-operated gearbox. Fig 5 shows a unit designed for operation from the steering position by a single lever which operates both the throttle and gearshift. Oil pressure in the gearbox is maintained at 4 bar (60 lb/in^2) by a relief valve. A selector valve is actuated by the throttle lever and in the 'ahead' position oil is fed through the annulus A to force the plates of the 'ahead' clutch together and engage forward drive. On return to 'neutral' fluid pressure is released and the clutch is freed. Movement of the lever to 'astern' feeds oil under pressure into annulus C and D to engage the 'astern' clutch. Release valves are included in both sides of the circuit to assist in rapid release of the clutches when the control lever is moved to the neutral position.

1 — release valve
2 — to valve
3 — neutral brake
4 — muff
5 — astern clutch
6 — release valve
7 — ahead clutch

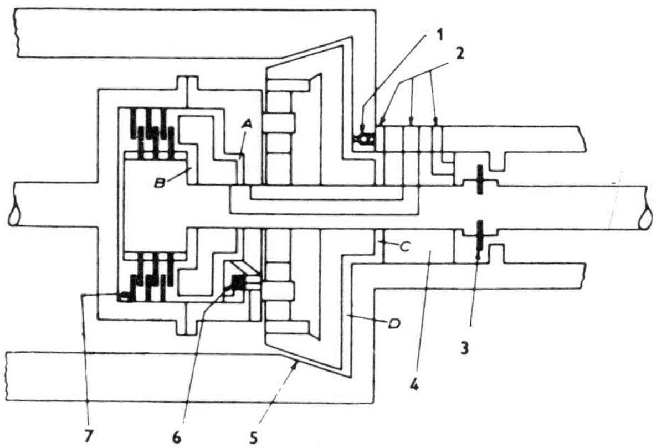

Fig 5 Clutch for reversing gearbox for small vessel. (Self-Changing Gears Ltd).

In the largest sizes of marine gearboxes where much greater powers have to be handled the clutch design is suitably modified. Typically, the clutch may consist of two grooved discs separated by a space into which oil can be pumped. The respective clutch plates are then engaged by applying hydraulic pressure to the outsides of the discs, and freed by applying pressure between them.

On larger vessels two or more engines are often connected to a common gearbox and final drive, this system being increasingly preferred to the employment of a single engine of the same total power or separate engines driving their own screws. In the latter case the same total power supplied through a single screw would normally show a substantially higher efficiency. Whilst this is mainly a matter of arranging suitable drives into the gearbox, hydraulic couplings can often be interposed between engine(s) and gearbox to advantage, particularly where diesel engines are involved. Such couplings enable the drive to be disengaged entirely and the load picked up progressively, where necessary.

Deck Machinery

Hydraulic windlasses for handling anchors are made in all sizes from gear designed as small yacht equipment up to the largest sizes required for ocean going vessels. They are produced both as independent units with their own hydraulic pump (electric motor-driven or engine-driven) and as pure hydraulic units for coupling to the main hydraulic system. In the latter respect it should be noted that in a 'complete' hydraulics system the demand for many of the services will seldom, if ever, be simultaneous. Thus the system power level can be proportioned to meet the maximum simultaneous demand likely to be encountered. In many cases it may be possible to reduce this demand to that of the single system with the greatest demand (for example, certain services like power-operated steering will only be required when the vessel is under way; and other services, like windlasses, when the vessel is stationary).

In the 'all-hydraulic system', or where a number of individual hydraulic services are operated from a common source of hydraulic power, an accumulator can materially reduce the size of power plant required and also provide 'reserve' power to meet peak demands, such as an unusual set of circumstances calling for a simultaneous demand greater than the design maximum operating level. This can apply particularly when a variety of deck machinery may be hydraulically operated off a common system.

A typical scheme for hydraulic drive for deck machinery on a medium size fishing vessel comprises a double pump driven directly by the main engine and feeding four separate hydraulic motors distributed over the deck. Each service is independently selected and controlled and can operate simultaneously, if necessary. Additional services can be added, or other combinations used, within the power-output capabilities of the pump.

A series system is shown in Fig 6 where a single pump is used to operate all the deck machinery, each machine being fitted with the same size of hydraulic motor. A second pump is incorporated as a standby or can be run simultaneously unloaded and brought in only when peak power

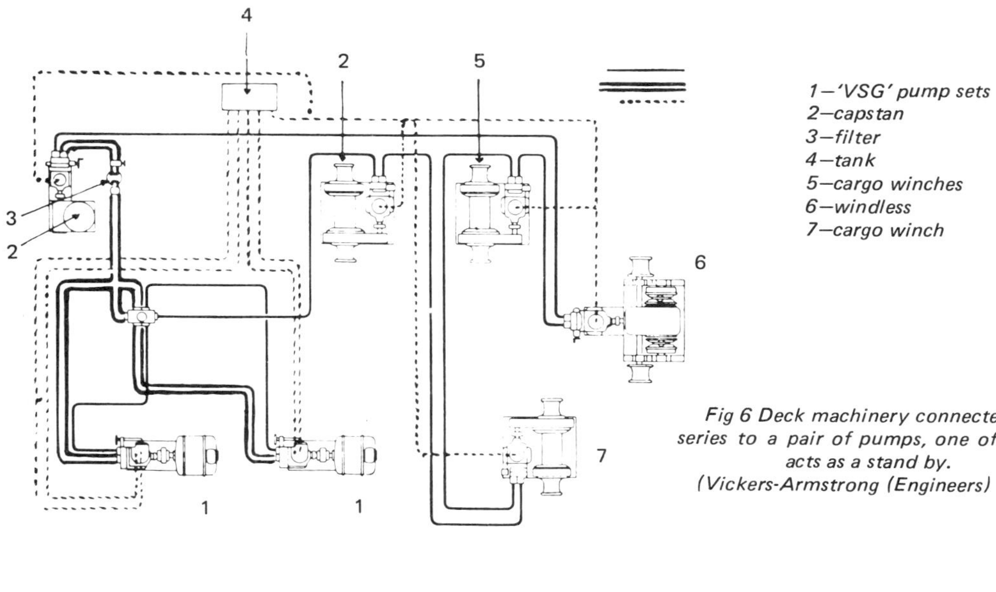

Fig 6 Deck machinery connected in series to a pair of pumps, one of which acts as a stand by. (Vickers-Armstrong (Engineers) Ltd).

demands have to be met. Advantage can be taken of the normal 'non-simultaneous' service requirements to keep the size of the pump and prime mover down to a minimum.

All the units in this circuit are connected in series *via* a ring main and each motor is controlled by a four-way combined input bypass valve. When this valve is operated the required amount of fluid can be diverted through the motor before the main flow is passed on to the next unit in the series connection. Each motor can be operated independently of the others and the pressure drop across each is proportional to the load on it. Normal maximum 'simultaneous' demand would be from the three winches, representing the peak system demand, and the pump size would be selected accordingly. A smaller pump may be specified on the basis that only two winches are likely to be used simultaneously (or any other two services together), when the standby pump could be run off-load and brought in should circumstances call for the operation of three services simultaneously.

Mooring Auxiliaries

These comprise capstans, aft warping winches and automatic mooring winches which would normally employ radial hydraulic motors and can be rendered in almost any size and capacity in neat and compact form. On larger vessels the pump is usually engine-driven (and thus located in the engine room), although an electrically driven pumping unit may be preferred in some installations. An electrically driven pumping unit has advantages for working deck machinery and similar services which may be required when the vessel is stationary since the main engines do not have to be run to provide hydraulic power.

Hatch Covers

A majority of modern cargo ship designs employ hydraulically operated hatch covers, the specific advantage being the very considerable saving in time compared with manual labour to do the same job. System design is specific to the size and type of hatch and the loads involved, but can be met with the simplest mechanical linkage connected to hydraulic jacks of suitable proportions.

Power demands for the actual operation may be relatively high, but since this service is only required at infrequent intervals accumulators are widely favoured as the power source. Accumulator size can be selected to cover a complete cycle of operation with sufficient reserve for emergencies, and then re-charged during the long idle periods of the system by a very small pump and motor.

Stabilizers

Stabilizing fins are now an accepted feature of modern ship designs for roll damping and are employed on naval as well as commercial passenger-carrying vessels. Stabilizers differ from bilge keels or similar fixed devices in that they are rotated to produce positive or negative 'lifting' forces, with their motion governed by a master roll controller (usually a gyroscopic roll detector). The necessary motions of the stabilizing fins are almost invariably produced by hydraulic power.

Pumps are used which give both variable and reversible delivery and these are controlled through a powerful servo-system capable of giving full effect to control impulses in a fraction of a second.

Arrangements are also provided for stowing the fins hydraulically when the ship enters port.

A similar system is available for smaller vessels and this is shown in detail in Fig 7. It has been produced by the co-operation of Saunders Roe (Anglesey) Ltd, William Denny & Brothers Ltd, Brown Brothers & Co Ltd, Muirhead & Co Ltd, and Automotive Products Co Ltd, who are responsible for the electrical and hydraulic systems.

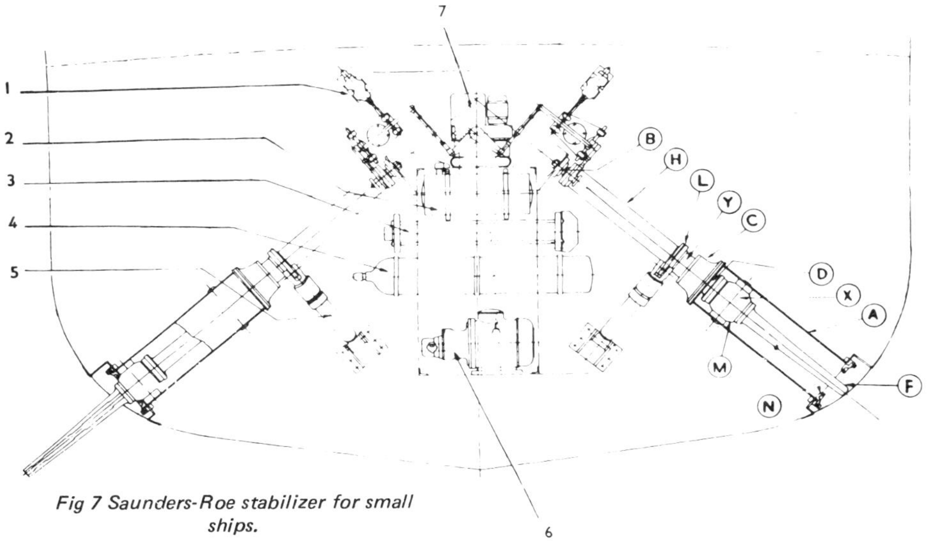

Fig 7 Saunders-Roe stabilizer for small ships.

1—control valve
2—supply tank
3—accumulator
4—air bottle
5—hydraulic jack
6—pump and motor
7—gyroscope

As will be seen from the drawing there are two fins which are extended when in operation and retracted when not in use. As in all Denny-Brown designs, the fins are in two parts — the main fin and the tail fin. The main fin is oscillated by power through an angle of 40 degrees and the tail fin automatically travels through a total of 100 degrees through a simple shackle device. Any tendency of the ship to roll is resisted by the forward motion of the ship acting on the fins, which are turned in such a way as to correct it. A roll to starboard would be corrected by turning the starboard fin anti-clockwise (looking from the outside) and the port fin clockwise.

A watertight box A is attached to the hull structure and to a hydraulic cylinder H which is pivoted at its upper end. The fin unit consists of the fin F, a crosshead X which runs in slots in the box A to transmit the turning motion and a double-acting piston operating a cylinder H. When extended, the crosshead makes a joint with the seating at the bottom of the box A, and by so doing all fin loads are transferred direct to the ship structure.

The fin assembly, including cylinder and water-box, is oscillated by the hydraulic jack, which is connected to the lever L. At the upper end the control valve spool is attached to a floating lever, one end of which is connected to an arm on the upper end of the cylinder H and the other to the gyroscopic control.

Hydraulics in Aircraft

HYDRAULICS IS virtually the standard choice for powered controls systems and utility services on modern aircraft, largely because of the highly favourable power/weight ratios attainable with system pressures of 210–280 bar (3 000–4 000 lb/in^2).

System pressures have become more or less standardized at 210 bar (3 000 lb/in^2) in the United States, although 280 bar (4 000 lb/in^2) are also now used – higher pressures being favourable to the power/weight ratio, although at the expense of increasing component stresses. Thus, light alloys have been largely excluded; jacks are fabricated from high tensile steels or titanium; and there is a preference for stainless steel over copper-based alloys for lines (although light alloy is still used for low-pressure return lines). Instead of coupling, welded joints have been adopted on many aircraft to reduce maintenance, again particularly with stainless steel lines.

A system pressure of 280 bar (4 000 lb/in^2) is considered by some authorities as the optimum with regard to equipment rate and response rate with servo-controls.

Power Levels

In the early days of powered flying controls applied to fairly large aircraft, input demand seldom exceeded 15 to 20 horsepower. By the late fifties 40 to 50 horsepower was quite common, and represents a typical level for modern high performance aircraft of smaller size.

Modern military aircraft of the fighter/bomber category require a hydraulic system output capability of some 100 horsepower or more. In the case of variable-sweep wing aircraft the demand may approach 200 horsepower, with about 70% of this power utilized for sweeping the wings. Size also plays an important part – thus in very large military aircraft up to 1 000 horsepower may be made available for the primary and utility systems (*eg* C5A Galaxy).

Flow rates required have also risen considerably – from about 9 to 22 litres (2 to 5 gallons) per minute with earlier powered control systems to 1 236 litres (250 gallons) per minute in the case of the Boeing 747 and 2 045 litres (450 gallons) per minute for the Galaxy.

Pumps

Radial or axial piston pumps are the normal choice, generally incorporating their own valves, to reduce delivery and off-load the power to a low level, when the pressure has reached some 90% of its maximum. Variable-delivery axial-piston units are generally preferred, because of their more flexible control characteristics, and current standard types can readily produce deliveries of 182–227 litres/min (40–50 gal/min) running at 5 500 rev/min.

With such pumps the use of a pressurized reservoir is essential. Typically this is provided by engine air bleed at 7 bar (100 lb/in^2), or by a detensified reservoir pressurized from the high-pressure supply. An example of this is shown in Fig 1, utilizing what is commonly known as a 'bootstrap' circuit.

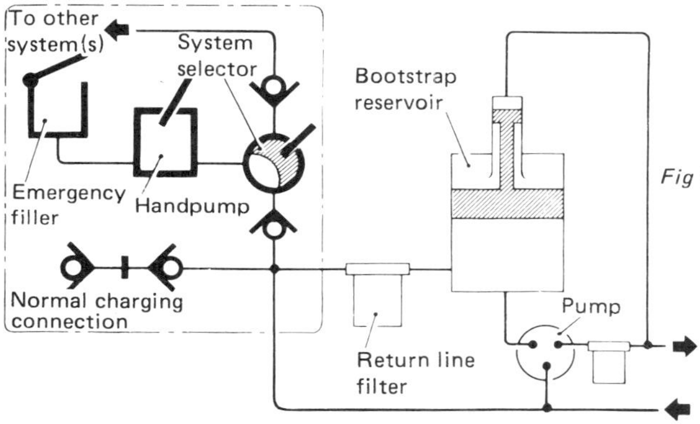

Fig 1 'Bootstrap' reservoir for system replenishment.

Power/Weight Ratio

From weight to power ratios of about 0.35–0.45 kilograms per horsepower (0.75–1 pound per horsepower), figures currently achieved with 280 bar (4 000 lb/in^2) systems are often better than 0.1 kilograms per horsepower (0.25 lb per horsepower). The very high power levels themselves, however, have added further problems, notably the necessity for adequate temperature control. Complex cooling systems may be required, which rely on both air and fuel heat-exchangers to maintain the hydraulic fluid at acceptable working temperatures.

Actual systems used have tended to become more and more complex, basically because of necessary system duplication on the primary controls. These cover all functions necessary to keep the aircraft in the air – *eg* at least elevator, rudder and aileron controls. In the case of smaller, low flying aircraft, where control surface loads are not excessive, reversion to manual control may be acceptable in the event of hydraulic failure. At higher input power levels, however, emergency conditions are normally provided for by a back-up system or systems based on a separate circuit or circuits and, where possible, a separate source of hydraulic power. Thus it is necessary to duplicate the hydraulic lines, valves and actuators, at least.

Triplex systems have, in fact, become commonplace on larger aircraft, providing two main systems with the third as standby. On a four-engined aircraft this may involve as many as six pumps in a combination which provides power from at least two pumps in the event of two-engine or two-pump failure.

On smaller, or less critical, aircraft, a logical solution in the case of multi-engine aircraft is to have two separate pumps, each driven by one engine. One pump can then be allocated to supply the primary system and the second the utility system. The utility system is then also the back-up system, and can be switched to power the primary system in an emergency, at the expense of isolating the utility services.

A preferred system is shown in Fig 2. This provides duplication of power for the flight control units, together with a separate source of power, in the form of a compressed air bottle, for emer-

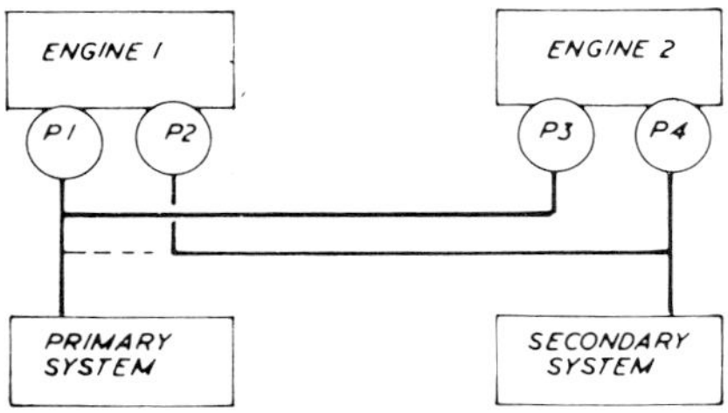

Fig 3

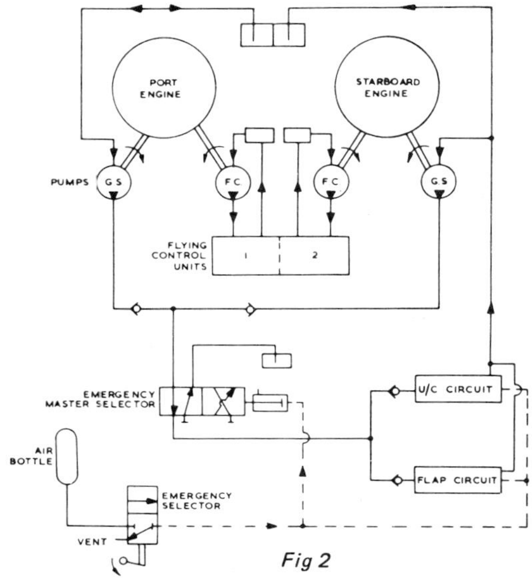

Fig 2

gency operation of flap and undercarriage. Shuttle valves are provided at the jacks, and release valves are included in the circuit to allow the return fluid to flow out of the alternate side. Flight control actuators are based on split cylinder construction, so that failure does not destroy its function.

An alternative system is shown in Fig 3. In the case of a twin-engine aircraft, each engine drives two pumps (or in a four-engine aircraft, each engine drives one pump). Pumps are parallel for each of the primary and utility systems. This arrangement gives better overall reliability, for it avoids complete loss of the secondary control system in the event of failure of one engine. However, partial isolation of the secondary services may be necessary, in order to maintain adequate performance in the main circuits, unless the pumps are so sized that a pair of pumps can meet maximum demand between them.

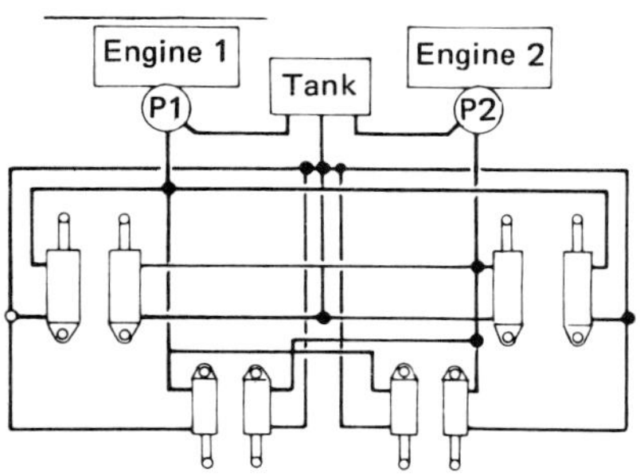

Fig 4

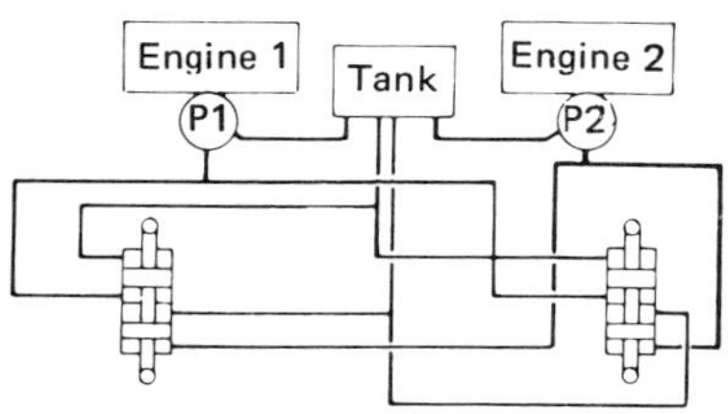

Fig 5

Physically, duplication normally involves the use of two complete and separate systems, each with its own actuators. These may be separate actuators (Fig 4) or tandem actuators (Fig 5). There is also the choice of using both actuators simultaneously for normal operation (that is, a control powered by both systems), or employing one of the pairs of actuators purely as a standby unit,

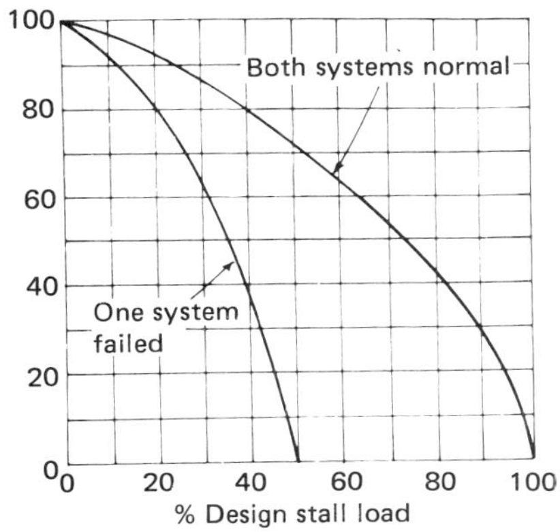

Fig 6

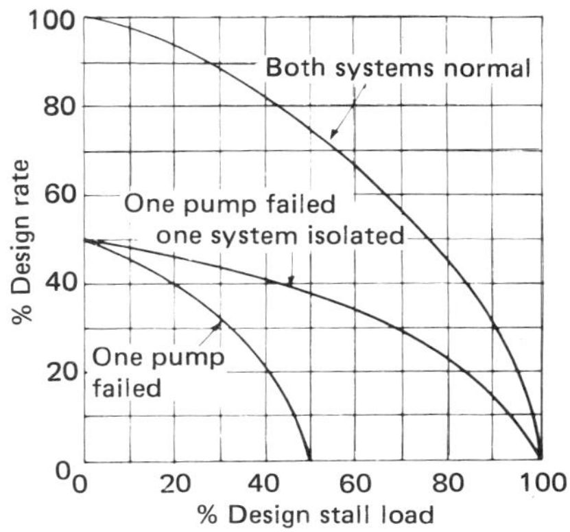

Fig 7

normally on open circuit, but to which backup power can be switched in an emergency. The choice affects both the attainable rate and the percentage of maximum output which can be achieved under emergency conditions.

Thus with a dual system (both systems live simultaneously under normal operation), characteristics may take the form shown in Fig 6. With a back-up system, the operating characteristics may be of the form shown in Fig 7. Again, depending on the specific arrangement installed, which will finally decide the most suitable configuration for the performance envelope required, there is flexibility of choice. In some cases, particularly in large aircraft, triple or even quadruple systems may be employed to provide back-up with maximum attainable rate and output. A typical duplicated system is shown in Fig 8.

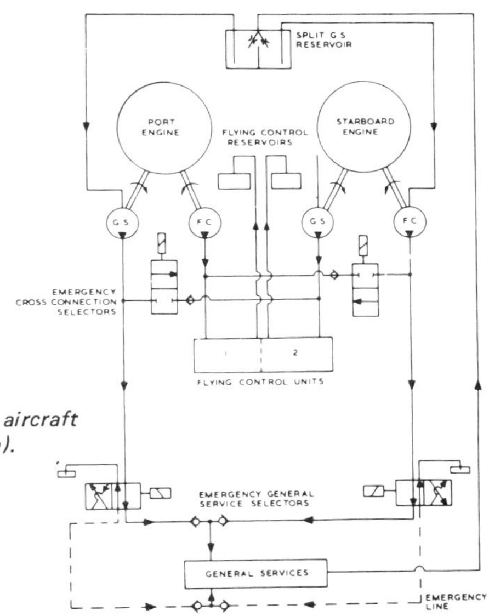

Fig 8 Typical allocation of aircraft services. (duplicated system).

Additional Groups

With the increasing complexity of controls on modern aircraft, the utilities may be sub-divided into further groups, depending on their importance. Thus the main flying controls remain Group 1, or the primary system to be safeguarded fully. The remaining controls are then allocated to additional groups, in order of significance. Hydraulic power is then developed by two (or more) separate and independent systems. The primary system is concerned solely with supplying power to Group 1.

These services are also connected to the utility system, which can also supply power to Group 1 services, or can act as a standby. The utility system also supplies power to Group II and Group III functions. Both these groups are automatically isolated in the event of loss of pressure in the primary system, when the utility system is directed solely to powering the essential flight services. Thus either the primary system, or the utility system, can supply power to all essential services, and with normal operation, both systems are connected to Group I services. The basic system layout is shown in Fig 9.

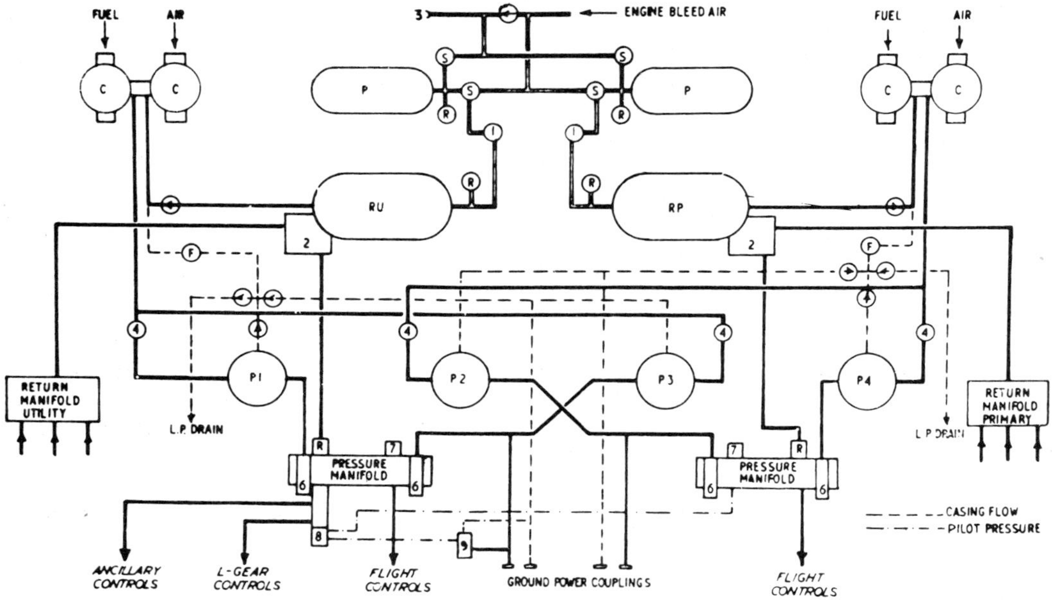

Fig 9 Simplified diagram of fluid power system.

P	—	*pneumatic bottle.*	*1*	—	*pressure regulator.*
RU	—	*hydraulic reservoir (utility service).*	*2*	—	*return filter and bypass assembly.*
RP	—	*hydraulic reservoir (primary service).*	*3*	—	*pneumatic charging point.*
S	—	*shuttle valve.*	*4*	—	*firewall shut-off valve.*
R	—	*relief valve.*	*6*	—	*pressure switch.*
P1	—	*left engine pump.*	*7*	—	*pressure transducer.*
P2	—	*left engine pump.*	*8*	—	*isolation valve.*
P3	—	*right engine pump.*	*9*	—	*manual isolation valve.*
P4	—	*right engine pump.*			

Electrical Signalling

The use of all-electrical signalling results in a control system which is easier to route through a densely packed airframe and is unaffected by airframe flexure. Most important of all, in a large supersonic aircraft, such a control system also overcomes the very serious problem arising from the change in relative lengths of the control linkage as compared with the airframe, which is subjected to very high temperatures arising from kinetic heating. Electrically transmitted control signals are not new, for they have long been the means of feeding in signals from an auto-stabilizer and sometimes from the auto-pilot also.

In the military aeroplane an electrical system need not have extremely high integrity as a carrier of auto-stabilizer signals; it need not be duplicated, for it were to fail the aircraft would simply carry on without an auto-stabilizer (and there are very few aircraft in which this would be really serious; it would merely mean a reduced flight envelope and possibly a curtailed mission). But for carrying primary control demands the signalling system must obviously have the same safety level, *via* duplication or triplication, as a comparable mechanical system. It must have built-in safety precautions to guard against any failure of a dangerous nature which would send out a signal calling for full control deflection about any axis. This type of situation has already been encountered generally in auto-stabilizer systems, and particularly in automatic terrain following. In the latter case the simplest answer is to arrange for any failure of the signalling while the terrain following mode is in operation to result in a safe departure from the original flight path – namely, to pull the aircraft up into a climb. Over enemy territory this would naturally increase vulnerability to defending weapons, but is considered preferable to leaving the pilot with inadequate time to take over and avoid hitting the next hill. With a supersonic aircraft, the human lag in the control loop is unacceptably great, and the only answer is a duplicated system with a safe preferred runaway direction.

For an airliner a more complex system is needed, in which the signalling system contains redundancy equalling that of the power control system. An enormous amount of work must be done before any electrically signalled system could be put into airline use. As the most advanced civil aircraft in full operation at the present time, the Concorde has a control system philosophy which almost certainly represents the latest stage in current development. In this particular aircraft layout the normal mode of control is electrical. Two independent electrical channels signal the commands to an electro-valve incorporated in each servo-control (lanes A and B). A mechanical channel (lane C) is provided in the event of a complete electrical failure. In normal operation, at any moment, the three modes send signals to the servo-control.

A monitoring system controls the operation of these modes and activates only one of them. In normal operation, control is ensured by electrical channel 1 (lane A). If an electrical or hydraulic failure occurs, the monitoring system cuts out electrical channel 1 and activates electrical channel 2 (lane B). As this channel is a standby one, the transient motion of the control surfaces is limited during switching. In the case of previous failure of the second channel, the monitoring system would directly switch to the mechanical control.

The servo-controls are dual tandem-cylinder, irreversible, hydraulically-powered servo-actuators. They are supplied by two independent main hydraulic systems, each being connected to one of the servo-control cylinders. In emergency, each cylinder can be supplied by a standby system.

To go beyond the Concorde philosophy to the wholly electrically signalled system, would require a research programme with an aircraft of suitable calibre being cleared by the ARB and FAA within three years on fully electrical signalling. It would mean much painstaking work. If a mechanical system fails, the result is 100% predictable and relatively easy to counter by modified design. Compared with a mechanical system, the electrical one may be lighter, and it would certainly lend itself better to generating complex non-linear responses, simply by inserting a micro-electronic filter network, weighing perhaps 28 gms (an ounce) but there would be far more parts to go wrong and the inherent reliability levels would start off well below those of the mechanical system.

Hydraulics in Mining

IT HAS been estimated that about a quarter of UK production of fluid power equipment has been used in mining. Compressed air is still widely used, but the higher output powers and forces available from more compact hydraulic equipment offers many advantages over air as well as opening up further fields of application (*eg* particularly for roof supports).

The only major problems associated with hydraulics in mines is the necessity of using a fire-resistant fluid and the exacting requirements necessary to avoid fluid contamination when working in dirty conditions.

Fire-resistant fluids have completely replaced mineral oil fluids in UK mines.

For mobile machines and rotary transmissions, invert emulsion (40% water) may be employed, permitting the use of most standard industrial hydraulic components, except that pumps and motors must only be of types able to tolerate the modest lubricating properties of this liquid. For the extended systems of powered roof supports, in which it is impossible to avoid some leakage, dilute emulsion (95% water) is used because it is so much cheaper, although it necessitates special valves and pumps. It is expected that the range of equipment able to run on dilute emulsion will be extended considerably in the future.

Hydro-kinetic fluid couplings of the self-contained type are filled with plain water, but those of the scoop-controlled design with an external circuit which cannot be pressurized, use phosphate ester.

Fluid Protection

The general design features adopted for mining hydraulic systems are those one would expect to combat the rugged and dirty conditions, the difficulty of ensuring that filter elements and other spares are available when needed, and, above all, the fire risk. Some early mining circuits were rather crude with little or no filtration, but on modern ones virtually the full range of precautions against contamination are taken. In spite of this, the state of the fluid usually corresponds to one of the worst of the standard industrial contamination classifications, or even dirtier.

Reservoirs are fully enclosed, with a filtered breather, and preferably the refilling fitting faces downwards so that fluid must be pumped in, since experience has shown that dirt is often introduced during open pouring. Reservoirs contain sludge weirs and magnetic filters. The pump suction is provided with a relatively coarse strainer, often of the rotary cleanable type. The finest filter in the system, down to 15 μm, is usually a replaceable element unit placed in the pressure line immediately following the pump where the pressure loss it causes will not be important. Local filters protecting sensitive components such as solenoid valves are also used. Reservoirs are fitted with safety devices such as low level and high temperature cut outs.

Exposed sliding surfaces are very liable to damage and corrosion. Where possible, rams are designed so that the rod is retracted in the rest position. Plating or hard chrome is employed on rods. Gaiters are sometimes used but are rather subject to mechanical damage. On surfaces, such as ram barrels, which are wholly immersed in the fluid, surprisingly little corrosion occurs even with water-based fluids, and bare steel is satisfactory. More critical components such as manifolds and valve bodies are often phosphated, whilst their moving parts are usually of stainless steel.

Underground Drilling Equipment

In almost all cases, the use of hydraulics represents a move to greater mechanization. Booms have been designed where both the boom and the feed can be swung quite freely. All the movements of the boom and feed are actuated by hydraulic cylinders. Unlike air, hydraulic oil is relatively incompressible, so that it is possible to position the feed or boom to drill a hole at a pre-set point and in a pre-set direction. The hydraulic fluid remains trapped between the valves when the position is fixed.

Hydraulic cylinders have the advantage over such mechanical alternatives as long screw feeds in that these are exposed to wear and disruptions. The hydraulic systems of booms are normally powered by a compressed air motor, although the rig may also feature an auxiliary diesel or electrically driven hydraulic pump, so that the booms can be swung when the rig has to negotiate curves.

An important advance over simple positioning by means of hydraulics is 'parallel holding', whereby a feed can be kept parallel to its original position when moved vertically and/or horizontally. Here hydraulic oil is transferred from one cylinder to another to compensate for the upwards, downwards, or sideways movement of the boom and keep the feed at the same angle to the tunnel face. On most tunnelling and drifting rigs the feeds are also hydraulically powered.

Hydraulic Pit Props

Hand-set individual props comprised the first major incursion of hydraulics into the mines, as a direct replacement for timber or rigid steel props with a great improvement in safety and ease of installation. They lend themselves readily to a variety of methods of mining and many millions have been produced at remarkably low cost. They are still extensively used, although now being superseded by powered supports.

A typical prop is shown in section in Fig 1. Basically, it embodies two steel cylinders fitted one inside the other, which can be extended by hydraulic pressure derived from a pump incorporated in the assembly. Provision is also made for quick release of pressure, whilst a relief valve in the hydraulic circuit also ensures that when the hydraulic pressure in the prop exceeds a specified figure the prop yields.

As the diagram shows, the main or pressure cylinder is enclosed by a guard tube. The inner cylinder, which slides in the pressure cylinder, forms a fluid reservoir as well as containing the pump and valve mechanisms. The pressure cylinder and inner cylinder operate together as a hydraulic ram, with a main bearing located at the open end of the pressure cylinder. This bearing also forms part of an outstop should the prop reach the limit of its travel. Included in the bearing assembly are a metal scraper ring and wiper ring to prevent dirt and water from entering the pressure cylinder.

A piston head is welded to the bottom of the inner tube and carries the pump cylinder, the main release valves and the relief valve capsule plus a gland ring which forms a high pressure seal, an anti-extrusion ring and a piston ring. Valves and rings are retained by a detachable plate.

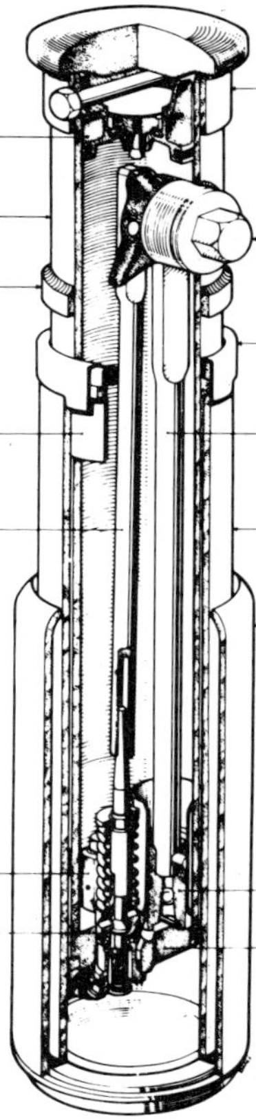

Fig 1 Dowty hydraulic pit prop.

A control shaft is mounted transversely in a housing at the top of the inner tube, linked *via* a crank to a connecting rod to the pump and a push rod to the valve assembly. An oscillatory motion imparted to the control shaft *via* a suitable handle operates the pump for setting the prop in position. Lifting the handle right up past the normal suction stroke position first compresses the spring in the valve assembly and then operates the relief valve mechanism for withdrawing the prop.

The pump is a single acting type with the pump cylinder projecting from the top face of the main piston head. The piston is pin-jointed to the connecting rod and a non-return valve forms part of the piston itself. Fluid is admitted to the pump cylinder *via* this non-return valve and then forced into the pressure cylinder by the pumping action. The prop thus extends a specific distance for each stroke of the pump, the setting up load being established by continued operation of the pump handle. This setting up load is normally from one quarter to one third of the maximum load for the prop. Thus once set the prop will continue to accept increasing roof loads up to the maximum figure, at which the relief valve is automatically operated and pressure is reduced by fluid passing from the pressure cylinder back to the reservoir. Immediately the load on the prop drops below a specified figure, the relief valve re-seats.

Under safe roof conditions withdrawal is accomplished by upward movement of the pump handle opening the release valve, after first overcoming the resistance of the spring. Prop closure can be controlled to fine limits since the relief valve allows for inching of the prop. Under difficult conditions – for example, where the roof is likely to be unstable – the prop can be released remotely by a rope and chain. In this case the handle is fitted pointing towards the bottom of the prop so that a horizontal pull will operate the release valve. The handle crank in this case should point inwards to prevent the prop from spinning.

Being fully enclosed units (except for a breather), hydraulic props may use mineral-oil fluid, although some may be designed specifically for operation on oil-in-water emulsions.

Powered Supports

In principle, a powered support unit consists of several props mounted on a common base which also contains a horizontal ram for advancing the whole assembly. A typical British unit, termed a 'chock' shown in Fig 2, shows a face equipped with similar supports. As with individual props, it is the trapped fluid within the legs which supports the roof, but powering permits the legs to be made heavier, and the rated thrust per leg is from 30 to 100 tones in current designs. Legs are interconnected in pairs, see Fig 3, each pair controlled by an isolating valve and a yield valve. The latter is set to open at a pressure of, say, 450 bar (6 500 lb/in^2).

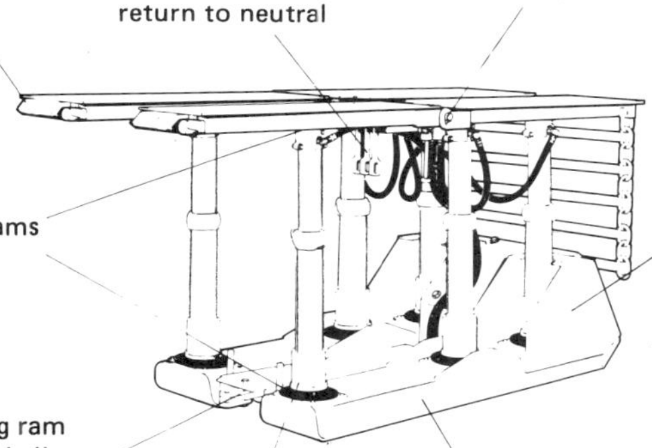

Fig 2 Six-leg powered support unit. (Dowty Mining Equipment Ltd).

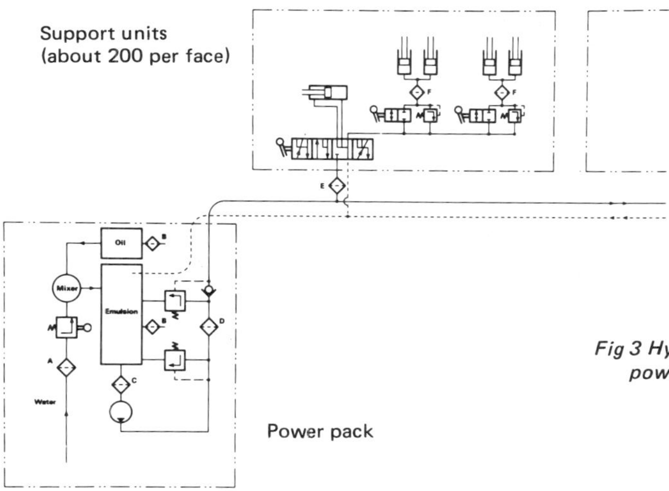

Fig 3 Hydraulic system of a coal-face powered support installation.

Setting of the legs is effected by taking fluid from the main supply hose which runs the whole length of the face. This also powers the advancing rams and any ancillary motions that may be provided, such as small rams to control the alignment of the chocks in an inclined seam. Control of all functions is by means of a selector valve, usually of the rotary face type. For safety, 'dead man's handle' action is required, and this is provided by combining the selector valve with a co-axial poppet valve controlling fluid admission. The valve handle is turned to select the service required and depressed to obtain power. There is of course also a connection from each chock to a main return hose.

Nominal supply pressure is 100–170 bar (1 500–2 500 lb/in^2) in the U.K. and the power pack is situated in the access tunnel near the end of the face. On the Continent it is customary for supply pressures to be up to twice as high, and often the pump installation is semi-permanent, feeding face through as much as 1.5 km (1 mile) of pipe. Since the fluid is dilute emulsion, the only suitable pumps are slow speed 3-cylinder types with oil-lubricated crankcase. These are fixed delivery pumps so that an unloader valve must be provided to control the output. Usually at least two pumps are provided to ensure reliability, the total output being about 2 lit/sec (30 gal/min), depending on the size of the chocks and the number of them to be moved per minute to keep pace with the other mining operations.

Underground Loaders

Increasing demands from all phases of mining and civil engineering have meant that loaders are being asked both to travel longer distances than before and to carry heavier loads. Haulage distances of 100–300 m make compressed air lines impractical. Air pressure is lost, the hose is exposed to wear and is generally difficult to handle. The necessary increase in hose diameter aggravates these problems.

Equally, the air cylinder, used in the past to raise the bin and load, took up most of the space between the wheels of the unit. Additional disadvantages were lack of speed and strength in certain circumstances. Since the hydraulic system, which replaced pneumatics, works at a much higher pressure, *eg* 150–250 bar, more speed and power became available and from much more compact equipment.

Once hydraulic power has been introduced for these machines, it can also be used close to each wheel, to provide four-wheel drive as well as good manoeuvrability. The diesel engine powers a series of hydraulic pumps; one for each pair of wheels and one for the bucket and bin. The number of hydraulic cylinders required can be kept low by the bucket and bin. A small extra cylinder can be added to increase the digging power of the bucket at the critical moment.

For cutting coal from the seam and loading it out, the main drive is normally electric, the inertia of the motors proving an advantage in coping with intermittent peak loads. Hydraulics are used, however, for all ancillary operations such as boom movements and for traction. Extreme compactness is required, so that proprietary pumps, motors and valves are seldom suitable and machine builders must develop their own. These components are often located within gearboxes so that mineral oil rather than fire resistant fluid is used, but this is tolerable since there is not much chance of an external high-pressure leak.

The most common longwall power loader is the shearer, of which an example is shown in Fig 4. The 200 kW water-cooled motor drives the two cutting drums through gear trains. A drive shaft passes through the haulage section of the machine where there is also a power take-off to the hydrostatic transmission. This comprises a variable-delivery swashplate pump delivering up to 1.8 lit/sec (23 gal/min) at 200 bar (3 000 lb/in^2) coupled in a closed circuit to the haulage motor, which in this model is of the radial piston type with a multi-lobe cam. This drives the haulage

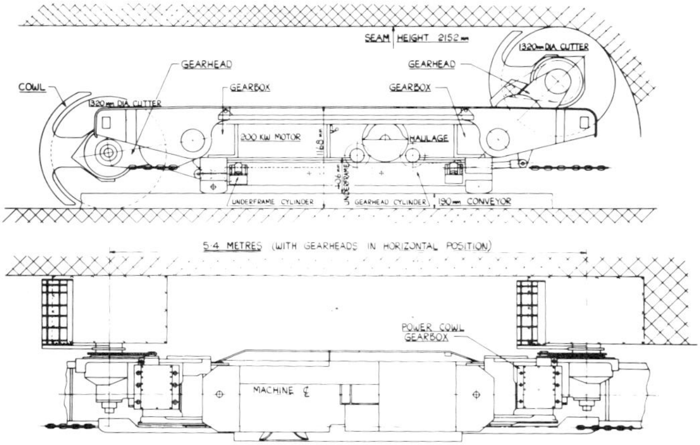

Fig 4 Double ended ranging drum shearer.

sprocket through 3.6:1 gearing. This propels the machine at up to 10 m/min (30 ft/min) by pulling on a fixed chain stretched along the face. Rack systems are also in use as an alternative to chain.

The pump swash angle may be varied to control the speed and direction of travel. There is also a torque motor sensing the load on the main electric motor so as to reduce haulage speed in case of overloading. The haulage section casting itself forms the reservoir and it also contains a priming pump, filter and oil cooler.

There are further pumps of 0.1 lit/sec (1.3 gal/min) capacity within the gearbox at each end of the machine for powering the ancillaries – boom lift, cowl rotation and machine tilting jacks.

Other longwall machines, such as the trepanner, cut and load the coal in a different manner but the haulage and other hydraulic aspects are basically similar.

For driving headings in coal and for pillar-and-stall mining in the American fashion, the machines are mainly mounted on caterpillar tracks. Machines such as the Joy Continuous Miner have a cutting head containing a broad mat of chains carrying picks to cut the coal and load it on to the internal conveyor of the machine. Raising, lowering, slewing and track drive are all powered by the integral hydraulic system.

Tunnelling

In tunnelling, the methods employed vary widely in accordance with the hardness of the rock. In the tunnelling associated with longwall coal mining, only rocks of medium hardness are encountered, but the need to keep pace with the relatively slow advance of the face renders the most sophisticated machines uneconomic, and also special problems are posed by the congestion of machinery.

The standard method for medium and hard rock tunnels is the use of explosives, for which the holes are drilled by pneumatic percussive drills. Retaining the pneumatic drills, performance may be greatly improved with a hydraulic boom rig or 'jumbo', which enables the drills to be quickly and accurately positioned and a much greater thrust to be applied to get optimum penetration rate. All movements are powered by rams, except advancing the drills, where a rotary hydraulic motor driving through a chain is used.

One of the obstacles in exploiting hydraulic oil as a power medium in rock drilling lies in its major difference from air – its incompressibility. Hydraulic oil can be compressed about 0.5%

under a pressure of 1 000 bar (14 500 lb/in^2.) If a piston mechanism is employed, the high shock pressures generated in the hydraulic fluid could break parts of the rock drill. In this case the inclusion of an accumulator can solve the problem.

Working pressures in hydraulic systems for rock drills are about 200 bar (2 900 lb/in^2) as compared to 5–8 bar (72.5–116 lb/in^2), or in certain cases up to 15 bar (217.5 lb/in^2), in a pneumatic system. This means that the size and weight of the rock drill can be reduced for the same amount of impact force, or the force can be increased without excessive demands for room and bearing power. At the same time the energy requirement for the same drilling capacity is only about one third as much. In addition, the higher impact force is achieved without increasing the stresses in the drill steel, which are a major factor in drill steel life.

Hydraulic pump units are smaller and can be located on the rig, facilitating changes in pressure to match particular drilling circumstances.

With the closed power transmission system of the hydraulic rock drill, there is no noise due to escaping 'power' and the noise level can be 10–15 dB(A) lower than for an air-powered drill. This is particularly important since the exhaust noise of a pneumatic drill is, to a great extent, low frequency sound, which is the most difficult to muffle. Finally, the absence of cold exhaust precludes the precipitation of humidity as mist and fog. Freezing problems are also completely eliminated.

Ripping

In coal mining, ripping is the operation of cutting down stone to enlarge a tunnel after the coal in the cross section has been removed. One of the types of machine used is that shown in Fig 5. This has direct electric drive to the rotary cutting head, but all other motions, including the drives to the scraper conveyor and the caterpillar tracks, are hydraulic. There is a multi-section gear pump supplying 5.2 lit/sec (70 gal/min) at 140 bar (2 000 lb/in^2) maximum pressure. The machine is capable of working on invert emulsion fluid.

Fig 5 Mk.IIa road header.
(Dosco Overseas Engineering Ltd).

Fig 6 Impact ripper.
(Gullick Dobson Ltd).

An alternative approach is the impact ripper, of which one version is shown in Fig 6. Its hydraulic impactor is capable of giving blows of 4 000 J (3 000 ft/lb) energy, at the rate of 600 per minute. In this design the force which accelerates the hammer piston is provided by nitrogen in an enclosed chamber acting as a large spring. This is alternately compressed and released by an annular hydraulic piston controlled by a pilot operated valve. In some other manufacturers' impactors, the hammer is a double-acting hydraulic piston and the very pulsatory flow is catered for by one or more external accumulators. Impact ripping is most effective in laminated strata where the existing fissures can be exploited by the chisel. Impactors are also proving very useful for surface applications such as concrete breaking.

To avoid confusion, it should be explained that 'hydraulic mining' means washing out a mineral by powerful water jets. It is used for surface working of sedimentary ores and occasionally for soft coal underground. It is also possible, with a water jet, to cut quite hard rock such as would be encountered when tunnelling, but this is only at the research stage at present. Exceptionally high pressures, *eg* 7 000 bar (100000 lb/in^2), are necessary, obtained by a hydraulic or pneumatic intensifier.

The broken rock obtained by any of these methods must be loaded out in some way. Only a few cutting machines have integral provision for this; otherwise a shovel loader must be used. In large tunnels and in some mines, diesel powered loaders of standard surface types may be used. For smaller workings, more compact mobile machines are available, chiefly powered by trailing cable to an electric motor on the machine. This drives a hydraulic pump which powers all motions including the crawler tracks. Some manufacturers formerly specializing in all-pneumatic loaders have found it impracticable to improve the lifting force without using unreasonably large air cylinders, and so have resorted to the unusual scheme of rotary air motors driving conventional pumps which in turn power hydraulic cylinders to carry out the actual motions.

If a cylindrical tunnel is required and its length is sufficient to warrant substantial capital cost, a mole tunnelling machine capable of both cutting and loading at a rapid rate may be used. The shields used for driving underground railway tunnels in clay are well known, but it is now possible to use the mole machines for at least medium-hard rocks. Roller cutters are used and hydraulic rams are essential to provide the very large thrusts required. For example, a 4 m (12 ft) diameter machine has a gross forward thrust of 400 tons reacted by gripping pads loaded to 440 tons on each side.

Underground Transport

Wherever possible the mineral product is brought out by conveyor, either to the shaft bottom or all the way to the surface up an inclined drift.

Some loading machines incorporate a small conveyor of the scraper-chain type. These are frequently jammed by large rocks and hydraulic drive is thus very suitable. Low speed motors requiring no gearbox are generally chosen. 'Armoured' face conveyors, also of the chain type, may be up to 300 m long with a drive of up to 180 kW at each end. Radial hydraulic motors have been used for this purpose, offering the useful safety feature of rapid stopping due to their low inertia, but the substantial power packs needed and the problem of synchronizing the drives at the two ends lead to this form of drive not being preferred to the standard one in which electric motors drive through traction-type fluid couplings and gearboxes.

Fluid couplings are also used for the main drives of belt conveyors, although in this case the scoop-controlled type is usual. The belt sags between the idler rollers when it is stationary, but this sag can be taken up without harm during the very gentle start made possible by the scoop regulation of torque. Hydrostatics may, however, be used for belt tensioning. The principle of this

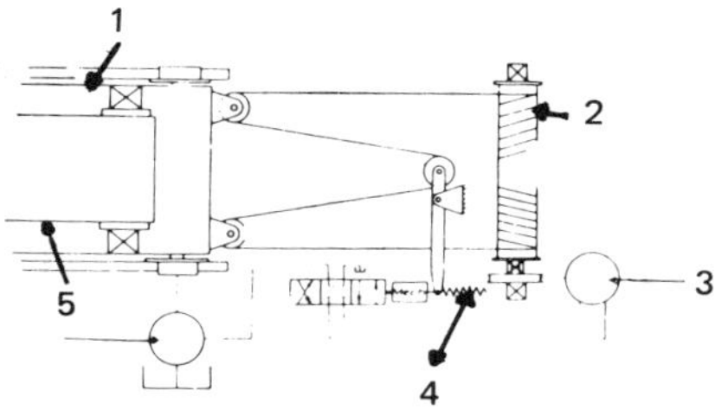

Fig 7 Hydraulic loop take-up for conveyor belts. Correct tension is automatically maintained whilst running and the belt does not sag between the rollers when it stops.
1—carriage. 2—winch. 3—motors. 4—Compression spring. 5—Conveyor belt.

is shown in Fig 7. The loop carriage is restrained by a wire rope from the winch which runs over a pulley mounted on a spring loaded lever. At the far end of the arm is a four-way valve, which returns to the central position when the pull from the belt and the spring are in equilibrium. Oil is supplied by a two-piston pump specially designed for this application. It will run in either direction and the impulses, combined with the general conveyor vibration, prevent any valve stick and ensure rapid response with immediate return of the valve to the neutral position once an adjustment has been made. When the conveyor stops, the pump, which may be driven from the main motor or a small motor linked in with it, stops too. Driving the winch, through a non-reversible worm gear, is a Deri Sine hydraulic motor, which rotates in the appropriate direction in response to the position of the control valve. In the neutral position the motor ports are locked and the pump output returns direct to the tank. With the belt stationary the winch is locked, and although the tension is greater than when running, this has no effect on the control as the pump is stopped. On starting up there is only about 19 mm (¾ inch) movement of the carriage compared with 1.12 m (3ft 8 ins) for the old method, and the belt moves normally in five seconds.

Hydraulics play a vital part in the discharge arrangements for underground storage bunkers. In some, hopper doors release the mineral, and these are operated by rams fed from a stationary power pack. In other cases a very robust chain conveyor forms the whole bottom of the bunker, and discharge is effected by driving the chains at slow speed. This may be done by automatically-reciprocating rams driving through pawls or there may be a chain sprocket fitted with a pair of Staffa motors.

Much of the transporting of men and materials is by means of rope haulage. A stationary haulage engine usually has electric drive, either direct or through a fluid coupling, but hydrostatic drives are being introduced for haulages where control of speed and torque is critical, for example, for winches intended for dragging coalface machinery into position. One winch, driven by a 5-cylinder Staffa motor, will give a pull of 2.9 t and will work satisfactorily on dilute emulsion if required.

For haulage over larger distances underground, trains running on track of conventional type are used, hauled by diesel, battery-electric or trolley-wire locomotives. The chief use of hydraulics here is that of fluid coupling transmission for diesel engines. Near the actual workings, overhead monorails and special channel track, on which the vehicles are trapped to prevent derailment, are used to some extent and on diesel-powered locomotives for these systems hydrostatic transmission has found an interesting application. To provide reliable tractive effort on inclines, the driving wheels are in the plane of the rails, gripping the webs. A gripping force proportional to the traction is provided by small rams fed from the same supply as the hydraulic driving motors.

A further use of hydraulics is in over-riding train brakes. The latest designs, for a fail-safe action, are spring-applied and taken off by hydraulic pressure. A centrifugal device is also fitted, to release the pressure in case of over-speeding.

For loading, tipping or docking, mine cars are being manipulated increasingly by hydraulic means. Several different methods are employed for this purpose, from small isolated units to comprehensive automatically controlled schemes.

To conserve power and prevent excessive oil heating when the cars are held stationary by the retarder, an unloading valve is provided which cuts out the main pump when the pressure rises to about 50 bar (700 lb/in^2) leaving a small subsidiary pump in circuit which blows off at 70 bar (1 000 lb/in^2) through a relief valve. As soon as the retarder is released the main pumps immediately pick up the load at the selected speed.

The retarder grips the wheel flanges between spring loaded bars to apply the braking force and is released hydraulically. When loading from a chute this is used in conjunction with a spotting creeper which pushes the cars forward when the retarder is released by the attendant. The creeper chain is furnished with tip-up horns at 8-link intervals and is driven by a Deri Sine pump which is controlled by a valve to give three motor speeds, 330, 660 and 1 000 rev/min, depending on whether the outputs of the small, large, or both sections of the pump are used.

Tub pullers are similar to creepers except that, being operated by rams, the pawls which engage with the dummy axles are reciprocated.

Shaft Winding

When no cars are used a bunker conveyor feeding into skips may be considered. A system of loading the skips employs an endless steel belt conveyor (Fig 8) about 9 m (30 ft) long which, whilst running at a slow speed of 5.6 m (18.8 feet) per minute, receives coal from a conveyor *via* a chute. The feed is loaded at this speed with the 6 tons necessary to fill one skip. When the skip is in position the belt speed is suddenly increased to 49 m (154 feet) per minute, so shooting the coal into the skip and filling it in 12 seconds compared with 15 seconds with bunker feed.

Basically the hydraulic system is simple, consisting of two pumps with capacities 15 lit/min (3.33 gal/min) and 127 lit/min (28 gal/min) driven by a 25 hp motor, a valve to unload the larger pump when on slow speed and a Deri Sine motor. The loaders are arranged in pairs, side by side, and are loaded alternately from a traversing chute. Provision is made for one pump to operate both loaders should the other pump fail for any reason. Control is very good and the instantaneous increase of speed necessary for discharging into the skip is obtained without shock. Safety features can be incorporated to prevent discharge when there is no skip in position.

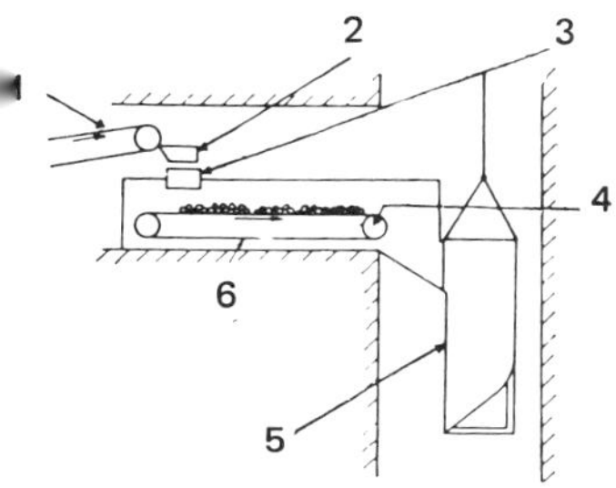

Fig 8 Skip loader with two-speed belt drive powered by twin pumps. 1—conveyor. 2—traversing chute. 3—fixed chute. 4—hydraulic motor. 5—skip. 6—twinfeeders.

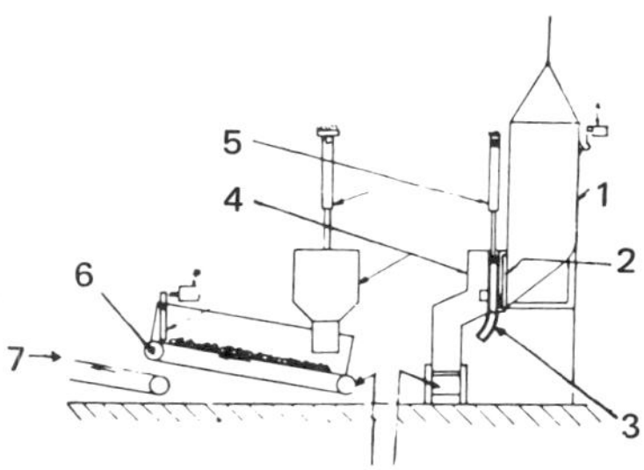

Fig 9 Skip unloader arranged to feed coal gradually into car conveyor. 1—skip. 2—skip door. 3—guide. 4—chute 5—skip door opening cylinder. 6—hydraulic motor. 7—main conveyor.

Skip unloading requires the reverse sequence to that described for loading and the not dissimilar arrangement shown in Fig 9 is used. As the skip comes to rest at the surface its position is proved by valve A and the door loading ram is actuated to lift the door in the side of the skip. Should the skip move more than 150 mm (6 inches) the ram is connected to exhaust and the door closes so there is no danger of the skip being held up by the door with the winding rope slack. The contents of the skip fall into the feeder whilst it is travelling at the fast speed of 52.5 m (175 feet) per minute, so distributing the coal in a layer over the feeder length. When the coal meets the gate at the far end it actuates a trip valve B which puts the feed motor into slow speed, 6.5 m (22 feet) per minute, by unloading the larger pump. The coal is then fed on to the main conveyor belt at the correct speed to give even distribution and no spilling.

The main winding engines of a mine may be of 2 000 kW or more and the hydraulic application here is that of the braking system, which must obviously be capable of delicate control and must be as safe as is humanly possible. Brakes are usually spring-applied and released by a single-acting ram which overcomes the spring force. Unlike a weight loaded brake, the system has practically no inertia and it is a simple matter to incorporate a time lag which reduces the initial rate of retardation, although normally this need not exceed a fraction of a second with this type of brake. In the event of pressure failure the brakes are applied automatically.

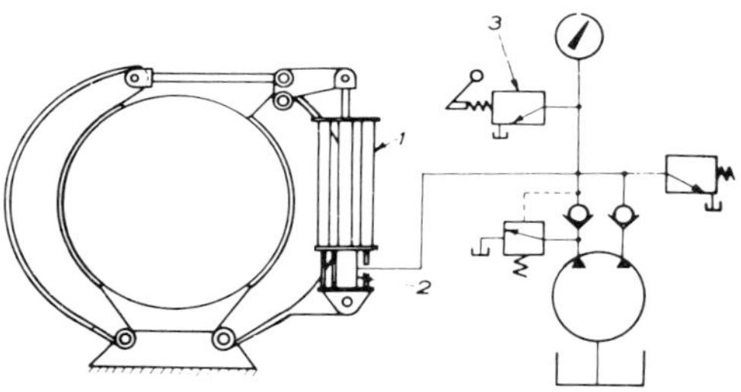

Fig 10 Principle of hydraulic winder brake. 1—nest of springs. 2—cylinder. 3—pressure control valve. (Black's Mining Equipment Ltd).

The usual arrangement (Fig 10) has a high pressure twin-delivery pump, the main delivery being unloaded when the 'brakes off' pressure is reached, leaving the minor supply to maintain pressure.

Because so much depends on local conditions, hydraulic systems often differ in detail. On sinking winding, for example, the brake must be able to discriminate between an unbalanced descending load and the same load ascending, and also compensate for the weight of rope paid out. These conditions will vary as sinking proceeds.

Exact positioning of the cage can be achieved by solenoid valves actuated by the mechanical parts. As the end of the wind approaches, the first solenoid valve is released, thus connecting the brake cylinder to a relief valve, and the pressure falls slowly to allow the brakes to contact the drum lightly. Just before the end of the wind the cage releases a second solenoid valve, introducing another relief valve set at a lower pressure than the first — the position when this occurs can be timed to suit the cage speed. If necessary a third solenoid valve applies full holding force when the cage comes to rest.

Centrifugal governors can be integrated with the hydraulic braking system and if desired, the speed controlled over the whole cycle including whilst stopping.

SECTION 7

Survey of Hydraulic Pumps and Motors

Abex Denison Limited

Abex Denison axial piston pumps operate at pressures up to 350 bar, with variable displacement models offering flow capacities up to 918 lit/min, or fixed displacement from 7 to 918 lit/min. Variable displacement pumps are controlled by cylinder, pressure compensator, linear servo- or electro-hydraulic servo methods.

Abex Denison Series 6, 7, 11 and 14 axial piston pumps in particular are combined with equivalent motors to produce compact, lightweight transmission packages. A special feature of these pumps is that all control functions — standard rotary servo- and pressure compensator over-ride, plus optional spring-positioning, brake-and-neutral bypass, torque over-ride *etc* — and the system replenishment auxiliary pump are contained within the pump package. Open-loop versions of the Series 6 and 7 pumps are also available.

It is planned to introduce these Denison transmissions as back-to-back units, with the pump close-mounted to the motor *via* an adaptor plate, for further space-saving. Another innovation will be a fully self-contained transmission, with the pump/motor combination enclosed within its own system replenishment reservoir complete with built-in cooler and filters.

Axial piston pumps newly available from 1983 include the PV Series, which are pressure-compensated variable displacement units offering discharge flows of from 25 to 112 lit/min at 206 bar maximum operating pressure. Also to be introduced, the HD61 Series axial piston pump will have a maximum discharge of 807 lit/min and offer the same control options as the transmission pumps mentioned previously.

Abex Denison fixed displacement balanced spring- and pin-vane pumps are available as single, double or triple units, *ie* incorporating one, two or three pump cartridges on the same drive shaft. Use of these hydro statically balanced replaceable cartridges permits considerable flexibility of pump outputs and combinations. Existing single pumps offer capacities ranging from 19 to 318 lit/min; dual pumps from 19 to 318 lit/min across the range of combinations; and triple pumps from 19 to 189 lit/min. Maximum continuous operating pressure is 175 bar.

New pin-vane pumps are to be introduced from 1983 designated T5S and T6. Based on the existing single and double pump capacities, the new designs will allow operating pressures up to 245 bar, and with higher speeds will be particularly suitable for mobile applications.

Robert Bosch Limited

A company with over 30 years experience in industrial and mobile hydraulics, Bosch produce a range of variable delivery piston pumps with the emphasis on long service life; also a range of quiet-running gear pumps.

Cessna Fluid Power Limited

The range of hydraulic pumps and motors produced by this company is primarily designed for application in mobile hydraulic equipment.

Commercial Hydraulics Limited

This company manufactures and markets an extensive range of hydraulic pumps and motors whose main applications are in mobile plant and mining equipment.

Standard gear pumps provide flows of up to 600 l/min at 173 bar. Developed from these pumps are units with hydrostatically-assisted plain bearings. Two versions of the plain-bearing design are available, one providing extended life when operating on fire-resistant fluids at pressures around 150 bar in mining applications, the other exploiting the improved bearing capacity to provide operating pressures on conventional hydraulic fluids up to 238 bar. Complementing the gear pumps is the Flowdyne series of variable-displacement axial-piston pumps, designed specifically for high pressure open-circuit mobile-hydraulic systems. Flowdyne pumps have a continuous operating pressure of 350 bar.

In addition to gear and axial-piston motors for use with the company's pumps, Commercial has also introduced a high-torque/low speed radial piston motor with a capacity of 1 312 cm^3/rev — equivalent to 100—150 hp.

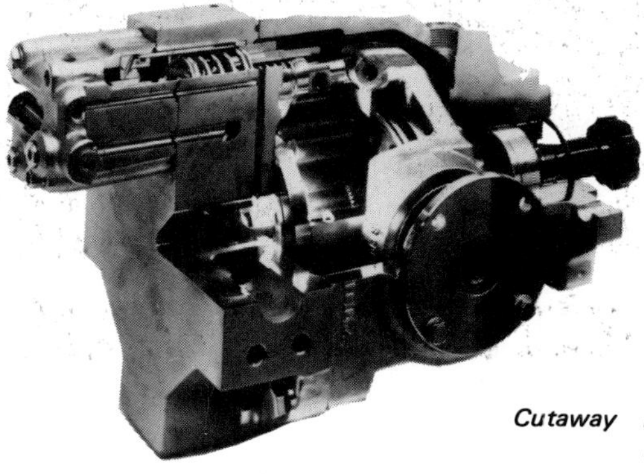

Cutaway Flowdyne variable displacement piston pump.

Hagglunds (UK) Limited

Hagglund hydraulic motors are of radial piston cam-curve design involving a rotating casing and fixed cylinder block, offering exceptionally high torque output for the motor's size. A two-speed

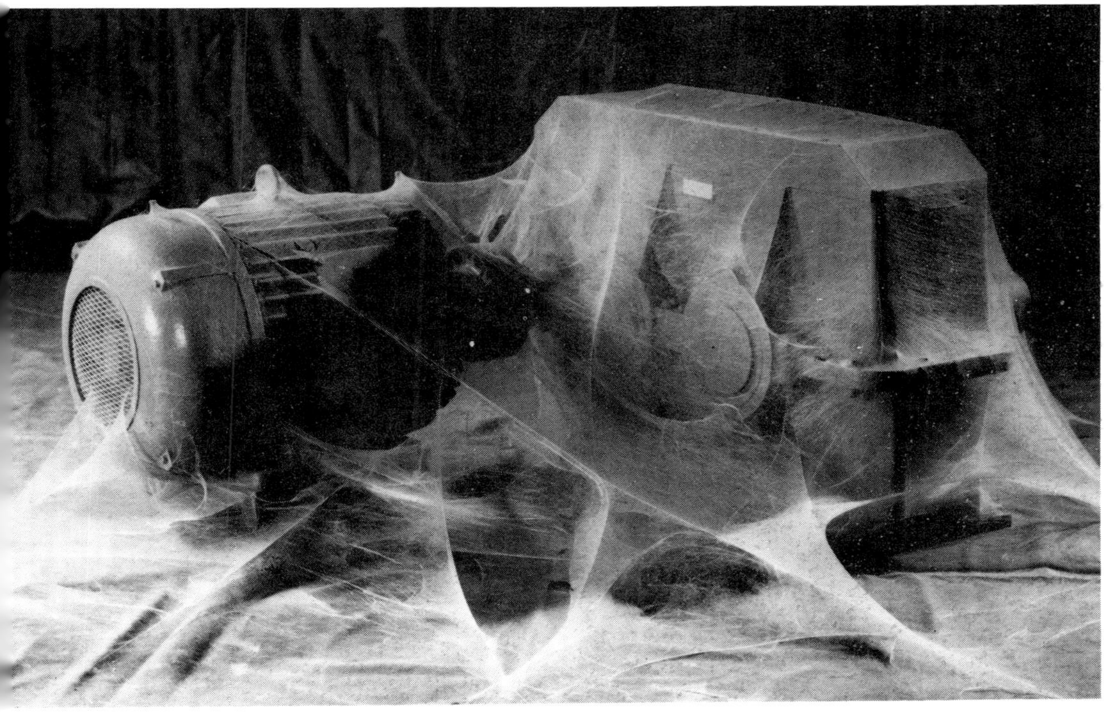

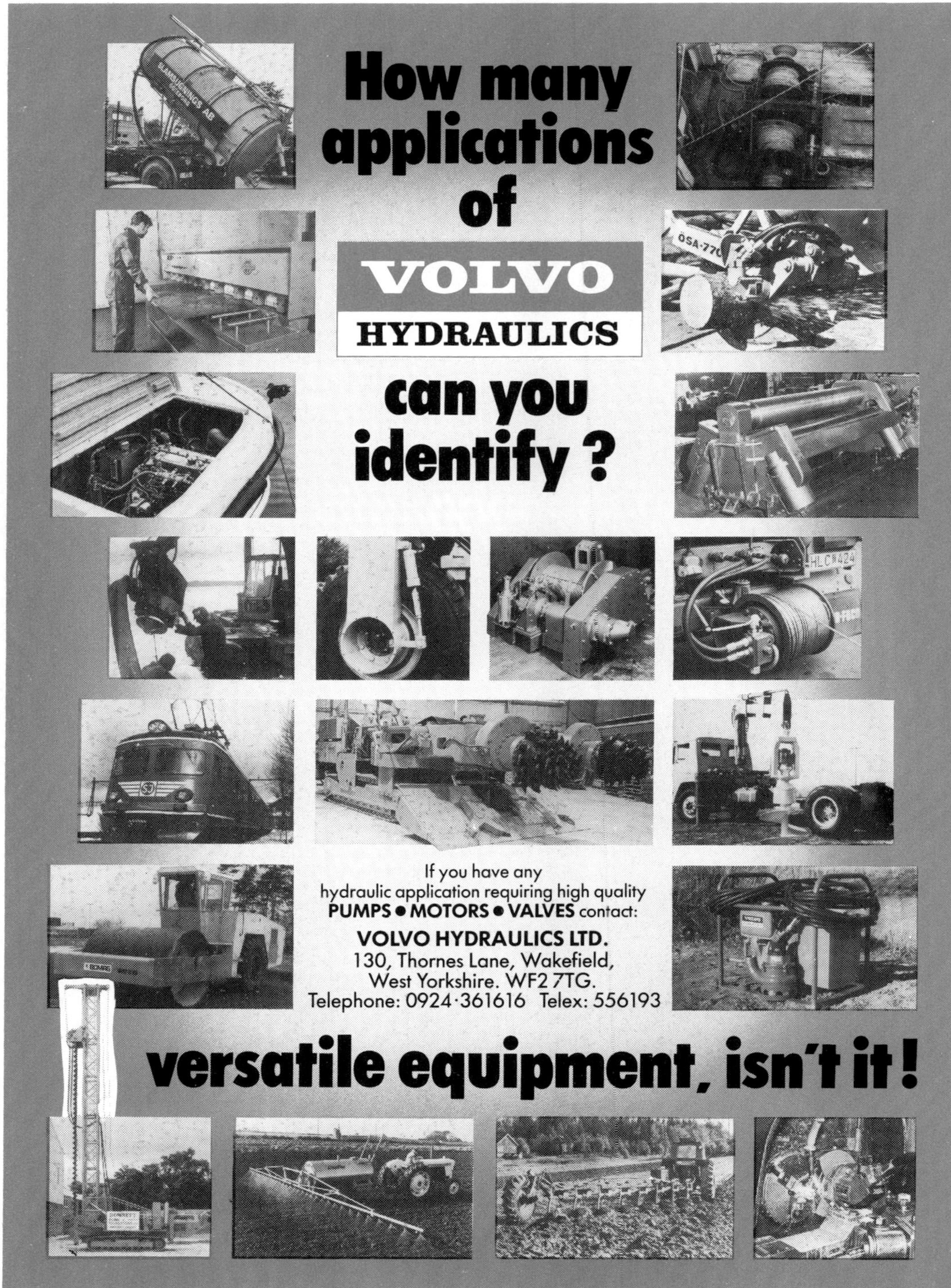
How many applications of
VOLVO
HYDRAULICS
can you identify ?
If you have any hydraulic application requiring high quality
PUMPS • MOTORS • VALVES contact:
VOLVO HYDRAULICS LTD.
130, Thornes Lane, Wakefield,
West Yorkshire. WF2 7TG.
Telephone: 0924·361616 Telex: 556193
versatile equipment, isn't it!

valve mounted on the motor ports provides a facility for operation at half rated displacement and because of its geometry the motor also has the capacity for freewheeling.

Hamworthy Engineering Limited

Hydraulic gear pumps and motors produced by this company are summarized in the Table. Dual pumps are also produced which may be of different capacity and operate at different pressures; also tandem pumps driven by one shaft.

HYDRECO HAMWORTHY HIGH PRESSURE GEAR PUMPS

Series	Model	Output per 1 000 rev/min		Maximum Pressure	
		Imp.gal/min	lit/min	lb/in^2	bar
1900	1905 1907 1909 1911 1913 1916	4.8 7.4 9.1 11.4 13.7 16.4	21.8 33.6 41.3 51.8 62.2 74.5	2 000 to 3 000	140 to 210
2200	2208 2210 2213 2215 2216	12.9 15.5 18.9 22.3 24.5	58.6 70.4 85.8 101.2 111.2	2 000 to 3 000	140 to 210
2400	2411 2413 2415 2416 2419	19.1 22.7 26.3 29.2 34.1	86.7 103.1 119.4 132.6 154.8	2 000 to 3 000	140 to 210

Industrial Hydraulics Division (Commercial Shearing)

The HD2 Series of variable displacement axial piston pumps produced by this company provide

HD2 Series hydraulic pump and motor with remote control from temperature sensing as used in cooling fan drives for locomotives, tracked vehicles and generator sets.

displacements up to 303 cm^3/rev with pressure ratings up to 410 bar maximum and 280 bar continuous. These pumps are designed for use in heavy-duty closed-circuit power transmissions. All HD2 units incorporate a high-power take-off shaft capable of transmitting 70 hp at 3 000 rev/min.

The HD2 Series motors are identical to the pumps.

Linde Hydraulics Limited

This company manufactures a wide range of axial piston high pressure hydraulic pumps and motors; also hydrostatic transmissions. Fixed capacity pumps are available in PF and MF models in 8 sizes offering deliveries from 20 to 296 cm^3/rev; the BMF models being fixed capacity motor equivalents in 6 sizes (35 to 186 cm^3/rev).

Variable delivery pumps include the B2PV double pump in 6 sizes (2x35 to 2x186 cm^3/rev); the PV bent-axis pump in 6 sizes (35 to 186 cm^3/rev); the BPV swashplate pump in 4 sizes (35 to 100 cm^3/rev); and the PR pressure-regulated or flow-regulated variable pumps in 6 sizes (35 to 186 cm^3/rev).

Lucas Fluid Power Limited

The Lucas Q range of internal gear pumps, noted for their quiet operation, produce pressures up to 300 bar.

Twenty basic models are available in one hundred and eighty different double-pump combinations. QT range has three model classifications for continuous pressures up to 250 bar. Tandem and triple combinations are available; also QA mobile gear pump for pressures to 200 bar. They are suitable for use with fire-resistant fluids.

The QT series is a further development using the Truninger system, available in six sizes and designed for pressure ratings of 80, 160 and 320 bar. The systematic graduation of the displacement volume from 5 to 500 cm^3/rev results in a very extensive range of 54 individual pumps and 1 300 tandem pumps. Adding the three and four circuit models gives a total of over 10 000 different pump combinations.

QT internal gear pump from Lucas Fluid Power offers wide choice of displacements and pump configurations.

Other pump types include:

External Gear – Five basic models for pressures to 210 bar. Available with integral relief, flow control and flow dividing valves. Wide range of mountings and drive configurations.

Vane – Fixed and variable displacement models with built-in pressure control features. Pressures up to 175 bar and flows up to 2 000 lit/min. Combination vane/gear units available. Fire resistant fluids capability.

Axial Piston – Fixed displacement pumps for continuous pressures to 180 bar, speeds to 3 000 rev/min. Swash-plate drive.

Lucas hydraulic motors are:

External Gear – Four basic models for pressures to 210 bar, speeds to 4 000 rev/min. Maximum torques to 216 Nm.

Axial Piston – Fixed displacement for pressures to 180 bar (nominal). Maximum torques to 118 Nm. Swash plate drive.

Radial Piston – Low noise, high torque. Speeds as low as 1 rev/min, maximum operating torques to 24 822 Nm. Wide model and specification range. Available with wide range of brakes and gearboxes.

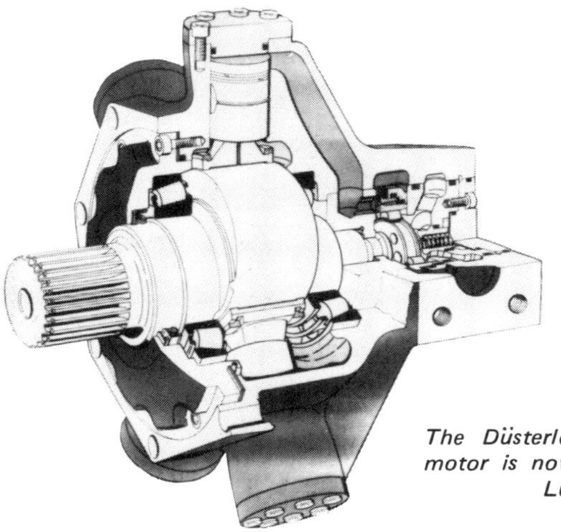

The Düsterloh radial piston hydraulic motor is now marketed in the UK by Lucas Fluid Power

Pedro Roquet S.A.

Hydraulic pumps manufactured by this company are of gear type, available in a comprehensive range of sizes.

Rexroth Limited

The G series of gear pumps manufactured by this company comprises nine different models each in a number of sizes (41 models in all). Displacements range from 0.82 cm^3/rev to 207.7 cm^3/rev. Maximum pressure ratings 210–250 bar (except the G5 series with built-in flow control where maximum pressure is 140 bar).

Fixed displacement 40° angle bent axis piston pump.

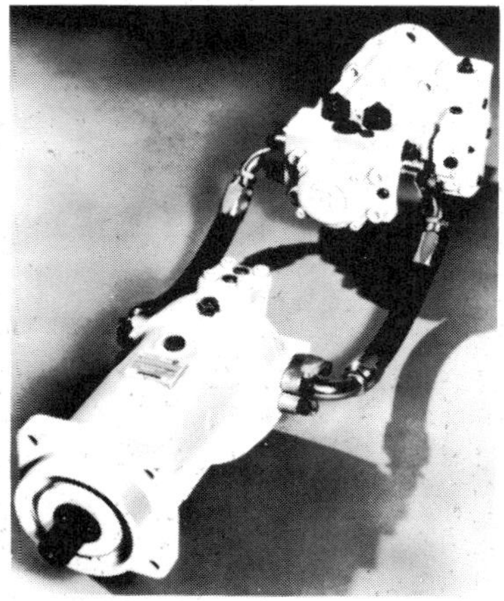

Hydrostatic transmission system. Pump type A4 with motor type A6.

Other pumps include the V2 fixed displacement vane pump (9 models, single and double); and the V3 and V4 series variable displacement vane pumps; the R4 series radial piston pumps in 3, 5 and 10 cylinder models with separate or combined output flows; and seven series of axial piston pumps.

Series A1, A4 and A5, and A10 are of swashplate design. A4 and A5 series are particularly suitable for mobile and transmission circuits. A10 series (3 models) are designed for hydrostatic transmissions in open circuit.

Bent-axis axial piston pumps are featured in the A2 series (18 models); and the A7 series (14 models) and A8 series (10 models).

RHL Hydraulics Limited

RHL's range of axial piston pumps and motors has recently been increased with the introduction of a 125 cm^3/rev unit, the A760. This brings the current range to five sizes: 11.5, 33, 62, 92 and 125 cm^3/rev.

The new pump has the same high performance specification as existing units, permitting operation at continuous pressures up to 275 bar and intermittently at 350 bar.

All units feature compact construction and excellent power/weight ratio. An extensive range of pump and motor controls is also produced.

Servotel Controls

As well as the S.I.G. range of screw pumps and servomotors, this company also produces Hartmann 'Rolvane' motors and servomotors.

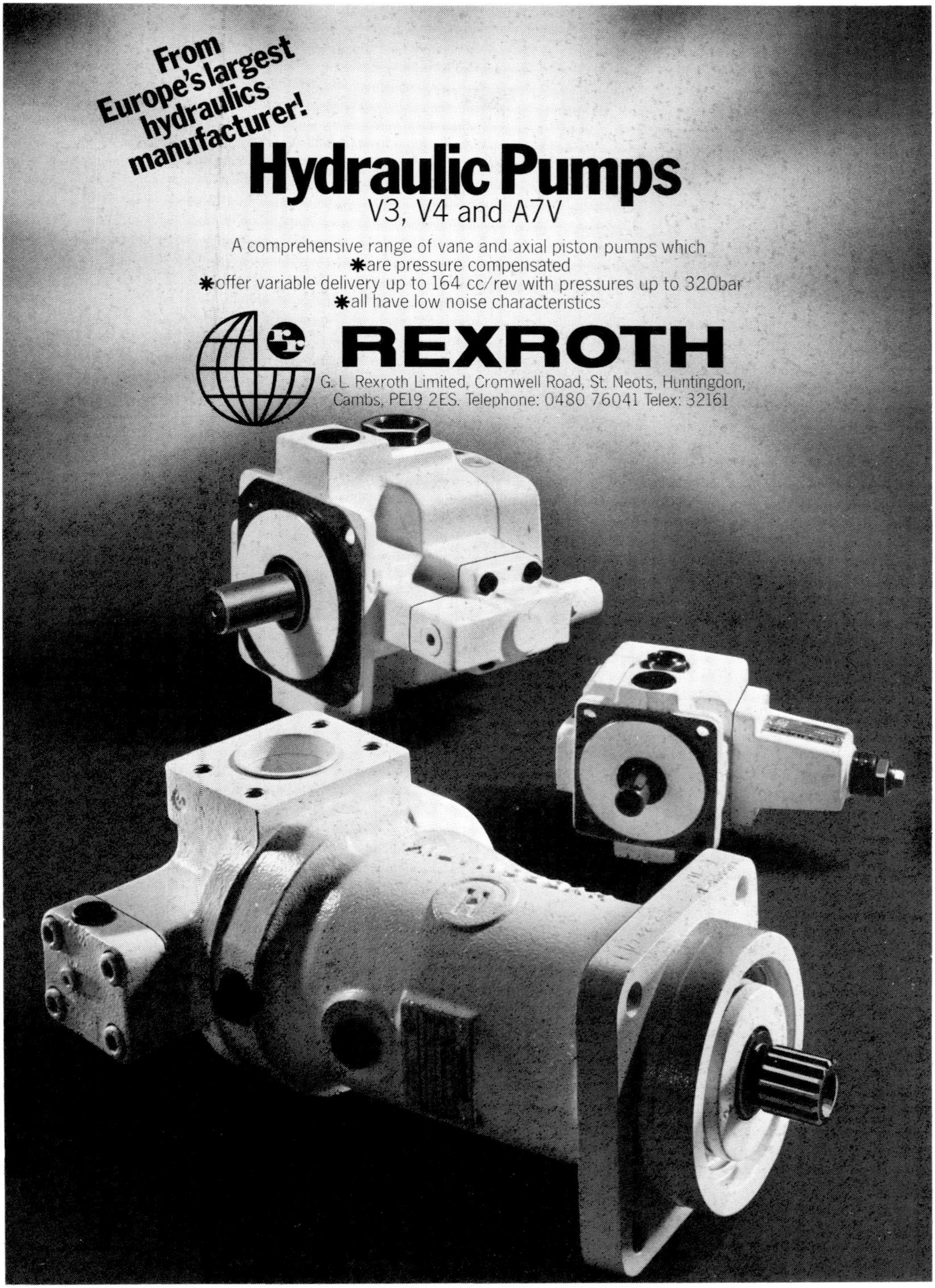
From Europe's largest hydraulics manufacturer!
Hydraulic Pumps
V3, V4 and A7V
A comprehensive range of vane and axial piston pumps which
*are pressure compensated
*offer variable delivery up to 164 cc/rev with pressures up to 320bar
*all have low noise characteristics
REXROTH
G. L. Rexroth Limited, Cromwell Road, St. Neots, Huntingdon, Cambs, PE19 2ES. Telephone: 0480 76041 Telex: 32161

RHL Hydraulics...
AXIAL PISTON PUMPS
& MOTORS
Performance, reliability and life are the requirements to satisfy demanding applications throughout modern industry – meet these confidently with RHL.
For further information phone (0632) 685505
From late 1983 091 268 5505
telex 537781.
RHL
RHL Hydraulics Limited
Planet Place,
George Stephenson Industrial Estate,
Killingworth,
Newcastle upon Tyne.
NE12 0EF.

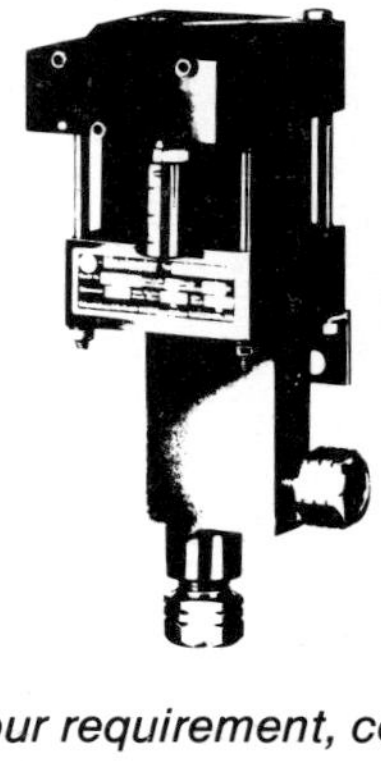

Motor pump unit P/5/12 pump. (Smiths Industries Hyd. Co).

Smiths Industries Hydraulics Company

This company produces the G series of high efficiency gear pumps. These pumps also form the basis for a series of hydraulic power packs, from 'mini' size upwards.

The complete range of variable- and fixed-delivery piston pumps and gear pumps gives deliveries up to 200 lit/min at pressures up to 280 bar (4 000 lb/in^2) as standard.

Sperry Vickers

Sperry Vickers produce an extremely wide range of hydraulic pumps and motors, summarzied as follows:

Fixed Dispacement Pumps:

Gear type — single, double, triple and multiple, with max flow rates 2.6—338 l/min; max pressures 100—250 bar

Piston type — 15—64 l/min, max pressure 210 bar.

Vane — single, double, triple and quadruple with a capacity range 6 to 289 l/min; max pressures up to 210 bar.

Variable Displacement Pumps:

Piston — max flow rates 15 to 296 l/min; max pressures up to 345 bar.

Vane — max flow rates 15 to 124 l/min; max pressures up to 175 bar.

Fixed Displacement Motors:

Gear — 0.80—0.39 Nm/bar torque; max pressure 250 bar.

Piston — High Speed 0.17—0.98 Mn/bar torque; speeds 2 400—3 600 rev/min.
High torque 3.07—71.70 Nm/bar torque; speeds 550—130 rev/min.

Gerotor type high-torque, low-speed motors; 0.98—9.54 Nm/bar torque at 790—200 rev/min; also variable speed type.

Vane — 0.38—5.05 Nm/bar torque; speeds 2 200—3 000 rev/min.

Towler Hydraulics (UK) Limited

The Towler Group of companies have a total system engineering capacity for complete 'packages' using pumps, valves, cylinders and electrical controls, as well as supplying individual components. The range of hydraulic pumps and motors available includes the following types.

Port Plate Pumps and Motors (Series P) – These are axial piston pumps with rotating body and are available as either infinitely variable with a tilting swash plate, or with fixed swash plate. The units can also be used as hydraulic motors and can stand full pressure on either port. Sizes range from 78 lit/min to 339 lit/min, with a normal pressure rating of up to 270 bar (4 000 lb/in^2).

Axial Piston Pumps (Series G, A, FBK, FCM and FCS) – These embrace the high pressure range of positively-sealed axial piston pumps with capacities from 2 lit/min to 640 lit/min. Maximum pressure ratings are 415 bar (6 000 lb/in^2) or 690 bar (10 000 lb/in^2). The largest sizes, FCM and FCS, are unique in having a built-in electrical device to warn of a malfunction. All bearings are of whitemetal, designed for infinite life with correct filtration levels.

Series HOA, DHOA, D, SJE and UL – These are a range of heavy duty positive displacement plunger pumps of extremely robust construction (originally manufactured by Andrew Fraser). Maximum capacities range from 1.16 lit/min to 5.5 lit/min. Pressure ratings up to 690 bar (10 000 lb/in^2). The D, SJF and UL offer multi-matched pump outlets up to twelve in number.

Cam Rotor Pumps – Originally manufactured by Andrew Fraser, this range of pumps/motors feature twin rotors phased at right angles and mounted coaxially on the same shaft. They are quiet in operation with an almost pulse-free delivery. Capacities range up to 132 lit/min and maximum pressure up to 170 bar (2 500 lb/in^2).

Voith Engineering Limited

Voith's well established range of high (IPH) and low (IPN) pressure, internal-gear pumps has recently been extended by the introduction of the medium pressure pump type IPR. With a continuous rating of 175 bar (210 bar peak) this pump, combining low noise and high efficiency, is

Voith IPH high pressure pumps including (from left) an IPH 6/5/4 triple pump, an IPH 5/4 double pump and an IPH 4 single pump.

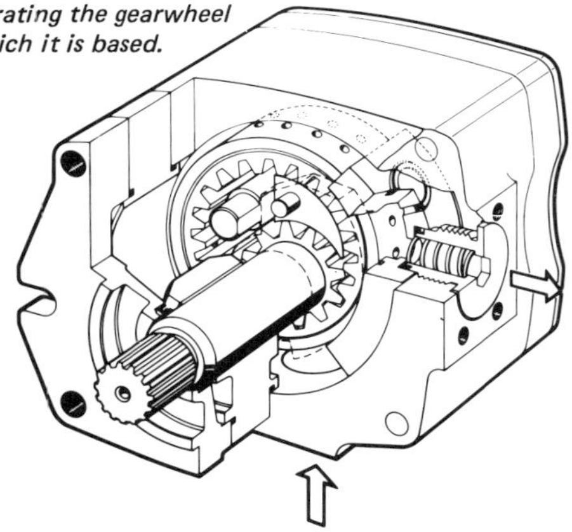

Sectional view of a Voith IPH high pressure pump, illustrating the gearwheel principle on which it is based.

in the same market sector as the vane pump. Nineteen displacement sizes are planned and nine are already in production, manufacture being split between Eckerle and Voith in Germany.

The IPR range will be integrated into the existing IPH and IPN programme so that multiple stage pumps will be possible.

Completing the range is the IPL miniature internal gear pump, rated at 70 bar.

Also available *via* Voith Engineering is the HAWE range of hydraulic pumps. Based upon radial-piston pumps with displacement sizes from 0.21 to 55.2 cm^3/rev and operating pressure up to 700 bar, the programme includes complete hydraulic units with oil submersible electric motors. An interesting feature of this range is the possibility of several equal or dissimilar outlets from one pump.

A further development is the introduction of a double pump combining a radial-piston unit and vane pump to give a high/low system.

Volvo Hydraulics Limited

The Volvo range of bent-axis piston pumps is now largely based on the F1 model, originally designed as a PTO pump for vehicles, but having a wide potential in other fields. It is notable for low weight, high efficiency and high power/weight ratio.

Another bent-axis model is the V11 variable motor in which the maximum angle of 40° can be varied by a choice of different control options. A speed range of 0–5 000 rev/min is possible, with continuous speed of 4 200 rev/min being available.

Volvo also produce the V30 variable swash plate axial piston pump, noted particularly for its quiet operation.

Survey of Hydraulic Cylinders

Bradford Cylinders Limited

Standard cylinders from Bradford — the largest manufacturer of power cylinders in the UK — cover the size range 40 to 400 mm and pressures up to 320 bar (4 600 lb/in^2). They include tie-bar and flanged cylinders together with two further ranges of thick-walled mill type cylinders. The company maintains quality assurance to the requirements of DEF-STAN05-29.

To meet special customer requirements, Bradford also offer hydraulic cylinders with bores up to 800 mm, strokes up to 10 metres, and working pressures up to 700 bar (1 100 lb/in^2). These include telescopic, double acting, constant velocity, rotary cylinders and distributors, water cooled, non-magnetic, bronze or stainless steel, rotary actuators, and piston accumulators with Lloyds, Det Norske Veritas and other authority approval.

Cessna Fluid Power Limited

This company specialises in the production of hydraulic components for mobile equipment, including a standard range of hydraulic cylinders. Additionally custom-built cylinders can also be supplied, designed and built from prototype stage onwards.

Denley Engineering Co (Heckmondwike) Limited

Denley cushioned and non-cushioned hydraulic cylinders are produced to CETOP standards, the RP 58H series embracing sizes from 25 mm to 160 mm; and the RP 73H series sizes from 200 mm to 320 mm. The company also manufactures proportional actuators and rotating hydraulic cylinders.

W.F.E. Hydraulics Limited

This company specializes in the design and manufacture of hydraulic cylinders and rotary distributors for mobile hydraulic applications; and also large rams for static plant. Sizes extend from 50 mm bore upwards, with working pressures up to 700 bar (10 000 lb/in^2); cylinders normally being individually designed for specific applications. These have included cylinders with 600 mm bores and others with 10 m strokes.

Two 320 mm bore marine quality hydraulic cylinders, with special rapid response capability, for operating a personnel services gangway on an Esso oil rig in the Norwegian sector of the North Sea, purpose made by W.F.E.Hydraulics Limited for Hudson-Wharton Limited.

Lucas Fluid Power Limited

Three different standard ranges of cylinders are produced by Lucas – The G.P. (general purpose) range in sizes from 40–200 mm bore; the M.S. fully metric range; and the T.D. range for heavy duty applications with mill type construction (designed to CETOP RP 58H specification).

LC3 Hydraulic cylinders.
(Lucas Fluid Power).

Parker Hannifin Corporation

The extensive range of standard cylinders produced by Parker Hannifin are of tie-rod construction.

Pedro Roquet S.A.

Hydraulic components manufactured by this company include a standard range of cylinders on metric sizes.

Servotel Controls

This company is noted for its semi-rotary torque actuators and servo-actuators, and linear servo-actuators.

Spenborough Engineering Company Limited

This company specializes in the manufacture of single- and double-acting telescopic cylinders, including special designs of telescoping boom cylinders for cranes (an example being one where a special cylinder allows oil to be supplied to it through a conventional cylinder, this eliminating the use of two hose reels and hoses).

Sperry Vickers Limited

Hydraulic cylinders produced by this company are the VSL single-acting range (22–180 mm bore); and the VDL double-acting range (40–250 mm bore). Both series are designed for a maximum pressure of 250 bar (3 500 lb/in^2).

Sterling Hydraulics Limited

Hydraulic cylinders produced by this company include both single- and double-acting types in bore sizes up to 160 mm, rated at 210 bar (3 000 lb/in^2). Special variations to the range can also be produced with larger bores and suitable for higher pressures.

Survey of Pipes and Pipe Fittings

Europower Hydraulics Limited

The Europower range of hydraulic fittings is one of the largest available and covers most application requirements.

Hyndburn Hydraulics Limited

Couplings produced by this company include the following ranges:

EMB Compression Fittings

Metric compression fittings to DIN 2353.

Bite type couplings from 4 mm to 8 mm in very light series, 6 mm to 42 mm in light series and 6 mm to 38 mm in heavy series. Available in steel, stainless steel to grade 1.4571 and brass grade SO MS 59. Weld nipples, weld nipple fittings and socket weld fittings also available in steel and stainless steel.

Optipress Pre-Assembly Machine

For pre-assembly of all types from 6 mm to 42 mm.

Fully automatic, simple and easy to use, not only takes all the hard work out of joint assembly, but gives the precise location and depth of cut required of the ring on the tube that is essential for satisfactory joint making.

Quick-Release Couplings

Manufactured by OBAC of Sweden a full range of couplings from 1/8 in BSP to 1/2 in BSP. Pressure rating to 1 500 lb/in^2. Available with all standard type finishes and in some cases in stainless steel. Fully interchangeable with all other makes, *eg* Parker, Pcl, CEJN, *etc.*

In addition the company also produces the 'Hydrazorb' range of steel pipe clamps with polypropylene inserts.

Lucas Fluid Power Limited

This company produces a wide range of 'bite' type couplings in phosphate-coated mild steel in three pressure ranges: LL for pressures up to 100 bar; L for pressures up to 250 bar; S for pressures up to 640 bar. The same couplings are also available in brass or stainless steel.

Bite type couplings.
(Lucas Fluid Power Ltd).

KR flange couplings are intended primarily for tube sizes 50 mm to 100 mm o.d.

Also the 'KR' range of re-useable compression couplings with O-ring seal and in stainless steel or cupro-nickel alloy. Sizes: Imperial range up to 7/8 in o.d. pressures up to 680 bar, and 1/2 in o.d. for pressures up to 500 bar. Metric range up to 22 mm for pressures up to 680 bar; and 25–50 mm for pressures up to 500 bar.

Parker Hannifin Corporation

Parker Hannifin produce an extremely wide range of pipe and tube fittings; also hydraulic hose and hose couplings and fittings covering virtually all requirements in the hydraulic industry.

Walther Couplings Limited

This company specializes in the production of low- and high-pressure quick-release couplings in a variety of materials. Extreme-pressure couplings are also available in the range with a burst pressure of 4 000 bar (58 000 lb/in^2).

Part of the Walther range of non-interchangeable couplings — to ensure correct connections.

Survey of Hose, Hose Couplings and Fittings

Adflow International

'Adlok' hermaphrodite couplings for hoses have been designed for use in high pressure pneumatics, hydraulics and fluid handling application, featuring a nitrile rubber seal and quick-action disconnection link. They are produced in four sizes from 10 to 25 mm.

Other sizes of couplings and pipe fittings in the 'Adlok' range extend up to 150 mm (6 in) diameter; and also include the 'Bosslock' — specially designed for use in hazardous environments.

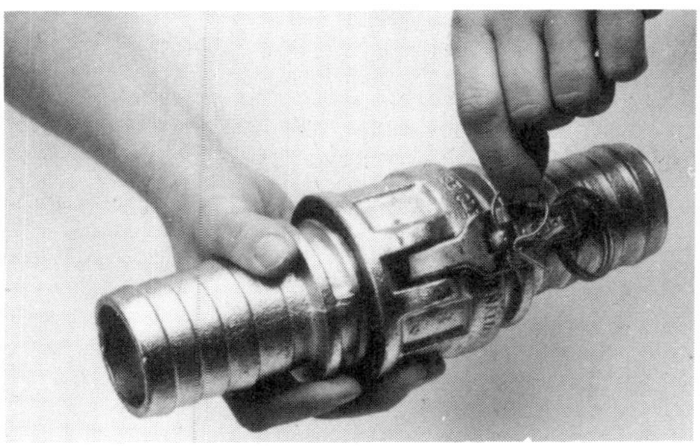

Adflow International 'Bosslock' quick action safety coupling.

Aeroquip (UK) Limited

The Aeroquip family of AQP hydraulic hose embraces the following types:

FC300 — with superior high temperature abrasion resistance and exceeding SAE 100R5 performance.

FC195 HI-IMPULSE® — rated for 40% greater than SAE 100R2.

FC510 — Hi-Pac® — exceeding SAE 100R2 performance.

FC194 HI-IMPULSE® — superior high temperature resistance with operating pressures up to 40% greater than SAE 100R1 hose.

FC350 — truck hose.

2661 — dual-purpose suction hose with superior fluid compatibility. (See also Table).

AEROQUIP AQP HOSE

Type	Standard	Reinforcement	Size Range i.d.	Pressure Range
FC300	Exceeds SAE 100R5	Polyester and wire braided	0.19 in to 1.81 in	3 000–300 lb/in^2 (210–21 bar)
FC195	Exceeds SAE 100R2A	Double wire braid	0.25 in to 2.0 in	5 750–1 500 lb/in^2 (400–105 bar)
FC510	Exceeds SAE 100R2	Patented wire reinforcement	0.25 in to 1.0 in	5 000–2 000 lb/in^2 (350–140 bar)
FC194	Exceeds SAE 100R1	Single wire braid	0.25 in to 1.25 in	3 250–900 lb/in^2 (225–60 bar)
FC390	Meets DOT FM V5S 106 Type H11	Polyester and single wire braid	0.19 in to 1.38 in	1 500–250 lb/in^2 (105–17.5 bar)
2661	Exceeds SAE 100R4	Helical wire between two textile braids	0.75 in to 4 in	300–50 lb/in^2 (23–3.5 bar)

FC300 and FC350 use standard Aeroquip re-usable fittings. FC194 and 2661 use standard crimped fittings. FC195 can be assembled with both standard crimped and re-usable fittings. FC510 (Hi-Pac) uses standard Hi-Pac re-usable fittings or the same crimped fittings as FC194 and FC195.

Aeroquip 2807 PTFE hose is a further type available with stainless steel outer braiding and PTFE tube.

Gates Hydraulics Limited

This company produces a wide range of hydraulic hose and hose fittings and couplings.

Intec Limited

Intec Limited, a member of the Tecalemit UK Group, specialize in thermoplastic extrusions both for L.P. tube and pressure hose. In addition to the hoses which meet the SAE 100R7 specification and, in all aspects other than materials, the SAE 100R1 specification, they produce mini-bore high pressure hose known as MS 40 for use with static and portable pressure-sensing and control systems and also a range of hoses recently developed and marketed known as KS 50, which incorporates Kevlar reinforcement and which meets the requirements of SAE 100R8 and technically SAE 100R2.

Intec have also recently reached an agreement with a European manufacturer of steel braided thermoplastic hoses to add such hoses to its range.

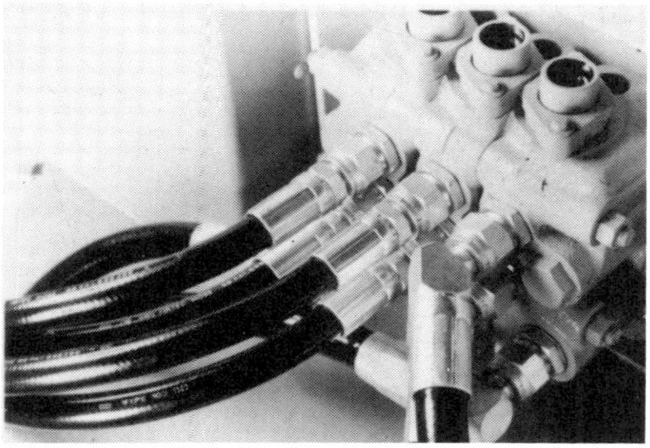

Typical application using Intec S40 hose and end fittings. (Intec Limited).

Lucas Fluid Power Limited

Lucas hose assemblies embrace flexible pressure hose in three main ranges:

(i) Standard, wire-braided ¾ in to 2 in bore, pressure-rated to 350 bar (5 000 lb/in^2);

(ii) Fabric braided, ¾ in to 1 in, pressure-rated to 350 bar (5 000 lb/in^2);

(iii) Multi-spiral, 3/8 in to 2 in bore, pressure-rated to 415 bar (6 000 lb/in^2);

Hoses are supplied with permanently swaged ends with BSP, JIC, NPTF or SAE threads. Hoses are also supplied loose with end fittings to be assembled by the user on matching machinery.

Plascoat International Limited (Griflex Creators)

Griflex Creators manufactures a range of reinforced and unreinforced PVC industrial hoses with excellent flexiblity and flow properties.

Surcoflex is reinforced with open mesh polyester fibre and has been designed to withstand higher pressures than other types of flexible PVC hose (up to 350 lb/in^2 working at 20°C).

Griflex Creators wire reinforced PVC hose.

Kanaflex hose is manufactured from PVC and incorporates a reinforcing spiral as an integral part of the hose wall. This method of construction results in a high performance hose with superior crush resistance.

The Kanaflex range which has a working temperature of $-15^{\circ}C$ to $+65^{\circ}C$ includes a blue oil-resistant hose specifically developed for hydraulic applications.

A range of nylon tubes both size 11 and 12 are available in flexible and semi-rigid grades.

Sterling Hydraulics Limited

Quick-release screw couplings produced by this company are available in flat face, pull-break, push-pull and trailer-brake configurations in sizes from 3/8 in to 2 in BSP. All couplings are produced in stainless steel.

Part of Sterling coupling range. (Sterling Hydraulics Ltd).

Survey of Hydraulic Valves and Selectors

Abex Denison Limited

Abex Denison manufactures a comprehensive range of hydraulic valves for directional flow, check and pressure control applications, including servo-controls.

Direct-operated or pilot-operated directional control valves are available for operating pressures up to 350 bar. Miniature 1/8 in direct-operated versions are designed for either subplate mounting or stacking systems and can be lever-, cam-, pneumatic- or electric solenoid-operated.

Larger sizes (3/8 in, 3/4 in and 1½ in) can be either direct-operated or pilot-operated with two or three spool positions, controlled by solenoids, lever, hydraulic pressure or stem. Subplate CETOP or manifold mounting is standard. Nominal flows are 180, 330 and 600 lit/min and maximum pressure is 240 or 350 bar for direct- and pilot-operation respectively.

Check valves are available in five sizes, *ie* 3/8, 3/4, 1, 1¼ and 1½ in and for threaded body, subplate or SAE 4-bolt flanged mounting. Cartridge type check valves are also available for manifold mounting.

In the pressure control series, pilot-operated valves are available in 3/8, 3/4 and 1½ in sizes for relief, reducing, sequencing and unloading applications in subplate mounting and cartridge versions. For rapid unloading, these valves can be used with the Denison solenoid-operated sandwich vent valve which is bolted directly between the main valve body and the pilot operator head. The Denison sandwich-type solenoid-operated proportional pressure control can be installed and used in a similar manner with these valves. A flange-mounting pilot-operated pressure control is also available in 3/4, 1 and 1¼ in sizes for relief, unloading and sequencing control.

Direct-operating pressure-relief valves are also available in 3/4 and 1½ in sizes for threaded body and subplate mounting.

An Abex Denison servo pressure-regulating valve can be used as a proportional 3-way control in a closed-loop or feedback circuit typically to control the speed, position and load on cylinders and servo motors. Alternatively it can be used for proportional pressure control, *ie* as a venting valve remotely to control larger valves in the relief, sequence or reducing range.

Pressure controls maximum system pressures range upwards from 7 to 350 bar with flows up to 600 lit/min.

The Abex Denison F5C solenoid-operated variable-flow control valve can be used non-compensated or in conjunction with two-port and three-port compensators. Operating at a maximum pressure of 210 bar, three sizes are available, (3/4, 1 and 1¼ in) covering a flow range from 0 to 380 lit/min. The flow control body is designed for flange mounting directly to pumps and motors. Smaller variable flow control valves in 1/4 and 3/8 sizes are also available.

Direct-seat valves with poppet-type cartridges are also available to control direction of flow either from Port A to Port V or *vice versa.* Manifold mounting cartridge valves in this series provide a compact, versatile method for replacing conventional directional-control valves. Subplate mounting and threaded-body versions using the same cartridge design supplement the range. Sizes range from 3/8 to 2½ in. Maximum operating pressure is 350 bar.

Robert Bosch Limited

Bosch hydraulic components include high flow-rated directional control valves; also directional, pressure and flow control proportional valves.

Cessna Fluid Power Limited

The range of control valves manufactured by this company is primarily intended for application in mobile hydraulic equipment such as cranes, earthmovers, fork lift trucks, harvesters and most other types of mobile machinery.

Commercial Hydraulics Limited

Latest addition to the series of control valves available from Commercial Hydraulics is the SLI range of electro-hydraulic proportional control valves providing flow rates up to 160 l/min.

Hamworthy Hydraulics Limited

Hydreco Hamworthy control valves are made in closed-centre and open-centre types. Open-centre valves are available in:

Tandem circuit — which is suitable for the majority of mobile equipment applications.

Parallel circuit — allowing simultaneous feed to a number of work ports.

Series circuit — allowing simultaneous operation of more than one circuit.

Combined circuit — allowing combinations of the Tandem, Parallel and Series circuits.

Tandem circuit valves are manufactured in a wide range of sizes. In multi-spool versions, the spool nearest the inlet takes priority of oil supply. Essentially, the tandem circuit valve is intended to feed the whole of the oil supply to one service at a time.

Parallel circuit valves have a common internal pressure gallery allowing simultaneous entry to any number of work ports. The service at the lowest static pressure will move first unless its spool is throttled to build up sufficient pressure to operate other services. The parallel circuit valve is able to divide the oil flow between any number of services, providing full pressure at any one or more, but with an operating speed less than the maximum for each. Manually controlled and variable simultaneous operation provided by parallel circuit valves gives the most effective type of work cycle for some machines.

Series circuit valves also allow simultaneous operation of more than one service. In this case, exhaust from the spool nearest the inlet is the supply for the next downstream valve operated. The operating pressure on the last service is the back pressure on the preceding upstream service and so on. Thus high speed is possible at simultaneously operated services, but the total pressure is divided between them.

Combined circuit valves are designed to suit individual machine systems and are evolved in close co-operation with the manufacturers of each machine.

Hytrol Hydraulic Valves

Hytrol design, manufacture and market a range of modular hydraulic valves, including lever-, oil-, and air-, plunger- and solenoid-operated models. All conform to CETOP 3 and 5 specifications and all, except the standard solenoid valves, are approved for NCB use. All are available for use with system pressures up to 350 bar (5 000 lb/in^2) and for flow rates up to 70 lit/min (15 gal/min). The solenoid valves are fitted with interchangeable coils for easy maintenance.

The size 6 CETOP 3 lever operated directional control valve. (Hytrol Hydraulic Valves).

The Hytrol range also includes intrinsically safe and explosion-proof solenoid valves.

Solenoid-operated 'cushioned' valves in the Hytrol range have hydraulically damped spools for smooth, controlled deceleration of actuators when the valve is de-energized.

A range of eight standard hydraulic powerpacks and systems is also available from Hytrol. These are suitable for use in machine tool, mechanical handling and other allied industries. The standard range is based on gear pumps with a variety of displacements and pressures supplied as standard. Normally supplied with a relief valve manifold, they are also vavilable with manifolds suitable for CETOP 3 and 5 control valves. In addition to the standard power packs, a variety of power packs and systems can be supplied to meet individual requirements.

Laser Engineering Limited

Laser directional control valves are based on 2 and 6 spool monoblocs and provide fine control, with low handle loads, plus the ability to mount extra valve sections and cover requirements from 2 to 8 services.

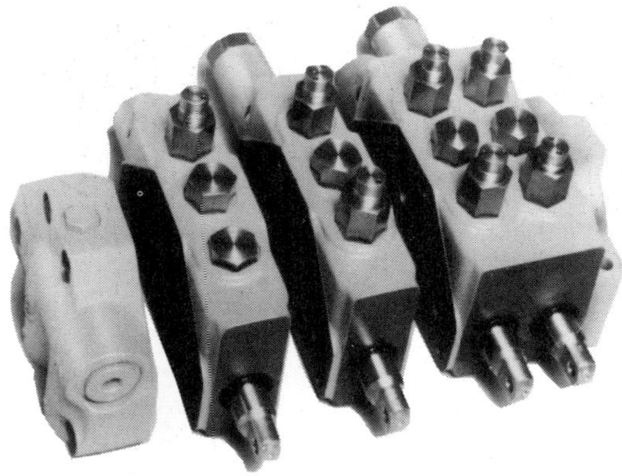

Laser 2-spool monobloc with two additional sections — plus system relief valve.

The valves are designed for flows up to 135 litres per minute. They are parallel connected and a series parallel option is available on the 2 spool monobloc. Float, motor and other spool types are available. Detents are mechanical or solenoid in either direction of spool movement.

The system relief valve is a Laser pilot-operated cartridge with flat characteristic and low hysteresis over the flow range 10—135 lit/min, from 150—210 bar. It possesses rapid response and low overshoot characteristics.

The circuit relief valves are Laser direct-acting cartridges that can be fitted to each of the work ports. They are suitable for pressures up to 330 bar.

Laser pressure-compensated fast-response flow regulators are available in three sizes covering flows up to 64, 114 and 218 lit/min, at pressures up to 315 bar. The valves provide accurate control in one direction and low pressure drop in the reverse free-flow direction.

Laser Engineering is also producing a range of lock-out and pilot-operated check valves, designed to meet the safety standards required for hose burst system protection and other site safety measures, for construction and mechanical handling machinery.

Pedro Roquet S.A.

Hydraulic components produced by this company include an extensive range of directional control valves, pressure control valves, flow control valves and solenoid valves.

Servotel Controls

This company produces the Pegasus range of servo valves.

Smiths Industries Hydraulics Company

The 'S' valve recently introduced by this company is a leakproof poppet type sealing in both directions (offering a cheaper alternative to multi-bank spool valves, or even directional valves in particular circuits).

2 port/2 position 'S' valve. Leakproof valve for pressures to 210 bar and nominal flow of 3 litres/min. Valve shown on the left is for banjo mounting and valve on the right is for line mounting. (Smiths Industries Hydraulics Company)

Sperry Vickers Limited

Sperry Vickers produce a very wide range of valves, most of which are available in alternative forms for modular mounting, pipe mounting or surface (gasket) mounting. Types include:

Pressure controls — reducing valves, relief valves, electro-modulated relief valves; also electrically selected pressure ranges and automatic unloading-type relief valves.

Sequence, unloading and back-pressure valves — deceleration valves (non-compensated).

Flow restrictor valves — compensated, compensated bypass type, also solenoid-operated, hydraulically operated and pneumatically operated flow-control valves.

Multiple valves — monobloc and sectional types.

Check valves

Cartridge valves

Sterling Hydraulics Limited

Hydraulic valves manufactured by this company include:

(i) A complete range of valves for pressure and flow control up to 350 bar at 300 litres.

(ii) Polyhydrons which utilize cartridge valves to form complete hydraulic systems in one complete assembly. Polyhydrons give great flexibility and can be designed to simplify and reduce the cost of most systems.

(iii) Directional-Control Valves designed for standard CETOP mounting, fitted to a polyhydron, offering complete hydraulic-control packages. The valves, most of which are suitable for 350 bar, are also available with a full range of subplates.

Towler Hydraulics (UK) Limited

The Towler Group produce a range of prefill and exhaust valves designed for use with hydraulic rams. Capacities range from 170 lit/min to 1 700 lit/min. Five different types of mounting are available.

Voith Engineering Limited

The HAWE programme, available from Voith Engineering Ltd in the UK, includes a comprehensive range of hydraulic valves suitable for flows up to 120 lit/min and 700 bar pressure.

The range of directional controls embraces both spool and ball-seated valves with electrical, pressure, mechanical and hand actuation for both in-line and manifold-mounted installations. Many designs are available in multiple valve banks. Apart from the normal range of controls, HAWE have a number of valves for special functions, *viz* line-rupture protection, pressure control and over-centre valves.

SECTION 8

Buyers' Guide

Please also refer to the addendum contained at the end of the Buyers' Guide section for additional companies whose entries arrived too late for inclusion in the main section.

Sub-section (a)

TRADE NAMES INDEX

AEROQUIP – Hydraulic hose and fluid seals – Aeroquip (UK) Ltd.
AQD – Hydraulic hose – Aeroquip (UK) Ltd.
AUTODRAULIC – Non-electric sequence control systems – Towler Hydraulics (UK) Ltd.

CEPAC – Subsea cylinder – Air Power & Hydraulics Ltd
COLORFLOW – Valves – Parker Hannifin (UK) Ltd.
C10 – Industrial cylinder – Air Power & Hydraulics Ltd

DENLEY – Hydraulic cylinders, hydraulic power units, hydraulic manifolds, hydraulic systems, electronic systems – Denley Engineering Co..

ELECTRAULIC – Electronically operated systems – Towler Hydraulics (UK) Ltd.
E.O. – Connectors – Parker Hannifin (UK) Ltd.
ERMETO – Connectors – Parker Hannifin (UK) Ltd.
EUROCRIMP – Swaging manuals – Europower Hydraulics Ltd.
EUROLOK – Staple fittings – Europower Hydraulics Ltd.
EUROSAFE – Safety valves – Europower Hydraulics Ltd.
EUROSWAGE – Swaging equipment – Europower Hydraulics Ltd.

HARTMANN – Rolvane motors – Servotel Controls Ltd.
HALLITE – Hydraulic and pneumatic seals – Hallite Seals Ltd.
HI-PAC – Hydraulic hose – Aeroquip (UK) Ltd.
HUNGER DFE – Seals, bearing elements and spherical bearings – Hunger Hydraulic UK Ltd.
HUNGER HYDRAULIK – Hydraulic cylinders – Hunger Hydraulic UK Ltd.
HUNGER SMG – Diamond hones and honing heads and hydraulic honing machines – Hunger Hydraulic UK Ltd.

KR COUPLINGS – High pressure hydraulic couplings – Lucas Fluid Power Ltd.
K10 – High pressure cylinder – Air Power & Hydraulics Ltd.

MANATROL – Valves – Parker Hannifin (UK) Ltd.
MAXIMATOR – Hydraulic intensifiers, hydraulic power plant, hydraulic test equipment – George Meller Ltd.

MODINA FILTRI – Simplex filters, Duplex filters, self-cleaning filters, filter elements – Modina Filtri SpA.

NO SKIVE – Hose – Parker Hannifin (UK) Ltd.

PACOMA – Hydraulic cylinders – Massey-Ferguson GmbH
PARFLEX – Hose – Parker Hannifin (UK) Ltd.
PEGASUS – Servo valves – Servotel Controls Ltd.
PIONEER – Quick action couplings – Parker Hannifin (UK) Ltd.
PNEUKO – Piston head – Freudenberg Simrit Ltd.
POLYON – Hydraulic hose – Aeroquip (UK) Ltd.

REXROTH – Hydraulics – G. L. Rexroth Ltd.
RHL – Axial piston pumps and motors – RHL Hydraulics Ltd.

SHM – Semi rotary actuators – Servotel Controls Ltd.
SIG – Screw pumps – Servotel Controls Ltd.
SIMKO – Piston packing – Freudenberg Simrit Ltd.
SIMMERING – Shaft seal – Freudenberg Simrit Ltd.
SIMRIT – Rubber compound – Freudenberg Simrit Ltd.
TDUO – Piston head – Freudenberg Simrit Ltd.
TELESIG – Power packs – Servotel Controls Ltd.
TELL-TALE – Filters – Parker Hannifin (UK) Ltd.
TORKO – Rotary actuatory – Air Power & Hydraulics Ltd.
TRIPLE-LOK – Connectors – Parker Hannifin (UK) Ltd.

VOLVO HYDRAULICS – Pumps, motors, valves – Volvo Hydraulics Ltd.

Sub-section (b)

CLASSIFIED INDEX OF HYDRAULIC EQUIPMENT AND COMPONENTS

Accumulators, Hydraulic
Henry Berry & Co Ltd
Robert Bosch Ltd
Bradford Cylinders Ltd
Cessna Fluid Power Ltd
Christie Hydraulics Ltd
Hunger Hydraulic UK Ltd
Lucas Fluid Power Ltd
Parker Hannifin (UK) Ltd
TI Reynolds Ltd
Towler Hydraulics (UK) Ltd

Accumulators, Hydro-Pneumatic
Christie Hydraulics Ltd
TI Reynolds Ltd

Actuators, Linear
Cessna Fluid Power Ltd
Bradford Cylinders Ltd
Denley Eng Co (Heckmondwike) Ltd
Hunger Hydraulic UK Ltd
Parker Hannifin (UK) Ltd
G. L. Rexroth Ltd
Servotel Controls Ltd
Sperry Vickers

Actuators, Rotary
Air Power & Hydraulics Ltd
Cessna Fluid Power Ltd
Commercial Hydraulics Ltd
Bradford Cylinders Ltd
Lucas Fluid Power Ltd
Parker Hannifin (UK) Ltd
Servotel Controls Ltd
Sperry Vickers

Adaptors
Adaptors (Engineering) Ltd

Bellows
Hallite Seals Ltd

Control Panels, Electric, *etc*
Commercial Hydraulics Ltd
Denley Eng Co (Heckmondwike) Ltd
Hunger Hydraulic UK Ltd
Towler Hydraulics (UK) Ltd

Control Panels, Hydraulic
Christie Hydraulics Ltd
Commercial Hydraulics Ltd
Denley Eng Co (Heckmondwike) Ltd
Hunger Hydraulics UK Ltd
Parker Hannifin (UK) Ltd
G. L. Rexroth Ltd
Servotel Controls Ltd
Sperry Vickers
Towler Hydraulics (UK) Ltd

Couplings, Flared
Parker Hannifin (UK) Ltd

Couplings, Quick Action
Parker Hannifin (UK) Ltd

Couplings, Self-Sealing
Adaptors (Engineering) Ltd
Aeroquip (UK) Ltd

Couplings, Swivel, Hydraulic
Adaptors (Engineering) Ltd

Europower Hydraulics Ltd
Parker Hannifin (UK) Ltd

Couplings, Tube, High Pressure
Europower Hydraulics Ltd
Hyndburn Hydraulics Ltd
Lucas Fluid Power Ltd
Parker Hannifin (UK) Ltd

Couplings, Tube, Hydraulic
Europower Hydraulics Ltd
Hyndburn Hydraulics Ltd
Lucas Fluid Power Ltd
Parker Hannifin (UK) Ltd

Cylinder Bodies, Extruded
TI Reynolds Ltd

Cylinders, Hydraulic
Air Power & Hydraulics Ltd
Commercial Hydraulics Ltd
Henry Berry & Co Ltd
Robert Bosch Ltd
Cessna Fluid Power Ltd
Bradford Cylinders Ltd
Dale (Mansfield) Ltd
Denley Eng Co (Heckmondwike) Ltd
Hunger Hydraulic UK Ltd
Lucas Fluid Power Ltd
Massey Ferguson GmbH
Parker Hannifin (UK) Ltd
Pedro Roquet S.A.
G. L. Rexroth Ltd
Servotel Controls Ltd
Sperry Vickers
TI Reynolds Ltd
Towler Hydraulics (UK) Ltd
Youngs Lifting Appliances Ltd

Cylinders, Hydraulic, Double-Acting
Air Power & Hydraulics Ltd
Henry Berry & Co Ltd
Robert Bosch Ltd
Cessna Fluid Power Ltd
Commercial Hydraulics Ltd

Cylinders, Hydraulic, Double-Acting
Air Power & Hydraulics Ltd
Henry Berry & Co Ltd
Robert Bosch Ltd
Bradford Cylinders Ltd
Cessna Fluid Power Ltd
Commercial Hydraulics Ltd
Dale (Mansfield) Ltd
Denley Eng Co (Heckmondwike) Ltd
Hunger Hydraulic UK Ltd
Lucas Fluid Power Ltd
Massey Ferguson GmbH
Parker Hannifin (UK) Ltd
Pedro Roquet S.A.
G. L. Rexroth Ltd
Servotel Controls Ltd
Sperry Vickers
Towler Hydraulics (UK) Ltd
Youngs Lifting Appliances Ltd

Cylinders, Hydraulic, Low Pressure
Henry Berry & Co Ltd
Robert Bosch Ltd
Bradford Cylinders Ltd
Cessna Fluid Power Ltd
Dale (Mansfield) Ltd
Denley Eng Co (Heckmondwike) Ltd
Lucas Fluid Power Ltd
Massey Ferguson GmbH
Parker Hannifin (UK) Ltd
Towler Hydraulics (UK) Ltd
Youngs Lifting Appliances Ltd

Cylinders, Hydraulic, Rotating
Bradford Cylinders Ltd

Cylinders, Hydraulic, Single-Acting
Air Power & Hydraulics Ltd
Henry Berry & Co Ltd
Robert Bosch Ltd
Bradford Cylinders Ltd
Cessna Fluid Power Ltd
Commercial Hydraulics Ltd
Dale (Mansfield) Ltd
Denley Eng Co (Heckmondwike) Ltd
Hunger Hydraulic UK Ltd
Lucas Fluid Power Ltd
Massey Ferguson GmbH
Parker Hannifin (UK) Ltd
Pedro Roquet, S.A.
G. L. Rexroth Ltd
Servotel Controls Ltd
Sperry Vickers
Towler Hydraulics (UK) Ltd
Youngs Lifting Appliances Ltd

Cylinders, Hydraulic, Telescopic
Air Power & Hydraulics Ltd
Henry Berry & Co Ltd
Bradford Cylinders Ltd
Cessna Fluid Power Ltd
Commercial Hydraulics Ltd
Howard Cook Ltd
Dale (Mansfield) Ltd
Hunger Hydraulic UK Ltd
Lucas Fluid Power Ltd
Pedro Roquet, S.A.
Towler Hydraulics (UK) Ltd
Youngs Lifting Appliances Ltd

Cylinders, Pneumatic
Bradford Cylinders Ltd

Cessna Fluid Power Ltd
Commercial Hydraulics Ltd
Dale (Mansfield) Ltd
Denley Eng Co (Heckmondwike) Ltd
Lucas Fluid Power Ltd
Parker Hannifin (UK) Ltd
Pedro Roquet S.A.
G. L. Rexroth Ltd
Servotel Controls Ltd
Sperry Vickers
Towler Hydraulics (UK) Ltd

Motors, Axial Piston
Abex Denison Ltd

Motors, Hydraulic
Abex Denison Ltd
Robert Bosch Ltd
Cessna Fluid Power Ltd
Commercial Hydraulics Ltd
Dale (Mansfield) Ltd
Hydroperfect International "HPI"
Linde Hydraulics Ltd
Lucas Fluid Power Ltd
Parker Hannifin (UK) Ltd
Pedro Roquet S.A.
RHL Hydraulics Ltd
G. L. Rexroth Ltd
Servotel Controls Ltd
Sperry Vickers
Towler Hydraulics (UK) Ltd
Volvo Hydraulics Ltd

Motors, Hydraulic, Cam-Rotor
Towler Hydraulics (UK) Ltd

Motors, Hydraulic, Rotary Abutment
G. L. Rexroth Ltd

Motors, Hydraulic, Vane Type
Abex Denison Ltd
Servotel Controls Ltd
Sperry Vickers
Towler Hydraulics (UK) Ltd

Motors, Radial Piston
Lucas Fluid Power Ltd

O-Rings
Aeroquip (UK) Ltd
Europower Hydraulics Ltd
Freudenberg Simrit Ltd
Hallite Seals Ltd
Hunger Hydraulic UK Ltd
Lucas Fluid Power Ltd
Parker Hannifin (UK) Ltd

Packings, Hydraulic, Laminated, *etc*
Freudenberg Simrit Ltd
Hallite Seals Ltd
Hunger Hydraulic UK Ltd
Parker Hannifin (UK) Ltd

Packings, Hydraulic, Leather
Aeroquip (UK) Ltd
Parker Hannifin (UK) Ltd

Packings, Hydraulic, Metallic
Aeroquip (UK) Ltd
Parker Hannifin (UK) Ltd

Packings, Hydraulic, Moulded
Freudenberg Simrit Ltd
Hallite Seals Ltd
Hunger Hydraulic UK Ltd
Parker Hannifin (UK) Ltd

Pipe Clamps
Hyndburn Hydraulics Ltd

Pipe Flanges
Hyndburn Hydraulics Ltd

Pipes, Hydraulic, Metal
Lucas Fluid Power Ltd
Parker Hannifin (UK) Ltd

Pipes, Hydraulic, Plastic
Intec Ltd

Plating, Hard Chrome, *etc*
Hunger Hydraulic UK Ltd

Power Packs, Hydraulic
Abex Denison Ltd
Air Power & Hydraulics Ltd
Robert Bosch Ltd
Cessna Fluid Power Ltd
Commercial Hydraulics Ltd
Dale (Mansfield) Ltd
Denley Eng Co (Heckmondwike) Ltd
Hunger Hydraulic UK Ltd
Hydroperfect International "HPI"
Lucas Fluid Power Ltd
George Meller Ltd
Parker Hannifin (UK) Ltd
Pedro Roquet, S.A.
G . L. Rexroth Ltd
Servotel Controls Ltd
Sperry Vickers
Towler Hydraulics (UK) Ltd
Volvo Hydraulics Ltd

Power Units, Hydraulic
Abex Denison Ltd
Air Power & Hydraulics Ltd
Robert Bosch Ltd
Commercial Hydraulics Ltd
Dale (Mansfield) Ltd
Denley Eng Co (Heckmondwike) Ltd
Hunger Hydraulic UK Ltd
Hydroperfect International "HPI"
George Meller Ltd
Parker Hannifin (UK) Ltd
Pedro Roquet, S.A.

Tees
Adaptors (Engineering) Ltd

Test Equipment, Hydraulic
Christie Hydraulics Ltd
Commercial Hydraulics Ltd
Denley Eng Co (Heckmondwike) Ltd
Europower Hydraulics Ltd
Lucas Fluid Power Ltd
George Meller Ltd
G. L.Rexroth Ltd
Servotel Controls Ltd
Towler Hydraulics (UK) Ltd

Test Facilities for Hydrostatic Pumps, Motors and Transmissions
National Engineering Laboratory

Test Units, Portable
Lucas Fluid Power Ltd

Torque Convertors
Volvo Hydraulics Ltd
Youngs (Lifting Appliances) Ltd

Transmissions, Hydrostatic
Abex Denison Ltd
Cessna Fluid Power Ltd
Commercial Hydraulics Ltd
Linde Hydraulics Ltd
G. L. Rexroth Ltd
Sperry Vickers
Towler Hydraulics (UK) Ltd
Volvo Hydraulics Ltd

Tube Benders, Hydraulic
Hyndburn Hydraulics Ltd
Parker Hannifin (UK) Ltd
Youngs (Lifting Appliances) Ltd

Tube Clamping
Lucas Fluid Power Ltd

Tubes, Hydraulic
Hunger Hydraulic UK Ltd
Lucas Fluid Power Ltd
Parker Hannifin (UK) Ltd
TI Reynolds Ltd

Valves, Cartridge Type, DIN 24342
Sperry Vickers

Valves, Direction, Flow and Speed Control
Abex Denison Ltd
Robert Bosch Ltd
Cessna Fluid Power Ltd
Commercial Hydraulics Ltd
Dale (Mansfield) Ltd
Hydroperfect International "HPI"
Laser Engineering Ltd, Hydraulics,
Linde Hydraulics Ltd
Lucas Fluid Power Ltd
Parker Hannifin (UK) Ltd
Pedro Roquet S.A.
G. L. Rexroth Ltd
Servotel Controls Ltd
Sperry Vickers
Towler Hydraulics (UK) Ltd
Volvo Hydraulics Ltd
Youngs (Lifting Appliances) Ltd

Valves, Flow Dividing
Robert Bosch Ltd
Cessna Fluid Power Ltd
Commercial Hydraulics Ltd
Hydroperfect International "HPI"
Lucas Fluid Power Ltd
Parker Hannifin (UK) Ltd
Pedro Roquet, S.A.
G. L. Rexroth Ltd
Towler Hydraulics (UK) Ltd

Valves, Locking
Cessna Fluid Power Ltd
Dale (Mansfield) Ltd
Laser Engineering Ltd, Hydraulics
Lucas Fluid Power Ltd
Parker Hannifin (UK) Ltd
G. L. Rexroth Ltd
Sperry Vickers
Towler Hydraulics (UK) Ltd
Youngs Lifting Appliances Ltd

Valves, Miniature
Dale (Mansfield) Ltd
Lucas Fluid Power Ltd
Parker Hannifin (UK) Ltd
Towler Hydraulics (UK) Ltd

Valve Testing Rigs
George Meller Ltd

Valves, Pilot
Abex Denison Ltd
Robert Bosch Ltd
Cessna Fluid Power Ltd
Commercial Hydraulics Ltd
Dale (Mansfield) Ltd
Hydroperfect International "HPI"
Lucas Fluid Power Ltd
Parker Hannifin (UK) Ltd
G. L. Rexroth Ltd
Sperry Vickers
Towler Hydraulics (UK) Ltd
Youngs (Lifting Appliances) Ltd

Valves, Pressure Regulating
Abex Denison Ltd
Robert Bosch Ltd
Cessna Fluid Power Ltd
Dale (Mansfield) Ltd
Hydroperfect International "HPI"
Laser Engineering Ltd, Hydraulics

Linde Hydraulics Ltd
Lucas Fluid Power Ltd
Parker Hannifin (UK) Ltd
Pedro Roquet, S.A.
G. L. Rexroth Ltd
Sperry Vickers
Towler Hydraulics (UK) Ltd
Volvo Hydraulics Ltd

Valves, Positional Control
Cessna Fluid Power Ltd

Valves, Proportional
G. L. Rexroth Ltd

Valves, Proportional Direction, Pressure and Flow Control, c/w Amplifiers
Robert Bosch Ltd

Valves, Proportional Electrically Modulated
Sperry Vickers

Valves, Reducing
Abex Denison Ltd
Robert Bosch Ltd
Cessna Fluid Power Ltd
Dale (Mansfield) Ltd
Laser Engineering Ltd, Hydraulics
Lucas Fluid Power Ltd
Parker Hannifin (UK) Ltd
Pedro Roquet, S.A.
G. L. Rexroth Ltd
Sperry Vickers
Towler Hydraulics (UK) Ltd
Youngs (Lifting Appliances) Ltd

Valves, Relief
Abex Denison Ltd
Robert Bosch Ltd
Cessna Fluid Power Ltd
Commercial Hydraulics Ltd
Dale (Mansfield) Ltd
Hydroperfect International "HPI"
Laser Engineering Ltd, Hydraulics
Lucas Fluid Power Ltd
George Meller Ltd
Parker Hannifin (UK) Ltd
Pedro Roquet, S.A.
G. L. Rexroth Ltd
Servotel Controls Ltd
Sperry Vickers
Towler Hydraulics(UK) Ltd
Volvo Hydraulics Ltd

Valves, Reversing
Robert Bosch Ltd
Dale (Mansfield) Ltd
Lucas Fluid Power Ltd
Parker Hannifin (UK) Ltd
G. L. Rexroth Ltd
Sperry Vickers
Towler Hydraulics (UK) Ltd
Youngs (Lifting Appliances) Ltd

Valves, Sequence
Abex Denison Ltd
Robert Bosch Ltd
Dale (Mansfield) Ltd
Laser Engineering Ltd, Hydraulics
Lucas Fluid Power Ltd
Parker Hannifin (UK) Ltd
Pedro Roquet, S.A.
G. L. Rexroth Ltd
Sperry Vickers
Towler Hydraulics (UK) Ltd

Valves, Servo
Abex Denison Ltd
Robert Bosch Ltd
Cessna Fluid Power Ltd
Commercial Hydraulics Ltd
Hydroperfect International "HPI"
Parker Hannifin (UK) Ltd
G. L. Rexroth Ltd
Servotel Controls Ltd
Sperry Vickers
Towler Hydraulics (UK) Ltd
Youngs (Lifting Appliances) Ltd

Valves, Solenoid Operated
Abex Denison Ltd
Robert Bosch Ltd
Cessna Fluid Power Ltd
Commercial Hydraulics Ltd
Dale (Mansfield) Ltd
Laser Engineering Ltd, Hydraulics
Lucas Fluid Power Ltd
Parker Hannifin (UK) Ltd
Pedro Roquet, S.A.
G. L. Rexroth Ltd
Servotel Controls Ltd
Sperry Vickers
Towler Hydraulics (UK) Ltd
Volvo Hydraulics Ltd

Valves, Stacking
Robert Bosch Ltd

Valves, Temperature Control
Parker Hannifin (UK) Ltd

Valves, Trailer Brake
Robert Bosch Ltd

Valves, Water, High Pressure
Henry Berry & Co Ltd

Winches, Hydraulic
G. L. Rexroth Ltd

Sub-section (c)

ALPHABETICAL LIST OF MANUFACTURERS WITH ADDRESSES, TELEPHONE AND TELEX NUMBERS, OF HEAD OFFICE, WORKS AND BRANCHES

ABEX DENISON LIMITED, Victoria Gardens, Burgess Hill, West Sussex RH15 9ND.
Telephone: Burgess Hill 04446 5121 Telex: 87168
Branches:
Thistle House, 34 Queen Street, Hitchin, Herts SG4 9TN.
Telephone: Hitchin 0462 31566 Telex: 826231
Crossford Court, Dane Road, Sale, Cheshire M33 1BJ
Telephone: 061 962 7674 Telex: 667421

ADAPTORS (ENGINEERING) LIMITED, 94 Tower Hamlets Road, Forest Gate, London E7 9DB.
Telephone: 01 534 6314/6324 Telex: 28480

AEROQUIP (UK) LIMITED, Industrial Division, Studley Road, Redditch, Worcs B98 7HQ.
Telephone: Redditch 20292 Telex: 339632
Branches:
Seals Division, 165/166 Birmingham Factory Centre, Kings Norton, Birmingham B30 3HG.
Telephone: 021 458 6841 Telex: 337374

AIR POWER & HYDRAULICS LIMITED, Hillington Industrial Estate, Glasgow G52 4PQ.
Telephone: 041 810 4511 Telex: 77:78 19 APHYDS G

HENRY BERRY & COMPANY LIMITED, Croydon Works, Leeds LS10 2BE, West Yorkshire.
Telephone: 0532 705481
Sales Office:
A.B.Barnett, 154 Redditch Road, Kings Norton, Birmingham 30
Telephone: 021 458 2980
R.J.P.Thomas, 7 Hamlet Road, Chelmsford, Essex.
Telephone: 0245 56682

BRADFORD CYLINDERS LIMITED, Allerton Road, Bradford, BD8 OBA.
Telephone: 0274 495611 Telex: 517528 Grams: Metallic Bradford

ROBERT BOSCH LIMITED, P.O.Box 166, Rhodes Way, Watford, WD2 4LB.
Telephone: Watford 44233

CESSNA FLUID POWER DIVISION, Eastfield Industrial Estate, Glenrothes, Fife, Scotland KY7 4NW
Telephone: 0592 771 771 Telex: 72305 Cessna G
Branches:
Cessna Fluid Power Division, P.O.Box 1028, Hutchinson, Kansas 67501, U.S.A.
Telephone: 316 663 5751 Telex: Dial 007 Key 255 910 740 1521

CHRISTIE HYDRAULICS LTD, Sandycroft Industrial Estate, Deeside, Clwyd, CH5 2QP.
Telephone: 0244 535515 Telex: 617026
Branches:
Wingate Hydraulics Ltd, 5 Manor Way, Old Woking, GU22 9JX, Surrey.
Telephone: 04862 27313 Telex: 8813487

COMMERCIAL HYDRAULICS LIMITED, Shuttleworth Road, Bedford
Telephone: 0234 50501 Telex: 82339

DALE (MANSFIELD) LIMITED, Rotherham Road, New Houghton, Mansfield, Nottinghamshire
Telephone: 0623 810659

DENLEY ENGINEERING CO (HECKMONDWIKE) LTD, Chapel Lane, Heckmondwike, W.Yorks.
Telephone: 0924 408234 Telex: 556384

EUROPOWER HYDRAULICS LIMITED, High Street, Market Weighton, Yorkshire YO4 3AD.
Telephone: 0696 72361
Branches:
Europower Hydraulics Limited, Holbrook Industrial Estate, Halfway, Sheffield.
Telephone: 0742 48316

FREUDENBERG SIMRIT LIMITED, Lutterworth, Leicestershire LE17 4DU.
Telephone: 04555 56521 Telex: 34125 Grams: Ceeflug G

HALLITE SEALS LIMITED, Oldfield Road, Hampton, Middlesex TW12 2HT.
Telephone: 01 941 2244 Telex: 21124

HYDROPERFECT INTERNATIONAL "H P I" 26, rue Condorcet B.P. 18, 99430 Chennevieres Sur Marne, France.
Telephone: (I) 576 60 22 Telex: 230 189 F

HUNGER HYDRAULIC UK LIMITED, 54 North Street, Emsworth, Hants PO10 7PJ.
Telephone: 02434 5308 Telex: 86495

HYNDBURN HYDRAULICS LIMITED, Arden Buildings, Bertha Street, Accrington, Lancs BB5 6TE.
Telephone: 0254 398204 Telex: 635431

INTEC LIMITED, Forder Valley Road, Plymouth, PL6 8LA.
Telephone: 0752 662844 Telex: 45561 Grams: Tecalemit Plymouth

LASER ENGINEERING LIMITED, Hydraulics, Victoria Road, Burgess Hill, Sussex.
Telephone: Burgess Hill 41571 Telex: Burgess Hill 877416

LINDE HYDRAULICS LIMITED, Nuffield Way, Abingdon, Oxon OX14 1RJ.
Telephone: 0235 22828 Telex: 837477

LUCAS FLUID POWER LIMITED, Torrington Avenue, Coventry CV4 9AJ.
Telephone: 0203 468111 Telex: 311801 Indhyd G.
Branches:
Unit 2, 148 Helen Street, Govan, Glasgow G51 3JS.
Telephone: 041 440 0662
Unit 23, Stretford Motorway Estate, Barton Dock Road, Stretford, Manchester M32 OZH.
Telephone: 061 865 9119
Unit 1, 154 Green Lane, Walsall WS2 8LE, Staffs.
Telephone: 0922 614357
Unit 10, Suttons Industrial Park, Earley, Reading, Berks RG6 1AZ.
Telephone: 0734 662313

MASSEY-FERGUSON GmBH, Postfach 520, 3440 Eschwege, West Germany.
Telephone: 49 565181 659. Telex: 993298 Grams: Masferg Eschwege
Branches:
Howard Cook Limited, 334 Meanwood Road, Leeds, Yorks LS7 2JF.
Telephone: 0532 624612 Telex: 557846

GEORGE MELLER LIMITED, Orion Park, Northfield Avenue, London W13 9SJ
Telephone: 01 579 2111 Telex: 27800

MODINA FILTRI SpA, Via San Chierico, 24060 Bolgare, Bergamo Italy
Telephone: 035 841250 Telex: 301589

NATIONAL ENGINEERING LABORATORY, East Kilbride, Glasgow G75 OQU, Scotland.,
Telephone: East Kilbride 20222 Telex: 777888

PARKER HANNIFIN (UK) LIMITED, P.O.Box 170, Greycaine Road, Watford, Herts WD2 4QA.
Telephone: 0923 44377 Telex: 23765

Branches:
Connectors Group, Haydock Park Road, Derby DE2 8JA.
Telephone: 0332 365631 Telex: 37427
Filter Division, Peel Street, Morley, Leeds, W.Yorkshire.
Telephone: 0532 537921
Mobile Division, Maylands Avenue, Hemel Hempstead, Herts.
Telephone: 0442 66691 Telex: 826577
Alenco Division, Riverside Road, Potterton Industrial Estate, Barnstaple, Devon.
Telephone: 0271 72591 Telex: 46185
PSI Division, P.O.Box 3, Stanmore, Middlesex
Telephone: 01 954 6288

PEDRO ROQUET, S.A., c/o Antonio Figueras, 91 Tona, Barcelona, Spain.
Telephone: 34 3 887 01 25 Telex: 57 644
G. L. REXROTH LIMITED, Cromwell Road, St. Neots, Cambridgeshire PE19 2ES.
Telephone: 0480 76041 Telex: 32161 Grams: Rexroth St. Neots
Branches:
Mitchelston Industrial Estate, Kircaldy, Fife.
Telephone: 0592 52781/2
12 Whitworth Road, South West Industrial Estate, Peterlee, Co.Durham
Telephone: 0783 860621 Telex: 537641
5-7 New York Road, Leeds LS2 7PL.
Telephone: 0532 441951/441952 Telex: 557614
Baystrait House, Station Road, Biggleswade, Bedfordshire SG18 8AL.
Telephone: 0767 315959 Telex: 925336
49-57 High Street, Droitwich, Worcs
Telephone: 0905 778151 Telex: 338388
Unit 3b, Hambridge Lane, Newbury, Berks RG14 5TU.
Telephone: 0635 41347 Telex: 848780
Linguaphone House, Beavor Lane, Hammersmith W6.
Telephone: 01 741 4356/7, 01 741 4091/2 Telex: 28871
RHL HYDRAULICS LIMITED, Planet Place, Killingworth, Newcastle-upon-Tyne
Telephone: 0632 685505 Telex: 537781
SERVOTEL CONTROLS LIMITED, 23 Eastville Close, Eastern Avenue, Gloucester
Telephone: 0452 28262 Telex: 43154
Branches:
31 Falstaff Avenue, Hollywood, Birmingham B47 5EL.
Telephone: 0564 823895
SPECIALIST HEAT EXCHANGERS LIMITED, Freeman Road, N.Hykeham, Lincoln LN6 9AP.
Telephone: 0522 683123 Telex: 56133
TI REYNOLDS LIMITED, P.O.Box 232, Hay Hall, Redfern Road, Tyseley, Birmingham B11 2BG.
Telephone: 021 706 3333 Telex: 338181
TOWLER HYDRAULICS (U.K.) LIMITED, Oaklands Road, Rodley, Leeds LS13 1LG, West Yorkshire.
Telephone: 0532 57721 Telex: 55413 Grams: Electrolic Leeds
Branches:
12 Sugarbrook Road, Aston Fields Industrial Estate, Bromsgrove, Worcs.
Telephone: 0527 78660
UCC INTERNATIONAL, International House, Mill Lane, Thetford, Norfolk IP24 3RT.
Telephone: 0842 4251 Telex: 81258
VOLVO HYDRAULICS LIMITED, 130 Thornes Lane, Wakefield, WF2 7TG.
Telephone: 0924 361616 Telex: 556193
Works:
Volvo Flygmotor A.B., Trollhattan, Sweden
Telephone: 0520 301 00 Telex: 420 40
YOUNGS LIFTING APPLIANCES LIMITED, 374–384 Moseley Road, Birmingham B12 9AY.
Telephone: 021 440 3131 Telex: 336988

Addenda to Buyers' Guide

TRADE NAMES

ARGUS — Hydraulic fittings — Argus Gesellschaft MBH
AQUABLAST — Water jetting hose and couplings — Hi-Flex International Ltd.
AQUACENT — Fire resistant hydraulic fluid — Century Oils International Ltd.

BOWMAN — Heat exchangers — E. J. Bowman (B'ham) Ltd.

DOUBLE A — Hydraulic pumps, valves and motors — Brown & Sharpe Fluid Power Ltd.
DUOCENT — Hydraulic and gear lubricants — Century Oils International Ltd.

E.P.E. — Cartridge filters — British Filters Ltd.

FIELDING — Hydrualic presses and ancillary equipment — Fielding & Platt Ltd.
FIRESAFE M64OR — Fire resistant hose and couplings — Hi-Flex International Ltd.

FLEXIFORGE — Flexible wire braid hose and couplings — Hi-Flex International Ltd.

GATES — V-belts and hose — Gates Europe N.V.
GEMINI — Pre-coat candle filter — British Filters Ltd.

HAGGLUND — Hydraulic motors — ASEA Hagglunds Ltd.

MULTITRAK — Multi-core hose assemblies — Hi-Flex International Ltd.

POWERDRAULIC — Multi-purpose hydraulic fluid — Century Oils International Ltd.
POWERGEAR — Multi-purpose gear oil — Century Oils International Ltd.
POWERTRAK — Multi-spiral hose and couplings — Hi-Flex International Ltd.

SLYDWAY — Bearing material — Shamban Europa UK Ltd.
STAFFA — Hydraulic motors — Brown & Sharpe Fluid Power Ltd.

TRACTAULIC — Agricultural PTO gearbox/pump unit — Joseph Young & Sons Ltd.
TURCITE — Seals and bearing materials — Shamban Europa UK Ltd.
TURCON — Seals and bearings — Shamban Europa UK Ltd.

CLASSIFIED INDEX OF HYDRAULIC EQUIPMENT AND COMPONENTS

Accumulators, Hydraulic
Fielding & Platt Ltd
Trans Nordic Ltd

Accumulators, Hydro-Pneumatic
Fielding & Platt Ltd

Actuators, Linear
Argus Gesellschaft MBH

Actuators, Rotary
Trans Nordic Ltd

Adaptors and Fittings, Hydraulic
B. & J. Light Engineering (Holdings) Ltd

Bearings, Ball and Roller
Shamban Europa UK Ltd

Bearing Materials
Shamban Europa UK Ltd

Control Panels, Electric, etc.
Brown & Sharpe Fluid Power Ltd
Fielding & Platt Ltd

Control Panels, Hydraulic
Fielding & Platt Ltd

Couplings, Self-Sealing
Argus Gesellschaft MBH
Hi-Flex International Ltd

Couplings, Swivel, Hydraulic
Argus Gesellschaft MBH

Couplings, Tube, High Pressure
Argus Gesellschaft MBH
B. & J. Light Engineering (Holdings) Ltd

Couplings, Tube Hydraulic
Argus Gesellschaft MBH
B. & J. Light Engineering (Holdings) Ltd

Cylinders, Hydraulic
Brown & Sharpe Fluid Power Ltd
Hi-Flex International Ltd
Trans Nordic Ltd

Cylinders, Hydraulic, Double-Acting
Brown & Sharpe Fluid Power Ltd
Hi-Flex International Ltd
Trans Nordic Ltd

Cylinders, Hydraulic, Low Pressure
Hi-Flex International Ltd
Trans Nordic Ltd

Cylinders, Hydraulic, Single-Acting
Hi-Flex International Ltd
Trans Nordic Ltd

Cylinders, Hydraulic, Telescopic
Hi-Flex International Ltd
Trans Nordic Ltd

Drives, Hydraulic, Infinitely Variable Speed
ASEA Hagglunds Ltd
Brown & Sharpe Fluid Power Ltd

Drives, Hydraulic, Synchronised Variable Speed
ASEA Hagglunds Ltd
Brown & Sharpe Fluid Power Ltd

Filters, Oil etc
British Filters Ltd
Hi-Flex International Ltd
Trans Nordic Ltd

Fluids, Hydraulic
Century Oils International Ltd

Fluids, Hydraulic, Fire Resisting
Century Oils International Ltd

Gauges, Pressure, Hydraulic
Trans Nordic Ltd

Gear Boxes
Brown & Sharpe Fluid Power Ltd
Trans Nordic Ltd

Gear Lubricants
Century Oils International Ltd

Heat Exchangers
E. J. Bowman (B'ham) Ltd
Serck Heat Transfer

Hose Assemblies, Hydraulic
Argus Gesellschaft MBH
Hi-Flex International Ltd

Hose, Hydraulic, High Pressure
Argus Gesellschaft MBH
Gates Europa N.V.
Hi-Flex International Ltd

Hose, Metal, Flexible, etc.
Hi-Flex International Ltd

Hose, Rubber, Flexible, etc.
Argus Gesellschaft MBH
Gates Europe N.V.
Hi-Flex International Ltd

Hose, Plastic
Argus Gesellschaft MBH
Gates Europe N.V.
Hi-Flex International Ltd

Hose Swaging Machinery, Portable
Hi-Flex International Ltd

Hose Swaging Machinery, Power Operated
Hi-Flex International Ltd

Lubricants
Century Oils International Ltd

Manifolds, Hydraulic
Brown & Sharpe Fluid Power Ltd
Fielding & Platt Ltd
Trans Nordic Ltd

Motors, Hydraulic
ASEA Hagglunds Ltd
Brown & Sharpe Fluid Power Ltd
Casappa Oleodinamica
Joseph Young & Sons Ltd

O-Rings
Shamban Europa UK Ltd
Trans Nordic Ltd

Power Packs, Hydraulic
Casappa Oleodinamica
Fielding & Platt Ltd

Trans Nordic Ltd
Joseph Young & Sons Ltd

Power Units, Hydraulic
Fielding & Platt Ltd
Trans Nordic Ltd
Joseph Young & Sons Ltd

Presses, Hydraulic, Belting
Fielding & Platt Ltd

Presses, Hydraulic, Bending
Fielding & Platt Ltd

Presses, Hydraulic, Extruding
Fielding & Platt Ltd

Presses, Hydraulic, Flanging
Fielding & Platt Ltd

Presses, Hydraulic, Forging
Fielding & Platt Ltd

Presses, Hydraulic, Joggling
Fielding & Platt Ltd

Presses, Hydraulic, Lamination
Fielding & Platt Ltd

Presses, Hydraulic, Moulding
Fielding & Platt Ltd

Pressure Gauges
Trans Nordic Ltd

Pump and Motor Units, Complete
Brown & Sharpe Fluid Power Ltd
Casappa Oleodinamica
Fielding & Platt Ltd
Trans Nordic Ltd
Joseph Young & Sons Ltd

Pumps, Hydraulic, Annular Piston Type
Trans Nordic Ltd

Pumps, Hydraulic, Gear
Brown & Sharpe Fluid Power Ltd
Casappa Oleodinamica
Trans Nordic Ltd
Joseph Young & Sons Ltd

Pumps, Hydraulic, Hand
Casappa Oleodinamica

Pumps, Hydraulic, High Speed Piston
Brown & Sharpe Fluid Power Ltd
Trans Nordic Ltd

Pumps, Hydraulic, Vane Type
Brown & Sharpe Fluid Power Ltd

Pumps, Hydraulic, Variable Delivery
Brown & Sharpe Fluid Power Ltd

Pumps, Hydraulic, Ram Type
Fielding & Platt Ltd

Pumps, Hydraulic, Variable Delivery
Trans Nordic Ltd

Rams, Hydraulic
Brown & Sharpe Fluid Power Ltd
Hi-Flex International Ltd
Trans Nordic Ltd

Remote Control, Hydraulic
Brown & Sharpe Fluid Power Ltd
Trans Nordic Ltd

Reservoirs, Hydraulic
Brown & Sharpe Fluid Power Ltd
Joseph Young & Sons Ltd

Sealing Plugs
Boneham & Turner Ltd

Sealing Rings, Square Section
Shamban Europa UK Ltd

Sealing Rings, Washers
Shamban Europa UK Ltd

Seals, Hydraulic
Boneham & Turner Ltd
Shamban Europa UK Ltd

Servo-Controls, Hydraulic
ASEA Hagglunds Ltd
Trans Nordic Ltd

Servo-Systems, Hydraulic
Fielding & Platt Ltd
Trans Nordic Ltd

Slide Tables, Hydraulic
Boneham & Turner Ltd

Solenoids
Trans Nordic Ltd

Transmissions, Hydrostatic
ASEA Hagglunds Ltd
Brown & Sharpe Fluid Power Ltd

Tube Benders, Hydraulic
Brown & Sharpe Fluid Power Ltd

Valves, Cartridge
Barmag Hydraulic UK

Valves, Check Hydraulic
Argus Gesellschaft MBH

Valves, Direction, Flow and Speed Control
Argus Gesellschaft MBH
ASEA Hagglunds Ltd
Barmag Hydraulic UK
Brown & Sharpe Fluid Power Ltd
Casappa Oleodinamica
Fielding & Platt Ltd
Trans Nordic Ltd
Joseph Young & Sons Ltd

Valves, Flow Dividing
Barmag Hydraulic UK

Brown & Sharpe Fluid Power Ltd
Trans Nordic Ltd

Valves, Hydraulic
Argus Gesellschaft MBH

Valves, Locking
Barmag Hydraulic UK
Trans Nordic Ltd
Joseph Young & Sons Ltd

Valves, Miniature
Brown & Sharpe Fluid Power Ltd
Trans Nordic Ltd

Valves, Pilot
ASEA Hagglunds Ltd
Barmag Hydraulic UK
Brown & Sharpe Fluid Power Ltd
Trans Nordic Ltd
Joseph Young & Sons Ltd

Valves, Pressure Regulating
ASEA Hagglunds Ltd
Barmag Hydraulic UK
Brown & Sharpe Fluid Power Ltd
Trans Nordic Ltd
Joseph Young & Sons Ltd

Valves, Proportional
Barmag Hydraulic UK

Valves, Reducing
ASEA Hagglunds Ltd
Barmag Hydraulic Ltd
Brown & Sharpe Fluid Power Ltd
Trans Nordic Ltd

Valves, Relief
ASEA Hagglunds Ltd
Barmag Hydraulic UK
Brown & Sharpe Fluid Power Ltd
Casappa Oleodinamica
Serck Heat Transfer
Trans Nordic Ltd
Joseph Young & Sons Ltd

Valves, Reversing
Brown & Sharpe Fluid Power Ltd
Trans Nordic Ltd
Joseph Young & Sons Ltd

Valves, Sequence
ASEA Hagglunds Ltd
Brown & Sharpe Fluid Power Ltd
Trans Nordic Ltd

Valves, Servo
ASEA Hagglunds Ltd
Barmag Hydraulic UK
Trans Nordic Ltd

Valves, Solenoid Operated
ASEA Hagglunds Ltd
Barmag Hydraulic UK
Brown & Sharpe Fluid Power Ltd
Serck Heat Transfer
Trans Nordic Ltd

Valves, Temperature Control
Barmag Hydraulic UK
Serck Heat Transfer

Winches, Hydraulic
ASEA Hagglunds Ltd
Trans Nordic Ltd

ALPHABETICAL LIST OF MANUFACTURERS WITH ADDRESSES, TELEPHONE AND TELEX NUMBERS, OF HEAD OFFICE, WORKS AND BRANCHES

ARGUS GESELLSCHAFT MBH, Goethestr. 15 D-7505 Ettlingen.
Telephone: 07243/103-1 Telex: 782 841
Branches:
BHS Birmingham Hydraulic Services, Wellington Works, Great Bridge Street, West Bromwich, West Midlands, B70, ODJ.
Telephone: 021 520 2444/5/6 Telex: 335 387
Powerite Limited, Poutney Street, Wolverhampton WV2 JN, West Midlands.
Telephone: Wolverhampton 27189, 24689 Telex: 339 507

ASEA HAGGLUNDS LTD, Milner Way, Ossett West, Yorkshire WF5 9JE.
Telephone: 0924 272581 Telex: 557252

B. & J. LIGHT ENGINEERING (HOLDINGS) LTD, 629 Spur Road, North Feltham Trading Estate, Feltham, Middlesex TW14 OSS.
Telephone: 01 89C 4282 Telex: 948227

BARMAG HYDRAULIC UK, Dunbeath Road, Elgin,
Telephone: Swindon (0793) 487414 Telex: 444485 Grams: Barmag G

BONEHAM & TURNER LTD, Nottingham Road, Mansfield, Nottinghamshire.
Telephone: 0623 27641 Telex: 37193

E. J. BOWMAN (B'HAM) LTD, Chester Street, Birmingham B6 4AP
Telephone: 021 359 5401 Telex: 339239 Grams: Intercool

BRITISH FILTERS LIMITED, Thames Industrial Estate, Fieldhouse Lane, Marlow, Bucks SL7 1TD.
Telephone: Marlow 73131 Telex: 849216 Grams: Filtersphone Marlow.

BROWN & SHARPE FLUID POWER LTD, Ernesettle, Plymouth, Devon PL5 2SA.
Telephone: 0752 364394 Telex: 45759

CASAPPA OLEODINAMICA, 43044 Cavalli di Collecchio, Parma, Italy.
Telephone: 804101 804147 Telex: 53364 Grams: Oleocasappa

CENTURY OILS INTERNATIONAL LTD, P.O.Box 2, Century Street, Hanley, Stoke-on-Trent, ST1 5HU.
Telephone: 0782 29521 Telex: 36213
Branches:
P.O.Box 20, Spelter Works Road, Hendon, Sunderland.
Telephone: 0783 79423
Century Wharf, Crayford Creek, Dartford, Kent.
Telephone: 03224 41114
Albert Road, Bristol
Telephone: 0272 779473
959 London Road, Leigh-on-Sea, Essex.
Telephone: 0702 78557
Units 13 & 14a, Central Trading Estate, Off Marine Parade, Southampton.
Telephone: 0703 334366
Aitkenhead Road, Tannochside, Uddingston, Glasgow.
Telephone: 0698 816126

FIELDING & PLATT LTD, P.O.Box 10, Atlas Works, Southgate Street, Gloucester GL1 5RF
Telephone: 0452 28611 Telex: 43187 Grams: Fieldplatt Gloucester A/B Fandpg.

GATES EUROPE N.V., Dr. Carlierlaan 30, B-9440 Erembodegem-Aalst, Belgium.
Telephone: 053 779999 Telex: 12322&12317
Branches:
Gates Europe S.A., 15a Market Place, Driffield, Yorkshire YO25 7AP.
Telephone: 0377 42771 Telex: 527501
Gates Europe S.A., Warehouse, Little End Road, Eaton Socon, St. Neots PE19 3JH.
Telephone: 0480 218018 Telex: 32169
Gates France S.A.R.L., Zone Industrielle, F-95380 Louvres, France.
Telephone: 3-4681990 Telex: 696946
Gates Deutschland GmbH, Haus Gravenerstr. 191-192 D-4018, Langenfeld 2, Germany.
Telephone: 02173 7950 Telex: 8515752

HI-FLEX INTERNATIONAL, P.O.Box 2, Salisbury, Wilts.
Telephone: 0722 6231 Telex: 47560
Branches:
Hi-Flex Northern Limited, Copse Industrial Estate, Fleetwood, Lancs.
Telephone: 03917 6711/6712
Hi-Flex Northern Limited, Cobden Street, Salford, Manchester M6 6WF
Telephone: 061 737 6978/9
Hi-Flex Northern Limited, National Industrial & Commercial Estate, Seeds Lane, Aintree, Liverpool 9.
Telephone: 051 523 4118
North Eastern Counties Hydraulics Limited, 2 Cowen Road, Blyadon, Tyne & Wear.
Telephone: 0632 447771

North Eastern Counties Hydraulics Limited, Wallis Road, Skippers Lane Industrial Estate, Middlesbrough, Cleveland.
Telephone: 0642 468500
North Eastern Counties Hydraulics Limited, 4 Maurice Road, Wallsend, Newcastle-upon-Tyne.
Telephone: 0632 620984
Hi-Flex Yorkshire Limited, Unit 5, Horbury Industrial Estate, Calder Vale Road, Horbury, Wakefield, Yorkshire WF4 5ER
Telephone: 0924 270070/271203
Hi-Flex Yorkshire Limited, Gilbey Road, Pyewipe, Nr.Grimsby, South Humberside.
Telephone: 0472 40813
Eastern Counties Hydraulics Limited, Highfield Lane, Orgreave, Sheffield S13 9NA
Telephone: 0742 697633 Telex: 54751
Hi-Flex Midlands Limited, Unit 3, 73 Aston Hall Road, Birmingham B6 7LP
Telephone: 021 327 1368
Hi-Flex (South Eastern) Limited, Unit 4, Somersham Road Industrial Estate, St. Ives, Cambs.
Telephone: 0480 65289 Telex: 32397
Hi-Flex Anglia Limited, Unit 2, Tayfen Road, Bury-St-Edmunds, Suffolk.
Telephone: 0284 62409
Hi-Flex Kent Limited, Playfair House, 158 Church Road, Southborough, Tunbridge Wells, Kent.
Telephone: 0982 37231
Hi-Flex Oxon Limited, Unit 30, Milton Trading Estate, Abingdon, Oxon OX14 4EF
Hi-Flex Surrey Limited, Unit 3, Goldsworth Road Industrial Estate, Woking, Surrey
Telephone: 04862 25065
Hi-Flex Southern Counties Limited, Unit 3, Easton Lane, Winnall Trading Estate, Winchester, Hants.
Telephone: 0962 60311
Hi-Flex Poole Limited, Unt
Hi-Flex Poole Limited, Unit 9, 43 Nuffield Road, Nuffield Trading Estate, Poole, Dorset
Telephone: 0202 683522
Western Counties Hydraulics Limited, Unit 22A, Barton Hill Trading Estate, Barton Hill, Bristol.
Telephone: 0272 552823/4
Hi-Flex South West Limited, Stonehouse Street, Plymouth PL1 3PN.
Telephone: 0752 669373/262268
Hi-Flex Scotland Limited, 4 Castlebank Crescent, Glasgow G11 6DQ.
Telephone: 041 3392332
Hi-Flex Scotland Limited, 2 Institution Street, Macduff, Scotland.
Telephone: 0261 32487
Hi-Flex Scotland Limited, West End No 9 Shed, Leith Docks, 35 Harbour Dock West, Edinburgh EH6 6LZ
Telephone: 031 553 4237
Hi-Flex Herts Limited, Unit RO8, Acton Workshops, School Road, Acton, London NW10
Telephone: 01 965 2345
Hi-Flex Cambs Limited, Unit 4, Somersham Road Industrial Estate, St. Ives, Cambs
Telephone: 0480 64657
Hi-Flex Essex Limited, 6 Bridge Close Estate, Old Church Road, Romford, Essex RM7 OA5
Telephone: 0708 754 523

SERCK HEAT TRANSFER, P.O.Box 598B, Warwick Road, Birmingham B11 2QY.
Telephone: 021 772 4353 Telex: 337341/339805/336487 Grams: Sercktrans
Branches:
60 Trafalgar Square, London WC2N 5DS.
Telephone: 01 839 2682 Telex: 919368
11 Sandy Ford Place, Glasgow G3 7NB
Telephone: 041 248 6991/2

SHAMBAN EUROPA UK LTD, Ewart House, 5 St. James Terrace, Nottingham
Telephone: 0602 411866 Telex: 337280 Telefax 0602 411278
Works:
W. S. Shamban Europa A/S, Fabriksvej 17, 3000 Helsingor, Dk. Danmark

TRANS NORDIC LTD, Vicarage Lane, Hoo, Rochester, Kent.
Telephone: 0634 252300 Telex: 964992 Nordic G
Branches:
112 Broomhill Road, Strood, Rochester, Kent.
Telephone: 0634 78396

JOSEPH YOUNG & SONS LTD, Turton Street, Bolton, Lancs BL1 2EY
Telephone: Bolton 21431 Telex: 63224 Grams: Hydraulics Bolton.

Index

T

Index to Advertisers

Acknowledgements

In pursuing my research, I have always felt the importance of having people around me who are psychologically supportive. I think this is the most precious kind of help that can be given to someone who is concerned with solving new problems, and is therefore totally focused on constructing a new framework.

I specifically want to thank the following people for trusting me in what I was doing.

First of all I want to thank Prof. Marvin Minsky, Toshiba Professor of Arts and Sciences at the Media Laboratory of the Massachusetts Institute of Technology. He has given me precious advice during my research. It was his adventurous approach to science that inspired me to find new solutions to new problems.

I want to thank Prof. Marco Somalvico, Founder and Director of the AI Group at the Politecnico in Milan. His enthusiasm has been a tremendous help during difficult times. By "difficult" I mean the problems a new idea encounters when people are not yet willing to accept new approaches.

I want to thank Prof. Antonio Carile, President of the Faculty of Humanities at the University of Bologna, for having supported my efforts.

I also want to thank Masoud Yazdani, President of Intellect, who has been an extremely valuable supporter of my work through his advice in editing some of my English books and papers; he also encouraged me to write this "peculiar" scientific book for multiple readers.

I want to thank Jim Richardson, writer and consultant, who has helped me in his expert role of technical writer to turn my three earlier Italian books into a summarized English version entitled "Writing as a Visual Art".

I also want to thank Lia Leonardi, who has helped me with the final editing of many of my texts and manuscripts. Both Jim and Lia have learned from me some basic CPP-TRS skills.

I want to thank Prof. Paul McKevitt for valuable comments upon reading this book and Dr Wendi Momen for her careful spelling and language revision.

I also want to thank Dr Luigi Turrino for having caught the real spirit of my work at the very beginning and for having given me valuable and practical support.

And last but not least, I want to thank my mother Lina, for her constant presence and precious help in everyday things that have always been a major concern for me, especially when I was trapped in writing, which basically means ... all the time!